Table of Formulas

Perimeter and Area

Rectangle

Perimeter: $P = 2l + 2w$

Area: $A = lw$

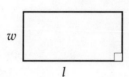

Parallelogram

Area: $A = bh$

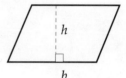

Square

Perimeter: $P = 4s$

Area: $A = s^2$

Trapezoid

Area: $A = \frac{1}{2}h(b_1 + b_2)$

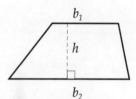

Triangle

Perimeter: P = the sum of the three sides

Area: $A = \frac{1}{2}bh$

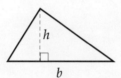

Right Triangle

Pythagorean Theorem:

$a^2 + b^2 = c^2$

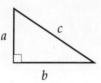

Circle

Circumference:

$C = 2\pi r$ or πd

Area: $A = \pi r^2$

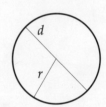

Volume and Surface Area

Rectangular Solid

Volume: $V = lwh$

Rectangular Prism

Surface Area:

$SA = 2(lh + wh + wl)$

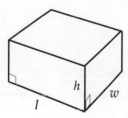

Cylinder

Volume: $V = \pi r^2 h$

Surface Area:

$SA = 2\pi r^2 + 2\pi rh$

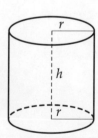

Cube

Volume : $V = s^3$

Surface Area:

$SA = 6s^2$

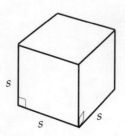

Sphere

Volume: $V = \frac{4}{3}\pi r^3$

Surface Area: $A = 4\pi r^2$

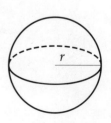

ELEMENTARY
ALGEBRA

ELEMENTARY ALGEBRA

Sandra Pryor Clarkson
Hunter College

Barbara J. Barone
Hunter College

with **Mary Margaret Shoaf** *Baylor University*

HOUGHTON MIFFLIN COMPANY **Boston New York**

*We dedicate this book
to our fathers for their complete faith in us,
to Shepherd G. Pryor III, and
in loving memory of Walter J. Primosch (1917–1995)
—S.P.C. and B.J.B.*

Sponsoring Editor: Maureen O'Connor
Development Editor: Dawn Nuttall
Project Editor: Cindy Harvey
Editorial Assistant: Michelle Francois
Production/Design Coordinator: Carol Merrigan
Senior Manufacturing Coordinator: Marie Barnes
Marketing Manager: Sara Whittern

Cover Design: Deborah Azerrad Savona
Cover Photo: Takashi Shima/Photonica
Interior Design: George McLean

Photo Credits: Chapter 1: David Witbeck/The Picture Cube; pages 2, 57, © Camerique/Picture Cube; page 3, Roy Morsch/The Stock Mkt; page 17, David Young-Wolff/Photo Edit; Chapter 2: © G.S.O. Images MCMXCIII/The Image Bank; Chapter 3: Dean Abramson/Stock Boston: Chapter 4: ITTC Productions/The Image Bank; Chapter 5: David M. Phillips/Science Source/Photo Researchers, Inc.; Chapter 6: Michael Melford/The Image Bank; Chapter 7: David Young-Wolff/PhotoEdit; Chapter 8: John Madere; page 467: The Chicago Historical Society; Chapter 9: Susan Van Etten/Photo Edit; Chapter 10: Glyn Kirk/Tony Stone Images.

Printed in the U.S.A.

Library of Congress Number: 97-72453

ISBNs:
Text: 0-395-74424-5
Instructor's Annotated Edition: 0-395-87940-X

123456789-VH-01 00 99 98 97

CONTENTS

PREFACE

Elementary Algebra serves as a bridge between the traditional and the reform text. It is a text that encourages students to explore mathematical ideas, formulate questions, and communicate mathematics using writing and graphing. This text embraces the best elements of the American Mathematical Association of Two-Year Colleges (AMATYC) standards and reform: motivating *applications* that drive the mathematics, including an abundance of *real-source data* graphs and tables; *technology* integrated where appropriate; *writing about mathematics*—in the exercises, as a new margin note feature, and as interactive chapter summaries; *group projects*, the majority of which involve real data; *variety in presentation* and illustration of mathematical concepts; and a *problem-solving approach* introduced early and integrated throughout.

Approach

Mastery   Many courses require students to demonstrate mastery of specific aspects of the material. Additionally, in this text, concepts and skills are integrated into a smoothly flowing text and not simply presented as isolated objectives. Section goals are identified, and the tests monitor achievement of those goals. Our exams give three questions for each objective. A student who answers two of the three questions correctly really knows the material.

Sensitivity to Learning Styles and Disabilities   Students have many opportunities to "see" mathematics in multiple formats—graphically, numerically, symbolically—and to use all these formats in solving problems. Additionally, Study Hints and Instructor Notes are given to aid students, including those with learning or perceptual difficulties. A section in the Instructor's Resource Manual is devoted to working with students with learning disabilities.

Motivation   We present problem situations from many areas of interest and concern that reflect the impact of mathematics in all aspects of modern life. Therefore, answers are not always artificially "nice" numbers but are often from real situations. This realism helps students develop genuine number sense.

Real-Life Applications

With respect to motivation and real data, newspapers, magazines, professional journals, and current and historical books are the source materials for the applications found in the following features.

Chapter and Section Applications   Each chapter has a *Chapter Lead-In* and a *Chapter Look-Back* and each section has a *Section Lead-In* and a *Section Follow-Up* that extends knowledge of mathematics in a real-life situation or in a mathematical exploration. These Lead-Ins relate peripherally to the section and chapter and help to interest students in the material to come.

Exercises and Examples Examples and exercises involve data gathered from contemporary business, sports, current events, the sciences, and virtually all real-life situations. They cover a diverse selection of topics, including biology, physics, astronomy, real estate, architecture, engineering, and even elements of popular culture, such as movies. They appear early in the text and throughout all chapters. The wide variety and realism of the applications will appeal to students. The areas of application are identified.

Excursions Excursions are extended project exercises that follow each section's exercise set. These problems can be assigned to individuals or groups. These exercises contain a wealth of real data, and a written answer or justification is often required. Many are also open-ended and encourage creative thinking and problem solving.

- *Class Act*   These problems are especially designed to stimulate discussion and cooperation as student groups formulate the best solution. These can be assigned to individuals but may require more time than for groups.
- *Data Analysis*   These problems encourage graphical and/or statistical analysis. They are appropriate for both groups and individuals.
- *Posing Problems*   Numerical information is given in table, graph, or paragraph format. Students are encouraged to pose their own problems and solve them and share with others. These activities often stimulate lively classroom discussion.
- *Exploring ...*   These problems often require some informed trial-and-error procedures to find a solution. Students will explore problems involving patterns, numbers, calculators, problem solving, and geometry.

Technology

Calculator Corners   All technology material has been specifically developed and written for this text. The Calculator Corners are integrated at appropriate places, providing detailed directions for how a graphing calculator can be useful—both in graphing and non-graphing topics. The Calculator Corners not only illustrate how a graphing calculator can make mathematical calculations easier, they also can take a mathematical idea one step further or present an alternate way to illustrate a mathematical concept. For classes not using calculators, the Calculator Corners are easy to identify and skip. Screen displays and directions are given for the Texas Instruments *TI 82/83* series, but the features discussed are found on all graphing calculator models.

The Calculator Corner examples and exercises were contributed by Dr. Mary Margaret Shoaf, who has given numerous workshops on the use of calculators in teaching mathematics. Additionally, her research and writing have focused on the use of calculators in enhancing student understanding. We are pleased to have her as a member of the writing team.

Writing/Group

Besides the Excursions and the section exercises identified with ✏, there are additional features that encourage student writing.

Margin Notes   In other texts, often student margin notes contain information for students to read only. We offer two types of student margin notes that ask students to give a written response, allowing students to interact immediately with the material.

- *Error Alert*   These can be assigned individually or used as a catalyst for group discussion. A problem is presented along with an incorrect or incomplete solution. Students are asked to identify and correct the errors. Common student errors are presented in this feature. We ultimately want students to be able to identify and correct the errors they might make in their own work. No answers are given in the text in order to encourage students to rely on themselves for these answers.
- *Writer's Block*   These questions require students to write about mathematics. Students explain, define, clarify, and interpret mathematical ideas, terms, and procedures.

Student Journal   The student text comes with a separate Student Journal booklet. This journal contains interactive chapter summaries in which students must write their own summaries as prompted by questions. Each chapter summary covers all the definitions and rules in that chapter and asks students to provide definitions, complete statements and diagrams, fill in blanks, and write explanations in their own words. Students are encouraged to give an example of each definition and rule and to make notes to help them remember the material. Students are also encouraged to keep a homework journal, with pages numbered, in order to note the page numbers of problems that they have worked that best "model" the chapter's main ideas. Overall, the Student Journal is designed to help students make the most efficient use of their time while studying the material.

Chapter Pedagogy

Section Goals   On the first page of each section, Section Goals list the terminal objectives for the section. These goals will be taught and practiced in that section, together with necessary "pre-skills."

Definitions and Rules   Key words are defined and rules (or procedures) are clearly delineated in boxes set apart from the rest of the material. These features build vocabulary and aid students in communicating mathematics. Each mathematical term appears in bold type when it first occurs in the text.

Study Hints   These student notes found in the margin aid students in learning the material.

Warm-Ups   Students are directed to a Warm-Up after each worked example. The Warm-Up parallels the example, reinforcing the concept just taught and building student confidence while also allowing students to interact immediately with the material. These special exercises appear right before the section exercises, and answers to all Warm-Ups are in the back of the text.

Connections to ...

- *Probability and Statistics*   Topics in probability and statistics are introduced throughout the text and reinforced periodically. Topics include mean, median, mode, standard deviation, range, line graphs, histograms, pie charts, and naming probabilities.

▪ *Geometry and Measurement*   Topics in geometry and measurement are introduced throughout the text and reinforced periodically. Topics include area and perimeter of plane figures, surface area and volume of rectangular solids, cylinders, spheres and pyramids, similar triangles, parallel lines cut by a transversal, changing units, and scientific notation.

Problem-Solving Preparation   Problem solving is emphasized throughout the text. A useful set of problem-solving strategies is developed in the text so that students learn to approach problems in an organized, efficient manner. Students are taught to reason mathematically using a four-step procedure modeled on Polya's problem-solving work.

Estimation   Throughout the text, estimation is taught and reinforced, where appropriate, to promote number sense.

Functions and Set Notation   Functions and set notation are introduced early and reinforced throughout the text. Students completing this text will be well prepared for intermediate algebra.

Assessment

We have incorporated a variety of assessment features into this text. Instructors might want to use a portfolio of these (and possibly also Warm-Ups, Excursions, Writer's Block, and Error Alert features) in evaluating student knowledge.

Skills Check   The Skills Check sections determine whether students have the prerequisite skills necessary for that chapter. Passing this test does not allow a student to "skip" the chapter; instead, it lets them know if they are ready to begin. Students who miss problems are referred to appropriate sections for review.

Section Exercise Sets   These exercises provide review and practice for all skills taught in the section. Exercises are arranged in pairs; answers to odd-numbered problems are at the back of the text.

Mixed Practice   In all sections except the first section in a chapter, there are mixed practice problems that review and reinforce skills previously taught in that chapter. Answers to the odd-numbered problems are at the back of the text.

Chapter Review   These exercises are divided into two sets. The first group of exercises provides a section-by-section review of the chapter's work. The second set is a mixed review that is not section referenced. All answers appear in the back of the text.

Chapter Test   This test follows the chapter review material. It is mastery-based and contains three questions for each section goal in a chapter. Questions are organized by section goal, and answers do *not* appear in the back of the text (solutions can be found in the Solutions Manual). A student who answers two questions correctly out of each group of three has most likely mastered the objective. The Chapter Test takes approximately one hour to complete. An additional Chapter Test as well as suggestions for Assessment

Alternatives—in keeping with the standards recommended by AMATYC and NCTM—appear in the Instructor's Resource Manual.

Cumulative Review   Beginning with Chapter 2, a Cumulative Review section appears at the end of each chapter. This review reinforces skills taught in all previous chapters including the current one. The exercises are mixed. Answers to all these problems are at the back of the text.

Supplements

FOR INSTRUCTOR USE

Instructor's Annotated Edition   The Instructor's Annotated Edition is an exact replica of the student text but also includes answers to all the exercises as well as helpful *Instructor Notes* in the margins. These Instructor Notes provide hints for teaching students, including those with learning or perceptual difficulties, and cautions about possible student misunderstandings.

Instructor's Resource Manual with Test Bank   The *Instructor's Resource Manual* contains information about how to organize a laboratory course using this text. Suggestions are also given about the use of cooperative learning groups and for working with students with certain types of perceptual or learning disabilities. There is also a section on how students can gain math confidence as well as a section listing the AMATYC standards. Finally, there is an additional Chapter Test for each chapter in the text. The questions are in random order and are slightly more challenging than the Chapter Tests in the text.

The *Test Bank* contains about 2000 test items. Items are grouped by section and goal and are also available in the Computerized Test Generator.

Computerized Test Generator   The Computerized Test Generator is the electronic version of the printed Test Bank. This user-friendly software permits an instructor to construct an unlimited number of customized tests from the 2000 test items offered. **On-line testing** and **gradebook** functions are also provided. It is available in Windows for the IBM PC and compatible computers.

Solutions Manual   The Solutions Manual contains full, worked-out solutions to all exercises in the text.

FOR STUDENT USE

Student Journal   This journal, which comes with the student edition, contains interactive chapter summaries in which students must write their own summaries as prompted by questions. Each chapter summary covers all the definitions and rules in that chapter and asks students to provide definitions, complete statements and diagrams, fill in the blanks, and write explanations in their own words.

Computer Tutor   The Computer Tutor is a text-specific, networkable, interactive, algorithmically driven software package. This powerful ancillary features full-color graphics, algorithmic exercises with extensive hints, animated solution steps, and a comprehensive classroom management system. It is available for the IBM PC and compatible computers and the Macintosh. A computer disk

icon ⌷ appears in the Section Goals box as a reminder that each section is covered in the tutor.

Videos   Within each section of the text, a videotape icon 📼 appears in the Section Goals box. The icon contains the reference number of the appropriate video, making it easy for students to find the extra help they may need. Each video opens with a relevant application which is then solved at the end of the lesson.

Student Solutions Manual   The Student Solutions Manual contains full, worked-out solutions to all the exercises whose answers are at the back of the text, namely all the Warm-Ups, the odd-numbered section exercises, all the Chapter Review exercises, and all the Cumulative Review exercises.

Acknowledgments

Many people have helped us directly with this project. In particular, we wish to thank Cindy Harvey, Kathy Deselle, George McLean, Dawn Nuttall, and Maureen O'Connor.

A special thanks must go to JoAnne Kennedy at LaGuardia Community College for authoring the Test Bank, Solutions Manual, and Student Solutions Manual, and to Mary Margaret Shoaf at Baylor University for authoring the Calculator Corners.

Through the years, each of us has people that inspire us, offer us opportunities, and teach us important lessons. The following have contributed to our growth in various ways: Mary Ellen Bohan; the late Ann Braddy; Mrs. Breiner; Mac Callaham; the late Tom Clarke; Tom Davis; the late Mary P. Dolciani Halloran and her husband James Halloran; Henry Edwards, Jr.; George Grossman; Eleanore Kantowski; Mrs. Leonardi; the late Sarah Anna Mathis; Ed Millman; the late Len Pikaart; Henry Pollack; Donna Shalala; June Smith; Andre Thibodeau; the late Mr. Towson; Zalman Usiskin; Bill Williams; Jim Wilson; and Gloria Wolinsky.

We would also like to thank the following reviewers for their suggestions:

Kathleen Burk, *Pensacola Junior College,* FL; Dennis Ebersole, *Northampton County Area Community College,* MA; Mitchel Fedak, *Community College of Allegheny County,* Boyce Campus, PA; Judy Godwin, *Collin County Community College,* TX; Karen Graham, *University of New Hampshire;* Rose Ann Haw, *Mesa Community College,* AZ; Edith Hays, *Texas Woman's University;* Georgia K. Mederer; William Radulovich, *Florida Community College,* Jacksonville; James Ryan, *Madera Community College,* CA; Lauri Semarne; Thomas Walsh, *City College of San Francisco;* Susan White, *DeKalb College,* GA; and Kenneth Word, *Central Texas College.*

And, of course, our families contributed most of all!

S.P.C. and B.J.B.

INDEX OF APPLICATIONS

INTRODUCTION TO THE REAL NUMBERS

*H*ow many bones are in your body? You would count your bones using the natural numbers, but you would need the whole numbers—the natural numbers and zero—to count the bones in a jellyfish or other invertebrates. You would need integers—the natural numbers, their opposites, and zero—to count the "bones" in a boneless breast of chicken.

■ *Describe the last time you used integers. Were they also natural numbers or whole numbers?*

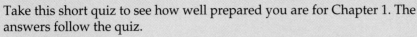

SKILLS CHECK

Take this short quiz to see how well prepared you are for Chapter 1. The answers follow the quiz.

1. Add: $2\frac{1}{3} + 4\frac{1}{6}$

2. Subtract: $3\frac{3}{4} - 1\frac{1}{8}$

3. Multiply: $8\frac{2}{5} \times 7\frac{1}{3}$

4. Divide: $9 \div 1\frac{1}{4}$

5. Rewrite $\left(\frac{1}{3}\right)^3 \left(\frac{2}{3}\right)^0$ without exponents.

6. Add: $6.28 + 7 + 0.48$

7. Subtract: $11 - 7.049$

8. Multiply: 76.45×2.03

9. Divide: $0.7\overline{)42.14}$

10. Simplify: $3 + 6 \times 2^3$

ANSWERS: **1.** $6\frac{1}{2}$ **2.** $2\frac{5}{8}$ **3.** $61\frac{3}{5}$ **4.** $7\frac{1}{5}$ **5.** $\frac{1}{27}$ **6.** 13.76 **7.** 3.951 **8.** 155.1935 **9.** 60.2 **10.** 51

CHAPTER LEAD-IN

- What happens when it becomes too difficult to count?
- We can count bones in the body but can we count the bodies in this bone?

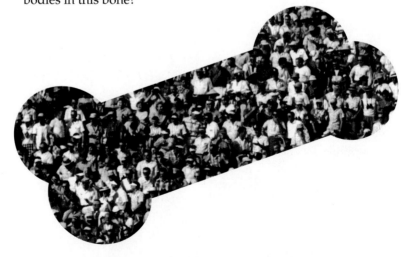

1.1 The Numbers of Algebra: Real Numbers

SECTION LEAD-IN

▪ Give three strategies for estimating the number of corn kernels you can see in this photograph.

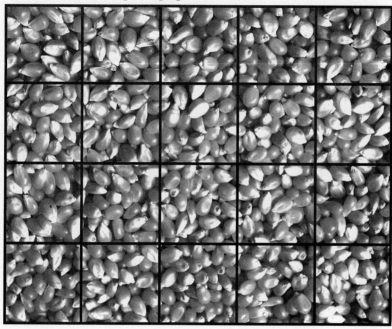

SECTION GOALS

▪ *To identify the whole numbers, integers, rational numbers, and real numbers*

▪ *To find the opposite of a real number*

▪ *To graph real numbers*

▪ *To compare and order real numbers*

▪ *To find the absolute value of a number*

Introduction to the Real Numbers

In arithmetic we work first with the **natural numbers,** which are the numbers we use to count with. By including zero with the natural numbers, we obtain the *whole numbers.*

Whole Numbers

The **whole numbers** consist of the natural numbers and zero.

However, there are also negative numbers—numbers that we can use to represent losses, to suggest decreases in temperature or speed, or to indicate the *opposite* direction.

The **opposite** of any non-zero number a is the number $-a$. If a is a positive number, then its opposite, $-a$, is a negative number.

For example, the opposite of 3 is -3, and the opposite of -3 is 3. Just as the number 3 can be located on a number line, its opposite, the negative number -3, can also be located on the number line. As you may have guessed by now, the number -3 lies to the *left* of zero.

Numbers to the right of zero on the number line are **positive numbers.**
Numbers to the left of zero on the number line are **negative numbers;** they have minus signs.
Numbers with signs are called **signed numbers.**

When we combine the positive whole numbers, their negatives, and zero, we get the set of *integers.*

Integers

The **integers** are the set of numbers that contain the positive whole numbers, their opposites (the negative numbers), and zero.

Integers are part of a larger set of numbers called *rational numbers.* Examples of rational numbers include $\frac{1}{2}$, -79, 37.007, and -0.65. We define the rational numbers as follows:

Rational Numbers

The **rational numbers** are the set of all numbers that can be represented as fractions, $\frac{a}{b}$, where a and b are integers and b is not equal to zero.

The integers are also rational numbers, because any integer a can be written as the fraction $\frac{a}{1}$. Therefore, such numbers as

$$0 = \frac{0}{1} \qquad 5 = \frac{5}{1} \qquad \text{and} \qquad -33 = \frac{-33}{1}$$

are also rational numbers. **Terminating decimals** such as

$$4.2 \qquad 6.134071 \qquad \text{and} \qquad 18$$

and **repeating decimals** such as

$$0.\overline{81} \qquad \text{The line over 81 here means that those}$$
digits repeat to give 0.81818181, and so on.

are rational numbers. But decimal numbers whose decimal parts do not terminate or repeat are *not* rational numbers. The number π (pi) is the ratio of the circumference of a circle to its diameter. However, it is not the ratio of two integers. In fact, this number is **irrational** (that is, not rational) and can only be estimated. It cannot be calculated. π estimated to 15 decimal places is

$$\pi = 3.14159\ 26535\ 89793$$

The irrational and rational numbers together make up the *real numbers*.

> **Real Numbers**
>
> The **real numbers** consist of the rational numbers and the irrational numbers.

▪▪▪
EXAMPLE 1

Which of the following real numbers are rational? Which are integers?

a. $\frac{1}{\pi}$ **b.** $-4\frac{1}{5}$ **c.** 6.91   **d.** $6.9\overline{1}$ **e.** 0   **f.** -7

SOLUTION

a. 1 is an integer but π is not. Because we cannot write this fraction as a ratio of integers, it is *not* a rational number and, therefore, not an integer either.

b. We can rewrite $4\frac{1}{5}$ as a fraction.

$$4\frac{1}{5} = \frac{4 \times 5 + 1}{5} = \frac{21}{5}$$

Thus $-4\frac{1}{5} = \frac{-21}{5}$. Because -21 and 5 are both integers, we classify $-4\frac{1}{5}$ as a rational number. It is not an integer.

c. 6.91 is a terminating decimal number. It is therefore a rational number. It is not an integer.

d. $6.9\overline{1}$ is a repeating decimal number, so it is a rational number. It is not an integer.

e. Zero is a rational number and an integer.

f. -7 is a rational number and an integer.

▶ CHECK **Warm-Up 1**

▪▪▪
WRITER'S BLOCK
Write a set of rules that you can use to determine what categories a number falls into.

Opposites of the Real Numbers

Like the integers, all non-zero real numbers have *opposites*.

> **Opposites of a Real Number**
>
> The **opposite** of any non-zero real number n is the real number $-n$.

Note that n may be positive or negative.

▪ ▪ ▪

EXAMPLE 2

Find the opposite of

a. $\dfrac{-5}{7}$ b. $-(-8.2)$ c. $0.39\overline{39}$

SOLUTION

a. The opposite of $\dfrac{-5}{7}$ is written $-\left(-\dfrac{5}{7}\right)$. The opposite of the negative number $-\dfrac{5}{7}$ is the positive number $\dfrac{5}{7}$.

b. From part (a), we see that $-(-8.2)$ is 8.2. So the opposite of $-(-8.2)$ is -8.2.

c. The opposite of $0.39\overline{39}$ is $-0.39\overline{39}$.

▶ *CHECK* **Warm-Up 2**

Calculator Corner

You can use the Home Screen of your graphing calculator to find the opposite of a number. Consider the following example.

Find the opposite of $-2/3$. (Note: the $\boxed{(-)}$ key means negative; the $\boxed{-}$ key means the operation of subtraction.)

   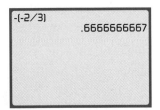

Press **ENTER.**

You can also obtain a fractional answer by telling the calculator to give the answer in fractional form.

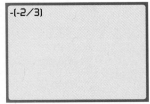   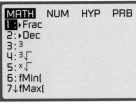   Press **ENTER.**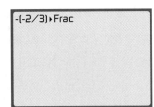

Press **ENTER.**

Some calculators will evaluate a "double negative sign" without the use of parentheses.

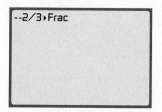

Press **ENTER.**

Graphing Real Numbers

Real numbers can be graphed on a number line.

> **To graph a real number that is not an integer**
> 1. Determine which two integers it lies between.
> 2. Approximate its position between those integers on a number line.

WRITER'S BLOCK

Explain to a friend how to graph a number that is an integer.

•••
EXAMPLE 3

Graph $\frac{-2}{3}$, 7.6, −6.29, and π on one number line.

SOLUTION

We survey the numbers we will be graphing.

$\frac{-2}{3}$ will lie between 0 and −1.
7.6 will lie between 7 and 8.
−6.29 will lie between −6 and −7.
$\pi \approx 3.14159$ will lie between 3 and 4.

We draw a number line and mark off divisions (usually integers). The line must be able to include all the numbers we are graphing.

To graph $\frac{-2}{3}$, we place a dot two-thirds of the way from 0 to −1. To graph 7.6, we place a dot slightly more than halfway from 7 to 8. We graph −6.29 and π similarly.

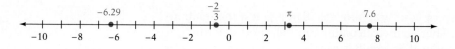

▶ CHECK **Warm-Up 3**

Ordering Real Numbers

A real number a is **less than** a real number b if a is to the left of b on the number line. In symbols, we write $a < b$.

A real number a is **greater than** a real number b if a is to the right of b on the number line. In symbols, we write $a > b$.

There is no greatest positive real number, and there is no least negative real number.

▪▪▪
EXAMPLE 4

Order 0.5, $-4\frac{3}{4}$, -730, and $4.11\overline{1}$ from least to greatest using the "is less than" sign, $<$.

SOLUTION

We survey the numbers as though we were graphing them.

0.5 is between 0 and 1.
$-4\frac{3}{4}$ is between -4 and -5.
-730 is at -730.
$4.11\overline{1}$ is between 4 and 5.

Placing these numbers in order as they would appear on a number line, we get

$$-730 < -4\frac{3}{4} < 0.5 < 4.11\overline{1}$$

If we order these numbers from greatest to least, we get

$$4.11\overline{1} > 0.5 > -4\frac{3}{4} > -730$$

▶CHECK **Warm-Up 4**

Calculator Corner

You can use the Home Screen of your graphing calculator to test the relationship between two numbers. That is, the calculator can help you determine if one number is greater or smaller than another number. As an example, consider the calculator work to accompany Example 4. The inequality signs can be found under the **2nd TEST** menu.

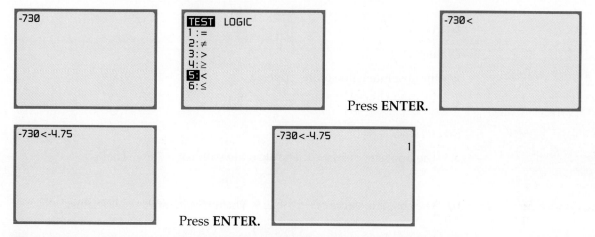

Press **ENTER.**

Press **ENTER.**

The calculator returns a response of 1 if the statement is true and a response of 0 if the statement is false. Notice the difference between the last calculator screen shown above and the one at the left below. Continuing to list the given numbers from least to greatest results in the screen at the right below.

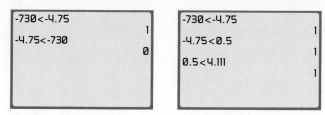

Now you should be able to write all four numbers from least to greatest.

Try the following problem on your own with the help of your graphing calculator.

Which of the following real numbers are greater than -3?

$$-2.999, \ -3.02, \ -17, \ -0.03, \ -2.00999$$

Absolute Value of Real Numbers

A real number and its opposite are the same *distance* from zero on a number line, but they lie on opposite sides of zero.

Absolute Value

The **absolute value** of a number a, symbolized $|a|$, is the distance of a from zero on the number line. Because absolute value represents distance, the absolute value of a number can never be negative.

Absolute values of real numbers are always positive numbers, so a number and its opposite have the same absolute value.

▪▪▪

EXAMPLE 5

Evaluate:

a.   the absolute value of the opposite of 7   b. $-|-(-9)|$

c. $-|4.56|$ d. $\left|-3\frac{1}{4}\right|$ e. $-|-2.16|$

SOLUTION

a.   The opposite of 7 is -7. The absolute value of -7 is 7. That is,
$$|-7| = 7$$

b.   The absolute value of $-(-9)$ is 9. The opposite of 9 is -9. That is,
$$-|-(-9)| = -(9) = -9$$

c.   The absolute value of 4.56 is 4.56. The opposite of that is -4.56. So
$$-|4.56| = -4.56$$

d.   The absolute value of $-3\frac{1}{4}$ is $3\frac{1}{4}$. So
$$\left|-3\frac{1}{4}\right| = 3\frac{1}{4}$$

e.   The absolute value of -2.16 is 2.16. The opposite of 2.16 is -2.16. So
$$-|-2.16| = -2.16$$

▶ *CHECK* **Warm-Up 5**

! ! !

ERROR ALERT

Identify the error and give a correct answer.

Simplify: $-|-2.5|$

Incorrect Solution:

By definition,
$-|-2.5| = 2.5$

Calculator Corner

You can use the Home Screen of your graphing calculator to evaluate the absolute value of a number. On most graphing calculators, the letters **ABS** refer to the Absolute Value function. For example, you can find the absolute value of -7.5 as follows:

| abs -7.5 |
| |

Press **ENTER.**

| abs -7.5 |
| 7.5 |

Explain why the following calculator result is different from the first example above. Why is the second answer a negative number while the first answer was positive?

```
abs -7.5
                      7.5
-abs -7.5
                     -7.5
```

Practice what you learned.

SECTION FOLLOW-UP*

- Which of your strategies involves counting?
- How effective are the counting strategies?
- What are the advantages and disadvantages of your strategies?

1.1 WARM-UPS

Work these problems before you attempt the exercises.

1. Which of the following real numbers are rational?

    $\frac{1}{7}$ $-4\frac{10}{5}$ 36.999991

2. Find $-(-5)$ and the opposite of 42.

3. Graph 4, -4, and -6 on a number line.

4. Order 10, -34, -3, and 4 from largest to smallest using the "is greater than" sign, $>$.

5. Evaluate $-|-(-97)|$ and the opposite of the absolute value of the opposite of 7.

*Each section's Follow-Up refers to topics or questions presented in the section's Lead-In.

1.1 EXERCISES

Note: Use your graphing calculator to check your results whenever possible.

In Exercises 1 through 8, determine which of the real numbers are rational (R) and which are integers (I).

1. -6

2. -3.5

3. $\frac{0}{6}$

4. 4

5. $3.18\overline{18}$

6. $\frac{-3}{-7}$

7. $\frac{\pi}{6}$

8. 0

In Exercises 9 through 16, find the opposite of each real number.

9. 6

10. -7

11. $-(-1933)$

12. $-(4.3)$

13. $-[-(5.86)]$

14. $5\frac{4}{7}$

15. $-0.9\overline{99}$

16. $-(-10.35\overline{3})$

In Exercises 17 through 38, evaluate each expression.

17. $-(-6)$

18. $-(-7)$

19. $-(-(-18))$

20. $-(-(-26))$

21. $|-6|$

22. $|-5|$

23. $\left|26\frac{7}{8}\right|$

24. $-|3.5|$

25. $|-2|$

26. $-|+17|$

27. $-|142|$

28. $-|45|$

29. $-(-|47|)$

30. $-|-9|$

31. the opposite of -8

32. the opposite of 14

33. the opposite of -122

34. the opposite of -43

35. the opposite of the absolute value of 355

36. the absolute value of the opposite of 9

37. the opposite of the absolute value of the opposite of 18

38. the opposite of the opposite of -10

In Exercises 39 through 44, graph and label each number on a number line.

39. $3, -6,$ and $|-2|$

40. $|4|, -3,$ and 0

41. $3, -1,$ and $-(-1)$

42. $-4, |-5|,$ and 6

43. $-1, 1,$ and $-(-3)$

44. $-(-2), -|-2|,$ and $|0|$

In Exercises 45 through 50, graph each number and show the two integers the number lies between.

45. $|-49|$

46. $-|-75|$

47. $-(-|54|)$

48. $-|10,000|$

49. $-(-90)$

50. $|-(-112)|$

That is, we can subtract a real number b from another real number a by adding the additive inverse, or opposite, of b to a.

In the subtraction $a - b = c$, a is called the **minuend,** b is the **subtrahend,** and c is the **difference.**

To subtract real numbers $(a - b)$

1. Replace the number being subtracted (b) with its additive inverse $(-b)$.
2. Rewrite the subtraction as addition by changing the operation sign.
3. Add $a + (-b)$ as indicated.

ERROR ALERT

Identify the error and give a correct answer.

Subtract: $(-27) - 22$

Incorrect Solution:

$(-27) + 22 = -5$

■ ■ ■

EXAMPLE 5

Subtract: $-16 - 12$

SOLUTION

To subtract, we first replace 12 with its additive inverse (-12) and rewrite the operation as addition.

$$-16 - 12 = -16 - (12)$$
$$= -16 + (-12)$$

Now we add. Because the signs are the same, we add absolute values.

$$|-16| + |-12| = 16 + 12 = 28$$

Because the number with the greater absolute value (-16) is negative, the result is negative: -28.

So $-16 - 12 = -28$.

▶ *CHECK* **Warm-Up 5**

■ ■ ■

EXAMPLE 6

Subtract: $-18\frac{1}{4} - \left(-10\frac{1}{2}\right)$

SOLUTION

To subtract, we first replace $-10\frac{1}{2}$ with its additive inverse $\left(10\frac{1}{2}\right)$ and rewrite the operation as addition.

$$-18\frac{1}{4} - \left(-10\frac{1}{2}\right) = -18\frac{1}{4} + \left(+10\frac{1}{2}\right)$$

Because the signs are different, we subtract absolute values to obtain the sum.

$$\left|-18\frac{1}{4}\right| - \left|10\frac{2}{4}\right| = 18\frac{1}{4} - 10\frac{2}{4}$$

WRITER'S BLOCK

Explain the difference between an additive identity and an additive inverse.

Because we cannot subtract $\frac{2}{4}$ from $\frac{1}{4}$, we must regroup (or borrow) from 18. We rewrite $18\frac{1}{4}$ as $17\frac{4}{4} + \frac{1}{4} = 17\frac{5}{4}$. Then, we subtract:

$$17\frac{5}{4} - 10\frac{2}{4} = 7\frac{3}{4}$$

The original number with the greater absolute value $\left(-18\frac{1}{4}\right)$ is negative, so the result is negative: $-7\frac{3}{4}$.

Thus $-18\frac{1}{4} - \left(-10\frac{1}{2}\right) = -7\frac{3}{4}$.

▶ *CHECK* **Warm-Up 6**

▪▪▪

WRITER'S BLOCK

Why isn't zero the identity element for subtraction?

EXAMPLE 7

Subtract: $2.75 - (-1.253)$

SOLUTION

First we replace -1.253 with its additive inverse $(+1.253)$ and rewrite the operation as addition.

$$2.75 - (-1.253) = 2.75 + (+1.253)$$

Now we add. Because the signs are the same, we add the absolute values.

$$|2.75| + |1.253| = 2.75 + 1.253 = 4.003$$

Because the original number with the greater absolute value (2.75) is positive, the result is positive: 4.003.

So $2.75 - (-1.253) = 4.003$

▶ *CHECK* **Warm-Up 7**

Practice what you learned.

SECTION FOLLOW-UP

- How are operations of real numbers used in your crowd estimation?
- **Research** Find some numbers that are estimates in newspapers or magazines. What procedures do you think were used to obtain the estimates?

1.2 WARM-UPS

Work these problems before you attempt the exercises.

1. Add: $-1703 + 294$

2. Add: $\left(-8\frac{2}{7}\right) + 7\frac{1}{6}$

3. What is the sum of (-5.62) and (-0.7) and 8?

4. Find the additive inverses of -3, $2\frac{1}{3}$, and $-|3.6|$.

5. Find the difference: $62 - (-18)$

6. Subtract: $\left(-86\frac{1}{6}\right) - \left(-23\frac{1}{3}\right)$

7. Subtract -1.6 from (-3.6).

1.2 EXERCISES

Note: Use your graphing calculator to check your results whenever possible.

In Exercises 1 through 50, add or subtract as indicated.

1. $(-27) - 63$
2. $23 - (-37)$
3. $+127 - (-94)$
4. $+136 - (-27)$

5. $67 - (-89)$
6. $(-37) - 28$
7. $(-18) - 75$
8. $-12 - 47$

9. $\left(-5\frac{6}{7}\right) + \left(-4\frac{6}{11}\right)$
10. $\left(-12\frac{2}{3}\right) + \left(-6\frac{7}{9}\right)$
11. $\left(-18\frac{5}{6}\right) + \left(-5\frac{3}{8}\right)$
12. $\left(-9\frac{5}{6}\right) + 3\frac{5}{8}$

13. $\left(-6\frac{2}{11}\right) + 5\frac{4}{7}$
14. $3\frac{4}{7} + \left(-15\frac{6}{7}\right)$
15. $\left(-4\frac{3}{4}\right) - \left(-1\frac{5}{8}\right)$
16. $\left(-6\frac{2}{3}\right) - 4\frac{1}{2}$

17. $\left(-9\frac{1}{2}\right) + \left(-4\frac{7}{12}\right)$
18. $\left(-8\frac{1}{3}\right) - 3\frac{2}{7}$
19. $\left(-14\frac{5}{7}\right) - 29\frac{6}{11}$
20. $7\frac{5}{9} + \left(-27\frac{7}{9}\right)$

21. $2.94 + (-1.8)$
22. $3.86 + (-2.6)$
23. $(-13.7) + 7.9$
24. $(-4.2) + (-9.8)$

25. $(-2.4) + (-7.36)$
26. $(-11.3) + (-8.45)$
27. $(-6.8) + (-9.46)$
28. $(-15.2) + 14.7$

29. $6.3 - (-14.7)$
30. $2.4 - (-6.2)$
31. $(-2.47) + (-3.21)$
32. $-12.6 - 2.51$

33. $(-90) + 47 + (-38)$
34. $(-232) + 76 + (-86)$
35. $(-56) + (+87) + (-54)$

36. $(-18) + (-24) + (-67)$
37. $(-54) + (-78) + (-58)$
38. $(-24) + (-56) + (-89)$

39. $97 + (-45) + 23$
40. $43 + (-28) + 47$
41. $63 + (-241) + (+75)$

42. $102 + (-98) + 77$
43. $(-89) + 90 + (-126)$
44. $(-32) + (-67) + (-41)$

45. $\left(-5\frac{2}{9}\right) + 1\frac{4}{9} + \left(-7\frac{5}{9}\right)$

46. $\left(-2\frac{1}{3}\right) + \left(-3\frac{2}{3}\right) + 5\frac{1}{3}$

47. $8.1 + (-4.6) + (-2.3)$

48. $7.7 + (-6.3) + 9.2$

49. $2.9 - 56 - 92.5$

50. $6.6 - 87 - 16.9$

In Exercises 51 through 58, perform the operations indicated.

51. Add 6.3 and the opposite of 14.7.

52. Add 2.4 and the opposite of 6.2.

53. Add the opposite of 12 to 4.7.

54. Add (-2.47) and the opposite of 0.21.

55. What is the total of $(-2.4) + (-7.06)$?

56. Subtract (-11.3) from (-8.45).

57. Find the sum of $\left(-2\frac{1}{8}\right)$ and $\left(-8\frac{2}{8}\right)$ and $5\frac{1}{8}$.

58. Find the sum of 9.7 and (-11.3) and (-15.9).

In Exercises 59 through 66, find the additive inverse.

59. -3

60. $-2\frac{1}{8}$

61. -7.9

62. $11\frac{2}{9}$

63. $(3 + (-2))$

64. $(-5 + -6)$

65. $(4 + |-9|)$

66. $-|2 + 7|$

In Exercises 67 through 74, fill in the blank.

67. _____ $+ (-15) = 0$

68. _____ $+ (-28) = 0$

69. $63 +$ _____ $= 0$

70. $102 +$ _____ $= 0$

71. $0 = \left(-7\frac{5}{6}\right) +$ _____

72. $0 =$ _____ $+ \left(-2\frac{1}{8}\right)$

73. $0 =$ _____ $+ \left(-17\frac{5}{7}\right)$

74. $0 = \left(-13\frac{1}{2}\right) +$ ___

In Exercises 75 through 78, find the missing digits.

75.
```
  1_4
  320
+ 993
 _417
```

76.
```
  153
  49_
+ 6_7
 1_59
```

77.
```
   34
   1_
   85
+  _6
  207
```

78.
```
   26
   _3
   5_
+ 45
  221
```

In Exercises 79 through 82, fill in the digits that make the problems correct.

79.
```
  8625_
- 9_35
 _6616
```

80.
```
  462_3
- _327
 3988_
```

81.
```
  5421_
- 18_2
 _2351
```

82.
```
  642_3
- _854
 6035_
```

MIXED PRACTICE

By doing these exercises, you will practice the topics in this chapter up to this point.

83. Subtract: $(-65) - \left(-94\frac{1}{2}\right)$

84. Add: $(85.8) + (-33.95)$

85. Find the sum: $(-123) + \left(46\frac{1}{6}\right)$

86. Arrange $-|-|-85||, -|37|$, and -42 in order from smallest to largest.

87. Subtract: $\left(11\frac{1}{4}\right) - \left(-52\frac{1}{2}\right)$

88. Insert $<$ or $>$ between the numbers $-|26|$ and $-|-63|$.

89. Find the sum: $(-56.85) + 79 + (-129.6)$

90. Arrange $|-27|$, $|-|32||$, and $-|-14|$ in order from largest to smallest.

91. Add: $35\frac{2}{3} + \left(-42\frac{1}{2}\right) + \left(-88\frac{1}{7}\right)$

92. Subtract 23.85 from (-27.5).

EXCURSIONS

Exploring Patterns

1. ✎ Do the first few of the following multiplications. Try to find a pattern and predict the answers to the last few. Use your calculator to find other patterns. Write a description of the pattern.

    a. $7 \times 11 \times 13$ **b.** $7 \times 11 \times 13 \times 2$ **c.** $7 \times 11 \times 13 \times 3$

    d. $7 \times 11 \times 13 \times 4$ **e.** $7 \times 11 \times 13 \times 7$ **f.** $7 \times 11 \times 13 \times 9$

2. ✎ Problems (a) through (d) exhibit a new pattern. What is it? Predict the results of (e) and (f). Justify your predictions and verify the results.

    a. 25×11 **b.** 25×111 **c.** 25×1111

    d. $25 \times 1,111,111,111$ **e.** $25 \times 111,111,110$ **f.** $111,111 \times 25$

3. ✎ Use your calculator to multiply each pair of numbers in turn. Stop multiplying when you discover a pattern to the answers. Indicate where you stopped multiplying, give the remaining answers, and describe in your own words the pattern you see.

a.	**b.**	**c.**	**d.**
1.01×11	9.9×123	1.23×555	1.05×7777
1.01×111	9.9×1234	1.23×5555	1.05×77777
1.01×1111	9.9×12345	1.23×55555	1.05×777777
1.01×11111	9.9×123456	1.23×555555	1.05×7777777
1.01×111111	9.9×1234567	1.23×5555555	1.05×77777777
1.01×1111111	9.9×12345678	1.23×55555555	1.05×777777777
		1.23×555555555	1.05×7777777777

CONNECTIONS TO *STATISTICS*

Finding the Range

In statistics we use numbers to describe the characteristics of sets of data. One of the measures that we use is that of "spread." That is, how far apart is the data. One basic measure of spread is the *range*. We use subtraction to find the range.

> **Range**
>
> The **range** of a set of data is the difference between the largest value and the smallest value in the set. It is given as a number with its unit, if it has a unit. A range is always positive or zero.

▪▪▪

EXAMPLE

Hourly readings on a pressure gauge were -1.80, -2.75, -2.25, and -1.25. The units are mb, millibars of mercury. What is the range of these readings?

SOLUTION

Our definition says that we must subtract the least value from the greatest. Because the least value is -2.75 and the greatest is -1.25, our problem becomes

$$-1.25 - (-2.75) = 1.5$$

1.5 mb is the range. Therefore, readings from -2.75 to -1.25 have a range of 1.5 millibars of mercury.

◢

PRACTICE

1.   The markings on a safe go from 31 to 158. What is their range?

2.   The following numbers of feet were recorded: -3.4, 8.9, 32.18, 45, -17.3, and -0.002. What is the range of these numbers?

3.   Markings from -17 to 101 are indicated along a number line. What is the range of these markings?

4.   What is the range of these test scores: 43, 25, 100, 18, 42, and 32?

5.   What is the range of these fractional measures: $2\frac{1}{3}$, $5\frac{4}{7}$, $8\frac{3}{8}$, and $1\frac{1}{5}$?

6.   A program randomly lists the numbers -102, 47, $188\frac{1}{2}$, 4.28, and -3.18. What is the range of these numbers?

7.   *Flying and Diving*   In 1961 people flew as high as 203.2 miles and dived as deep as 728 feet below sea level, using a gas mixture for breathing. What is the range of these record achievements?

8.   *Record Highs and Lows*   Asia has a high point of 29,028 feet and a low point of 1302 feet below sea level; North America has a high point of 20,320 feet and a low point of 282 feet below sea level. Which continent has the greater range in elevation? What is its range?

9. *Height and Depth Records* The following heights above and below sea level were recorded by a scientific team: 35 feet below, 46.3 feet above, 19 feet below, 25.17 feet above, 0.3 feet above, 135.4 feet below, 35 feet below, and 6.24 feet above. What is the range of these readings?

10. *Height and Depth Records* In 1954 a man flew 93,000 feet high in a rocket plane. Another dived 13,287 feet below sea level in a bathyscaphe. What is the range of these records?

1.3 Operations on Real Numbers: Multiplication and Division

SECTION GOALS

▪ To find the product of two real numbers

▪ To find the quotient of two real numbers

SECTION LEAD-IN

1. a. Take your pulse.

   b. Estimate how long it would take for your pulse to beat a million times.

   c. Write four more questions that involve estimation.

2. a. Have one person in the group look at the second hand on a watch. See if you can estimate when exactly one minute has passed.

   b. Write a paragraph describing your estimation method. Did it work well?

   c. How would you alter your procedure next time? In estimation do you believe that "practice makes perfect"?

Multiplication

Multiplicative Identity

When 1 is multiplied by any real number, the result is always that real number. Because of this property, 1 is called the **multiplicative identity,** or the **identity element for multiplication.**

Identity Element for Multiplication

If a is any real number, then

$$(a)(1) = (1)(a) = a$$

Multiplication by Zero and by −1

Any number multiplied by zero gives zero.

WRITER'S BLOCK

Can zero be the multiplicative identity? Why not?

> **Multiplication Property of Zero**
>
> If a is any real number, then
> $$(a)(0) = (0)(a) = 0$$

Whenever we multiply a real number by −1, we obtain its opposite. To accomplish the multiplication, we can merely change the sign of the number we are multiplying.

> For all real numbers n,
> $$(-1)(n) = -n$$
> $$(-1)(-n) = n$$

Multiplying Real Numbers

In the multiplication $(a)(b) = c$, a and b are called **factors,** and c is the **product.** Here is a general multiplication procedure that works for all real numbers:

ERROR ALERT

Identify the error and give a correct answer.

Multiply: $(-9)(-12)$

Incorrect Solution:

$(-9)(-12) = -108$

> **To multiply two real numbers ($a \times b$)**
>
> Multiply the absolute values of the factors.
>
> **1.** If the factors have the same sign, the product is positive.
> **2.** If the factors have different signs, the product is negative.

▪▪▪

EXAMPLE 1

Multiply: $(-93)(43)$

SOLUTION

First we multiply the absolute values.
$$|-93||43| = (93)(43) = 3999$$

Because the signs are different, the product is negative: -3999.

So $(-93)(43) = -3999$.

▶ *CHECK* **Warm-Up 1**

Calculator Corner

```
(-93)(43)
            -3999
-93*43
            -3999
```

Multiplication can be done on the Home Screen of your graphing calculator using parentheses or the multiplication sign. Example 1 could be done either way on the Home Screen, as shown at the right.

▪▪▪ EXAMPLE 2

Multiply: $\left(-2\frac{1}{3}\right)\left(-7\frac{2}{5}\right)$

SOLUTION

We first must multiply their absolute values.

$$\left|-2\frac{1}{3}\right|\left|-7\frac{2}{5}\right| = \left(2\frac{1}{3}\right)\left(7\frac{2}{5}\right)$$

Changing to an improper fraction: To multiply, we have to change these mixed numbers to improper fractions.

$$2\frac{1}{3} = \frac{2(3)+1}{3} = \frac{7}{3}$$
$$7\frac{2}{5} = \frac{7(5)+2}{5} = \frac{37}{5}$$

Multiplying fractions: To multiply fractions, we multiply their numerators and their denominators.

$$\left(\frac{7}{3}\right)\left(\frac{37}{5}\right) = \frac{(7)(37)}{(3)(5)} = \frac{259}{15}$$

Changing to a mixed number: Next we simplify the result by changing it to a mixed number.

$$\frac{259}{15} = 17\frac{4}{15}$$

Because the signs of the original factors are the same, the product is positive. So $\left(-2\frac{1}{3}\right)\left(-7\frac{2}{5}\right) = 17\frac{4}{15}$.

▶ CHECK **Warm-Up 2**

We can multiply more than two numbers by multiplying two at a time. We will do that next with decimal numbers.

▪▪▪ EXAMPLE 3

Multiply: $(-3.1)(-4.2)(-0.2)$

! ! !
ERROR ALERT

Identify the error and give a correct answer.

Multiply: $\left(-6\frac{2}{3}\right)\left(-5\frac{1}{4}\right)$

Incorrect Solution:

$$\left(-6\frac{2}{3}\right)\left(-5\frac{1}{4}\right) = 30\frac{2}{12}$$
$$= 30\frac{1}{6}$$

SOLUTION

We begin by multiplying any two of these numbers together; suppose we pick the first two.

$$(-3.1)(-4.2)(-0.2)$$

We multiply their absolute values.

$$|-3.1||-4.2| = (3.1)(4.2) = 13.02$$

Because both factors have the same sign, the product is positive: 13.02. We still have to multiply this result by the remaining factor: $(13.02)(-0.2)$. Multiplying 13.02 by 0.2 we obtain 2.604.

The factor 13.02 has two decimal places, and 0.2 has one decimal place, so the result has three places. We count these from the right:

$$2.6\underset{\smile\smile\smile}{04}$$

Because the factors have different signs, the product is negative: -2.604. So

$$(-3.1)(-4.2)(-0.2) = -2.604$$

▶ CHECK **Warm-Up 3**

Division

Multiplicative Inverse

The *multiplicative inverse* of any non-zero real number $\frac{a}{b}$ is the real number $\frac{b}{a}$. Thus the multiplicative inverse of $\frac{5}{4}$ is $\frac{4}{5}$, and the multiplicative inverse of $-\frac{6}{7}$ is $-\frac{7}{6}$. The multiplicative inverse of 3 is $\frac{1}{3}$ (because we can write 3 as $\frac{3}{1}$). The product of any real number and its multiplicative inverse is 1.

Multiplicative Inverse

The non-zero real numbers $\frac{a}{b}$ and $\frac{b}{a}$ are **multiplicative inverses,** and their product is 1. That is,

$$\left(\frac{a}{b}\right)\left(\frac{b}{a}\right) = 1 \qquad \text{and} \qquad \left(\frac{b}{a}\right)\left(\frac{a}{b}\right) = 1$$

The multiplicative inverse is also called the **reciprocal.** Zero does not have a reciprocal.

▪▪▪

EXAMPLE 4

Find the multiplicative inverses of

a. -2 **b.** $\frac{2}{3}$ **c.** $4\frac{3}{5}$ **d.** 2.1

SOLUTION

a. $-2 = \frac{-2}{1}$. The multiplicative inverse is $-\frac{1}{2}$.

b. $\frac{2}{3}$. The multiplicative inverse is $\frac{3}{2}$.

c. First we rewrite the mixed number as an improper fraction.

$$4\frac{3}{5} = \frac{[4(5) + 3]}{5} = \frac{23}{5}$$

Then we "invert" the fraction:

$$\frac{5}{23}$$

Thus the multiplicative inverse of $4\frac{3}{5}$ is $\frac{5}{23}$.

d. The multiplicative inverse of 2.1 is $\frac{1}{2.1}$, which we write without a decimal as

$$\frac{1}{2.1} = \frac{1 \times 10}{2.1 \times 10} = \frac{10}{21}$$

▶ *CHECK* **Warm-Up 4**

The multiplicative inverse enables us to write a fraction in three different ways:

> For real numbers a and b with b not equal to zero,
>
> $$\frac{a}{b} = a \times \frac{1}{b} = \frac{1}{b} \times a$$

Dividing Real Numbers

The result of dividing one number by another is called their **quotient.** The number we divide by is the **divisor,** and the number we are dividing is the **dividend.**

> To divide two real numbers ($a \div b$, where b is not zero)
>
> Divide the absolute value of a by the absolute value of b.
>
> **1.** If a and b have the same sign, the quotient is positive.
> **2.** If a and b have different signs, the quotient is negative.

▪▪▪

EXAMPLE 5

Divide: $-1908 \div (-9)$

SOLUTION

First we divide the absolute values of the two numbers.

$$|-1908| \div |-9| = 1908 \div 9$$

Because the original numbers have the same sign, the quotient is positive.

So $-1908 \div (-9) = 212$.

▶ *CHECK* **Warm-Up 5**

In our next example, we have to divide one fraction by another. To do so, we make use of the multiplicative inverse.

▪▪▪
EXAMPLE 6

Divide: $\left(-3\frac{3}{5}\right) \div \left(1\frac{1}{2}\right)$

SOLUTION

According to our division procedure, we first divide the absolute values of these mixed numbers.

$$
\begin{aligned}
\left|-3\frac{3}{5}\right| \div \left|1\frac{1}{2}\right| &= 3\frac{3}{5} \div 1\frac{1}{2} \\
&= \frac{18}{5} \div \frac{3}{2} \quad \text{Rewriting as improper fractions} \\
&= \left(\frac{18}{5}\right)\left(\frac{2}{3}\right) \quad \text{Multiplying by the reciprocal of } \frac{3}{2} \\
&= \frac{36}{15} \\
&= 2\frac{2}{5} \quad \begin{array}{l}\text{Reducing to lowest terms and} \\ \text{writing as a mixed number}\end{array}
\end{aligned}
$$

Because the signs of the original numbers were different, the quotient is negative: $-2\frac{2}{5}$.

So $\left(-3\frac{3}{5}\right) \div \left(1\frac{1}{2}\right) = -2\frac{2}{5}$.

▶ *CHECK* **Warm-Up 6**

Calculator Corner

Example 6 could be worked on your graphing calculator as follows. (Note: See the Calculator Corner on page 6 to review how to get an answer in fractional form.)

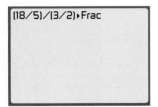

Press **ENTER.**

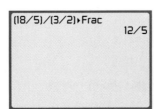

Question: Does it really matter that the parentheses are used in this problem? What would happen if the parentheses were left out?

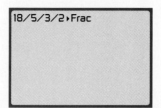

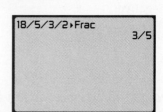

Press **ENTER.**

How did the calculator arrive at this answer? Write out the sequence of steps that the calculator did to arrive at the answer 3/5.

When we divide one decimal number by another, the procedure we use is similar to the one we use for integers.

▪▪▪

EXAMPLE 7

Divide: $0.364 \div (-0.02)$

SOLUTION

We need to divide the absolute values of the dividend and divisor.

$$|0.364| \div |-0.02| = 0.364 \div 0.02$$

But before we can divide, we must remove the decimal point from the divisor. We do this by multiplying the dividend and the divisor by a number that makes the divisor an integer. Using 100 gives us

$$\frac{0.364}{0.02} = \frac{(0.364)(100)}{(0.02)(100)} = \frac{36.4}{2}$$

We could also accomplish this step by "moving" the decimal point out of the divisor in the actual division. We would write

$$.02\overline{)36.4}$$

being careful to move the decimal point the same number of places in both divisor and dividend. We would then divide, as follows:

```
       18.2    Place the decimal point.
 .02) 36.4
      2         The result of 1 × 2
      16        Subtract and bring down 6.
      16        The result of 8 × 2
       04       Subtract and bring down 4.
        4       The result of 2 × 2
        0       No remainder
```

We place the decimal point in the quotient directly over the decimal point in the dividend. Because the signs of the original numbers were different, the result is negative:

$$0.364 \div (-0.02) = -18.2$$

▶ *CHECK* **Warm-Up 7**

Zero as the Divisor or Dividend

You must be careful when zero appears as a divisor or as a dividend. Remember the following rule:

■■■
WRITER'S BLOCK

Explain in your own words why $n \div 0$ does not have an answer.

> For all non-zero real numbers a,
> $$\frac{0}{a} = 0$$
> $$\frac{a}{0} \text{ is undefined}$$

When zero is divided by any non-zero number, the quotient is zero. Division by zero, on the other hand, has no meaning in the real numbers.

Signs in Fractions

> For all real numbers a and b,
> $$-\left(\frac{a}{b}\right) = -\frac{a}{b} = \frac{-a}{b} = \frac{a}{-b}$$

Using Cancellation to Multiply and Simplify

There is a short-cut way to multiply and simplify in the same step. Follow along in this example. Multiply and simplify as you go:

$$\frac{2}{3} \times \frac{9}{10} \times \frac{1}{18}$$

Begin by setting up the problem.

$$\frac{2}{3} \times \frac{9}{10} \times \frac{1}{18} = \frac{2 \times 9 \times 1}{3 \times 10 \times 18}$$

Next look and find which numbers have common factors and remove those factors. Here, we can divide the 2 in the numerator by 2 and the 10 in the denominator by 2. We cross out the numbers that have been divided, and we place the quotients above or below the numbers.

$$\frac{2 \times 9 \times 1}{3 \times 10 \times 18} = \frac{\overset{1}{\cancel{2}} \times 9 \times 1}{3 \times \underset{5}{\cancel{10}} \times 18}$$

This indicates the division of both numerator and denominator by 2. Next we can divide the 9 in the numerator and the 18 in the denominator by 3, yielding 3 in the numerator and 6 in the denominator; then we can divide the 3 and the 6 by 3 again, yielding 1 and 2 respectively. Note that we continue to cross out dividends and place the resulting quotient above or below the numbers.

$$\frac{\overset{1}{\cancel{2}} \times 9 \times 1}{3 \times \underset{5}{\cancel{10}} \times 18} = \frac{\overset{1}{\cancel{2}} \times \overset{3}{\cancel{9}} \times 1}{3 \times \underset{5}{\cancel{10}} \times \underset{6}{\cancel{18}}} = \frac{\overset{1}{\cancel{2}} \times \overset{\overset{1}{\cancel{3}}}{\cancel{9}} \times 1}{3 \times \underset{5}{\cancel{10}} \times \underset{\underset{2}{\cancel{6}}}{\cancel{18}}} = \frac{1 \times 1 \times 1}{3 \times 5 \times 2} = \frac{1}{30}$$

We could have shortened the process by dividing the numerator 9 and denominator 18 by 9 immediately.

Cancelling in this manner reduces the possibility of errors because it gives smaller multipliers. You may use this process on any multiplication or division problem in this chapter.

Practice what you learned.

SECTION FOLLOW-UP

Compare your estimation procedures in your group.

- Which procedure worked the best?
- Which procedures involved multiplication?

1.3 WARM-UPS

Work these problems before you attempt the exercises.

1. Multiply: $(-15)(493)$

2. Multiply: $\left(-6\frac{1}{4}\right)\left(-5\frac{2}{5}\right)$

3. Multiply: $(0.5)(-0.6)(-8.7)$

4. Find the multiplicative inverses of -4, $2\frac{1}{3}$, and -3.2.

5. Divide: $(-18) \div (9)$

6. Divide: $\left(-3\frac{1}{4}\right) \div \left(6\frac{1}{3}\right)$

7. Find the quotient: $(-0.3) \div (-5)$

1.3 EXERCISES   Note: Use your graphing calculator to check your results whenever possible.

1. Find two integers whose product is -7.

2. Find an integer that when multiplied by $16\frac{1}{3}$ results in a product of 0.

3. Find a real number that when multiplied by 9.3 results in a product of -9.3.

4. Find the missing integer x: $(-1)(x) = +5$

In Exercises 5 through 50, multiply.

5. $(6)(-1)$

6. $(-9)(-1)$

7. $(0)(12)$

8. $(9)(0)$

9. $(-6.7)(-1)$

10. $(-1)(-8.3)$

11. $(-18.7)(0)$

12. $(0)(-14.16)$

13. $(-5)(7)$

14. $(8)(-9)$

15. $(9)(-6)$

16. $(-7)(8)$

17. $(-112)(7)$

18. $(6)(-73)$

19. $(5)(-74)$

20. $(-8)(32)$

21. $(1)\left(5\frac{2}{6}\right)$

22. $\left(-3\frac{1}{2}\right)(0)$

23. $\left(\frac{2}{3}\right)(-1)$

24. $(-1)\left(4\frac{3}{10}\right)$

25. $(-6)\left(3\frac{5}{12}\right)$

26. $(-9)\left(8\frac{4}{15}\right)$

27. $(-7)\left(-4\frac{1}{3}\right)$

28. $\left(7\frac{5}{7}\right)(-7)$

29. $(-74)(0.08)$

30. $(-0.72)(-0.03)$

31. $(-0.42)(-0.02)$

32. $(-63)(0.08)$

33. $(-25)(-2)(6)$

34. $(-9)(3)(-4)$

35. $(-22)(2)(-8)$

36. $(7)(-5)(-13)$

37. $(-15)(8)(-9)$

38. $(-32)(5)(-3)$

39. $(-1674)(-18)(-1)$

40. $(23)(-8)(-10)$

41. $\left(-\frac{5}{11}\right)\left(\frac{2}{3}\right)\left(\frac{22}{25}\right)$

42. $\left(-\frac{1}{2}\right)\left(\frac{8}{11}\right)\left(\frac{3}{4}\right)$

43. $\left(-\frac{2}{5}\right)\left(\frac{5}{9}\right)\left(-\frac{1}{2}\right)$

44. $\left(-\frac{2}{3}\right)\left(-\frac{6}{7}\right)\left(-\frac{21}{32}\right)$

45. $(0.1)(-2)(0.3)$

46. $(0.5)(-6)(0.4)$

47. $(0.7)(-5)(0.8)$

48. $(0.3)(9)(0.2)$

49. $(-0.2)(-4)(-0.01)$

50. $(-2.1)(-1)(-0.02)$

In Exercises 51 through 58, find the multiplicative inverse.

51. $\frac{2}{7}$

52. $-\frac{3}{5}$

53. $-3\frac{4}{5}$

54. 0

55. -6

56. -5.8

57. $3\frac{1}{3}$

58. $-2\frac{1}{9}$

In Exercises 59 through 94, divide:

59. $-6 \div 0$

60. $-5.4 \div 0$

61. $0 \div 3.5$

62. $0 \div 2\frac{1}{2}$

63. $-2\frac{1}{8} \div 2\frac{1}{8}$

64. $9\frac{2}{7} \div \left(-9\frac{2}{7}\right)$

65. $-7.6 \div 1$

66. $-8.7 \div (-1)$

67. $6 \div (-73)$

68. $5 \div (-74)$

69. $(-25) \div (-2)$

70. $(-38) \div (-2)$

71. $230 \div (-10)$

72. $(-1674) \div (-18)$

73. $\left(-8\frac{2}{5}\right) \div \left(-1\frac{6}{7}\right)$

74. $\left(3\frac{3}{5}\right) \div \left(-2\frac{7}{10}\right)$

75. $\left(1\frac{1}{2}\right) \div \left(-3\frac{1}{4}\right)$

76. $(-3) \div \left(-1\frac{4}{7}\right)$

77. $\left(-\frac{4}{7}\right) \div \frac{1}{2}$

78. Divide (-74) by 0.08.

79. Divide (0.3) by (-9).　**80.** $(-0.16) \div (-0.2)$　**81.** Divide (-112) by 7.　**82.** Divide (-190) by (-38).

83. $\left(-6\frac{1}{3}\right) \div \left(-1\frac{4}{9}\right)$　**84.** $\left(-\frac{5}{11}\right) \div \frac{2}{3}$　**85.** $\left(-\frac{1}{2}\right) \div \frac{8}{11}$　**86.** $(-7) \div \left(-4\frac{1}{3}\right)$

87. Divide (-1) by $4\frac{3}{10}$.　**88.** $-11 \div 8\frac{4}{5}$　**89.** $2\frac{2}{5} \div \left(-6\frac{2}{5}\right)$　**90.** $7\frac{1}{7} \div \left(-7\frac{1}{2}\right)$

91. $(-82) \div 0.04$　**92.** $(-63) \div 0.07$　**93.** $(-2) \div (-14)$　**94.** $(-17) \div (-6)$

MIXED PRACTICE

By doing these exercises, you will practice the topics up to this point in the chapter.

95. Graph $-\frac{5}{6}$, $-\left(-\frac{7}{8}\right)$, and $\frac{4}{9}$, and show what two integers they fall between.

96. Simplify: $-|-3.6|$

97. Multiply: $\left(6\frac{1}{2}\right)\left(-8\frac{1}{2}\right)$

98. Simplify: $-\left|-\left|3\frac{3}{4}\right|\right|$

99. Arrange $-2\frac{6}{7}$, $-12\frac{5}{8}$, and $-(-2.64)$ in order from greatest to least using the symbol $>$.

100. Find the product: $(2.3)(0.002)$

EXCURSIONS

Class Act

1. A page from the 1988 *Manual on Uniform Traffic Control Devices,* published by the U.S. Department of Transportation, is shown below. For each arrow, find the size that may be used under low-speed urban conditions.

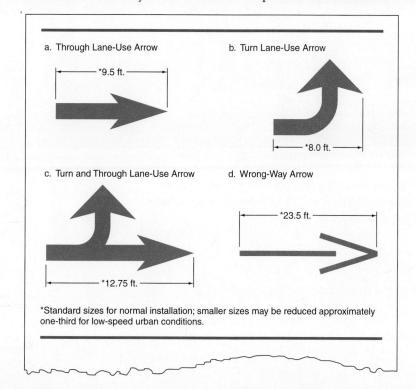

a. Through Lane-Use Arrow

*9.5 ft.

b. Turn Lane-Use Arrow

*8.0 ft.

c. Turn and Through Lane-Use Arrow

*12.75 ft.

d. Wrong-Way Arrow

*23.5 ft.

*Standard sizes for normal installation; smaller sizes may be reduced approximately one-third for low-speed urban conditions.

2. In parts (a) through (d), each of the problems has missing digits. Find them.

a.
```
    2__
 _7)74__
   ==
   __
   00
    0
```

b.
```
    ____
 4)92__
   92
   __
   00
    0
```

c.
```
    0_
 7)829_
   9
  __5
  3__
   0
```

d.
```
    0_
 8_)92_5
   5
  _6_
  7__
   0
```

e. Make up four missing digit problems. Swap them with another group.

f. ✏ Explain to a friend how to solve the missing digit problems (a) to (d).

3. The following table lists the average weight of various animals.

a. What is the range of the weights?

b. What is the total weight of the three lightest animals? the three heaviest?

Animal	Weight
hedgehog	1.88 pounds
chinchilla	1.5 pounds
ferret	2.04 pounds
beaver	58.5 pounds
guinea pig	1.54 pounds
wild boar	302.5 pounds

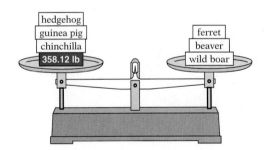

The balance scale above shows that the difference in weight between the three heaviest animals and the three lightest is 358.12 pounds. Rearrange the animals in groups so that the difference is

c. 354.04 pounds

d. 364.2 pounds

e. Arrange four animals so that the difference is 0.12 pounds.

f. Arrange three animals so that the difference is 1 pound.

g. ✏ Make up four other problems using this table. Solve them or swap with a classmate.

1.4 Other Operations on Real Numbers

SECTION LEAD-IN

a. Estimate the number of tickets available for each section in Yankee Stadium.

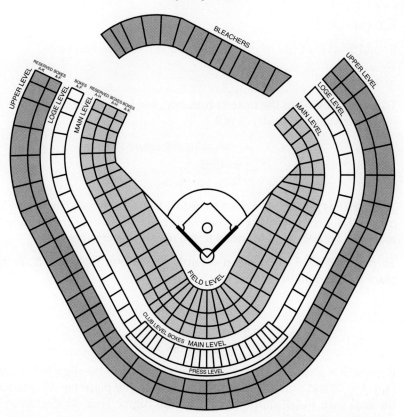

**Yankee Stadium
Capacity: 57,545**

Source: Diagram of Yankee Stadium as found in the Bronx Nynex Yellow Pages®. Reprinted by permission of Yankee Stadium.

b. ✏ Write a paragraph describing your estimation procedure.

For the most part, operations that can be applied to whole numbers can also be applied to real numbers.

Rounding Numbers

When we are computing, and particularly when we are solving application problems, we often need to **round** a decimal number to a certain place value.

> **To round a decimal number to a given place value**
>
> Locate the digit *to the right* of the place you are rounding to.
>
> 1. If the digit is less than 5, change it and all digits to the right to zero.
> 2. If the digit is greater than or equal to 5, add one to the digit *to its left,* and change all digits to the right of that digit to zero.
> 3. Drop any unnecessary zero digits.

Many people like to start by underlining the place they are rounding to.

•••

EXAMPLE 1

Round 98.4563 to the nearest hundred and to the nearest tenth.

SOLUTION

To round 98.4563 to the nearest hundred, we look at the digit in the *tens* place (to the right of hundreds). It is 9.

Rounding to the nearest hundred
$$\underline{\ \ }98.4563$$
Look at this digit.

The digit 9 is greater than 5, so we add 1 to the hundreds digit (0 + 1 = 1) and make all digits to the right 0. The result is 100.0000. We can drop the zero decimal part. Thus 98.4563 rounded to the nearest hundred is 100.

To round 98.4563 to the nearest tenth, we look at the digit in the *hundredths* place (to the right of tenths). It is 5.

Rounding to the nearest hundred
$$98.\underline{4}563$$
Look at this digit.

Because 5 is greater than or equal to 5, we add 1 to the digit in the tenths place (4 + 1 = 5) and make all digits to the right 0. The place we rounded to is in the decimal part, so we simply drop the zero digits. Thus 98.4563 rounded to the nearest tenth is 98.5.

▶ CHECK **Warm-Up 1**

Calculator Corner

Most graphing calculators have the capability of rounding a result to a specified number of decimal places. The calculator can be instructed to give answers in tenths as was asked for in Example 1.

Press **ENTER.**

Sometimes we use decimal numbers to express very large numbers! This might sound impossible, but the next example will illustrate it.

■■■

EXAMPLE 2

The following is a list of the oil-carrying capacities, in barrels, of four super-tankers. Express each amount to the closest tenth of a million barrels.

Bellamya	3,304,205
Pierre Guillaumat	3,307,317
Globtik London	2,820,658
Globtik Tokyo	2,796,883

SOLUTION

We are interested in rounding to the nearest tenth of a million. To do so, we first rewrite the numbers as millions by placing a decimal point immediately after the millions place. Then we can round them exactly as we rounded decimal numbers in the last few examples. We first get

Bellamya	3.304205 million
Pierre Guillaumat	3.307317 million
Globtik London	2.820658 million
Globtik Tokyo	2.796883 million

Rounding each to the nearest tenth, we obtain

3.304205 rounded is 3.3 million
3.307317 rounded is 3.3 million
2.820658 rounded is 2.8 million
2.796883 rounded is 2.8 million

Our list with rounded values looks like this:

Bellamya	3.3 million
Pierre Guillaumat	3.3 million
Globtik London	2.8 million
Globtik Tokyo	2.8 million

> **WRITER'S BLOCK**
> Describe two situations where you would want to use a rounded number.

▶ *CHECK* **Warm-Up 2**

Raising a Real Number to a Power

Real numbers can be raised to powers just as whole numbers and fractions can. In the exponential notation a^n, a is the **base** and n is the **exponent.** The exponent indicates how many times the base is used as a factor.

$$\text{Exponent} \longrightarrow$$
$$\text{Base} \longrightarrow 3^5 = \underbrace{(3)(3)(3)(3)(3)}_{\text{5 factors of 3}}$$

WRITER'S BLOCK

Explain to a classmate why it makes sense for a^0 to equal one and not zero.

To raise a real number to a power

For a real number a and a positive integer n,

$$a^n \text{ means } \underbrace{a \times a \times \ldots \times a}_{n \text{ factors}}$$

Also,

$$a^1 = a \quad \text{and} \quad a^0 = 1$$

Thus, for example, $35^3 = 35 \times 35 \times 35$ and $4^1 = 4$ and $5^0 = 1$. We have special names when 2 and 3 are used as exponents. We call 10^2 "ten squared" and 4^3 "four cubed" instead of saying "ten to the second power" and "four to the third power."

▪▪▪

EXAMPLE 3

Evaluate: $\left(-\frac{5}{7}\right)^2 (-5)^0$

SOLUTION

To raise $-\frac{5}{7}$ to the second power, we multiply it by itself.

$$\left(-\frac{5}{7}\right)^2 = \left(-\frac{5}{7}\right)\left(-\frac{5}{7}\right) = \frac{(-5)(-5)}{(7)(7)} = \frac{25}{49}$$

-5 raised to the zero power is 1.

So $\left(-\frac{5}{7}\right)^2 (-5)^0 = \left(\frac{25}{49}\right)(1) = \frac{25}{49}$.

▶ *CHECK* **Warm-Up 3**

Calculator Corner

Raise $-3/5$ to the fourth power.

```
(-3/5)^4▸Frac
```

Press **ENTER.**

```
(-3/5)^4▸Frac
                81/625
```

Finding Square Roots and Cube Roots of Real Numbers

The opposite process of raising a number to the second power, or squaring it, is finding the *square root* of a number.

> **Square Root**
>
> A **square root** of a positive real number a is another real number b that, when squared, results in a.

Thus if you had to find a square root of 49, you would need to ask, "What number when squared results in 49?" One such number is 7, because $7^2 = 49$. Another square root of 49 is -7, because $(-7)^2 = 49$.

Every positive real number has a positive square root and a negative square root. The positive square root of a real number a is written $\sqrt{a}$ and is called the **principal square root.** Thus $\sqrt{100} = 10$ because $10^2 = 100$. In words, the positive square root of 100 is 10 because the square of 10 is 100.

In a similar way, we say that a real number b is the **cube root** of a real number a if $b^3 = a$. As an example, we know that $2^3 = 2 \times 2 \times 2 = 8$, so 2 is the cube root of 8. We denote the cube root of a real number a by $\sqrt[3]{a}$. We will work more with square roots and cube roots later.

▪▪▪

EXAMPLE 4

Find:

a. the square of 9   **b.** the positive square root of 64

c. the cube of -1 **d.** the cube root of -125

> **! ! !**
> *ERROR ALERT*
>
> Identify the error and give a correct answer.
>
> Simplify: $(0.3)^2$
>
> *Incorrect Solution:*
>
> $(0.3)^2 = 0.9$

SOLUTION

a. $9^2 = 9 \times 9 = 81$

b. Because $8 \times 8 = 64$, we can write $\sqrt{64} = 8$.

c. $(-1)^3 = (-1)(-1)(-1) = -1$

d. Because $(-5)(-5)(-5) = -125$, we can write $\sqrt[3]{-125} = -5$.

▶ *CHECK* **Warm-Up 4**

Calculator Corner

a. Find the square of 9.

Press **ENTER.**

b. Find the positive square root of 64.

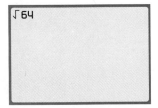
Press **ENTER.**

c. Find the cube of -1.

Press **ENTER.**
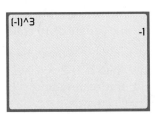

d. Find the cube root of -125.

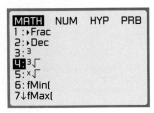
Press **ENTER.**

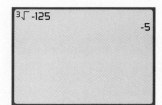

Press **ENTER.**

> **To find the principal square root of a perfect square**
>
> **1.** Find which two "tens" the square root lies between.
> **2.** Find the possible last digits of the square root.
> **3.** Test the possibilities.

...

EXAMPLE 5

Find: $\sqrt{7569}$

SOLUTION

The square root of 7569 is smaller than 90 because $90^2 = 8100$. It is larger than 80 because $80^2 = 6400$. Thus we know that $80 < \sqrt{7569} < 90$. Therefore, the first digit of the square root is 8.

The last digit of 7569 is 9, so the last digit of its square root must be 3 or 7. Why? Because $3 \times 3 = 9$ and $7 \times 7 = 49$, and no other digit has a square ending in 9. Try both.

$$83 \times 83 = 6889 \quad \text{Not what we are looking for}$$
$$87 \times 87 = 7569 \quad \text{Yes!}$$

Because $87 \times 87 = 7569$, the square root of 7569 is 87.

▶ *CHECK* **Warm-Up 5**

Estimating Answers

In some calculations, an approximate result is enough. **Front-end estimation** is useful when the numbers in a calculation include multidigit whole numbers or decimal numbers with whole number (or integer) parts.

> **To use front-end estimation**
>
> **1.** For each number, identify the place value of the first non-zero digit.
> **2.** Round the number to that place value.
> **3.** Use these rounded numbers in computation.

▪ ▪ ▪

EXAMPLE 6

Estimate the following sum, and then find the exact total.

$$
\begin{array}{r}
341{,}287 \\
274{,}384 \\
12{,}967 \\
+\,986{,}586 \\
\end{array}
$$

SOLUTION

We round to the greatest place in each number and add the rounded numbers.

341,287 rounds to	300,000
274,384 rounds to	300,000
12,967 rounds to	10,000
986,586 rounds to	1,000,000
Estimate of sum:	1,610,000

The actual sum is 1,615,224.

▶ *CHECK* **Warm-Up 6**

▪ ▪ ▪

EXAMPLE 7

a. Use front-end estimation to determine an estimate for:
$(1947)^2 + (3{,}287{,}210 \times 475)$

b. Estimate the result of the computation: $(0.00495)(4.3786)(0.00000999)$

WRITER'S BLOCK

Write a set of rules to use when you estimate the answer to a computation.

SOLUTION

a. We round 1947 to the nearest thousand: 2000. We round 3,287,210 to the nearest million: 3,000,000. And we round 475 to the nearest hundred: 500.

Our estimated computation becomes

$$(2000)^2 + (3{,}000{,}000 \times 500)$$
$$= 4{,}000{,}000 + 1{,}500{,}000{,}000$$
$$= 1{,}504{,}000{,}000$$

When some of the numbers in a calculation have only decimal parts, or a decimal number with a whole-number part less than 10, we use the same estimation procedure as before.

b. (0.00495) is rounded to 0.005.
(4.3786) is rounded to 4.
(0.00000999) is rounded to 0.00001.

Then the computation is estimated as: $(0.005)(4)(0.00001) = 0.00000020$.

▶ *CHECK* **Warm-Up 7**

Using Grouping Symbols

When we compute with signed numbers, we often use parentheses and other **grouping symbols** to set off, or group, certain numbers and operations. In addition to parentheses, the symbols we use for grouping include brackets [ ] and braces { }. When you encounter grouping symbols within grouping symbols, always work *from the inside out.*

Using the Standard Order of Operations

The **standard order of operations** is another concept that applies to all real numbers. It tells us the order in which to perform operations, so that each mathematical expression will represent the same value to everyone.

Standard Order of Operations

1. Simplify within grouping symbols, working from the innermost out.
2. Remove exponents.
3. Multiply and divide from left to right in order.
4. Add and subtract from left to right in order.

Let's use the standard order of operations to simplify an expression.

∎∎∎

EXAMPLE 8

Simplify: $14[5 - (2)(3) \div 7]$

SOLUTION

The grouping symbols here are used to show multiplication. There are no exponents. So we begin within the brackets by multiplying and dividing in order from left to right.

$$14[5 - (2)(3) \div 7]$$
$$14[5 - 6 \div 7] \qquad \text{Multiplying } 2 \times 3$$
$$14\left[5 - \frac{6}{7}\right] \qquad \text{Dividing } 6 \div 7$$

Then we subtract, still within the brackets.

$$14\left[4\frac{7}{7} - \frac{6}{7}\right] \qquad \text{Regrouping}$$
$$14\left[4\frac{1}{7}\right] \qquad \text{Subtracting } 5 - \frac{6}{7}$$

We rewrite $4\frac{1}{7}$ as an improper fraction.

$$14\left[4\frac{1}{7}\right] = 14\left[\frac{29}{7}\right]$$

■■■
WRITER'S BLOCK

Explain to a friend
how we can say that
$$\frac{14(29)}{7} = (2)(29)$$

We now multiply this fraction by 14.

$$14\left[\frac{(29)}{7}\right] = \frac{14(29)}{7} = (2)(29) = 58$$

So $14[5 - (2)(3) \div 7] = 58$.

▶ *CHECK* **Warm-Up 8**

The next example contains grouping symbols and exponents.

❗ ❗ ❗
ERROR ALERT

Identify the error and
give a correct answer.

Simplify:
$12 - 3 \times 2 + 4$

Incorrect Solution:

$12 - 3 \times 2 + 4$
$= 9 \times 2 + 4 = 22$

■ ■ ■
EXAMPLE 9

Simplify: $-(2.8 - 1.4)^2 + 3.2 - \left[\frac{5 + 7}{9 - (-3)}\right] \div (-2)$

SOLUTION

We work inside the parentheses first.

$$-(2.8 - 1.4)^2 + 3.2 - \left[\frac{5 + 7}{9 - (-3)}\right] \div (-2)$$
$$-(1.4)^2 + 3.2 - \left(\frac{12}{12}\right) \div (-2) \qquad \text{Subtracting, adding}$$
$$-(1.4)^2 + 3.2 - 1 \div (-2)$$

Then we eliminate the exponent.

$$-(1.4)^2 + 3.2 - 1 \div (-2)$$
$$-[1.96] + 3.2 - 1 \div (-2) \qquad \text{Squaring}$$

Because the opposite of 1.96 is -1.96, we rewrite this as

$$-1.96 + 3.2 - 1 \div (-2)$$

■■■
WRITER'S BLOCK

Explain to a friend
why we can write
$$-1 \div (-2)$$
as
$$-[1 \div (-2)]$$

Next we do the division.

$$-1.96 + 3.2 - [1 \div (-2)]$$
$$-1.96 + 3.2 - [-0.5] \qquad \text{Dividing}$$

Then we do the additions and subtractions from left to right.

$$-1.96 + 3.2 - [-0.5]$$
$$-1.96 + 3.2 + 0.5 \qquad \text{Rewriting } -[-0.5]$$
$$1.24 + 0.5 \qquad \text{Adding } -1.96 + 3.2$$
$$1.74 \qquad \text{Adding } 1.24 + 0.5$$

So $-(2.8 - 1.4)^2 + 3.2 - \left[\frac{5 + 7}{9 - (-3)}\right] \div (-2) = 1.74$.

▶ *CHECK* **Warm-Up 9**

Practice what you learned.

SECTION FOLLOW-UP

- What operations and mathematical procedures did you use in estimating the tickets available? Make an ordered list (the order in which you used them) of these procedures.

1.4 WARM-UPS

Work these problems before you attempt the exercises.

1. Round 1.9345 to the nearest unit (ones place), tenth, and thousandth.

2. Rewrite each of the following to the closest tenth of a thousand dollars.

   $4567
   $6477
   $8190
   $8566

3. Write $\left(-\frac{3}{7}\right)^3 (-1)^3$ without exponents.

4. Find the square of 16 and the principal square root of 81.

5. Find $\sqrt{576}$ and $\sqrt{961}$.

6. Round to the nearest hundred thousand and add:

$$
\begin{array}{r}
273{,}613 \\
985{,}274 \\
673{,}780 \\
+\ 11{,}427 \\
\end{array}
$$

7. Use front-end estimation to determine a reasonable estimate of:

   9876553 + 298 + 5763002035

8. Simplify: $5 - 6.2 \div 20 - 10$

9. Simplify: $-4.5\left[1.8 \div \left(-\frac{3}{5}\right)\right]^2 - 2.1$

1.4 EXERCISES

Note: Use your graphing calculator to check your results whenever possible.

In Exercises 1 and 2, round each number to the nearest ten, tenth, and thousandth.

1. a. 262.0024   **b.** 76.96318

    c. 984.339   **d.** 24.5908

2. a. 3.4814   **b.** 10.77249

    c. 0.3933249   **d.** 4.0132

In Exercises 3 and 4, the average personal income per capita in 1984 is given for several major cities. Round each number to the given place.

3. a. New York–New Jersey metropolitan area: $15,957 (nearest tenth of a thousand dollars)

    b. Chicago metropolitan area: $14,456 (nearest hundred dollars)

    c. Philadelphia metropolitan area: $13,785 (nearest tenth of a thousand dollars)

    d. Boston metropolitan area: $15,932 (nearest tenth of a thousand dollars)

    e. Washington, D.C. metropolitan area: $17,724 (nearest ten thousand dollars)

4. a. Los Angeles metropolitan area: $14,591 (nearest ten thousand dollars)

    b. San Francisco metropolitan area: $17,171 (nearest tenth of a thousand dollars)

    c. Detroit metropolitan area: $13,984 (nearest hundred dollars)

    d. Houston metropolitan area: $14,374 (nearest tenth of a thousand dollars)

    e. Dallas metropolitan area: $15,272 (nearest tenth of a thousand dollars)

The following table shows the population of the United States from 1810 to 1900. Use this information to answer the questions in Exercises 5 through 8.

U.S. Population from 1810 to 1900

1810	7,239,881	1860	31,443,321
1820	9,638,453	1870	39,818,449
1830	12,866,020	1880	50,155,783
1840	17,069,453	1890	62,947,714
1850	23,191,876	1900	75,994,575

5. In which year did the population, rounded to the nearest million, first reach forty million?

6. Which three years have the same population, rounded to the nearest ten million?

7. When did the population, rounded to the nearest hundred million, first reach one hundred million?

8. When did the population, rounded to the nearest ten million, first reach ten million?

In Exercises 9 through 22, round each of the underlined numbers to the indicated place.

9. In 1920 there was $5,698,214,612 in circulation. (millions)

10. The number of ways to get dealt three of a kind from a deck of 52 playing cards is 54,912. (thousands)

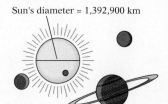
Sun's diameter = 1,392,900 km

11. The diameter of the sun is 1,392,900 kilometers. (hundred thousands)

12. The number of Americans injured by needles or pins in one year is 40,150. (hundreds)

13. Each year Americans eat about sixty-two billion fifty million pounds of canned tuna. (hundred millions)

14. Each year Americans buy about one hundred million, three hundred seventy-five thousand pounds of yarn. (hundred thousands)

15. Each day Americans eat about eight hundred fifteen billion calories of food. (billions)

16. The population of the United States in 1986 that was under 5 years of age was one hundred eighty-one million. (hundred millions)

17. It is estimated that by 2000, there will be 34,882,000 people over the age of 65 in the United States. (hundred thousands)

18. The federal acreage of the national parks system in 1986 was 75,862,484 acres. (millions)

19. In 1985 the population of Arkansas was 2,286,435. (thousands)

20. Monument Valley is 29,816 acres in size. (ten thousands)

21. Arizona's highest mountain, Humphreys Peak, is 12,655 feet high. (thousands)

22. Alaska has an area of five hundred eighty-six thousand square miles. (millions)

In Exercises 23 through 38, rewrite each expression without exponents.

23. $\left(1\frac{1}{5}\right)^2$

24. $\left(1\frac{2}{3}\right)^3$

25. $(-1.3)^3$

26. $-(0.00071)^0$

27. $(0.25)^2(-1)^3$

28. $(-0.38)^0(-1.5)^2$

29. $-(0.37)^2(-3)^3$

30. $(-0.005)^3(-1.2)^0$

31. $(0.04)^3(10)^3$

32. $-(-0.03)^2(3)^2$

33. $(0.021)^0(-1.9)^2$

34. $(0.23)^2(-1)^8$

35. $-(-0.0007)^4(0)^5$

36. $-(-0.005)^3(0)^3$

37. $-\left(-2\frac{1}{3}\right)^3(-9)^0$

38. $-\left(1\frac{3}{4}\right)^2\left(-\frac{1}{2}\right)^2$

In Exercises 39 and 40, the number on the left has been rounded. Of the other three numbers, circle those that it could have been before rounding.

39. **a.** 6.9   6.85   6.94999   6.9049

  b. 7.23   7.239   7.23   7.2349875

  c. 2.0   2.093   1.99   1.957867

40. **a.** 3.04   3.0399876   3.04499   3.045

  b. 2.19   2.1949   2.18997   2.1854

  c. 3.611   3.61107   3.61195   3.610997

In Exercises 41 through 44, find the requested power or root.

41. **a.** the square of 144   **42.** **a.** the principal square root of 9

  b. the principal square root of 36   **b.** the square of 25

  c. the cube root of 8   **c.** the cube of 3

43. **a.** the cube of 10   **44.** **a.** the cube root of 64

  b. the square of 16   **b.** the principal square root of 81

  c. the cube of 8   **c.** the cube root of 1000

In Exercises 45 through 48, find the square root of each perfect square.

45. **a.** 4356   **46.** **a.** 2704   **47.** **a.** 7056   **48.** **a.** 8836

  b. 3969   **b.** 3249   **b.** 7569   **b.** 9409

  c. 5041   **c.** 6241   **c.** 7921   **c.** 8464

  d. 4096   **d.** 5476   **d.** 6561   **d.** 9801

In Exercises 49 through 52, use your calculator to determine whether each of these numbers is a perfect square.

49. **a.** 7228   **50.** **a.** 5776   **51.** **a.** 22,504   **52.** **a.** 26,244

  b. 6089   **b.** 8286   **b.** 20,739   **b.** 14,404

Use estimation to solve Exercises 53 through 60.

53. ***Motor Vehicle Registration***   In 1987 total motor vehicle registrations in British Columbia, Québec, and Ontario were 2,175,032, 2,974,099, and 5,179,918, respectively. Estimate how many million vehicles were registered.

54. ***Revenue in China***   The 1987 estimated revenue for the Republic of China was $16,349,000,000. Estimated expenses were $16,329,200,000. Estimate the difference to the nearest ten million.

55. ***College Graduates***   There were 953,000 college graduates in the academic year 1981–1982 and 930,684 in the academic year 1971–1972. Estimate the change in the number of college graduates to the nearest ten thousand. Use your calculator to check your answer.

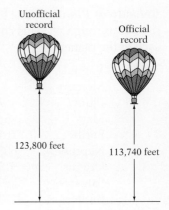

Unofficial record

Official record

123,800 feet

113,740 feet

56. *Hot Air Balloon*   The greatest official altitude reached by an occupied balloon is 113,740 feet, whereas the greatest unofficial altitude reached by an occupied balloon is 123,800 feet. How much higher is the unofficial record, estimated to the nearest thousand? Use your calculator to check your answer.

57. *Population*   Estimate the total population of these three cities to the nearest hundred thousand: Atlanta 426,090; Austin 397,001; Dallas 974,234. Use your calculator to check your answer.

58. *Movie Attendance*   The 18-and-over movie-viewing audience showed in one year that it had the following preferences:

> Drama: 12,810,000
> Suspense: 10,350,000
> Comedy: 14,220,000
> Adventure: 8,150,000

To the nearest million, estimate the greatest and the least number of people who could have been surveyed, if everyone chose at least one category. (*Hint:* What if each person could choose only one category? What if each person could choose as many as four categories?)

59. *A Marathon*   The fraction of women finishers of the New York City marathon in 1986 was 3323 of 19,689 total finishers, or $\frac{3323}{19,689}$. In 1991, 5204 women and 20,593 men finished. What was the fraction of women finishers in 1991? Did the fraction of women finishers increase or decrease from 1986 to 1991?

60. *Land Area*   Estimate the total land area of these five cities in the United States. Give your answer to the nearest ten square miles.

Columbus, Ohio	189.272 square miles
Honolulu, Hawaii	617 square miles
Washington, D.C.	68.25 square miles
Seattle, Washington	144.6 square miles
Cleveland, Ohio	79 square miles

In Exercises 61 and 62, use what you have learned about rounding to find the numbers.

61. *Volcano*   Mount Haleakala is a volcano with a crater large enough to hold Manhattan. It is 10,000 feet tall. If this number was rounded to the nearest ten thousand before being reported, what is the tallest and shortest height it could be?

62. *Parks*   Waimea Falls Park, in Oahu, is eighteen hundred acres in size. If this number was rounded to the nearest hundred before being reported, what is the least area and the most area the park could have?

Velocity of Planets   Use the following chart to estimate the answers to Exercises 63 through 66. This chart gives the rotational velocity, at the equator, of each of the planets.

Venus	4.05	miles per hour
Mercury	6.73	miles per hour
Pluto	76.56	miles per hour
Mars	538	miles per hour
Earth	1040	miles per hour
Neptune	6039	miles per hour
Uranus	9193	miles per hour
Saturn	22,892	miles per hour
Jupiter	28,325	miles per hour

63. About how many time faster is Pluto's rotation than that of Venus? Is it closer to:

20 times?   2 times?   200 times?   0.2 times?

64. About how many times faster is Earth's rotation than that of Pluto? Is it closer to:

13 times?   1.3 times?   130 times?   0.13 times?

65. About how many times faster is Jupiter's rotation than that of Mercury? Is it closer to:

400 times?   4000 times?   4 times?   0.4 times?

66. About how many times faster is Neptune's rotation than that of Mars? Is it closer to:

1.2 times?   12 times?   120 times?   0.12 times?

Of the following numbers, find the pair whose quotient is closest to each estimate given in Exercises 67 through 70. Use a calculator.

1598762	59887	1217788	1968777
312161	883260	32597	18212

67. 88   **68.** 4   **69.** 27   **70.** 33

In Exercises 71 through 90, simplify each expression.

71. $\left(8 + 1\frac{1}{3}\right)(-3)$

72. $26 + 1.8(6)$

73. $3.5 + (-1.7)(5)$

74. $(4 + 0.8)^2 + \left(1\frac{3}{5}\right)^2 - 15^0$

75. $[1.1 + (-5)](1.4)^2 - 2$

76. $(1.5 + 3)^2(1.3) - 11$

77. $3.3 + [(-1.3)(2)] - 1.7$

78. $[(1.1)^2(6)] + 15 \div 1.5$

79. $[(3)(0.5)]^2 + 4 \div 2.5 - 5^0$

80. $-(1.2)(4) + 1.6 \div (-0.4)^2$

81. $-(5.1)(2) + 2.7^2 \div 1.0$

82. $(-2.9 + 1.4)(2^5)^0 + (-0.5)$

83. $(1.3)^2 + 0.9 (3.7) + 16^0$

84. $-2\frac{1}{5} + 3\left(1\frac{1}{5}\right)^3 + 5 \div 5^2$

85. $\frac{-2}{9} + \frac{1}{3}(9) + 3\left(1\frac{1}{3}\right)^2$

86. $-0.9 + 13.5(2)$

87. $[(-1)^2 + (-1+1)^4 + (-1)(-1)^4 + (-1)^{11}]^9$

88. $[1 + (-1)^{11} - 1]^{13} + [(-1)^{18} + (1)^9]^2$

89. $[[1 - (1)]^3 - 1] + (-1)^{15} + (-1)^{18}$

90. $[(-1)^{18} + 1^{10} - (-1)^{16} - 1]^3 - 1$

MIXED PRACTICE

By doing these exercises, you will practice the topics up to this point in the chapter.

91. Divide: $25\frac{1}{2} \div 3$

92. Simplify: $\frac{1}{3} - 6^2 \div 3^3 + (-4)$

93. Round 16.77932 to the nearest tenth, hundredth, and thousandth.

94. Subtract: $-18\frac{1}{3} - \left(-9\frac{1}{18}\right)$

95. Simplify: $[1 \div (2 \div 3) \times 6]^2 - 18 \div 3 + 2$

96. Evaluate: $-|3-(-2)|$

97. Arrange $-(-3)$, $-\left|3\frac{1}{2}\right|$, and $-2\frac{1}{2}$ in order from greatest to least.

98. Evaluate: $-(-(-|-3|))$

99. Multiply: $\left(-\frac{1}{3}\right)\left(2\frac{1}{5}\right)\left(-3\frac{1}{2}\right)$

100. Is $\frac{(56)(493)(74,222)}{(69,999)(989)}$ closest to 3, to 30, or to 300?

EXCURSIONS

Class Act

1. Use the formula for relative humidity and the table, to answer parts (a) and (b).

$$\text{Relative humidity} = \frac{\text{partial pressure of } H_2O}{\text{saturated vapor pressure of } H_2O} \times 100$$

Saturated Vapor Pressure of Water

Temperature (°C)	torr (= mmHg)	Temperature (°C)	torr (= mmHg)	Temperature (°C)	torr (= mmHg)
−50	0.030	20	17.5	70	234
−10	1.95	25	23.8	80	355
0	4.58	30	31.8	90	526
5	6.54	40	55.3	100	760
10	9.21	50	92.5	120	1489
15	12.8	60	149	150	3570

a. On one very hot day, the temperature is 25°C and the partial pressure of water vapor in the air is 21.0 torr. What is the relative humidity?

b. When the temperature was 10°C, the relative humidity was 65%. Find the partial pressure of water vapor in the air.

2. Place +, −, ×, and ÷ signs between the numbers in each row to give the indicated answers. Use a calculator. Compare your results with other groups, and find each result as many ways as possible. Be sure to use ( ) and [ ] where needed.

a. 0 1 2 3 4 5 6 7 8 9 = 0   **b.** 0 1 2 3 4 5 6 7 8 9 = 1

c. 10 9 8 7 6 5 4 3 2 1 = 0   **d.** 0 1 2 3 4 5 6 7 8 9 = 2

e. 0 1 2 3 4 5 6 7 8 9 = −3   **f.** 0 1 2 3 4 5 6 7 8 9 = −4

g. 0 1 2 3 4 5 6 7 8 9 = 10   **h.** 10 9 8 7 6 5 4 3 2 1 = 1

i. 0 1 2 3 4 5 6 7 8 9 = 9   **j.** 0 1 2 3 4 5 6 7 8 9 = 100

Posing Problems

3. Use the following data about the 1992 Summer Olympics to ask and answer four questions. Exchange your questions with another group.

1992 Summer Olympics Numbers

2 Athletes on the teams from Albania, Cook Islands, Mali and Monaco.
5 Olympic appearances for British javelin thrower Tessa Sanderson and American runner Francie Larrieu Smith.
8 Olympic teams made by Michael Plumb of the U.S. equestrian team.
13 Age of Chinese diver Fu Mingxia.
30 World records in pole-vault set by Sergei Bubka.
64 Tons of fireworks for opening and closing ceremonies.
107 Medals won by the U.S. in Mexico City in 1968.
132 Medals won by the Soviet Union at 1988 Seoul Olympics.
144 U.S. athletes from California.
172 Countries in the Olympics.
257 Events at the Olympics.
610 U.S. athletes at the Olympics.
626 Medals won previously by American track and field athletes.
1,691 Total medals to be awarded.
1,768 U.S. medals won in all Summer Olympics.
9,172 Spaniards who helped carry the Olympic torch through the country.
2,350,000 Watts of electricity for opening and closing ceremonies.
401,000,000 Dollars paid by NBC for rights fees for the Olympics.
920,000,000 Dollars spent on Barcelona's new phone system.
1,800,000,000 Dollars spent to refurbish Barcelona's waterfront.
8,000,000,000 Dollars spent on refurbishing all of Barcelona.

Source: Associated Press.

Exploring with Calculators

4. **a.** Input each of the following expressions into your graphing calculator Home Screen and obtain the result. Then place parentheses around the operation that should be performed first.

$$2 + 3 - 1 \qquad 2 \times 3 - 1 \qquad 2 + 3 \times 4$$
$$2 \times 3 + 4 \qquad 2 - 3 \times 4 \qquad 2 + 3 - 4$$
$$5 \times 6 + 3 \times 2 \qquad 5 + 6 + 3 \times 2 \qquad 5 + 6 \times 3 + 2$$
$$5 + 6 \times 3 \times 2 \qquad 5 + 6 - 3 + 2$$

b. Insert $+$, $-$, or $\times$ one time each in each of the following sequences of numbers in order to obtain the given result.

$$4 \quad 5 \quad 3 \quad 2 = 21 \qquad 4 \quad 5 \quad 3 \quad 2 = 17 \qquad 4 \quad 5 \quad 3 \quad 2 = 19$$
$$4 \quad 5 \quad 3 \quad 2 = 3 \qquad 4 \quad 5 \quad 3 \quad 2 = -9 \qquad 4 \quad 5 \quad 3 \quad 2 = 5$$

CHAPTER LOOK-BACK

Whenever large numbers of people gather, the exact count is difficult to determine.

1. Describe three methods that you believe you could use to determine the number of people in this picture.

2. Which method in part (1) would give the answer closest to the correct number? Justify your answer.

3. Are crowd estimates likely ever to be exact? Give some reasons why you believe this.

4. What questions would you want someone to answer before you would believe their estimate?

CHAPTER 1
REVIEW PROBLEMS

The following exercises will give you a good review of the material presented in this chapter.

SECTION 1.1

1. Simplify: $|-34|$

2. Find the opposite of -25.

3. Evaluate: $|-(-24)|$

4. Graph -7, -8, and 2 on a number line.

5. Arrange $-\frac{23}{4}$, $\frac{13}{5}$, and $\left|-\frac{19}{6}\right|$, in order from least to greatest.

6. Find the opposite of -5.128, $-|-5.12|$, and $|-5.1|$.

7. Graph $-\left|\frac{9}{5}\right|$, $-\left|1\frac{1}{3}\right|$, and 2.4 on a number line.

8. Between what two integers does $-[-(-9.999)]$ lie?

SECTION 1.2

9. Add: $(-5.8) + (-0.9)$

10. Find the sum of 3.004 and (-1.6).

11. What is the total of (-3.7) and (-6.5) and 3.9?

12. Add: $\left(-9\frac{1}{8}\right) + \left(-16\frac{2}{5}\right)$

13. Add: $-2 + 4 + (-7)$

14. Add: $\left(-5\frac{3}{4}\right) + \left(-2\frac{1}{5}\right)$

15. Subtract: $35.4 - (-18.6)$

16. Subtract: $6\frac{6}{7} - \left(-4\frac{3}{5}\right)$

17. Subtract 12.4 from (-67.2).

18. Subtract: $\left(-13\frac{3}{4}\right) - \left(-9\frac{4}{7}\right)$

19. Subtract: $-26\frac{3}{7} - 4\frac{19}{21}$

20. What is the additive inverse of $-\frac{4}{5}$?

SECTION 1.3

21. Multiply: $(-12.5)(4.3)$

22. Multiply: $(-24.6)(12.5)$

23. Multiply: $(19)(-18)(26)$

24. Multiply: $\left(3\frac{15}{16}\right)\left(-4\frac{2}{3}\right)$

25. Multiply: $\left(-2\frac{8}{11}\right)\left(-3\frac{2}{5}\right)$

26. Multiply: $(-26.5)(-7.3)$

27. Divide (-90.8) by 0.02.

28. Divide $7\frac{4}{5}$ by $\left(-6\frac{3}{8}\right)$.

29. What is the multiplicative inverse of $8\frac{1}{5}$?

30. What is the multiplicative inverse of -0.01?

31. Divide -3069 by (-15).

32. Divide $\left(-16\frac{2}{3}\right)$ by $3\frac{5}{6}$.

SECTION 1.4

33. Write $(0.005)^3$ without exponents.

34. Simplify: $-\{-[(-6)^3]\}$

35. Simplify: $2 - \left[\left(\frac{1}{4}\right) \div \left(\frac{2}{3}\right) \div \left(-\frac{6}{8}\right)\right]^2 \times \left(3\frac{1}{2}\right)$

36. Write -7^2 without exponents.

37. Simplify: $[18 + 2(1 - 4)^3] \div 3 + 7^2$

38. Write $(-9)^2$ without exponents.

39. Round 7150.009 to the nearest thousand.

40. Round 0.0034999 to the nearest thousandth.

MIXED REVIEW

41. Arrange $-|2.4|, \left|-\left|\frac{8}{3}\right|\right|$, and $-|-9.5|$ in order from greatest to least.

42. What is the multiplicative inverse of -1?

43. Simplify: $\left(\frac{3}{5}\right) + \left(\frac{1}{4}\right) \div \left(-\frac{5}{9}\right) - \left(4\frac{2}{3}\right)$

44. Add: $\left(-6\frac{1}{2}\right) + 3\frac{1}{4}$

45. Simplify: $-|\{-[-(0.012)]\}|$

46. Divide: $\left(-6\frac{4}{7}\right) \div 8\frac{2}{5}$

47. What is the additive inverse of -1?

48. Simplify: $13 - 6(2) \div 9 + 12$

49. Subtract: $\left(-3\frac{3}{7}\right) - \left(-6\frac{1}{2}\right)$

50. Multiply: $\left(5\frac{3}{7}\right)\left(-6\frac{1}{8}\right)$

51. Subtract $42 - (19.602)$

52. Between what two integers does $-\left|-\left|-13\frac{1}{2}\right|\right|$ lie?

53. Divide (-11.46) by (-0.004).

54. Divide 7.008 by (-0.002).

CHAPTER 1 TEST

This exam tests your knowledge of the material in Chapter 1.

1. Simplify:

    a. $|-6.7|$ **b.** $-|-5|$ **c.** $-\left|-\left|-3\frac{1}{2}\right|\right|$

2. Arrange each group of numbers in order from least to greatest.

    a. 3.042, 3.04, 3.4   **b.** $\left|\frac{3}{8}\right|, 0.374, 0.37532$ **c.** $\frac{2}{5}, -\frac{5}{12}, \frac{3}{7}$

3. Add:

    a. $(-84) + (-96.5)$ **b.** $\left(-3\frac{2}{5}\right) + 4\frac{1}{2}$ **c.** $\left(-6\frac{1}{5}\right) + 7\frac{2}{3} + \left(-11\frac{1}{2}\right)$

4. Subtract:

    a. $\left(-41\frac{3}{8}\right) - \left(-12\frac{1}{8}\right)$ **b.** $0.34 - (-2.7)$ **c.** $\left(-11\frac{2}{9}\right) - 6\frac{3}{7}$

5. Multiply:

   a. $\left(-\frac{8}{9}\right)\left(-\frac{7}{12}\right)$ **b.** $(0.3)(-7.2)(0.09)$ **c.** $\left(16\frac{5}{7}\right)\left(-3\frac{8}{9}\right)$

6. Divide:

   a. $(-0.6) \div 1.8$ **b.** $\left(-\frac{26}{35}\right) \div \left(-\frac{4}{7}\right)$ **c.** $\left(11\frac{2}{3}\right) \div \left(-2\frac{1}{9}\right)$

7. Simplify, using the standard order of operations:

   a. $2(6-3)^2 + (9-8)^8$ **b.** $-4(6.4-1.8)^2 - 3.4(-9.6 \div 2)$ **c.** $\frac{1}{4}\left(1\frac{1}{3} - 2\frac{2}{3}\right)^2 \div \frac{3}{4}(-1)^6$

INTRODUCTION TO SETS AND ALGEBRAIC NOTATION

*H*ow many meteorites have struck Earth in the last 500 million years? Say one impact occurs every 10,000, or 10^4, years. Let the variable y stand for "number of years ago." Using the algebraic expression $\left(\frac{1}{10^4}\right) y$, you can estimate that there have been 50,000 impacts.

■ *Use a variable to describe a familiar situation.*

SKILLS CHECK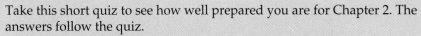

Take this short quiz to see how well prepared you are for Chapter 2. The answers follow the quiz.

1. Add: $(-5) + (-3)$

2. Add: $-7\frac{1}{6} + 2\frac{5}{7}$

3. Add: $3.5 + 6$

4. Subtract: $6 - 4.32$

5. Subtract: $(-87) - (-28)$

6. Multiply: $\left(-\frac{1}{2}\right)(2)$

7. Multiply: $\left(6\frac{2}{3}\right)\left(5\frac{1}{4}\right)$

8. Multiply: $(-2.5)(-7.36)$

9. Simplify the expression: $6 \div 2 + 4 - 7$

ANSWERS: **1.** -8 [Section 1.2]   **2.** $-4\frac{19}{42}$ [Section 1.2]   **3.** 9.5 [Section 1.2]   **4.** 1.68 [Section 1.2]
5. -59 [Section 1.2]   **6.** -1 [Section 1.3]   **7.** 35 [Section 1.3]   **8.** 18.4 [Section 1.3]   **9.** 0 [Section 1.4]

CHAPTER LEAD-IN

Many problems have to be translated into language that we understand before we can solve them.

- About how many miles have you walked in the last year?
- About how many feet high is your school building?
- About how long does your hair grow in a minute?
- About how many times does your heart beat in 10 years?

2.1 The Language of Sets and Problem Solving

SECTION LEAD-IN

Before you work a word problem, you must read and interpret the given information and determine what question is being asked.

"There are 3 times as many students as there are professors. . . ." So starts a famous problem that many students find difficult to express using algebraic notation.

1. Please fill in the following blanks:

   a. There are 90 students, so there are _____ professors.

   b. There are 18 professors, so there are _____ students.

2. Complete these sentences:

   a. The number of students is _____ the number of professors.

   b. The number of professors is _____ the number of students.

The language of a problem situation is extremely important.

An Introduction to the Language of Sets

A **set** is a collection of objects. The objects are called **members,** or **elements,** of the set. We name sets with capital letters, and we identify the members of a set by enclosing them in braces { }. Thus the statement $A = \{a, *, 1, 2\}$ tells us that A is the set that contains the elements a, $*$, 1, and 2.

The set that contains no elements is called the **empty set** and is denoted by either empty braces or the symbol $\emptyset$ (but not both). The set E of whole numbers between 1 and 2 is an empty set, because there are no whole numbers between 1 and 2. We write $E = \{ \ \}$ or $E = \emptyset$.

Describing Sets

A set is correctly described when we can tell, from the description, whether any given object is a member of the set. Two methods are most often used. In **roster notation,** the elements of the set are listed. Set A above is given in roster form, and so is $N = \{1, 2, 3, 4, \ldots\}$. You should recognize N as the set of natural numbers. Set A is a finite set (its members can be counted), whereas set N is an infinite set (its members cannot be counted). We designate an infinite set by including the ellipsis symbol ($\ldots$).

In **set-builder notation,** a typical element is named (usually with a letter) and carefully described. Thus

$$N = \{x \mid x \text{ is a natural number}\}$$

is a set-builder description of the set of natural numbers. The thin vertical line is read as "such that," so the entire set-builder statement is read, "N is the set of all elements x such that x is a natural number."

WRITER'S BLOCK

How do we read the set-builder notation for T and X in Example 1?

EXAMPLE 1

a. Write the following sets in roster notation:

$$T = \{t \mid t \text{ is an even whole number less than } 18\}$$
$$X = \{x \mid x \text{ is an integer greater than } -6 \text{ and less than } 4\}$$

b. Write the following sets in set-builder notation:

$$Y = \{1, 3, 5, 7, \ldots\} \qquad Z = \{-2, -1, 0, 1\}$$

SOLUTION

a. Set T contains all the even whole numbers less than 18. We list them within braces.

$$T = \{0, 2, 4, 6, 8, \ldots, 16\}$$

And the description of set X tells us that

$$X = \{-5, -4, -3, -2, -1, 0, 1, 2, 3\}$$

b. The elements of Y are all the odd whole numbers, so

$$Y = \{y \mid y \text{ is an odd whole number}\}$$

The elements of Z are the integers greater than -3 and less than 2, so

$$Z = \{a \mid a \text{ is an integer } and \text{ } a \text{ is greater than } -3 \text{ and less than } 2\}$$

▶ *CHECK* **Warm-Up 1**

Subsets

Set A is said to be a **subset** of set B when every element of set A is also an element of set B. The empty set is a subset of every set.

Set Operations

Two operations on sets, *union* and *intersection,* produce new sets.

Set Union

The **union** of two sets A and B is the set of all elements contained in either A or B or in both.

The union of A and B is written $A \cup B$.

Set Intersection

The **intersection** of two sets A and B is the set of all elements that are contained in both A and B.

The intersection of sets A and B is written $A \cap B$.

▪ ▪ ▪

EXAMPLE 2

a. Let $A = \{1, 2, 5\}$ and $B = \{2, 3, 4, 5\}$. Find $A \cup B$.

b. Given $E = \{x \mid x$ is a whole number$\}$ and $F = \{x \mid x$ is an integer less than 3$\}$, find $E \cap F$.

SOLUTION

a. $A \cup B$ contains the elements that are in either A or B.

We start by listing the elements in A:

$$\{1, 2, 5 \qquad \}$$

Then we include those additional elements from set B:

$$\{1, 2, 5, 3, 4\}$$

We usually give the answer as a set with the elements arranged in order.

$$A \cup B = \{1, 2, 3, 4, 5\}$$

b. We are looking for elements that these two infinite sets have in common. We can see those elements more clearly if we write the sets in roster form, one above the other. We get

$$E = \{0, 1, 2, 3, 4, \dots\}$$
$$F = \{\dots, -2, -1, 0, 1, 2\}$$

Now we can see that E and F have only the elements 0, 1, and 2 in common. So $E \cap F = \{0, 1, 2\}$.

▶ CHECK **Warm-Up 2**

> **STUDY HINT**
>
> *Each element in a set is listed only once.*

Set Complements

> Two sets A and A' are **complements** of each other when the intersection of the two sets is $\varnothing$ and their union is U, the universal set.

> ▪▪▪
>
> **WRITER'S BLOCK**
> Distinguish between
> $\{\ \}$ and $\{0\}$.

▪ ▪ ▪

EXAMPLE 3

Let $U = \{1, 3, 5, 7, 9\}$ and $A = \{1, 5\}$. Find A'.

SOLUTION

Start with the universal set $\{1, 3, 5, 7, 9\}$. The set A' will have all the same elements except for 1 and 5:

$$\{\cancel{1} \quad 3 \quad \cancel{5} \quad 7 \quad 9\}$$

So $A' = \{3, 7, 9\}$.

▶ CHECK **Warm-Up 3**

▪▪▪

WRITER'S BLOCK

Explain when two sets will have the following property:

$A \cup B = A$
and
$A \cap B = \emptyset$

Complements have certain properties. If A and A' are complements, then

$$A \cap A' = \emptyset$$
$$A \cup A' = U$$

An Introduction to Problem Solving

We will be applying algebra to solve a variety of problems—especially word problems—in this text. The best way to attack such problems is with a systematic method, step by step. Here's one that works for many people.

Interpret the problem. You must read and understand the problem. Determine what information is given, what you are actually asked to find, and what information you need to supply.

Decide on a method for solving the problem. Do you have to add, subtract, multiply, or divide? There are sometimes several different ways to approach a problem. Decide which way will work best. Can you solve the problem directly, or must you solve for a piece of information first and then use it to find the answer?

Apply the method. Perform the actual computations in the proper order to obtain an answer.

Look back and reason. Once you have an answer, go back and check two things:

1. Check your calculations to make sure that you have not made any computational errors.
2. Reread the original question to make sure your answer is reasonable and valid. (An answer of $2\frac{1}{2}$ might be computationally accurate, but if you were asked for the number of children in a family, it would not be a reasonable answer.)

All together, these four steps are:

Four Steps for Problem Solving

1. INTERPRET the problem.
2. DECIDE on a method for solving the problem.
3. APPLY the method.
4. Look back and REASON.

▪▪▪

EXAMPLE 4

Blood flows through the kidneys at a rate of 1.3 liters per minute. How much blood flows through the kidneys in 1 hour?

SOLUTION

INTERPRET We are given the amount of blood that flows each minute (1.3 liters). We are asked how much blood flows in 1 hour. We know there are 60 minutes in an hour.

DECIDE   To find the amount of blood flowing in 1 hour, we must multiply the amount of blood that flows in 1 minute by 60 minutes.

APPLY   Multiplying yields

$$\frac{1.3 \text{ liters}}{\text{minute}} \times \frac{60 \text{ minutes}}{\text{hour}} = 78 \frac{\text{liters}}{\text{hour}}$$

REASON   Rereading the problem, we see that the blood flows at slightly more than 1 liter per minute, and our result shows the amount per hour to be slightly more than 60 liters per hour (60 minutes). The answer is reasonable. We can check the result by dividing 78 by 60. The answer checks.

So 78 liters of blood flow through the kidneys each hour.

▶ CHECK **Warm-Up 4**

...

EXAMPLE 5

A certain roller coaster ride lasts $1\frac{5}{6}$ minutes. If the entire ride is 3250 feet long, what is the average speed in feet per second?

SOLUTION

INTERPRET   The ride is 3250 feet long and takes $1\frac{5}{6}$ minutes to complete. We want the speed in feet traveled per second. We know there are 60 seconds in a minute. (From here on, we will not indicate the units throughout the computations; however, we will always include them in our answer.)

DECIDE   To find the time the ride takes in seconds, we must multiply $1\frac{5}{6}$ minutes times 60 seconds per minute. Then we can divide the distance traveled (3250 feet) by the time traveled in seconds. We will do the computations and then indicate the answer in feet per second.

APPLY   First we find the time in seconds.

$$\left(1\frac{5}{6}\right) \times 60 = \left(\frac{11}{6}\right)(60) = 110 \text{ seconds}$$

Then we divide the length of the ride (3250 feet) by the time needed to travel it (110 seconds).

$$\frac{3250}{110} = 29\frac{60}{110} = 29\frac{6}{11}$$

The roller coaster travels at $29\frac{6}{11}$ feet per second.

REASON   According to the original problem statement, the roller coaster travels about 3000 feet in about 100 seconds. Our result, then, should have been about 30 feet per second. The answer, $29\frac{6}{11}$ feet per second, is reasonable.

▶ CHECK **Warm-Up 5**

In the next example, you will need to know that the **markup rate** is the percent by which a dealer increases the cost to obtain the selling price. Note that the

! ! !
ERROR ALERT

Identify the error and give a correct answer.

$6000 divided between 2 people:
$$\frac{\$6000}{2} = \$3000$$
This answer is incomplete.

discount rate is the percent by which the original cost is reduced. The **percent of increase or decrease** is the ratio of the amount of change to the original amount.

■ ■ ■
EXAMPLE 6

A coat costs retailers $180. It is sold for $230. What is the markup rate?

SOLUTION

INTERPRET   We are being asked, "What percent of the cost is the given markup amount?" Because the amount the coat has been marked up is $230 − $180, or $50, we must find

"What percent of $180 is $50?"

Here, the *percent* is the missing value, the *base* is $180, and the *amount* is $50.

DECIDE   We substitute the known values into the **percent equation.**

$$\text{Percent (as a decimal)} \times \text{base} = \text{amount}$$
$$n \times 180 = 50$$

A good estimate of the answer is $50 \div 200$, or 0.25. So 25% is the estimated markup rate.

APPLY   We solve the equation by dividing both sides by 180.

$$\frac{n \times 180}{180} = \frac{50}{180}$$
$$n = 0.277\overline{7} \quad \text{Dividing}$$

Rounded to the ten-thousandths place, n is 0.2778.

REASON   The solution thus far is a decimal, but we are asked for a *percent*. So we must move the decimal point two places to the right and add a percent sign.

Thus the markup rate is 27.78%.

▶ CHECK Warm-Up 6

In the next example, we have to find a missing piece of information first and then use it to solve the problem.

■ ■ ■
EXAMPLE 7

> **WRITER'S BLOCK**
>
> What do you have to INTERPRET to work the problem in Example 7?

An inventory of spare parts showed that 168 had rusted. This represented 56% of the total number of parts. How many parts had not rusted?

SOLUTION

DECIDE   Solving this problem requires two steps. First, we must find the total number of parts. Then we can subtract the number of rusted parts from that to find the number that did not rust. We know that 168 is 56% of the total number of parts.

The *percent* is 56%, which gives us the ratio $\frac{\text{part}}{\text{whole}} = \frac{56}{100}$. The *base* (or *whole*) is the missing value, and the *amount* (or *part*) is 168; they give us the ratio $\frac{168}{n}$.

APPLY   These fractions are equal. We set up a **proportion** and solve.

$$\frac{56}{100} = \frac{168}{n} \qquad \text{The proportion (equal fractions)}$$

$$56 \times n = 100 \times 168 \qquad \text{Cross-multiplying (If } \tfrac{a}{b} = \tfrac{c}{d}, \text{ then } ad = bc.)$$

$$56 \times n = 16{,}800 \qquad \text{Multiplying out}$$

$$\frac{56 \times n}{56} = \frac{16{,}800}{56} \qquad \text{Dividing by 56}$$

$$n = \frac{16{,}800}{56} = 300 \qquad \text{Simplifying}$$

REASON   The total number of parts is 300.

DECIDE   Now we must subtract the number of rusted parts from the total.

APPLY $300 - 168 = 132$

REASON   So 132 parts were not rusted.

▶*CHECK* **Warm-Up 7**

Practice what you learned.

SECTION FOLLOW-UP

Let x be the number of professors and y be the number of students. Then ordered pairs (x, y) can be written that satisfy the equation

Number of students = 3 times number of professors

$(30, 90)$ is one ordered pair that shows this relationship.

1. Write three ordered pairs that fit this relationship.

2. Fill in the blanks to make ordered pairs that fit this relationship.

   {(35, _____), (100, _____), (_____ , 432), (_____ , 1101)}

3. Finish this statement using the variables as given in part 2:

   $x = $ _____ y

4. Finish this statement using the variables as given in part 2:

   $y = $ _____ x

2.1 WARM-UPS

Work these problems before you attempt the exercises.

1. Describe the set of integers greater than -2 and less than 5, using both roster and set-builder notation.

2. Let $A = \{1, 3, 5, 7\}$
   $B = \{2, 3, 4, 5\}$
   $C = \{1, 4, 5\}$

   a. Find $A \cup B$.   **b.** Find $(A \cup B) \cup C$.

3. Let $A = \{1, 2, 3, 4, 5\}$
   $B = \{1, 3, 5\}$
   $U = \{1, 2, \ldots, 10\}$

   a. Find A'.   **b.** Find B'.   **c.** Find $(A \cap B)'$.

4. An astronaut can circle Earth in a space shuttle in 1.5 hours. How many complete times can the astronaut circle Earth in exactly 1 day?

5. If the roller coaster from Example 5 traveled a mile at its average speed, how long would it take to complete the trip? (Give your answer in minutes.)

6. A price ticket on a suit said "PRICE REDUCED $75." If the original price was $250, what is the discount rate (based on original price)?

7. On a certain day, 22% of the workers called in sick. If the company employs 150 people, how many people did not call in sick?

2.1 EXERCISES

Note: Use your graphing calculator to check your results whenever possible.

In Exercises 1 through 4, determine whether the sets are empty.

1. $\{0\}$

2. $\{y \mid y$ is an even whole number between 3 and 6$\}$

3. $\{x \mid x$ is even and odd$\}$

4. $\{y \mid y$ is a natural number less than 1$\}$

In Exercises 5 through 12, write the sets in roster form.

5. $\{x \,|\, x$ is an even whole number$\}$

6. $\{y \,|\, y$ is an integer greater than -5 and less than 3$\}$

7. $\{t \,|\, t$ is a prime number (a number with exactly two factors—itself and one) less than 30$\}$

8. $\{$the odd numbers between 4 and 6$\}$

9. $\{v \,|\, v$ is a positive-integer factor of 12$\}$

10. $\{$positive even-number factors of 50$\}$

11. $\{y \,|\, y$ is an integer greater than -5 and less than 10$\}$

12. $\{w \,|\, w$ is a whole number between 17 and 21$\}$

In Exercises 13 through 16, answer true or false.

13. The set of integers is a subset of the set of real numbers.

14. The set of real numbers has the set of rational numbers as a subset.

15. The set of natural numbers is a subset of the set of integers.

16. The empty set is a subset of the irrational numbers.

In Exercises 17 through 20, find the indicated result when $A = \{0, 2, 3, 4, 5\}$; $B = \{0, 3, 5, 9\}$; and $C = \{0\}$.

17. $(A \cup B) \cup C$

18. $(A \cap B) \cap C$

19. $(A \cup B) \cap C$

20. $C \cap (A \cup B)$

In Exercises 21 through 24, find $A \cup B$ and $A \cap B$.

21. $A = \{$whole numbers$\}$
$B = \{$integers$\}$

22. $A = \{$rational numbers$\}$
$B = \{$integers$\}$

23. $A = \{x \,|\, x$ is an even whole number$\}$
$B = \{y \,|\, y$ is an odd whole number$\}$

24. $A = \{t \,|\, t$ is an integer multiple of 3$\}$
$B = \{z \,|\, z$ is an integer multiple of 6$\}$

25. *High Jump*   In the 1988 summer Olympics in Seoul, Korea, the American who won the heptathlon, Jackie Joyner-Kersee, had come to the high jump with a previous personal-best jump of 6 feet 4 inches. In the Olympic event, however, she jumped 6 feet $1\frac{1}{4}$ inches. By how much did she miss her personal best?

26. *High Jump*   In the 1996 summer Olympics in Atlanta, the winner of the heptathlon, Ghada Shouaa from Syria, jumped 6 feet 1.23 inches. By how much did she miss Jackie Joyner-Kersee's 1988 Olympic jump? her personal best jump? (Refer to Exercise 25.)

27. *Caves*   The Cuyaguatega cave system in Cuba is $32\frac{7}{10}$ miles long. The Flintridge cave system in the United States is $148\frac{7}{10}$ miles longer. How long is the Flintridge system?

28. *Glaciers*   The Antarctic glacier flows $84\frac{3}{5}$ yards in a week; the Greenland glacier flows $236\frac{9}{10}$ yards in the same time. How much farther does the Greenland glacier flow in a week?

29. *Baseball*   The distance from the pitcher's mound to the home plate on a baseball diamond is $60\frac{1}{2}$ feet. If a player's stride is $2\frac{1}{2}$ feet long, how many steps would he take to walk from the mound to home plate?

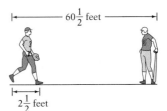

30. *Travel*   The road distance (in miles) between Memphis and Pittsburgh is approximately $1\frac{1}{4}$ times the air distance (in statute miles) between Pittsburgh and Memphis. If the air distance is approximately six hundred sixty statute miles, how many miles is the road distance?

31. *Counterfeit Money*   During the Civil War, it was estimated that nearly one-third of all the currency in circulation was counterfeit. Approximately 1600 state banks were designing and printing their own money, and there were about 7000 different varieties of genuine bills. If there were 2 million bills in circulation, about how many were counterfeit? (*Hint:* Beware of unneeded information.) Use 0.333 as an estimate for $\frac{1}{3}$.

32. *Classic Cars*   Only six 1930 Bugatti Royale open cars were produced. If the total sales from these cars was $270,000 and they all sold for the same amount, about how much did each car sell for? Round your answer to the nearest whole number.

33. *Football*   The most points ever scored by one team over another in a football game was scored by Georgia Tech in 1916. The total number of points scored was 222. If a touchdown gives a team 6 points, what is the greatest number of touchdowns that could have been scored by Tech?

34. *Weight Lifting*   The heaviest weight ever lifted by a woman was lifted by Josephine Blatt in 1895. She lifted the equivalent of 26 women of average weight. If a woman of average weight weighs 135 pounds, how much weight did Blatt lift?

35. *Frog Long Jump*   A trained frog can jump 17 feet $6\frac{3}{4}$ inches. If a man can jump about $\frac{3}{4}$ that distance, how far can he jump?

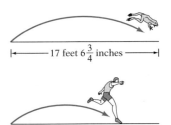

36. *Horse High Jump*   A horse has jumped 8 feet $1\frac{3}{4}$ inches high. The 1994 pole vault record is twice that plus $46\frac{1}{6}$ inches. About how high is the 1994 pole vault record jump?

37. *Ocean Liner*   The Verrazano Narrows Bridge is how many times longer than the width of the Queen Elizabeth ocean liner if the bridge is 4260 feet long and the ocean liner is 119 feet wide? Round your answer to the nearest whole number.

38. *Roller Coaster Ride*   Six Flags Over Georgia, an amusement park, boasts one of the world's largest and tallest wooden roller coasters, the Scream Machine. It is 105 feet tall, and the cars travel 57 miles per hour (0.95 miles per minute). It takes about 1.5 minutes to complete one trip. How many miles long is the trip?

39. *Mortgage Rate*   A mortgage company advertises that its current home mortgage rate is 11.75%, which represents a drop of 0.6% from the previous rate. What was the old rate?

40. *Stock Prices*   A certain stock has fluctuated as follows: up $\frac{4}{8}$, down $\frac{3}{8}$, up $\frac{2}{8}$, up $\frac{1}{8}$. If the opening price was $3\frac{5}{8}$, what is the current price?

41. *Heart Rate*   Your heart rate changes with the amount of energy that you exert. If your heart rate was 70 beats per minute (bpm) before you exercised, increased 40 bpm during exercise, and then dropped 10 bpm after exercise, what was your heart rate after exercise?

42. *Sea Level*   An airplane is at an altitude of 29,000 feet above sea level. It flies over a valley that has a low point 25 feet below sea level. How far above the valley "floor" is the airplane?

43. *Temperature Range*   The temperatures in a certain part of the world have been known to range from $-23°F$ to $102°F$. What is the range of these temperatures?

44. *Plant Growth*   A certain plant can grow at any altitude from 35 feet below sea level to 72 feet above sea level. What is the range of its "growth zone"?

45. *Stock Prices*   The stock for a certain company closed yesterday at $31\frac{1}{8}$. Today it opened $3\frac{3}{8}$ points lower than yesterday's closing, and then it went to a level $6\frac{2}{8}$ points higher than yesterday's closing price. What is the range of these fluctuations?

46. *Sea Level*   If Mt. Everest's base were at the bottom of the Marianas Trench, deep under the Pacific Ocean, its peak would be 7169 feet below sea level. Mt. Everest is 29,028 feet tall. How deep is the Marianas Trench?

47. *Scale Readings*   The changes in a reading on a scale were -2, $3\frac{1}{2}$, 10.5, -16, -15.5, and -4.5. What does the scale read now in relation to its starting point?

48. *Tsunami*   The highest recorded tsunami (tidal wave) was 220 feet high. The Statue of Liberty is 305 feet tall with its pedestal. How much taller is the statue than this wave?

49. *Hammer Throw*   In the 1996 summer Olympics, Lance Deal, the silver medal winner from the United States, threw the hammer 76.94 meters, 75.62 meters, 77.26 meters, and 81.12 meters. By how much did his longest throw exceed his shortest throw?

50. *Ocean Size*   There are 139,500,000 square miles of salt water oceans in the world. This is 139,170,000 more than the number of square miles of fresh water lakes. How much area is covered by fresh water lakes?

51. *Bowling*   In 1984 there were 8,401,000 members in the American Bowling Congress. This was 868,000 fewer than in 1983. What was the enrollment in 1983?

52. *Rent Increases*   Your rent increases 15% after the first year and 10% more the second year. If your rent before any increases was $200, what is your rent after the two increases?

53. *Sales Tax*   Find the sales tax on a dress that costs $124.80 if the tax rate is 5.5%.

54. *Discount Prices*   A coat is marked "20% off." How much money will you save if the coat was originally priced at $175?

55. *Price Reduction*   A bicycle originally priced at $220 was reduced by $52.80. What was the percent reduction?

56. *Land Area*   Ice covers 10% of the land area of Earth. If Earth's land area is approximately 58,433,000 square miles, how much is covered by ice?

57. *Carbon Weight*   Anthracite is 95% carbon by weight. If a specimen weighs 24 ounces, how much of it is carbon?

58. *Fish Harvest*   95% of the 75 million tons of fish harvested each year come from the oceans. How many millions of tons of fish come from other water sources?

59. *Ocean Area*   The surface area of Earth is approximately 196,949,970 square miles. Approximately 147,712,470 square miles are covered by oceans. Round these numbers to the nearest million, and determine what percent of Earth is covered by oceans. Round to the nearest whole percent. What percent is land?

60. a. *Grade Point Average* Su's freshman GPA was 3.15. It went up by 0.34, down by 0.15, and up by 0.08 in her next three years. What was her final GPA?

b. ✏ How is GPA calculated at your school?

61. *Sea Level* In Africa the highest point of elevation is Mt. Kilimanjaro at 19,340 feet above sea level. The lowest point is a spot in Egypt that is 436 feet below sea level. A man is standing at the lowest spot. He is 6 feet $3\frac{1}{3}$ inches tall. Another man, 5 feet $8\frac{1}{4}$ inches tall, is standing on top of Mt. Kilimanjaro. Find the difference in the elevations of the tops of these men's heads. Draw a picture showing this problem situation.

62. *Meter Reader* A meter is checked and is found to read over or under the actual level by these amounts: $+0.004$, -0.36, -0.004, $+0.5$, and -0.003. What is the range of these readings?

63. a. *Tip Size* A bill for a meal in a restaurant is $56. If you decide to leave a tip of 18%, what is the total cost of the meal?

b. ✏ In some locations, people "double the tax" to determine the size of a tip. Determine the tax on food in your location. Would you consider a "tax-doubled" tip sufficient? Explain.

64. a. *Realtor Commission* How much of a commission will a Realtor in a large city receive on a sale of a $246,000 house if the rate of commission is 12%?

b. ✏ Write a plan that a Realtor can use to find what price house she would need to sell to earn a given commission.

65. a. *Hockey* When David Williams was one of the San José Sharks, he scored his first game-winning NHL goal on February 6, 1992, against the Chicago Blackhawks. The Sharks eventually won 5 to 2. What fraction of the Sharks' goals did Williams score? What fraction of the total goals did he score? What fraction of the total goals did the Blackhawks score? Write these fractions in order from smallest to largest.

b. *Research* What does the phrase "game-winning goal" mean in hockey?

66. *Cross-Country Runners* According to *Runner's World* magazine (March 1992), the first successful cross-country run (coast to coast) was in 1890. It took John Ennis 80 days and 5 hours to complete the run. The fastest crossing, in 46 days, was accomplished 90 years later.

a. In what year was the fastest crossing?

b. *Research* Find the distance of a "cross-country" run.

67. **Trip Reimbursement** Dr. Callaham drives from Dahlonega to Athens for a class (70 miles), then to Atlanta for a meeting (65 miles), and then back through Athens to Dahlonega. If he gets $0.23 reinbursement per mile driven for travel expenses when he drives to meetings and classes, how much should he be paid for this trip?

68. **a. Crossword Puzzles** A crossword puzzle appeared for the first time in the *New York Times* magazine section in 1942. If a puzzle appeared each Sunday after that, how many puzzles appeared in the next 51 years? (Disregard leap years and their effect on your solution.)

　b. ✏ Explain to a friend how to find the results if leap years are included.

69. **Motorcycles** Motorcycles often have a two-cylinder engine. To determine the engine size, multiply the displacement of the cylinder by the number of cylinders. The displacement (D) of one cylinder is found by using the formula

$$\text{displacement} = \pi \times \left(\frac{\text{bore}}{2}\right)^2 \times \text{stroke}$$

where

$$\pi \approx 3.14$$
$$\text{bore} = \text{diameter of the cylinder in centimeters}$$
$$\text{stroke} = \text{distance a piston moves in centimeters}$$
$$\text{displacement} = \text{volume of a cylinder in cubic centimeters}$$

Find the displacement for a two-cylinder Harley Davidson® motorcycle with a bore of 88.8 millimeters and a stroke of 108.0 millimeters.

70. **a. Pottery Sale** Peggy and Michael have a pottery sale the first Sunday in December each year. They feed breakfast to everyone who comes to the sale. It costs an average of $1.39 for each person, and they spent $300 on breakfast this year. How many people attended?

　b. ✏ Plan a breakfast for a family or a group of friends. How much will it cost per person? Describe how you found your estimated cost per person.

Use the following chart to answer Exercises 71 through 74.

Menu Planning

Serving size	Calories	Serving size	Calories
beef, rib, lean roasted (4 oz)	273	rice, boiled (4 oz)	400
chicken, mixed meat, roasted (4 oz)	206	potatoes, baked (4 oz)	115
carrots, cooked (4 oz)	20	apple pie (1 slice)	350
cabbage, cooked (4 oz)	16	chocolate chip cookie (1)	50

71. Plan a meal that consists of beef or chicken, a vegetable, potato or rice, and a dessert and that has fewer than 390 calories.

72. Plan a meal that consists of beef or chicken, a vegetable, potato or rice, and a dessert and that has more than 1000 calories.

73. Plan a meal that consists of beef or chicken, a vegetable, potato or rice, and a dessert and that has fewer than 1000 calories and more than 950 calories.

74. ✏ Write your own question that can be answered using the Menu Planning table. Answer your question.

75. *Travel Time*   Len Pikaart used to drive his Corvette all over the state of Georgia to give math workshops for teachers. He usually averaged 695 miles a week during the school year, and each semester was 15 weeks long. How many semesters would it take him to travel 150,000 miles?

76. *Text Author*   Professor Nakahara writes mathematics books for his students. He can write 7 problems every 10 minutes. How many hours will he have to work to write 6000 math problems?

77. *Elevator Operator*   Loraine runs an elevator at a college. The elevator holds 23 people. How many trips must Loraine make to transport 123 people from floor three to floor twelve?

78. *Runner*   Ruth runs around the lower loop in the park. The distance is 1.7 miles. How many times must she run the loop in order to have run a total distance of at least 26 miles?

79. *Newspaper Prices*   Each week, Babatunde buys the *New York Times* 5 times, the *New York Post* 6 times, the *Wall Street Journal* twice, and *USA Today* once. In addition, he buys the Sunday *New York Times*. The cost of each paper is shown in the following table. What does Babatunde spend in a year for papers?

	Daily	Sunday
New York Times	$0.60	$2.00
New York Post	$0.50	$1.00
Wall Street Journal	$0.75	—
USA Today	$0.75	—

80. *Long-Distance Charges*   One long-distance phone company advertises that on the weekend, it costs 11 cents a minute to call anywhere in the United States. Stacie and A.J. speak long-distance from Twentynine Palms to Sacramento for 1 hour and 28 minutes one weekend, using this phone plan. How much does this call cost?

81. *Sea Level*   Albuquerque is 4944 feet above sea level. Mexico City is about 1.463 times as high. How much higher is Mexico City? Give your answer to the nearest 10 feet.

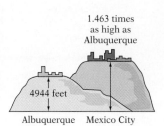

82. *Creek Widths* Black Creek in Mississippi ranges in width from 6.67 yards to 33.33 yards. What is the difference between these two widths? A duck in the center of this creek is how far from the shore at the widest part?

83. *Kitty Treats* Each year, on my cat Rail's birthday, I feed her 10 treats for each year old she is. When she is 10 years old, how many birthday treats will she have eaten altogether?

84. *Landscaping* A woman wants to arrange 36 stones into the shape of a square with the same number on each side to surround a flower bed. How many stones will there be on each side?

EXCURSIONS

Class Act

1. **a.** The tallest skyscrapers in the United States are: in Chicago—the Sears tower (1454 feet tall, 110 stories), the Standard Oil Building (1136 feet tall, 80 stories), and the John Hancock Center (1127 feet tall, 100 stories); and in New York—The World Trade Center (1350 feet tall, 110 stories) and the Empire State Building (1250 feet tall, 102 stories).

   Ask four questions that can be answered using the skyscraper data. Answer those questions.

   b. Find some data that you can use to make up some word problems. Swap your word problems with another group.

Posing Problems

2. Use the following data to ask and answer four questions. Try to ask questions that other students would want to answer.

Manufacturers' Dollar[1] Shipments of Recordings (in millions)

	1991	1992	1993	1994
Vinyl singles	63.9	66.4	51.2	47.2
LP's/EP's	29.4	13.5	10.6	17.8
CD's	4,337.7	5,326.5	6,511.4	8,464.5
Cassettes	3,019.6	3,116.3	2,915.8	2,976.4
Cassette singles	230.4	298.8	298.5	274.9
CD singles	35.1	45.1	45.8	56.1
Music videos	118.1	157.4	213.3	231.1

1. List price value. *Source:* Recording Industry Association of America, Inc., as cited in the *1996 Information Please® Almanac* (©1995 Houghton Mifflin Co.), p. 741. All rights reserved. Used with permission by Information Please LLC.

3. Ask and answer four questions about the data in the following table.

National Recreation Areas

Name and location	Total acreage	Name and location	Total acreage
Amistad (Tex.)	58,500.00	Gauley River (W. Va.)	10,300.00
Bighorn Canyon (Wyo.-Mont.)	120,296.22	Glen Canyon (Ariz.-Utah)	1,236,880.00
Chattahoochee River (Ga.)	9,259.91	Golden Gate (Calif.)	73,179.90
Chickasaw (Okla.)	9,930.95	Lake Chelan (Wash.)	61,886.98
Coulee Dam (Wash.)	100,390.31	Lake Mead (Ariz.-Nev.)	1,495,665.52
Curecanti (Colo.)	42,114.47	Lake Meredith (Tex.)	44,977.63
Cuyahoga Valley (Ohio)	32,524.76	Ross Lake (Wash.)	117,574.59
Delaware Water Gap (Pa.-N.J.)	67,204.92	Santa Monica Mountains (Calif.)	150,050.00
Gateway (N.Y.-N.J.)	26,310.93	Whiskeytown-Shasta-Trinity (Calif.)	42,503.46

Source: 1996 Information Please® Almanac (©1995 Houghton Mifflin Co.), p. 578. All rights reserved. Used with permission by Information Please LLC.

CONNECTIONS TO *PROBABILITY*

Introduction to Probability

> The **probability** of an event is the numerical likelihood that the event will occur.

In an **experiment,** we gather data. We can list all possible outcomes in a universal set called a **sample space,** S. Any subset of this sample space is called an **event.**

We can describe any event by using a rule (such as set-builder notation) or by listing the outcomes (such as roster notation). Each outcome in a sample space is a **simple event.**

When we toss two coins, we have four possible **outcomes.** Let H stand for heads and T for tails, then the possible outcomes can be written as HH, HT, TH, TT. The sample space of the experiment is $\{HH, HT, TH, TT\}$. Experience shows us that each of these events is **equally likely** and has a **probability** of

$$p = \frac{1}{\text{total number of outcomes in sample space}} \quad \text{or} \quad \frac{1}{n(S)}$$

So the probability of HH, written $p(HH)$, is $\frac{1}{4}$; $p(HT) = \frac{1}{4}$; $p(TH) = \frac{1}{4}$, and $p(TT) = \frac{1}{4}$.

The sum of the probabilities of the simple events in a sample space is 1.

To give the probability of any event, we use the rule:

Probability of event E ⟶ $p(E) = \frac{n(E)}{n(S)}$ ←Number of outcomes in E
←Number of outcomes in S

> **WRITER'S BLOCK**
>
> Explain why *HT* and *TH* are both listed.

▪▪▪

EXAMPLE

Toss two coins. Find

a. the probability of getting exactly one head.

b. the probability of getting at least one tail.

c. the probability of getting two heads or two tails.

SOLUTION

a. The event of getting exactly one head is $\{HT, TH\}$.

$$\text{Let } A = \{HT, TH\}$$

The probability is

$$p(A) = \frac{n(A)}{n(S)} = \frac{2}{4}$$

b. $B = \{HT, TH, TT\}$, so

$$p(B) = \frac{n(B)}{n(S)} = \frac{3}{4}$$

c. $C = \{HH, TT\}$, so

$$p(C) = \frac{n(C)}{n(S)} = \frac{2}{4}$$

We may reduce the fractions or leave them in this form.

◢

Just as with sets, two events A and B can be complementary, if $A \cap B = \varnothing$ and $A \cup B = S$.

▪▪▪

EXAMPLE

Let $S = \{HH, HT, TH, TT\}$. Let event A be no more than one head. Let event B be no tails.

a. List A and B.

b. Find the probability of each event.

c. Use the rules $A \cap A' = \varnothing$ and $A \cup A' = U$, and show that A and B are complements.

SOLUTION

a. $A = \{HT, TH, TT\}$

$B = \{HH\}$

b. $p(A) = \frac{n(A)}{n(S)} = \frac{3}{4}$

$p(B) = \frac{n(B)}{n(S)} = \frac{1}{4}$

c. $A \cap B = \{HT, TH, TT\} \cap \{HH\} = \emptyset$

   $A \cup B = \{HT, TH, TT\} \cup \{HH\} = \{HH, HT, TH, TT\}$

   Since $\{HH, HT, TH, TT\} = U$, we can say

   $$A \cup B = U$$

◢

Note that in part (b), $p(A) + p(B) = 1$. This property is true for all complementary sets.

If A and A' are complements of each other,

$$p(A) + p(A') = 1$$
$$1 - p(A) = p(A')$$
$$1 - p(A') = p(A)$$

▪▪▪

EXAMPLE

Ten balls with numbers from 0 to 9 are placed in a container. The chance, or probability, of choosing the ball with the number 6 on it, if you draw just one ball, is 1 out of 10, or $\frac{1}{10}$. Under these same conditions, what are the chances of *not* drawing the ball with the number 6 on it?

SOLUTION

The number required by the problem is the chance that the number 6 will *not* be drawn. This number is equal to 1 minus the chance that it will be drawn. So we write 1 as $\frac{10}{10}$ and subtract $\frac{1}{10}$.

$$\frac{10}{10} - \frac{1}{10} = \frac{9}{10}$$

Thus there is a $\frac{9}{10}$ probability of *not* drawing the ball with a 6 on it.

◢

PRACTICE

1. Toss one die. Find the probability of getting a 3. Find the probability of getting an even number.

2. Toss two coins. Find the probability of getting at least one head. Find the probability of getting three heads.

3. Show that the events {2, 4, 6} and {1, 3, 5} are complements when tossing one die is the experiment.

4. Flip three coins. Are the events {all heads} and {no heads} complements of each other? Explain.

5. The probability that it will rain tonight is $\frac{1}{3}$. The probability that it will not rain is what?

6. The probability that I will get tickets for the Collective Soul concert is $\frac{3}{7}$. The probability that I will not get tickets is what?

In the experiment of tossing three coins, let $S = \{HHH, HHT, HTH, THH, HTT, THT, TTH, TTT\}$.

Event A is the event of exactly one tail.
Event B is the event of two or more heads.
Event C is the event of exactly two heads or exactly two tails.
Event D is no more than one head.
Event E is no less than two heads.

7. **a.** Find $p(A)$ **b.** Find $p(B)$

8. **a.** Find $p(C')$ **b.** Find $p(E')$

9. **a.** Give an event F different from A, B, C, D, or E. List the elements and give the probability of F.

 b. Make up a different sample space and define four events of that sample space. List the elements of S and your four events.

10. **a.** You have probably heard the expression, "You're one in a million." What is the probability that you are *not* one in a million?

 b. The chances of winning a certain scratch-off game are 3 out of 35. What is the probability of losing?

2.2 Algebraic Expressions

SECTION LEAD-IN

In each of the following problems, ask and answer a question that can be answered using the given data.

a. In 1987, the U.S. Post Office sold $31,028,300,000 worth of stamps.

b. In 1997, the cost of first class postage is $0.32 for each letter up to one ounce and then 20 cents for each additional ounce.

c. Fresh tuna steak costs $4.49 a pound.

d. Two hundred twenty people paid 75 cents each to buy a *Wall Street Journal*.

e. One-half cup of three-bean salad has 120 calories.

    f. One hundred percent cotton flannel costs five-and-a-half dollars a yard.

    g. One-fourth of a 9-inch pie is 405 calories.

    h. One-half cup of strawberry ice cream has 232 calories and 0.75 cup of strawberry frozen nonfat yogurt has 135 calories.

The Language of Algebraic Expressions

In arithmetic, we use the digits 0 through 9 to write the symbols for numbers. Each of these symbols $\left(\text{such as 23 and } 151\frac{2}{3}\right)$ always represents the same number, so they are called *constants*.

> **Constant**
>
> A **constant** is a symbol that stands for a certain number. It has a fixed value.

In algebra, we also use symbols that represent *unknown numbers*. We call them *variables*.

> **Variable**
>
> A **variable** is a symbol that stands for a number that is yet to be determined. Its value can vary.

Almost any written symbol can be used to represent a variable, but we generally use letters. Because a variable stands for a number, we can multiply it by itself, obtaining such products as x^2, a^5, and w^n. Sometimes letters are also used to represent constants. We are usually told when a constant is represented by a letter.

> **Term**
>
> A **term** consists of variables and/or constants connected by multiplication or division.

A term has two parts—a *variable part* and the *coefficient* of that variable. Some examples of terms are $\frac{1}{n}$, 3, x, $4.7x^3y^2$, and $4n$. These terms are also called algebraic expressions.

> **Algebraic Expression**
>
> An **algebraic expression** consists of numbers and variables linked together by any operation.

Thus $x + 4$, y^2, and $x^2 - 3x + 4$ are all algebraic expressions.

The sign that precedes each term in an expression is considered part of that term. We usually write each term with the variables in alphabetical order. So we would write xy instead of yx.

■ ■ ■
EXAMPLE 1

Identify the terms in the algebraic expression $2y^2 + 3 - 7xy + 5x^2y$.

SOLUTION

There are four terms in this expression. The terms, with their signs, are

$$+2y^2 \qquad +3 \qquad -7xy \qquad +5x^2y$$

We usually don't write the plus signs, so we can write these terms as $2y^2$, 3, $-7xy$, and $5x^2y$.

▶ *CHECK* **Warm-Up 1**

Remember that a term has two parts—a variable part and its coefficient.

> **Variable Part**
>
> The **variable part** of a term consists of all the variables, together with their exponents.

> **Coefficient**
>
> The **coefficient** of a term is the number, along with the sign, that multiplies the variable part. A term with no number has the coefficient 1 or -1, depending on its sign.

A term that consists only of a constant is sometimes called a **constant term.** Remember that its variable part is simply raised to the zero power. So, 3 can be thought of as $3x^0$ and x as $1x$.

■ ■ ■
EXAMPLE 2

Name the coefficient and the variable part in each term of the algebraic expression $2x^2 - xy + y^2 + 5$.

SOLUTION

The coefficient of a term is the number with its sign, so the coefficients are, in order, 2, -1, 1, and 5.

The variable parts of the first three terms are, in order, x^2, xy, and y^2. The fourth term has no variable part that we need list.

▶ *CHECK* **Warm-Up 2**

Evaluating Algebraic Expressions

To "evaluate an algebraic expression" means to find its numerical value. We can do this only when we know the values represented by its variables.

> **To evaluate a given algebraic expression**
> 1. Substitute the numerical value for each variable.
> 2. Use the standard order of operations to simplify the expression.

▪ ▪ ▪
EXAMPLE 3

Evaluate $-fn(a^2 - g \div n)$ when $a = 4, f = 3, g = 0,$ and $n = -2.$

SOLUTION

First we substitute the values.

$$-fn(a^2 - g \div n)$$
$$= -(3)(-2)[4^2 - 0 \div (-2)] \quad \text{Substituting}$$

Next we use the order of operations to simplify, clearing the parentheses first.

$$= -(3)(-2)[16 - 0 \div (-2)] \quad \text{Squaring}$$
$$= -(-6)[16 - 0] \quad \text{Multiplying/dividing}$$
$$= -(-6)(16) \quad \text{Subtracting}$$
$$= 6(16) \quad \text{Multiplying}$$
$$= 96$$

Thus $-fn(a^2 - g \div n)$ is 96 when $a = 4, f = 3, g = 0,$ and $n = -2.$

▶ *CHECK* **Warm-Up 3**

Calculator Corner

The graphing calculator can evaluate any expression containing one or more variables. The variables *do not* always have to be the commonly used x and y. As long as you have **STO**red a numerical value for each variable that is used in the expression, the calculator will find the result. This is an excellent way to check your own work.

Evaluate $x - a^2$ when $x = 20$ and $a = -4$ using the Home Screen and the **STO**re features of your graphing calculator. On Texas Instruments Graphing calculators, press the $\boxed{\text{STO▶}}$ key to produce the arrow you see on screen. The calculator "**STO**res" the value 20 for x and -4 for A and then uses them to evaluate $x - A^2$, as shown in the accompanying figure.

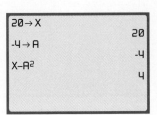

▪ ▪ ▪

EXAMPLE 4

Evaluate $\dfrac{-6n(2x^3 - t)^2 + 8n}{n - t}$ when n is $\frac{1}{2}$, t is -3, and x is -2.

SOLUTION

First we substitute the values.

$$\frac{-6n(2x^3 - t)^2 + 8n}{n - t} = \frac{-6\left(\frac{1}{2}\right)[2(-2)^3 - (-3)]^2 + 8\left(\frac{1}{2}\right)}{\frac{1}{2} - (-3)}$$

Then we use the order of operations to simplify. We start with the numerator. Working within the brackets, we simplify the exponent first and then do the multiplication and subtraction.

$$= \frac{-6\left(\frac{1}{2}\right)[2(-8) - (-3)]^2 + 8\left(\frac{1}{2}\right)}{\frac{1}{2} - (-3)} \qquad \text{Clearing exponent}$$

$$= \frac{-6\left(\frac{1}{2}\right)[-16 + 3]^2 + 8\left(\frac{1}{2}\right)}{\frac{1}{2} - (-3)} \qquad \text{Simplifying}$$

$$= \frac{-6\left(\frac{1}{2}\right)[-13]^2 + 8\left(\frac{1}{2}\right)}{\frac{1}{2} - (-3)} \qquad \text{Adding}$$

Next we simplify the exponent and multiply as indicated.

$$= \frac{-6\left(\frac{1}{2}\right)(169) + 8\left(\frac{1}{2}\right)}{\frac{1}{2} - (-3)} \qquad \text{Squaring}$$

$$= \frac{-507 + 4}{\frac{1}{2} + 3} \qquad \text{Multiplying}$$

We then do additions.

$$= \frac{-503}{3\frac{1}{2}} \qquad \text{Adding}$$

This is the same as saying -503 divided by $3\frac{1}{2}$.

$$-503 \div 3\frac{1}{2} = -503 \div \left(\frac{7}{2}\right) = -503 \cdot \left(\frac{2}{7}\right)$$

$$= -\frac{1006}{7} \text{ or } -143\frac{5}{7}$$

So when n is $\frac{1}{2}$, t is -3, and x is -2,

$$\frac{-6n(2x^3 - t)^2 + 8n}{n - t} = -143\frac{5}{7}$$

▶ CHECK **Warm-Up 4**

Practice what you learned.

SECTION FOLLOW-UP

Now go back and ask another question for each problem situation that uses a different operation.

2.2 WARM-UPS

Work these problems before you attempt the exercises.

1. Identify the terms in the algebraic expression $-4y^3 - 2y^2 + 5$.

2. Name the coefficient and the variable parts in each term: $3x, 4x, -3xy$.

3. Evaluate $-f(c^2 - g \div h)$ when $f = -3, c = 4, g = 1$, and $h = -2$.

4. Evaluate $\dfrac{-3x(2x - 9)^2 + 3t}{t + x}$ when $x = \frac{1}{4}$ and $t = \frac{1}{2}$.

2.2 EXERCISES

Note: Use your graphing calculator to check your results whenever possible.

In Exercises 1 through 8, identify the terms, coefficients, and variable parts in each algebraic expression.

1. $3x - 2x + 3y + 8x$

2. $-3x^2 + 7y - 8y - 5z$

3. $-9y^3 + 7y^2 + 7y + 4$

4. $3t^4 + 2t^3 + 3t^2 + 12t - 9$

5. $-7ef - 8ef + 7ef + 7ef^2$

6. $9ac - 5ac + 4ac - ac + ac^2$

7. $7fg - 8fg + 6fg - 6f + 7f^2g^2$

8. $-x + 2x - y + 3y - 3y + x$

In Exercises 9 through 20, evaluate the expression when

$$f = 2 \qquad h = -1 \qquad c = 0 \qquad t = 3 \qquad e = -2$$

9. cf^2

10. f^2h

11. $f^2 + hc - t$

12. $t^2e + t^2 \div h^2$

13. $c \div fh$

14. cht

15. eft

16. $ef \div h$

17. $fh + c$

18. $fh - t$

19. $ef + h$

20. $ft - e$

In Exercises 21 through 32, evaluate the expression when

$$c = 2 \qquad e = -\frac{1}{2} \qquad f = -\frac{3}{4} \qquad h = 4 \qquad t = -2$$

21. $(ce^2 \div f^2)htf^2$

22. $e^2f^2h^3$

23. $fh^2 - t^2$

24. $(c - f^2t^2)h$

25. $(et^3 - te \div f)f$

26. $(-f + tf \div h)ht$

27. $(c \div t - f \div e) + e$

28. $(ef - tf^3)c^5$

29. $eh - e^2h$

30. $fh - fh^2$

31. $c^2f^2h^2$

32. $c^2h^2t^2$

In Exercises 33 through 44, evaluate the expression when

$$f = -3.2 \qquad g = -5 \qquad h = 1.4 \qquad x = 2.5 \qquad y = -4$$

33. $\dfrac{hy^2(g + 2x)}{g - h}$

34. $\dfrac{gx(xy - f)}{g - y}$

35. $\dfrac{y(h + 2f) + g}{2x}$

36. $-3(y - g)^2 + (2x)^2$

37. $x(y \div g) - gh$

38. $4y + (2x + gh)$

39. $6fg - (yx - g)$

40. $8(2h - f)x \div g$

41. $-4(y^2 - g)x$

42. $(7g(h - y)) \div 3$

43. $(-gf)^2(y - g)$

44. $\{[-5x^2(g - y)] \div g^5\}(2g)^2$

In Exercises 45 through 50, use a calculator and evaluate the expression when

$$x = 3.1 \qquad y = -5 \qquad t = 2.4 \qquad u = 7$$

45. $3xy - 2xt + u^3$

46. $-4(xy - 5x^2t) - u$

47. $5xt - 7x^2y^2 + 8(x - y)$

48. $5x \div y + tu$

49. $3x^2 - 2y$

50. $tu \div y + 2$

In Exercises 51 through 56, evaluate each expression.

51. $(c^2 + e)^3$, when $c = -2$ and $e = 4$

52. $(e^2 - 2ce)^2$, when $e = 5$ and $c = -1$

53. $[-2c(f^2 - 2gh)]^2$, when $f = -3$, $c = -4$, $g = -1$, and $h = 2$

54. $[-f(2c^2) \div h]^4$, when $f = -3$, $c = 4$, $g = 1$, and $h = -2$

55. $\dfrac{-3x(2x - 9)^2 + 3t}{xt}$, when $x = 0.2$ and $t = 0.5$

56. $\dfrac{7r(5x + 8r) - 3n}{2n - r}$, when $n = 0.5$, $r = 1.6$, and $x = 2$

MIXED PRACTICE

By doing these exercises, you will practice the topics up to this point in the chapter.

57. Simplify: $-\frac{5}{6}\left[\frac{3}{4} + \left(-\frac{7}{8}\right)\right]$

58. Evaluate $(r - t)v^2$ when $r = -2$, $t = -3$, and $v = 1$.

59. Evaluate $4fx + cf^3$ given $c = -4$, $x = 7$, and $f = 3$.

60. $A = \{0, 2, 4\}$ and $B = \{1, 3, 5\}$. Find $A \cap B$.

61. Evaluate $(v - t)^2 + (r - v)^3$ when $r = 2$, $t = 5$, and $v = 2$.

62. Let $A = \{0, 1, 2\}$; let $B = \{0, 2, 4, 6\}$; and let $C = \{2, 3, 4, 5, 6\}$. Find $(A \cup B) \cap C$.

EXCURSIONS

Class Act

1. a. Let $A = B$
Let $B = 3$
Then $A = ?$

b. Let $A = B$
Let $B = C + 1$
Let $C = D \div 3$
Let $D = 27$
Find A, B, and C

c. Let $A = B - 6$
Let $B = C + 8$
Let $C = D \div 4$
Let $D = E + 10$
Let $E = F$
Let $F = 6$
Find A, B, C, D, and E

d. Let $C = 5$
Let $A = B + 7$
Let $B = 24 \div D$
Let $D = C - 2$
Find A, B, C, and D

e. Write four number rules. Share them with another group.

2. Each person in the group should toss a coin 20 times and keep a record of heads and tails. Put all your data together and answer the following questions.

a. What is the group total of heads and of tails?

b. ✏ With a fair coin, the probability of a head is $\frac{1}{2}$ and the probability of a tail is $\frac{1}{2}$. Discuss your results. Is your coin fair?

c. ✏ Would it be possible for someone tossing a fair coin to get ten heads in a row? Justify your answer.

d. ✏ What do you think is meant by a "fair" coin?

Exploring with Calculators

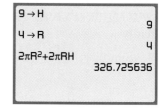

3. The surface area of a cylinder is calculated using the formula $SA = 2\pi r^2 + 2\pi rh$, where h is the height of the cylinder and r is the radius of the circular base of the cylinder. Use your graphing calculator to find the surface area of the following four cylinders. Remember that your calculator has π already stored in memory. (Note: See the Calculator Corner on page 85 to review how to store values for variables.)

   a. $h = 9$ inches $r = 4$ inches   (See the accompanying figure.)

   The **2nd ENTRY** feature of your graphing calculator will "recall" the formula for the surface area so that you do not need to re-enter it for each new problem!

   b. $h = 11.9$ inches $r = 1.6$ inches

   c. $h = 23.06$ meters $r = 16.73$ meters

   d. $h = 579.001$ feet $r = 801.9003$ feet

4. Suppose you receive 1 cent on Monday, 2 cents on Tuesday, 3 cents on Wednesday, 4 cents on Thursday, and so on. How much money would you have at the end of twenty days?

CONNECTIONS TO *GEOMETRY*

Introduction to Geometric Figures

In this text we will be investigating the uses of real numbers and algebra. Many real-life applications occur in geometry. But first we need some definitions before we can explore those applications.

Our approach to geometry and measurement is informal. This is a review of basic ideas and terminology. Some basic ideas—point, line, angle, and degree—are left undefined.

A **polygon** is a closed figure made up of three or more line **segments,** or parts of lines. (See the accompanying figure.) The segments are called the **sides** of the polygon. The point where two sides meet is a **vertex** and the sides form an **angle** at the vertex.

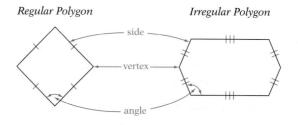

Regular Polygon   *Irregular Polygon*

side

vertex

angle

The measure of a side is called its **length.** If the sides of a polygon all have the same length and its angles all have the same measure, the polygon is called a

regular polygon. If the lengths or angles differ, the polygon is **irregular.** Some polygons that you are probably familiar with, and some new ones, are shown here.

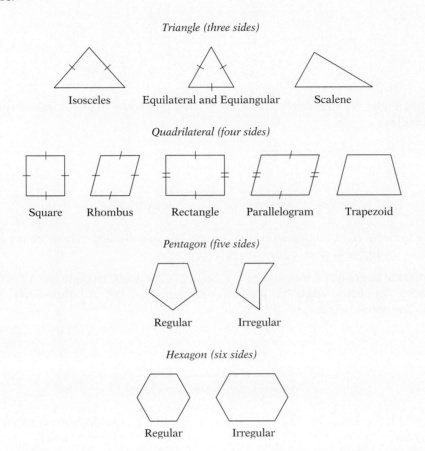

Triangle (three sides)

Isosceles   Equilateral and Equiangular   Scalene

Quadrilateral (four sides)

Square   Rhombus   Rectangle   Parallelogram   Trapezoid

Pentagon (five sides)

Regular   Irregular

Hexagon (six sides)

Regular   Irregular

Triangles that have *two* sides of equal length are called **isosceles triangles.** Triangles that have *three* sides of equal length are called **equilateral triangles;** their angles are all of equal measure too. Those that have *no* sides equal in length are called **scalene triangles.**

Quadrilaterals are a little more difficult to classify. **Rectangles** are quadrilaterals that have two pairs of opposite sides of equal length and all four angles of equal measure; the angles are called **right angles,** or 90° angles. The sides that form right angles are said to be **perpendicular** to each other. **Squares** are quadrilaterals that have all four sides equal in length; their angles too are all right angles. If the angles are not right angles, the figure is called a **rhombus.** **Parallelograms** are quadrilaterals that have opposite sides equal in length and parallel. **Parallel** means that the perpendicular distance between the two lines is the same no matter where on the line the comparison is made. Parallel lines never intersect. (See the accompanying figure.)

Trapezoids are quadrilaterals that have one pair of opposite sides parallel.

All the common polygons shown in these figures are **plane figures;** they exist only on a flat surface. Note that we use tick marks to show equal measure. Lines with the same number of marks have equal lengths.

Parallel   Non-parallel

The "angle" measure of a **circle** is 360°. A **right angle** has a measure of 90° (that is, $\frac{360}{4} = 90°$).

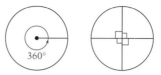

Right angles are found in squares, rectangles, and some triangles (called **right triangles**).

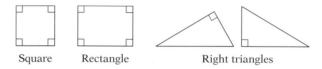

Square   Rectangle   Right triangles

We indicate right angles with little boxes. We usually leave those boxes out when the figure is clearly a square or a rectangle.

An **acute angle** has a measure of less than 90°. An **acute triangle** has all acute angles. An **obtuse angle** has a measure of more than 90°. An **obtuse triangle** has one obtuse angle.

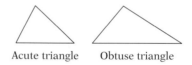

Acute triangle   Obtuse triangle

You will use many of these definitions and ideas in the geometry sections in this text.

2.3 Real Numbers and Algebra

SECTION GOALS

- *To identify and use the associative, commutative, and distributive properties*
- *To identify and combine like terms*

SECTION LEAD-IN

The number of calories in a cup of homemade popcorn depends on how the corn is popped and whether you add butter. The following numbers are approximately correct:

air popped, butter and salt	45 cal/cup
oil popped, with salt	35 cal/cup
oil popped, no salt	35 cal/cup
oil popped, with butter and salt	55 cal/cup
air popped, no salt	25 cal/cup

a. How many calories per cup does butter add to air-popped popcorn?

b. How many calories would be in $8\frac{1}{2}$ cups of oil-popped popcorn that has butter and salt?

c. You eat buttered, salted, oil-popped popcorn that has $192\frac{1}{2}$ calories. How much popcorn did you eat?

d. ✏ Write the plan you used to answer questions (a) through (c).

e. Ask and answer four questions using this data.

Properties of Real Numbers

Certain properties of operations with real numbers are useful in the study of algebra. The **commutative property of addition** tells us that we can add two numbers in any order.

Commutative Property of Addition

If a and b are real numbers, then

$$a + b = b + a$$

Remember: To *commute* means to *interchange*.

Similarly, when we add more than two real numbers, we can group them together in any way without altering the sum. This is called the **associative property of addition.**

Associative Property of Addition

If a, b, and c are real numbers, then

$$(a + b) + c = a + (b + c)$$

Remember: To *associate* means to *group*.

Similar properties hold for the multiplication of real numbers.

Commutative Property of Multiplication

If a and b are real numbers, then

$$ab = ba$$

Associative Property of Multiplication

If a, b, and c are real numbers, then

$$a(bc) = (ab)c$$

> ■ ■ ■
> ***WRITER'S BLOCK***
>
> Give an example of a real-life "operation" that is not commutative. One example is putting on socks and putting on shoes. I wouldn't want to put my socks on last!

...

EXAMPLE 1

What properties are used in each numbered line of the following computation?

$$[(-986) + 21] + (-18)$$
$$= [21 + (-986)] + (-18) \quad (1)$$
$$= 21 + [(-986) + (-18)] \quad (2)$$

SOLUTION

1. The commutative property (to change the order in the first addition)

2. The associative property (to regroup)

▶ *CHECK* **Warm-Up 1**

One further property—the **distributive property**—affects both addition and multiplication. It tells us how to find a product when one factor is a sum.

Distributive Property

If a, b, and c are real numbers, then

$$a(b + c) = ab + ac$$

Remember: We can *distribute* multiplication over both addition and subtraction, because a subtraction $a - b$ is defined as the addition $a + (-b)$.

...

EXAMPLE 2

Solve the following problem using the distributive property. Then check your answer by solving with the standard order of operations.

$$26(18 - 2) + (-13)(-2 + 3) = ?$$

SOLUTION

We use the distributive property to multiply by 26 and -13, respectively.

$$26(18 - 2) + (-13)(-2 + 3) = ?$$
$$= (26)(18) + (26)(-2) + (-13)(-2) + (-13)(3) \quad \text{Distributive property}$$
$$= 468 + (-52) + (26) + (-39) \quad \text{Multiplying}$$

Next we use the associative and commutative properties so that we can add numbers with the same signs.

$$468 + (-52) + 26 + (-39)$$
$$= 468 + 26 + (-52) + (-39) \quad \text{Commutative property}$$
$$= 494 + (-91) = 403 \quad \text{Adding}$$

Check: $(26)(18 - 2) + (-13)(-2 + 3) = (26)(16) + (-13)(1)$
$$= 416 + (-13) = 403$$

▶ *CHECK* **Warm-Up 2**

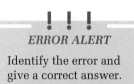

! ! !

ERROR ALERT

Identify the error and give a correct answer.

$(6 \times 2) \times 3$
$= (6 \times 3) \times (2 \times 3)$
$= 18 \times 6$
$= 108$

The results are the same, so unless we made mistakes in both procedures, the answer is probably correct.

Combining Like Terms

In algebra, we are often concerned with terms that are "alike" in an important way. We call them *like terms.*

Like Terms

Like terms have the same variables raised to the same powers.

Therefore,

$2x$ and $3x$ are like terms and $3rs$ and $4sr$ are like terms.

$5x^2$ and $5y^2$ are *not* like terms—the variable parts differ.

$3n$ and $2n^2$ are *not* like terms—the variable parts are not the same because the exponents differ.

One way to simplify an algebraic expression is to combine like terms. In order to do that, we can use the properties of operations with real numbers. Suppose you were asked to combine $2n + 4n$. You would apply the distributive law, obtaining

$$2n + 4n = (2 + 4)n = 6n$$

We use the associative and commutative properties to group like terms. Then we use the distributive property to combine their coefficients. Therefore,

---

To combine like terms

Group like terms, and then combine their coefficients.

▪▪▪

EXAMPLE 3

Combine like terms: $-6x^2y - 18xy^2 + 14xy^2 - 8x^2y^2 - 5x^2y$

SOLUTION

Close inspection of this expression shows that there are only two sets of like terms.

$$\underset{\text{Like terms}}{-6x^2y \underbrace{- 18xy^2 + 14xy^2} - 8x^2y^2 - 5x^2y}$$

ERROR ALERT

Identify the error and give a correct answer.

$(12 \div 6)(-3)$
$= (12)(-3) \div (6)(-3)$
$= -36 \div -18$
$= 2$

Rearranging and combining coefficients, we get

$$-6x^2y - 18xy^2 + 14xy^2 - 8x^2y^2 - 5x^2y$$
$$= -6x^2y - 5x^2y - 18xy^2 + 14xy^2 - 8x^2y^2$$
$$= (-6 - 5)x^2y + (-18 + 14)xy^2 - 8x^2y^2$$
$$= -11x^2y - 4xy^2 - 8x^2y^2$$

So $-6x^2y - 18xy^2 + 14xy^2 - 8x^2y^2 - 5x^2y$ is simplified to $-11x^2y - 4xy^2 - 8x^2y^2$.

▶ *CHECK* **Warm-Up 3**

In our last example, we must first apply the distributive property.

▪▪▪
EXAMPLE 4

Simplify: $2(3x - 4) - 4x + 6 + x(r - 2)$

SOLUTION

We first use the distributive property to clear all parentheses.

$$2(3x - 4) - 4x + 6 + x(r - 2)$$
$$= 2(3x) + 2(-4) - 4x + 6 + x(r) + x(-2)$$
$$= 6x - 8 - 4x + 6 + xr - 2x$$

ERROR ALERT

Identify the error and give a correct answer.

$12 \div (6 - 3)$
$= 12 \div 6 - 12 \div 3$
$= 2 - 4$
$= -2$

Then we combine like terms.

$$6x - 4x - 2x - 8 + 6 + xr = 0x - 2 + xr$$

So $2(3x - 4) - 4x + 6 + x(r - 2) = xr - 2$

▶ *CHECK* **Warm-Up 4**

Practice what you learned.

SECTION FOLLOW-UP

Using the popcorn data, ask and answer a question that can be solved by the following:

a. addition

b. subtraction

c. multiplication

d. division

c. any two operations

f. any three operations

2.3 *WARM-UPS*

Work these problems before you attempt the exercises.

1. What properties are used in this computation?

$$[(-26) + 185] + (-96)$$
$$= [185 + (-26)] + (-96)$$
$$= 185 + [(-26) + (-96)]$$

2. Solve this problem by using the distributive property.

$$-3\tfrac{1}{2}\left(\tfrac{2}{7} + \tfrac{6}{7}\right) = ?$$

3. Combine like terms: $9r^2t - 7r^2t + 5rt^2 - 5t^2r$

4. Simplify: $5(2a - 4) + 6a$

2.3 EXERCISES

Note: Use your graphing calculator to check your results whenever possible.

In Exercises 1 through 8, determine whether each statement is true or false. For each true statement, give the name of the property that explains why it is true. For each false statement, explain in your own words why it is false.

1. $(-4 + 5) + (-3) = -4 + [5 + (-3)]$

2. $12(4 + 6) = (4)(12 + 6)$

3. $3 + 5(132 + 321) = 5(132 + 321) + 3$

4. $425 + [75 + (-1183)] = (425 + 75) + (-1183)$

5. $(-38)(10) + (-3) = -38 + (10)(-3)$

6. $(-264 + 512) + (-439) = -263 + [512 + (-439)]$

7. $17 + 5(132 + 121) = 670 + 17 + 605$

8. $8^2(5^5 + 3^6) = 5^5(8^2 + 3^6)$

In Exercises 9 through 14, state which property or properties discussed in this section support each statement.

9. $12 + \{31 + [14 - 81(9 + 5)]\} = (12 + 31) + [14 - 81(9 + 5)]$

10. $2 + \{3 + [4 - 8(9 + 5)]\} = 2 + \{[4 - 8(9 + 5)] + 3\}$

11. $6[2(9 + 33) - 18] = 6[(9 + 33)2 - 18]$

12. $(806)(38^2 - 5) = 806(38^2) + 806(-5)$

13. $6[2(9 + 33) - 18] = 6[2(33 + 9) - 18]$

14. $[(-14)(-45)](-3) = (-3)[(-14)(-45)]$

In Exercises 15 through 22, rewrite the given expression using the property that is indicated.

15. $\frac{86}{5} + (38 + 5)$
associative property of addition

16. $(340)(98)(-209 - 39)$
commutative property of multiplication

17. $97 + [3 + 4(876 + 723)]$
distributive property

18. $(76 + 56) + (-34)(23)(-19)$
commutative property of addition

19. $45 - 8[(24)(23)]$
associative property of multiplication

20. $5[16 - (715 - 9)]$
distributive property

21. $(27 - 18) + 3(6)(-2)$
associative property of addition

22. $607 + (38 - 5)$
associative property of addition

In Exercises 23 through 30, use the distributive property to simplify each expression. Then use the standard order of operations to check your answer.

23. $-336 + 5(24 + 18)$

24. $98[23 + (-34)]$

25. $3\{[14 + (-12)] + (-27)\}$

26. $4 + 3\{[-2 + (-8)]6\}$

27. $-9[(18) + (-8)]$

28. $11\{7[-2(-16) + (-5)]\}$

29. $-87\{2[(-47) - 38] + -5[(-23) + 12]\}$

30. $2\frac{1}{2}\left\{5\frac{3}{4} + 6\frac{7}{8}\left[2\frac{1}{4} + \left(-3\frac{1}{8}\right)\right]\right\}$

In Exercises 31 through 46, determine whether each statement is true or false. If the statement is false, give one numerical example that illustrates this fact. Here, a, b, and c are real numbers. Numerical examples may differ.

31. $a - 0 = 0 - a$

32. $g + 0 = 0 - g$

33. $1 \div (-c) = (-c) \div 1$

34. $a \div 1 = 1 \div a$

35. $a + b = b + a$

36. $(-a) \div b = a \div (-b)$

37. $-a - b = -b - a$

38. $a - b = b - a$

39. $a \div (b + c) = (a \div b) + (a \div c)$

40. $(-a + b) \div c = (-a \div c) + (b \div c)$

41. $-a \div b = -(a \div b)$

42. $-(a + b) = -a + b$

43. $-(a - b) = a + b$

44. $-(a + b) = -a - b$

45. $-(a + b - c) = c - a - b$

46. $-(a - b - c) = c - b - a$

In Exercises 47 through 50, identify the like terms.

47. $14x^2y - 11x^2y + 8xy^2 - 2xy^2 + 6x - 23y$

48. $9rs - 8sr + 3r^2s - 6s^2r + 7s^2r + 11r^2s$

49. $3rt^2 - 4t^2r - 8t^2 + 4rt^2$

50. $2x^2y + 5x^2y - 3xy^2 - 2yx^2 - 9yx^2$

In Exercises 51 through 62, combine like terms.

51. $3u + v + 5v + 7u$

52. $3r - 4r + 5r - 6r + 8r$

53. $4y + 9y - 8y + 12y - 18y$

54. $9f - 3f + 6f - 8f + 4$

55. $n^3 + 4n^2 - n + n^2 + n^3$

56. $-3 + r - 3r - 3r^2 - 3r^3$

57. $4x^2 + 3x - 6x + 9x^2 - 7x$

58. $12y^2 + 8y - 6 - y^2 + y - 3$

59. $6s - 7s^2 + 9s^2 + 4s - 3s$

60. $5n + 3n^2 - 4n^2 - 8n$

61. $3x^2 - 7x + 6x^2 - 2x + x - x^2$

62. $7rt - 5tr - 3rt + 2tr + 4tr - 7rt - r + 6t$

In Exercises 63 through 78, apply the distributive law and simplify.

63. $-9(t - 5) + 3t$

64. $\left(-\frac{1}{2}\right)(x + 3) - 4x$

65. $5(2a - 4) + 6a$

66. $-6x - 3(2x + 5) + 5$

67. $7x - 4(x - 3) + 8$

68. $4(n - 3) - 10n - 4$

69. $-1.2(4x - 5) + 2x - 3 + 6(x - 9)$

70. $0.2(x + y) - 1.3(x + y) + 4.5y$

71. $\frac{2}{5}x^2 - \frac{3}{4}(x^2 - y) + 1.3(y - x^2)$

72. $-\frac{1}{4}u^2 + \frac{1}{2}(u + x^2) + \frac{5}{6}(x + u^2)$

73. $2.4(t^2n + 1.2t^2n) - 0.2(t^2n + 0.4t^2n) - 2.9tn^2 + 9$

74. $\frac{3}{4}(x^3y - 8x^3y) - (6xy^3 + 5) - 12(x^3y + 2x^3y)$

75. $-7[3(n^2y + 5n^2y) - 9(n^2y - 3n^2y)] - 15ny^2 - 9$

76. $6ty^2 - 5 - 5[3(2t^2y - 5t^2y) + 5(t^2y - 5t^2y)]$

77. $x(y - 3) + y(x - 3) - 4y + 3x$

78. $-r(t - 4) - t(r + 5) + 5r - 3t$

MIXED PRACTICE

By doing these exercises, you will practice the topics up to this point in the chapter.

79. What property is illustrated?
$336 + [24 + (-5)] = [24 + (-5)] + 336$

80. $A = \{5, 6, 7\}$; $B = \{6, 13\}$; and $C = \{0, 5, 6\}$.
Find $(A \cup B) \cap C$.

81. Show that $-3 - 6 \div 5 = -\left(\frac{6}{5} + 3\right)$.
Justify each step.

82. Rewrite $\left[4 + \left(-\frac{3}{4}\right)\right] \times 8$ using the
distributive property.

83. What property is illustrated by the statement
$\left(\frac{1}{4}\right)(4 + 8) = 1 + 2$?

84. True or false: Some rational numbers are
whole numbers.

85. Simplify: $63t - 36c - 5(t - 3) + 8(2t - 7)$

86. True or false: All integers are whole numbers.

87. Combine like terms: $-(-2a) + (-3a) - 5(2a + 4)$

88. Identify the terms in the algebraic expression
$$2xy^2 + 3x - 5y^3 + 2x$$

89. Which of the following are equivalent expressions?
  a. $-3[7 + (-5 + 8)]$
  b. $7(-3) + (-5)(-3) + (8)(-3)$
  c. $(-3)(7) + (8) + (-5)$

90. The world's deepest mine is Western Deep in South Africa. It reaches 12,600 feet deep. The world's deepest drilling site, the Kola peninsula in the former Soviet Union, is 31,911 feet deep. How much deeper is the Kola site?

91. Show that $5 + 6 \times 7 = 7 \times 6 + 5$. Justify each step.

92. Simplify $-9[4 + (-5) + (-8)]$ using the distributive property.

93. Is $\frac{5}{6} + 2\left(\frac{8}{9} + \frac{1}{2}\right)$ the same number as $2\left(\frac{5}{6} + \frac{8}{9}\right) + \frac{1}{2}$? Justify your answer.

94. Rewrite $-3[(2)(-5)]$ using the associative property.

EXCURSIONS

Class Act

1. In order for a law to be considered applicable to a set of numbers, it has to work two ways. For example, the distributive law for multiplication over addition (or subtraction) works because

$$c(a + b) = (a + b)c$$

The commutative law does not work for division because

$$c \div a \neq a \div c$$

What other laws can you find that will work for one operation but not for all? Show by giving a "non-example" (an example that doesn't work).

Posing Problems

2. Ask and answer four questions using the data in the following tables.

Audience Composition by Selected Program Type[1] (average minute audience)

	General drama	Suspense and mystery drama	Situation comedy	Informational[2] 6–7 p.m.	Feature films	All regular network programs 7–11 p.m.
Women (18 and over)	7,610,000	10,220,000	8,882,000	6,910,000	8,810,000	8,270,000
Men (18 and over)	4,560,000	6,870,000	5,930,000	5,200,000	5,620,000	5,950,000
Teens (12–17)	1,010,000	760,000	1,610,000	320,000	1,220,000	1,180,000
Children (2–11)	1,180,000	990,000	2,620,000	540,000	1,820,000	1,640,000
Total persons (2+)	14,360,000	18,840,000	18,990,000	12,980,000	17,480,000	17,040,000

Weekly TV Viewing by Age

	Time per week			Time per week	
	Nov. 1994	Nov. 1993		Nov. 1994	Nov. 1993
Women 18–24 years old	26 h 23 min	25 h 42 min	Men 55 and over	38 h 38 min	38 h 28 min
Women 25–54	30 h 55 min	30 h 35 min	Female teens	20 h 20 min	20 h 50 min
Women 55 and over	44 h 11 min	44 h 11 min	Male teens	21 h 59 min	21 h 10 min
Men 18–24	22 h 41 min	22 h 31 min	Children 6–11	21 h 30 min	19 h 59 min
Men 25–54	27 h 13 min	28 h 04 min	Children 2–5	24 h 42 min	24 h 32 min

Source: Nielsen Media Research, copyright 1995, Nielsen Media Research, as cited in the *1996 Information Please® Almanac* (©1995 Houghton Mifflin Co.), p. 743. All rights reserved. Used with permission by Information Please LLC.

CONNECTIONS TO *STATISTICS*

Reading a Table

You may sometimes have to refer to a complex table or chart to obtain information you need. Table 1 is such a table, and it can be confusing. The color notations have been added to help you interpret the information in the table, above and below it, and at the left. Note, especially, the *"total"* and *"average"* rows, which can get in the way of reading the table. See the key on the following page.

1 TABLE 1

2 **U.S. Travel to Foreign Countries—Travelers and Expenditures: 1975 to 1985**

3 Travelers in thousands; expenditures in millions of dollars, except as indicated. Covers residents of United States and Puerto Rico.

Item and Area	1975	1979	1980	1983	1984	1985	
Total overseas travelers	**6,354**	**7,835**	**8,163**	**9,628**	**11,252**	**12,309**	
Region of destination:							
Europe and Mediterranean	3,185	4,068	3,934	4,780	5,760	6,457	7
Caribbean and Central America	2,065	2,533	2,624	2,989	3,313	3,497	
South America	447	434	594	535	557	553	
Other	657	800	1,011	1,324	1,622	1,802	
Total Expenditures abroad	**6,417**	**9,413**	**10,397**	**13,556**	**15,449**	**16,482**	
Canada	1,306	1,599	1,817	2,160	2,416	2,694	
Mexico	1,637	2,460	2,564	3,618	3,599	3,531	
Total overseas areas	3,474	5,354	6,016	7,778	9,434	10,257	
Europe and Mediterranean	1,918	3,185	3,412	4,201	5,171	5,857	8
Average per trip (dollars)*	602	783	867	882	897	(NA)	
Caribbean and Central America	787	1,019	1,134	1,428	1,786	1,830	
South America	242	288	392	408	357	365	
Japan	131	142	185	276	400	458	
Other	396	720	893	1,465	1,720	1,747	

The "Item and Area" column header row also carries a **4** marker at the right, and bracket markers **5** and **6** appear at the left grouping the region/expenditure subrows.

9 NA Not available in dollars.
10 Adapted from *Statistical Abstracts of the United States, 1988.*

1	table number
2	title: describes the information in the table
3	units
4	years for which data was compiled
5	places visited
6	places visited
7	numbers of travelers (in thousands)
8	travel costs (in millions of dollars); *Average per trip is in dollars
9	explanatory note
10	source of data

■ ■ ■

EXAMPLE

a. Using Table 1, determine the total amount spent by people traveling from the United States to Canada during 1983.

b. How many people traveled to South America from the United States in 1979?

SOLUTION

a. Look across the *row* that starts "Canada" to where it *intersects* the *column* headed "1983." The number in that intersection is 2160. This is the number of millions of dollars spent, so the answer is 2160 million dollars, or $2,160,000,000.

b. In the section that refers to travelers, look across the row that starts "South America" until you are under the column headed "1979." The number there is 434, so the answer is 434 thousand (or 434,000) people.

Travelers from the United States who visit other countries may need to determine the money exchange rates to budget for their trips. The accompanying table shows various exchange rates—that is, how much $1 was worth in foreign currencies—for late September 1989.

WRITER'S BLOCK

Research Find up-to-date exchange rates and compare them to those shown in Table 2. Which have gone up? Which have gone down? What reasons can you give for these changes?

■ ■ ■

EXAMPLE

A Mexican hotel advertises rooms at 225,000 pesos per night. How much is this in U.S. dollars?

SOLUTION

Using Table 2, we first find the exchange rate for Mexico: 2387 pesos per $1. We then set up a proportion with the information we have and solve for the unknown value.

TABLE 2 Foreign Exchange Rates

Country	Foreign exchange for $1	Country	Foreign exchange for $1	Country	Foreign exchange for $1
Argentina (austral)	550.96	Holland (guilder)	2.00	Singapore (dollar)	1.82
Australia (dollar)	1.20	Hong Kong (dollar)	7.37	South Africa (rand)	2.53
Austria (schilling)	12.50	India (rupee)	14.97	South Korea (won)	602.05
Belgium (franc)	37.17	Indonesia (rupiah)	1,515	Spain (peseta)	109.91
Brazil (cruzado)	4.05	Ireland (pound)	0.67	Sweden (kroner)	6.07
Britain (pound)	0.59	Israel (shekel)	1.65	Switzerland (franc)	1.55
Canada (dollar)	1.13	Italy (lira)	1,287	Tahiti (franc)	111.37
Chile (peso)	259.81	Japan (yen)	133.98	Taiwan (dollar)	23.65
Colombia (peso)	359.32	Jordan (dinar)	0.60	Thailand (baht)	23.65
Denmark (krone)	6.89	Mexico (peso)	2,387	Turkey (lira)	1,634
Ecuador (sucre)	328.08	New Zealand (dollar)	1.58	Venezuela (bolivar)	33.06
Egypt (pound)	2.14	Norway (kroner)	6.52	W. Germany (mark)	1.78
Finland (mark)	4.03	Philippines (peso)	18.74	Yugoslavia (dinar)	18519
France (franc)	6.09	Portugal (escudo)	145.77		
Greece (drachma)	145.65	Saudi Arabia (riyal)	2.39		

$$\frac{2387 \text{ pesos}}{1 \text{ dollar}} = \frac{225{,}000 \text{ pesos}}{n \text{ dollar}}$$

$$\frac{2387}{1} = \frac{225{,}000}{n} \qquad \text{Dropping units}$$

$$2387 \times n = 225{,}000 \times 1 \qquad \text{Cross-multiplying}$$

$$\frac{2387 \times n}{2387} = \frac{225{,}000}{2387} \qquad \text{Dividing by 2387}$$

$$n = 94.26 \qquad \text{Simplifying and rounding}$$

So 225,000 pesos is $94.26.

PRACTICE

Use Table 1 to answer Questions 1 through 5.

1. How much more was spent for U.S. travel to Japan in 1985 than was spent 10 years before that?

2. How many people traveled from the United States to the Caribbean and Central America in 1985?

3. What were the expenditures by travelers in Japan during 1984?

4. What was the average amount spent per trip in Europe and the Mediterranean during 1984?

5. An average vacation to Europe and the Mediterranean would have cost how much more in 1983 than in 1975?

Use Table 2 to answer Questions 6 through 10.

6. In which countries was the dollar worth less than one of the foreign currency units?

7. A tourist brought back 9000 Japanese yen from a trip. How much is this in U.S. currency?

8. A visitor to London, England, bought jewelry for 250 pounds. How much did she pay in U.S. currency?

9. A trip to Norway resulted in the purchase of an embroidered blouse for 260.8 kroner. That same blouse was available in New York for $92. Was it more or less expensive in New York?

10. Money that we had invested in Canada was exchanged for 118 U.S. dollars. About how many Canadian dollars is this?

CONNECTIONS TO *GEOMETRY*

Parallel Lines

A straight angle has a measure of 180°. It is called a **line.** Two angles are complements of each other, or **complementary,** if their measures total 90°. Two angles are supplements of each other, or **supplementary,** if their measures total 180°.

Two distinct lines either **intersect** or do not. When they intersect, they intersect in just one point. The angles formed opposite each other are called **vertical angles.** Vertical angles have equal measure. Angles formed next to each other are **adjacent angles.** Adjacent angles formed by two intersecting lines are supplementary.

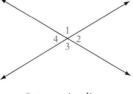

Angles 1 and 3 are vertical.
Angles 2 and 4 are vertical.
Angles 1 and 2 are adjacent.
Angles 2 and 3 are adjacent.
Angles 3 and 4 are adjacent.
Angles 1 and 4 are adjacent.

Intersecting lines

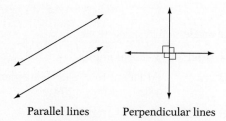

Parallel lines   Perpendicular lines

Lines in a plane that do not intersect are called **parallel lines.** Lines that intersect in right angles are called **perpendicular lines.**

When a line intersects two or more lines in different points, it is called a **transversal.**

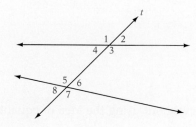

We give names to the angles:

Exterior angles	1, 2, 7, 8
Interior angles	3, 4, 5, 6
Corresponding angles	1 and 5, 2 and 6, 4 and 8, 3 and 7
Alternate exterior angles	1 and 7, 2 and 8
Alternate interior angles	3 and 5, 4 and 6

When two parallel lines are cut by a transversal,

1. corresponding angles are equal in measure.
2. alternate exterior angles are equal in measure.
3. alternate interior angles are equal in measure.

▪ ▪ ▪

EXAMPLE

Lines a and b are parallel and angle 2 is 60°. Find the measure of all other angles.

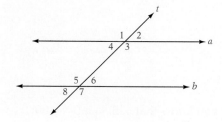

SOLUTION

We use $\angle$ to indicate angle.

$\angle 2 = \angle 4 = 60°$	Vertical angles
$\angle 1 = 120°$	Adjacent to $\angle 2$ and supplementary: $180° - 60° = 120°$
$\angle 1 = \angle 3 = 120°$	Vertical angles
$\angle 2 = \angle 8 = 60°$	Alternate exterior angles
$\angle 8 = \angle 6 = 60°$	Vertical angles
$\angle 3 = \angle 5 = 120°$	Alternate interior angles
$\angle 5 = \angle 7 = 120°$	Vertical angles

So $\angle 2 = \angle 4 = \angle 6 = \angle 8 = 60°$
$\angle 1 = \angle 3 = \angle 5 = \angle 7 = 120°$

PRACTICE

1. Redo the example without using the idea of equality of vertical angles to justify a result.

2. Redo the example using only vertical angles, supplementary angles, or alternate interior angles to justify each result.

For Questions 3 through 8, use the following figure.

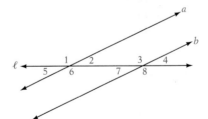

Lines a and b are parallel.
ℓ is a transversal.

Find the measure of each angle when

3. $\angle 6 = 150°$

4. $\angle 8 = 125°$

5. $\angle 5 = 32°$

6. $\angle 7 = 20°$

7. ✏ Show how you know that $\angle 5 + \angle 3 = 180°$.

8. ✏ Show how you know that $\angle 1 + \angle 4 = 180°$.

2.4 Translating Between English and Algebra

SECTION LEAD-IN

Let $S = NV + X$, where

   S is the score on an exam
   N is the number of questions correct
   V is the value for each question
   X is the number of extra credit points

Choose numbers to fill in the blanks in the following statements to give four different problems. Use the given formula to solve each problem.

a. Find the score on an exam when 27 questions are answered correctly at _____ points each, and an additional 5 points are given for extra credit.

b. Find the number of extra credit points when there are 30 questions, each question counts _____ points, there are _____ questions correct, and the total score is 178.

c. Find the value for each question when 10 points are given for extra credit, 98 is the total score, and there are _____ questions on the exam.

d. Find the number of questions on the exam when there were _____ points counted per question, the total score was _____, and there were _____ points given for extra credit.

Translating English to Algebra

The symbols and expressions of algebra enable us to state information simply and efficiently. For that reason, algebra is very useful in solving real problems and problems from other areas of mathematics. But to use algebra, we need to be able to translate English phrases into algebraic expressions.

To translate an English phrase into algebra

1. Identify the quantities and express them as symbols.
2. Identify the operation.
3. Connect the symbols with the operation sign.

The following lists contain some words and phrases that will help you in translating each of the operations.

Addition and Multiplication

Because addition and multiplication are commutative, the order of the terms in each phrase does not matter.

Addition   Algebraic expression: $c + 8$

the sum of c and 8 c increased by 8
8 more than c c added to 8
8 plus c the total of 8 and c

Multiplication   Algebraic expression: $\frac{1}{2}y$ or $\left(\frac{1}{2}\right) \cdot y$

the product of $\frac{1}{2}$ and y $\frac{1}{2}$ times y
$\frac{1}{2}$ multiplied by y $\frac{1}{2}$ of y

Subtraction and Division

Because subtraction and division are not commutative, the order in which the terms are written is critical.

Subtraction   Algebraic expression: $x - 9$

the difference of x and 9   9 less than x
9 fewer than x x decreased by 9
9 subtracted from x

Division   Algebraic expression: $12 \div x$ or $\frac{12}{x}$

the quotient of 12 and x 12 divided by x
the ratio of 12 to x 12 over x

▪▪▪

EXAMPLE 1

Rewrite each phrase as an algebraic expression:

a. 11 times Bob's salary

b. the difference between a number and 13

c. the ratio of men to women

SOLUTION

a. The quantities are "11" and "Bob's salary." We choose the variable n to represent Bob's salary. The word "times" indicates multiplication. So "11 times Bob's salary" is written as $(11)(n)$ or $(n)(11)$ or $11n$.

b. The quantities are "a number" and "13." We choose the variable n to represent "a number." "The difference between" signals subtraction, so we rewrite this phrase as $n - 13$.

c. The quantities are "men" and "women." Let x represent "men" and let y represent "women." The word "ratio" indicates division. So the phrase is written $x \div y$ or $\frac{x}{y}$.

! ! !

ERROR ALERT

Identify the error and give a correct answer.

Translate into algebra: 5 less than q

Incorrect Solution:

$5 - q$

▶ *CHECK* **Warm-Up 1**

More complex symbols may require that you translate phrases involving several operations and quantities. Again, you must identify the quantities, assign them symbols, and link them with the proper operation sign.

■ ■ ■
EXAMPLE 2

Write "half the difference of thirteen and a number" as an algebraic expression.

SOLUTION

There are three quantities: "half" $\left(\frac{1}{2}\right)$, 13, and "a number." Let the number be n. Because 13 and n are linked by "the difference of," subtraction is indicated. This gives us $13 - n$.

Now, "half the difference" translates as multiplication between $\frac{1}{2}$ and "the difference," which is $13 - n$. The complete translation is then $\frac{1}{2}(13 - n)$, or $\left(\frac{13 - n}{2}\right)$.

Note: Parentheses are added to clarify the quantities. $\frac{1}{2}(13 - n)$ is *not* the same as $\left(\frac{1}{2}\right)13 - n$.

So "half the difference of thirteen and a number" is $\frac{1}{2}(13 - n)$.

▶ *CHECK* **Warm-Up 2**

■ ■ ■
EXAMPLE 3

Express "the cost of two pears and three apples" as an algebraic expression if a pear costs 10 cents less than an apple.

SOLUTION

There are two unknown quantities: the cost of pears and the cost of apples. However, the problem does give additional information—pears cost 10 cents less than apples. If the cost of one apple is n, then the cost of one pear is $n - 10$.

We are asked for the cost of two pears and three apples. We must multiply the number of each fruit by its price.

$$3 \text{ apples cost } (3)(n) = 3n.$$
$$2 \text{ pears cost } (2)(n - 10) = 2n - 20.$$

Simplifying, we find that the combined cost is

$$3n + (2n - 20) = 5n - 20$$

So the cost of two pears and three apples is $5n - 20$ cents.

▶ *CHECK* **Warm-Up 3**

Often, we need to symbolize a statement that requires additional information—information that is not contained in the statement. Look at the following example.

■ ■ ■

EXAMPLE 4

Write as an algebraic expression "the value in cents of the coins in a stack of quarters."

SOLUTION

We can let q be the number of quarters in the stack. Following the Study Hint directions, we write it down.

$$q = \text{number of quarters}$$

Now we want their value. To find it, we need the additional information that each coin (a quarter), has a value of 25 cents.

Then we find the total value of the stack by repeated addition of 25 cents or (much more readily) by multiplication. The value is $(25)(q)$, or $25q$.

So the value of the quarters in cents is $25q$.

▶ *CHECK* **Warm-Up 4**

STUDY HINT

When identifying the quantities in a problem, read the problem all the way through before you try to name any of them. Then write down what you are going to let your variable stand for, and define all quantities in terms of that variable.

Translating Algebra to English

On occasion, you will have to write an algebraic phrase in words.

> **To translate an algebraic phrase into English**
>
> 1. Find the operation and choose appropriate words to indicate it.
> 2. Identify the quantities connected by this operation sign.
> 3. Write the phrase.
>
> *Note:* There may be more than one way to express a given algebraic phrase.

■ ■ ■

EXAMPLE 5

Rewrite the algebraic expression $26a + 15$ in words.

SOLUTION

There are two quantities linked by a plus sign: $26a$ and 15. $26a$ is a way of writing "the product of 26 and a." So we are adding the product of 26 and a to 15.

We can express this in several ways: "the sum of the product of 26 and a, and 15," "fifteen added to the product of 26 and a," "fifteen added to $26a$," or "the product of 26 and a, increased by 15."

▶ *CHECK* **Warm-Up 5**

Practice what you learned.

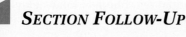

SECTION FOLLOW-UP

Write a problem similar to this section's Lead-In. The formula you use must differ from that given in the Lead-In. Swap problems with a classmate.

2.4 WARM-UPS

Work these problems before you attempt the exercises.

1. Rewrite "5 more than a number" and "12 fewer than a number" as algebraic expressions.

2. Write "three-quarters of the quotient of a number and five" as an algebraic expression.

3. Write "the cost of four books and five magazines" if the cost of a book is $5.50 more than the cost of a magazine.

4. Write "the number of days in y weeks" as an algebraic expression.

5. Write $14b - 8$ in words.

2.4 EXERCISES

Note: Use your graphing calculator to check your results whenever possible.

In Exercises 1 through 68, write each verbal expression as an algebraic expression. When a variable is not given, use n.

1. 13 decreased by f

2. x decreased by 11

3. four less than a number

4. x less than 8

5. the square of t

6. 5 fewer than m

7. the value of d dimes in dollars

8. the number of dimes in d dollars

9. the cost of 15 items at n cents each

10. half the sum of r and 1.5

11. twice the difference of 8 and x

12. one and one-fourth times as expensive as gold

13. the number of coins in n dimes and y nickels

14. the annual rent budgeted in weekly payments

15. eight days after David's birthday

16. three and three-quarter pounds more than her sister's birth weight

17. the monthly payment on a year's insurance bill

18. three-fourths as much money per hour as Shepherd earns

19. one and five-tenths the size of a golf ball

20. the value of t half-dollars in dollars

21. half my age if I am y years old

22. the cost of 13 items at n cents each

23. the number of coins in n dimes and y quarters

24. My weight is z. Express my weight after I lose 15 pounds.

25. Express the distance traveled from New York to San Bernardino, if you stop over in Chicago and the distance from New York to Chicago is x and the distance from Chicago to San Bernardino is y.

26. the product of 3 and the sum of a and 5

27. the sum of the product of 11 and x, and 4

28. the difference of the product of 3 and x, and 5

29. the sum of a number and the number decreased by 2

30. twice the difference of 4 and x

31. $1.94 less than the price of butter

32. thirteen years older than Susan

33. one hundred twenty-six more calories than lean beef

34. the quotient of fifty-two and a number

35. two hundred eighty-five calories less than chicken

36. half as expensive as leather

37. one-third the product of a number increased by four and that number

38. the quotient of 4 times a number and 26

39. the quotient of a number and that number less 4

40. the product of five hundred twenty-eight and a number

41. the difference of one quarter of a number and nine

42. the sum of eighty-five hundredths and a number

43. the cost of two pears and three apples if pears cost 2 cents more than apples

44. the difference of a number and eleven and five-tenths

45. the sum of Julio's age and Anne's age if Julio is half Anne's age

46. The first integer is n. Express the next two larger integers in terms of n.

47. ten less than a number

48. a number less twenty-five

49. the quotient of a number and sixty-five

50. Lenora's age ten years ago

51. $1.94 less than the original price

52. the sum of Maxwell's age and Lisa's age if Maxwell is $\frac{1}{3}$ Lisa's age

53. twice Catherine's age three years from now

54. the difference of one-quarter of a number and twelve

55. the sum of a number and the number decreased by one-fifth of the number

56. the sum of Marie's age and Jason's age if Marie is 2 years older than Jason

57. the square of a number plus the number increased by two

58. the cube of the product of a number and 7

59. the quotient of the square of a number, and 23

60. one-fourth the product of a number increased by five and that number

61. one and one-fourth times as expensive as silver

62. one monthly installment on an annual rent

63. the difference between a number and six more than that number

64. the sum of one-third of a number increased by four and that number

65. the sum of eight less than a number, and a number

66. the quotient of a number and that number decreased by 4

67. the difference between a number and thirty more than that number

68. the product of a number and fifty-two more than that number

In Exercises 69 through 76, write each algebraic expression in words.

69. $n + 6$

70. $6 + n$

71. $n \div 7$

72. $18 \div n$

73. $6n$

74. nt

75. $n - 10$

76. $n - b$

In Exercises 77 through 92, use the given variable and represent the English expression in symbols.

77. When r rooms are rented at $155 per room, the amount collected is _____.

78. When 180 rooms are rented at s per room, the amount collected is _____.

79. You had d dollars and you spent \$4.15. How much do you have left? _____

80. When you had a \$20 bill and you spent d dollars, the amount you received in change was _____ .

81. The cost of n bagels at 50 cents and 6 coffees at p cents is _____ .

82. The cost of 16 books at p dollars each and c pencils at 25 cents each is _____ .

83. Each of n people have h hamburgers. How many hamburgers are there? _____

84. My age is a years. Last year my age was _____ .

85. I am mailing c cards with one stamp on each. How many stamps do I need? _____

86. When there is 6% tax, the cost of n items at \$1 each is $n +$ _____ n dollars.

87. What is the number of people when n rooms have p people in each room? _____

88. What is the total number of people on a bus that is half full if the bus holds b people? _____

89. When n tickets are sold at \$6 each and 80 tickets are sold at d dollars each, the total collected is _____ .

m passengers

x passengers

90. When x people go to Atlanta by train and m people travel by air, how many more travel by air than by train? _____

91. I am n years old and my daughter is d years younger. Our total age in two years is _____ .

92. I earn d dollars and my sister earns s dollars. How much less than I does she earn? _____

MIXED PRACTICE

By doing these exercises, you will practice the topics up to this point in the chapter.

93. Write "A less than 36" as an algebraic expression.

94. Evaluate $4t + r + (t - v)^2$ when $r = -2$, $t = 0$, and $v = -3$.

95. Write $16a + 24$ as an English phrase.

96. Simplify: $6x - 9x + 56y + 16x$

97. Evaluate $(-3r - t)^2 + (t - s)$ when $r = 4$, $t = -3$, and $s = 2$.

98. Determine whether $-3.4[(-7.1) + (-5.4)]$ and $(-7.1)(-3.4) + (-5.4)(-7.1)$ are equivalent.

99. Write "the quotient of the product of 6 and a number and 56, and 12" as an algebraic expression.

100. Write $15a - 12$ as an English phrase.

EXCURSIONS

Posing Problems

1. What information would you like that you can find in this table? Ask and answer four questions using these data.

Resident Population, by Age Group, Race, and Hispanic Origin, 1995[1] (in thousands)

Age	White	Black	Hispanic origin[2]	American Indian, Eskimo & Aleut	Asian & Pacific Islanders	All persons
Under 5	12,524	2,932	3,191	176	770	19,593
5–9	12,630	2,665	2,658	191	670	18,814
10–14	12,641	2,739	2,424	198	711	18,713
15–19	12,291	2,689	2,280	171	636	18,067
20–24	12,184	2,507	2,334	158	690	17,873
25–29	13,126	2,447	2,488	150	752	18,963
30–34	15,686	2,679	2,533	157	821	21,876
35–39	16,508	2,661	2,158	152	797	22,276
40–44	15,370	2,288	1,729	136	713	20,236
45–49	13,666	1,780	1,313	111	585	17,455
50–54	10,850	1,326	962	84	413	13,635
55–59	8,844	1,095	782	65	320	11,102
60–64	8,141	954	636	53	266	10,050
65–69	8,220	889	543	42	228	9,922
70–74	7,544	677	405	34	174	8,834
75–79	5,802	498	258	23	104	6,685
80–84	3,911	311	189	15	58	4,484
85–89	2,046	160	64	8	22	2,300
90–94	864	80	38	4	8	994
95–99	229	22	9	1	3	264
100 and over	42	7	2	1	1	53
All ages	193,519	31,606	26,978	1,931	8,743	262,777
16 and over	152,755	22,496	18,237	1,328	6,454	201,270
18 and over	147,835	21,422	17,332	1,258	5,199	193,046
65 and over	28,658	2,644	1,508	128	598	33,536

1. July 1, 1995. 2. Persons of Hispanic origin may be of any race. The information on the total and Hispanic population shown in this table was collected in the 50 states and the District of Columbia, and, therefore, does not include residents of Puerto Rico. *Source:* U.S. Bureau of the Census, as cited in the *1996 Information Please*® *Almanac* (©1995 Houghton Mifflin Co.), p. 830. All rights reserved. Used with permission by Information Please LLC.

2. Ask and answer four questions using the data in the following table. Share your questions with a classmate.

Currency Exchange Rate per New York Dollar			
	Today	6 mos. ago	1 yr. ago
Australian dollar	1.2870	1.3200	1.3446
Austrian schilling	**10.440**	**10.388**	**9.900**
Belgian franc	30.58	30.38	28.99
British pound	**.6481**	**.6503**	**.6246**
Canadian dollar	1.3720	1.3713	1.3558
French franc	**5.0605**	**5.0725**	**4.8560**
German mark	1.4848	1.4755	1.4062
Greek drachma	**237.10**	**243.55**	**226.64**
Hong Kong dollar	7.7340	7.7309	7.7377
Irish punt	**.6224**	**.6326**	**.6105**
Israeli shekel	3.1495	3.1412	3.0210
Italian lira	**1518.00**	**1572.50**	**1587.60**
Japanese yen	107.90	106.95	91.60
Mexican peso	**7.5050**	**7.4550**	**6.1350**
Dutch guilder	1.6640	1.6548	1.5795
Norwegian krone	**6.4095**	**6.4485**	**6.2140**

Exploring Problem Solving

3. The senior class officers at the College of Wakefield have arranged for class members to attend a performance of the *Nutcracker Ballet* in New York City in December. The bus has a capacity of 75 people and will cost $725 plus $40 for each passenger. Each ticket for the performance costs $50. The class officers have decided to charge each participant $125 for the trip and ballet performance. How many people must go on the trip so that the college does not lose money? Use your graphing calculator to help you construct a table that shows the costs for 20 to 75 students in increments of 5.

CHAPTER LOOK-BACK

Most word problems encountered in your mathematics class are presented in a form so that you can work in a straightforward manner. Real-world problems, however, are not always so neatly presented.

Problems such as

About how many miles have you walked in the last year?

must be approached first by formulating the problem in a solvable way. That is, we must INTERPRET this problem—what does it mean? We must make a plan and DECIDE on a strategy. We APPLY this strategy, and then, after obtaining an answer, we check to see that the answer makes sense. We check it, and then we look back. We use REASON to determine the validity of our solution.

1. Explain how you approached each problem at the beginning of Chapter 2.

2. Ask a question, answer it, and tell how you can check your answer.

    a. In 1987, 1.2 million high-school seniors spent $14,549,475 to take the SAT exams.

    b. There is a sound-activated burglar alarm that mimics the barking of a dog. You can buy it for $69.95.

    c. It costs about ten thousand dollars to freeze Wollman rink in Central Park at the beginning of each winter.

    d. It costs $1.5 billion dollars a year to replace stolen road signs.

    e. P. T. Barnum's famous six and one-half ton elephant, Jumbo, ate two hundred pounds of food daily.

    f. On a standard badminton shuttlecock, there are 14 to 16 feathers, each 2.5 to $2\frac{3}{4}$ inches long.

    g. The Suez Canal is 105 miles long.

    h. The leaves in Central Park were raked by volunteers last year. It took 4546 people-hours to rake all these leaves.

CHAPTER 2
REVIEW PROBLEMS

The following exercises will give you a good review of the material presented in this chapter.

SECTION 2.1

In Exercises 1 through 3, answer true or false.

1. All integers are whole numbers.

2. Some rational numbers are whole numbers.

3. Zero is a whole number, an integer, a rational number, and a real number.

Use the following information to answer Exercises 4 through 6.

$$A = \{5, 6, 7\}; B = \{6, 13\}; \text{ and } C = \{0, 5, 6\}$$

4. Find $(A \cup B) \cap C$.

5. Find $A \cap C$.

6. Find $B \cup C$.

SECTION 2.2

7. Identify the variable parts and the coefficients in the algebraic expression $5v - t - rv - tr$.

8. Evaluate $(4f - h)^3 - ch^3$ when $c = 4, f = -3$, and $h = 1$.

9. Evaluate $(fx - cx) + cf$ given $c = -1, f = 0$, and $x = 6$.

10. Evaluate $4x^2 - 3(y + 6xy)^2$ when $x = 4$ and $y = -3$.

11. Evaluate $eh^2 - f(h - e)^3$ when $e = 4, f = -3$, and $h = 3$.

12. Evaluate $r^2 + 3t - v^3$ when $r = -2, t = 3$, and $v = 4$.

SECTION 2.3

13. Verify that $2[3 + (-4)] = 2(3) + 2(-4)$.

14. Which property allows us to say $[(-3)(4)](7) = (-3)[(4)(7)]$?

15. Determine whether $[(-3x) + (-4x)] + 5x = (-3x) + [(-4x) + 5x]$.

16. Simplify: $-2\{(-8)[(-3) + (-5)]\}$

17. Combine like terms: $14y - 52y + 16y + 18y - 32y$

18. Combine like terms: $4x - 5 + 6x + 8 - 2x + 1x - 5$

19. Combine like terms: $3t^2 - 4t + 5tx + 7t - tx + 3t^2 - 8t$

20. Combine like terms: $-12a + 8(2a - 9)$

21. Simplify: $2x - 5x - 3(x - 5)$

22. Simplify: $-8(x - 5y) + 2(-3x - 9y)$

SECTION 2.4

23. Write $4x - 7x$ in words.

24. Write an algebraic expression for "the difference of a number and the product of twelve and the number."

25. Write an algebraic expression for "the sum of a number and the quotient of the number and ten."

26. Write $-9(18 + a)$ in words.

27. Write an algebraic expression for "the number of students in the class" if there are fourteen more boys than girls.

28. Write algebraic expressions for "the previous two consecutive integers" if the starting integer is x.

MIXED REVIEW

29. Evaluate $5(hf - e)^3$ when $e = -1, f = 4$, and $h = 3$.

30. Is $(-3) - (-5)$ equivalent to $(-5) - (-3)$? That is, does the commutative property hold for subtraction of real numbers?

31. Write an algebraic expression for "the sum of a number and 10."

32. Evaluate $e^3 - 2f + h$ when $e = -1, f = 2$, and $h = -2$.

33. Combine like terms: $19a - 3(a^2 - 12) - 8a$

34. Write "four years less than the age of my car" as an algebraic expression.

35. Write algebraic expressions for "the next two consecutive whole numbers" if the first whole number is symbolized by n.

36. Evaluate $xy^2 - v^3x$ when $x = -1$, $y = 3$, and $v = 7$.

37. Write "sixty-seven times a number" as an algebraic expression.

38. Simplify: $-3.4(x - 5y) + 4.8(x + 3.4y) - (x - y)$

39. Write $x - 7$ in words.

40. Write an algebraic expression for "the cost of 3 pencils and two pens" if pens cost 50 cents more than pencils.

41. Combine like terms: $-6(2x + 3) + x$

42. Simplify $-(-3xy) + 4 - y(-2x - 3) + y(-5x)$ by collecting like terms.

43. Write an algebraic expression for "the quotient of 4 and a number."

44. Write an algebraic expression for "the quotient of the sum of a number and 10, and twenty-five."

45. Write $3x$ in words.

46. Write "one and one-half times the size of a coconut" as an algebraic expression.

CHAPTER 2 TEST

This exam tests your knowledge of the material in Chapter 2.

1. Let $A = \{0, 1, 2\}$; let $B = \{0, 2, 4, 6\}$; and let $C = \{2, 3, 4, 5, 6\}$.

    a. Find $(A \cup B)$.   **b.** Find $(B \cap C)$.   **c.** Find $(A \cap B) \cup C$.

2. Evaluate the following algebraic expressions, given that $x = -1$, $y = 2$, and $t = -3$.

    a. $6(x - y)t$ **b.** $5x - y + x^3$ **c.** $3(xy)^2 - t^2$

3. Determine which property is illustrated.

    a. $3(2 + 4) = 3(2) + 3(4)$ **b.** $(-7)[(1)(-9)] = [(-7)(1)](-9)$ **c.** $(-6) + 3 = 3 + (-6)$

4. Simplify by combining like terms.

    a. $-5x + (-6x)$ **b.** $-8y - (-4y)$ **c.** $6r + (-3r) - 2t$

5. Simplify by using he distributive law.

    a. $-5(7x + 4)$ **b.** $-3(-6x + 8)$ **c.** $-x(3y - 5)$

6. Simplify by using the distributive law and then combining like terms.

    a. $-3(2n - 5) - 7n$ **b.** $8(-3x + 4) - 8x$ **c.** $(-4y + 9)6 + 3(11y - 6)$

7. Write each algebraic expression in words.

   a. $23(x + 18)$

   b. $2(27 - w)$

   c. $\dfrac{x + 8}{3}$

8. Write each English statement as an algebraic expression.

   a. the number of coins in y nickels and t pennies

   b. the quotient of 11 and x

   c. your age $3\frac{1}{2}$ years ago

CUMULATIVE REVIEW

CHAPTERS 1–2

The following exercises will help you maintain the skills you have learned in this and previous chapters.

1. Multiply: $(0.3)(-7.2)(0.09)$

2. Divide: $\left(11\frac{2}{3}\right) \div \left(-2\frac{1}{9}\right)$

3. Simplify $\frac{2}{3} + \left(7 - 4\frac{3}{5}\right)^2 \div [6(4.8 \div 9.6)^2]$ using the standard order of operations.

4. Simplify $-7(-3mt - 3) + 4mt$ by combining like terms.

5. Write the quotient of 25 and t as an algebraic expression.

6. Find the sum: $(-6.7) + 82.4 + 78 + (-96.4)$

7. Find the difference: $(-25.4) - (-19.9)$

8. Find the quotient: $(-30.6) \div (-0.03)$

9. Add the sum of $27\frac{1}{3}$ and $19\frac{3}{5}$ to the difference of $-36\frac{1}{2}$ and $45\frac{2}{3}$.

10. Evaluate: $16 \div 12 - (2 + 5)^2 - 4^2$

11. Simplify: $2(x + 5) - 3(x - 4)$

12. Simplify: $-4(y - 7) + 6(3 - y) + 9(2 + y)$

13. Compare $|-26.3|$, $-|23.9|$, and $-(-25.7)$, and arrange them in order from least to greatest.

14. Simplify: $\left(42\frac{1}{2}\right)\left(18\frac{2}{5}\right) + \left(-27\frac{5}{6}\right)$

15. Simplify: $-8(m + 4) - 9(5 - m)$

16. Evaluate $x^2 - wz - y$ when $w = 2$, $x = -3$, $y = -4$, and $z = 0$.

17. Evaluate $xy^2 - 3xy + \frac{x}{y}$ when x is -11 and $y = -4$.

18. Express $2^5 \cdot 3^2$ as a whole number.

19. Express $(-3)^4(-1)^5$ as a whole number.

20. Write in algebraic notation: the quotient of five more than an unknown number and eleven.

LINEAR EQUATIONS AND INEQUALITIES

*H*ow can one determine whether one engine is more powerful than another? The power of an engine, often given in horsepower, is the rate at which it produces work. Power can be calculated using the formula $P = W \div t$, where $P =$ power, $W =$ the amount of work produced, and $t =$ time. Depending on which variables are known, and following the rules and properties of literal equations, the formula can be rewritten: for example, $t = W \div P$ solves for time when W and P are known, or $W = Pt$ determines the work done when the values for P and t are known.

■ *Think of an equation and then rewrite it to solve for a different variable.*

SKILLS CHECK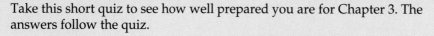

Take this short quiz to see how well prepared you are for Chapter 3. The answers follow the quiz.

1. Simplify:
   $2x + 8 - 3x - 4x + 7$

2. Evaluate $2x^2 - 3x + 1$ when $x = 4$.

3. State the meaning of $<, >,$ $\leq,$ and $\geq$.

4. Simplify: $-3(x + 5)$

5. Add: $-8 + -10$

6. Subtract $-5\frac{1}{2}$ from $6\frac{3}{4}$.

7. Multiply $-2\frac{1}{2}$ by $-5\frac{3}{4}$.

8. Divide -1.2 by 0.4.

ANSWERS: **1.** $-5x + 15$ [Section 2.3] **2.** 21 [Section 2.2] **3.** $<$ less than; $>$ greater than; $\leq$ less than or equal to; $\geq$ greater than or equal to [Section 1.1] **4.** $-3x - 15$ [Section 2.3] **5.** -18 [Section 1.2] **6.** $12\frac{1}{4}$ [Section 1.2] **7.** $14\frac{3}{8}$ [Section 1.3] **8.** -3 [Section 1.3]

CHAPTER LEAD-IN

In forensic science, it is sometimes important to be able to tell something about a person from investigating a bone. The height of a person can be calculated from knowing the lengths of certain major bones: the femur (F), the tibia (T), the humerus (H), and the radius (R). When the length of one of these bones is known, one of the following formulas is used to determine the height. (All measurements are in centimeters.)

Male	Female
$h = 69.09 + 2.24F$	$h = 61.41 + 2.32F$
$h = 81.69 + 2.39T$	$h = 72.57 + 2.53T$
$h = 73.57 + 2.97H$	$h = 64.98 + 3.14H$
$h = 80.41 + 3.65R$	$h = 73.50 + 3.88R$

After the age of thirty, the height of a person begins to decrease at the rate of approximately 0.06 centimeter per year.

Using these linear equations, ask and answer three questions. Compare your questions with those of your classmates. See what we asked at the end of this chapter.

3.1 Linear Equations in One Variable

SECTION LEAD-IN

Given

$$F = c + 40$$

where F is the Fahrenheit temperature and c is the number of cricket chirps in fifteen seconds, answer the following questions.

1. As the temperature gets colder, what happens to the number of cricket chirps?

2. As the number of chirps goes up, how has the temperature changed?

3. What do you think happens to crickets in the winter?

4. Think of another real-life situation that can be described by an equation such as

$$x = a + b$$

SECTION GOALS

- To solve equations using the addition and/or multiplication principles

- To solve equations by combining like terms

- To solve equations that involve terms in parentheses

Solving Linear Equations with One Operation

> **Equation**
>
> An **equation** is a mathematical statement in which two expressions are set equal to each other.

When an equation contains only constants and a single variable with an exponent of 1, it is called a **linear equation in one variable.**

> **Solution**
>
> A **solution of an equation in one variable** is a number that makes the equation true when the number is substituted for the variable.

The Addition Principle of Equality

Consider the equation

$$x + 3 = 5$$

You can probably see that the solution is 2, because $2 + 3 = 5$. However, when the solution is not quite so obvious, we can find the solution to an equation of this form by applying the **addition principle of equality.**

> **Addition Principle of Equality**
> For real numbers a, b, and c,
> $$\text{If } a = b, \text{ then } a + c = b + c$$

In words, if we add the same real number to both sides of an equation, the solution of the equation remains unchanged. We will use this procedure to isolate the variable on one side of the equation.

▪▪▪
EXAMPLE 1

Solve for x: $x + 12 = 5$

SOLUTION

To isolate x, we add -12 (the additive inverse of 12) to both sides of the equation and then simplify.

$$
\begin{aligned}
x + 12 + (-12) &= 5 + (-12) &&\text{Adding } -12 \text{ to both sides} \\
x + 0 &= 5 + (-12) &&\text{Additive inverse property} \\
x &= -7 &&\text{Simplifying}
\end{aligned}
$$

The solution is $x = -7$.

To check that -7 is a solution, we substitute -7 for x in the original equation.

Check:
$$
\begin{aligned}
x + 12 &= 5 \\
(-7) + 12 &= 5 \\
5 &= 5 \quad \text{True}
\end{aligned}
$$

So $x = -7$ is the solution to the equation $x + 12 = 5$.

▶ *CHECK* **Warm-Up 1**

▪▪▪
EXAMPLE 2

Solve for x: $x - 4 = 10$

SOLUTION

To isolate x, we add 4 to both sides of the equation.

$$
\begin{aligned}
x - 4 + 4 &= 10 + 4 &&\text{Adding 4 to both sides} \\
x + 0 &= 10 + 4 &&\text{Additive inverse property} \\
x &= 14 &&\text{Simplifying}
\end{aligned}
$$

Check:
$$
\begin{aligned}
x - 4 &= 10 \\
(14) - 4 &= 10 \\
10 &= 10 \quad \text{True}
\end{aligned}
$$

So $x = 14$ is the solution to the equation $x - 4 = 10$.

▶ *CHECK* **Warm-Up 2**

▪▪▪
WRITER'S BLOCK

Explain to a friend how to solve an equation such as

$$16 = 10 + x$$

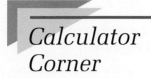

Calculator Corner

You can easily check your solution to an equation involving one operation using your graphing calculator. For instance, Example 2 reads:

$$x - 4 = 10$$

Let Y1 equal the left side of this equation and let Y2 be the right side of the equation. Then use the **TRACE** feature of your graphing calculator to find the point where Y1 and Y2 intersect. This is the point where the two sides of the equation are equal. We simply traced the graph of the slanted line to where $y = 10$ to find the value of x that makes this true.

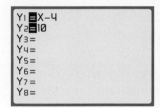 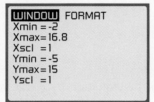 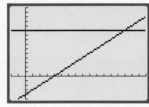 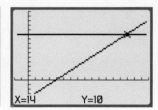

The point of intersection is (14, 10). The result obtained through algebra was $x = 14$. These two lines intersect at $x = 14$, therefore, confirming the answer that was found using algebra. This simple method of graphing both sides of an equation and finding the point where the two equations intersect can be used to check and confirm your paper-and-pencil work. We will learn more about graphing in Chapter 4.

The same procedure is used to solve an equation that has the variable on the right side of the equation.

•••
EXAMPLE 3

Solve for n: $-2.1 = n + 0.6$

SOLUTION

Here we will add the additive inverse of 0.6 to both sides of the equation.

$$-2.1 = n + 0.6$$
$$-2.1 + (-0.6) = n + 0.6 + (-0.6) \quad \text{Adding } -0.6 \text{ to both sides}$$
$$-2.7 = n \quad\quad\quad\quad\quad\quad \text{Simplifying}$$

So the solution is $n = -2.7$. We leave the check for you to do.

▶ *CHECK* **Warm-Up 3**

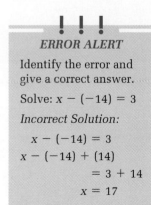

! ! !
ERROR ALERT

Identify the error and give a correct answer.

Solve: $x - (-14) = 3$

Incorrect Solution:

$$x - (-14) = 3$$
$$x - (-14) + (14)$$
$$= 3 + 14$$
$$x = 17$$

The Multiplication Principle of Equality

To solve an equation such as $3x = 21$, we must make the coefficient of x equal 1. To do so, we can use the **multiplication principle of equality.**

> **Multiplication Principle of Equality**
>
> For real numbers a, b, and c,
> $$\text{If } a = b, \text{ then } ca = cb$$

In words, if we multiply both sides of an equation by the same value, the solution remains unchanged. To use this principle, we multiply both sides of an equation by the multiplicative inverse (reciprocal) of the coefficient of x.

• • •

EXAMPLE 4

Solve for x: $5x = 65$

SOLUTION

We multiply both sides of the equation by $\frac{1}{5}$ (the multiplicative inverse of 5) and then simplify.

$$5x = 65 \qquad \text{Original equation}$$
$$\left(\tfrac{1}{5}\right)(5x) = \left(\tfrac{1}{5}\right)(65) \qquad \text{Multiplying both sides by } \tfrac{1}{5}$$
$$1x = \tfrac{65}{5} \qquad \text{Multiplicative inverse property}$$
$$x = 13 \qquad \text{Simplifying}$$

So $x = 13$ in the equation $5x = 65$. We leave the check for you to do.

▶ CHECK **Warm-Up 4**

In the the next example, the coefficient is both negative and a fraction. Its reciprocal is also negative and a fraction.

• • •

EXAMPLE 5

Solve for t: $-\frac{3t}{16} = 27$

SOLUTION

First we rewrite the equation so that we can see the coefficient more easily.

$$-\frac{3t}{16} = 27$$
$$\left(-\frac{3}{16}\right)t = 27$$

Next we multiply both sides of the equation by $-\frac{16}{3}$ (the multiplicative inverse of $-\frac{3}{16}$) and simplify.

$$\left(-\frac{3}{16}\right)t = 27 \qquad \text{Original equation}$$

$$\left(-\frac{16}{3}\right)\left(-\frac{3}{16}\right)t = \left(-\frac{16}{3}\right)(27) \qquad \text{Multiplying by } -\frac{16}{3}$$

$$t = -\frac{16}{3}(\overset{9}{27}) \qquad \text{Multiplicative inverse property}$$

$$t = -144 \qquad \text{Simplifying}$$

Remember to check your work.

▶ *CHECK* **Warm-Up 5**

Solving Linear Equations with Two Operations

In solving linear equations of the type $x + b = c$, we use the addition principle. To solve equations of the type $ax = c$, we use the multiplication principle. To solve an equation of the form $ax + b = c$, we must use *both* the addition principle and the multiplication principle.

We apply the addition principle first, to isolate the term that contains the variable. Then we use the multiplication principle to obtain the coefficient 1 for the variable.

More formally,

> **To solve a linear equation**
> 1. Simplify both sides of the equation using the properties of real numbers.
> 2. Use the addition principle to isolate the term that contains the variable.
> 3. Use the multiplication principle to give the variable a coefficient of 1.

Let's look at an example.

■ ■ ■

EXAMPLE 6

Solve for x: $2x + 4 = 8$

SOLUTION

To isolate the variable term $2x$, we add -4 to both sides of the equation.

$$2x + 4 = 8 \qquad \text{Original equation}$$

$$2x + 4 + (-4) = 8 + (-4) \qquad \text{Adding the additive inverse of 4}$$

$$2x = 4 \qquad \text{Simplifying}$$

Then, to obtain a coefficient of 1 for x, we multiply both sides of the equation by $\frac{1}{2}$ (the multiplicative inverse of 2) and simplify.

$$\left(\frac{1}{2}\right)(2)x = \left(\frac{1}{2}\right)(4) \qquad \text{Multiplying both sides by } \frac{1}{2}$$

$$x = \left(\frac{1}{2}\right)(\overset{2}{4}) \qquad \text{Multiplicative inverse property}$$

$$x = 2 \qquad \text{Simplifying}$$

! ! !
ERROR ALERT

Identify the error and give a correct answer.

Solve: $\frac{3}{4} = 12 + x$

Incorrect Solution:

$$\frac{3}{4} = 12 + x$$

$$\frac{3}{4}(4) = 12 + x(4)$$

$$3 = 12 + 4x$$

$$3 + (-12) = 12 + 4x$$

$$\qquad\qquad + (-12)$$

$$-9 = 4x$$

$$\left(\frac{1}{4}\right)(-9) = \left(\frac{1}{4}\right)(4x)$$

$$-\frac{9}{4} = x$$

To check, we substitute 2 for x in the original equation.

$$2x + 4 = 8$$

$$2(2) + 4 = 8$$

$$4 + 4 = 8$$

$$8 = 8 \quad \text{True}$$

So $x = 2$ in the equation $2x + 4 = 8$.

▶ *CHECK* **Warm-Up 6**

Calculator Corner

Example 6 can also be solved by graphing both sides of the equation and finding the point of intersection.

 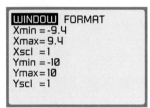 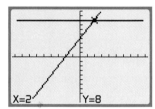

When $y = 8$, the side $2x + 4$ is true only when $x = 2$.

In the next example the variable has a fractional coefficient.

▪▪▪

EXAMPLE 7

Solve for n: $\frac{2n}{3} + 2 = 9$

SOLUTION

To isolate the term that contains the variable, we add -2 to both sides of the equation.

$$\frac{2n}{3} + 2 = 9$$

$$\frac{2n}{3} + 2 + (-2) = 9 + (-2) \quad \text{Adding the additive inverse}$$

$$\frac{2n}{3} = 7 \quad \text{Simplifying}$$

Next we rewrite $\frac{2n}{3}$ as $\left(\frac{2}{3}\right)n$, obtaining

$$\left(\frac{2}{3}\right)n = 7$$

Because the coefficient of n is $\frac{2}{3}$, we multiply both sides of the equation by $\frac{3}{2}$. This will result in a coefficient of 1 for n.

$$\left(\frac{3}{2}\right)\left(\frac{2}{3}\right)n = \left(\frac{3}{2}\right)7 \qquad \text{Multiplying by } \tfrac{3}{2}$$

$$1n = \frac{21}{2} \qquad \text{Multiplicative inverse property}$$

$$n = \frac{21}{2} = 10.5 \qquad \text{Simplifying}$$

You should check the solution by substituting in the original equation or by using a graphing calculator.

Look Again

We could have solved this problem by first multiplying both sides of the equation by 3 to clear the fractions.

$$3\left(\frac{2n}{3} + 2\right) = (9)(3)$$

$$3\left(\frac{2n}{3}\right) + 3(2) = (9)(3) \qquad \text{Distributive property}$$

$$2n + 6 = 27 \qquad \text{Multiplying}$$

$$2n + 6 + (-6) = 27 + (-6) \quad \text{Adding } -6$$

$$2n = 21 \qquad \text{Simplifying}$$

$$\left(\frac{1}{2}\right)(2n) = (21)\left(\frac{1}{2}\right) \qquad \text{Multiplying by } \tfrac{1}{2}$$

$$n = \frac{21}{2}$$

▶ *CHECK* **Warm-Up 7**

Calculator Corner

You must be careful when entering equations into your calculator when parentheses are needed in the equation. For instance, how would you enter the equation in Example 7 using two operations (division and addition) into your graphing calculator? (Note: Remember that you must use x's and y's as the variables with your graphing calculator. So instead of using an n as is shown in Example 7, you will use an x when you enter the left side into your graphing calculator.)

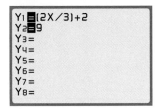 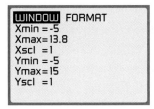 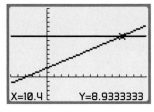 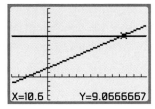

What do you notice happens when you **TRACE** on this graph? The **TRACE** feature does not land on $x = 10.5$. It goes from $x = 10.4$ to $x = 10.6$. So how can you confirm your answer obtained using algebra when **TRACE** only *suggests* the correct answer? Some graphing calculators are able to calculate points of intersection using a **CALC**ulate feature.

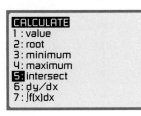 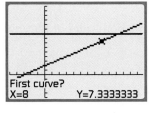 Press **ENTER.**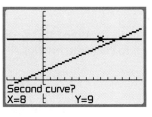

Press **ENTER.** 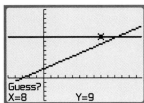 Press **ENTER.**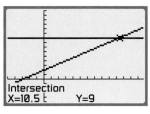

The **INTERSECT** utility of your graphing calculator *confirms* your answer $x = 10.5$.

In the next example, the variable is on the right side. We proceed as usual.

▪▪▪

EXAMPLE 8

Solve for x: $-1.6 = 2.3x + 0.7$

SOLUTION

We first isolate the variable on one side of the equation.

$$-1.6 = 2.3x + 0.7 \qquad \text{Original equation}$$
$$-1.6 + (-0.7) = 2.3x + 0.7 + (-0.7) \qquad \text{Adding } -0.7 \text{ to both sides}$$
$$-2.3 = 2.3x \qquad \text{Simplifying}$$

Multiplying by the reciprocal of the coefficient and simplifying yield:

$$\left(\tfrac{1}{2.3}\right)(-2.3) = \left(\tfrac{1}{2.3}\right)(2.3)x \qquad \text{Multiplying both sides by } \tfrac{1}{2.3}$$
$$-1 = 1x \qquad \text{Simplifying}$$

or $\qquad\qquad\quad -1 = x$

You should check $x = -1$ in the equation $-1.6 = 2.3x + 0.7$.

LOOK AGAIN

$$-1.6 = 2.3x + 0.7$$

We can solve this equation by multiplying by 10 first.

$$-16 = 23x + 7$$
$$-16 - 7 = 23x + 7 - 7$$
$$-23 = 23x$$
$$\left(\tfrac{1}{23}\right)(-23) = \left(\tfrac{1}{23}\right)23x$$
$$-1 = x$$

▶ *CHECK* **Warm-Up 8**

Solving More Complex Linear Equations

Suppose you are asked to solve an equation such as $4x + 16 = 12x$, where the variable x appears in two terms. Your first step must be to isolate the variable term on one side of the equation. For that you can use the addition principle.

▪▪▪
EXAMPLE 9

Solve for x: $4x + 16 = 12x$

SOLUTION

First, we use the addition principle to move the variable term $4x$ to the right side of the equation.

$$4x + 16 = 12x \qquad\qquad\qquad \text{Original equation}$$
$$4x + 16 + (-4x) = 12x + (-4x) \quad \text{Adding } -4x \text{ to both sides}$$
$$16 = 12x - 4x \qquad\qquad \text{Simplifying}$$
$$16 = (12 - 4)x \qquad\qquad \text{Distributive property}$$
$$16 = 8x \qquad\qquad\qquad \text{Simplifying}$$
$$\left(\tfrac{1}{8}\right)16 = \left(\tfrac{1}{8}\right)8x \qquad\qquad \text{Multiplying by } \tfrac{1}{8}$$
$$2 = x \qquad\qquad\qquad \text{Simplifying}$$

Check your answer in the original equation.

▶ *CHECK* **Warm-Up 9**

■ ■ ■

EXAMPLE 10

Solve for n: $3n - 4 - 9 - 4n = 5n - 6$

SOLUTION

Simplifying the left side gives us

$3n - 4 - 9 - 4n = 5n - 6$	Original equation
$3n - 4n - 4 - 9 = 5n - 6$	Commutative property
$(3 - 4)n - 13 = 5n - 6$	Distributive property
$-1n - 13 = 5n - 6$	Simplifying
$-1n - 13 + 1n = 5n - 6 + 1n$	Adding n to both sides
$-13 = 6n - 6$	Combining terms
$-13 + 6 = 6n - 6 + 6$	Adding 6 to both sides
$-7 = 6n$	Simplifying

Finally, we multiply by the reciprocal of 6, which is $\frac{1}{6}$, to get n alone.

$\left(\frac{1}{6}\right)(-7) = \left(\frac{1}{6}\right)(6n)$	Multiplying both sides by $\frac{1}{6}$
$-\frac{7}{6} = n$	Simplifying

Checking will show that $n = -\frac{7}{6}$ in the equation $3n - 4 - 9 - 4n = 5n - 6$.

▶ *CHECK* **Warm-Up 10**

Calculator Corner

Your graphing calculator can also help you to check your answers for more complex linear equations. For instance, Example 10 can be worked on the calculator as follows. If you use the **TRACE** feature of your calculator, you may not land on the *exact* point of intersection. If your calculator has a **CALC**ulate **INTERSECT** feature, use it to confirm your result. (Note: See the Calculator Corner on page 128 to review how to find the intersection points of two graphs.)

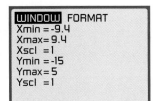
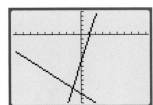
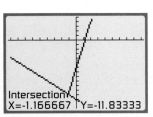

The calculator gives the answer in decimal form. On many graphing calculators, you can use the Home Screen and easily convert this decimal result into a fraction. Go to your Home Screen by pressing **2nd QUIT**; then press the x-variable key followed by the **MATH** key and choose the **FRAC** option. Now you can see the answer as a fraction equal to the result that was obtained by the traditional paper-and-pencil method.

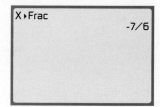

▪▪▪

EXAMPLE 11

Solve for k: $0.2(6 + 3k) = 9$

SOLUTION

First we apply the distributive property to the left side of the equation to remove the parentheses.

$$0.2(6 + 3k) = 9 \qquad \text{Original equation}$$
$$0.2(6) + 0.2(3k) = 9 \qquad \text{Distributive property}$$
$$1.2 + 0.6k = 9 \qquad \text{Simplifying}$$

Then we use the addition and multiplication principles, in that order, and simplify.

$$1.2 + (-1.2) + 0.6k = 9 + (-1.2) \qquad \text{Adding } -1.2 \text{ to both sides}$$
$$0.6k = 7.8 \qquad \text{Simplifying}$$
$$10(0.6)k = (7.8)10 \qquad \text{Multiplying both sides by 10}$$
$$6k = 78 \qquad \text{Simplifying}$$
$$\left(\tfrac{1}{6}\right)6k = \left(\tfrac{1}{6}\right)78 \qquad \text{Multiplying both sides by } \tfrac{1}{6}$$
$$k = 13 \qquad \text{Simplifying}$$

A check will show that $k = 13$ in the equation $0.2(6 + 3k) = 9$.

▶ CHECK **Warm-Up 11**

> ▪▪▪
> ### WRITER'S BLOCK
> Explain to a friend how we knew to multiply both sides of the equation $0.6k = 7.8$ by 10.

In this next example, we must first remove parentheses and then combine like terms.

▪▪▪

EXAMPLE 12

Solve for n: $5(3n + 4) - n = -2(n + 6)$

SOLUTION

Using the distributive property to remove the parentheses gives us

$5(3n + 4) - n = -2(n + 6)$ Original equation
$5(3n) + 5(4) - n = -2(n) + (-2)(6)$ Distributive property
$15n + 20 - n = -2n - 12$ Simplifying

Then we combine like terms on the left side of the equation.

$(15 - 1)n + 20 = -2n - 12$ Commutative and distributive properties
$14n + 20 = -2n - 12$ Simplifying

Then we use the addition principle twice to isolate the variable.

$14n + 20 + (-20) = -2n - 12 + (-20)$ Adding -20 to both sides
$14n = -2n - 32$
$14n + 2n = -2n - 32 + 2n$ Adding $2n$ to both sides
$16n = -32$ Simplifying

Finally, we use the multiplication principle to give the variable a coefficient of 1.

$16n = -32$
$\left(\frac{1}{16}\right)(16)n = \left(\frac{1}{16}\right)(-32)$ Multiplying both sides by $\frac{1}{16}$
$n = -\frac{32}{16} = -2$ Simplifying

A check will show that $n = -2$ in the equation $5(3n + 4) - n = -2(n + 6)$.

▶ *CHECK* **Warm-Up 12**

Calculator Corner

We can work Example 12 on the graphing calculator. (Note: See the Calculator Corner on page 128 to review how to find the points of intersection of two graphs.)

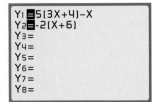 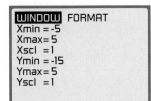 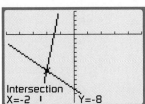

Practice what you learned.

SECTION FOLLOW-UP

Here are answers to the questions asked in the Section Lead-In using the equation

$$F = c + 40$$

where F is the Fahrenheit temperature and c is the number of cricket chirps in fifteen seconds.

1. The chirps are less frequent.

2. The temperature has gone up.

3. According to this equation, they don't chirp. They probably die or are protected from the cold somehow.

4. Answers may vary.

3.1 WARM-UPS

Work these problems before you attempt the exercises.

1. Solve for n: $n + 9 = -4$

2. Solve for m: $m - 27 = 16$

3. Solve for n: $3.7 = n + 4.9$

4. Solve for x: $12x = 144$

5. Solve for t: $-\frac{4t}{5} = 484$

6. Solve for y: $16y - 13 = 3$

7. Solve for n: $\frac{3n}{2} + 4 = -8$

8. Solve for x: $9 = 0.3x + 5.4$

9. Solve for y: $8 - y = 12y$

10. Solve for t: $6t - 4 + 5t = -9 + 7t$

11. Solve for r: $-0.8(6 + 3r) = 12$

12. Solve for n: $6(2n + 8) - n = 3(n + 8)$

3.1 EXERCISES

Note: Use your graphing calculator to check your results whenever possible.

In Exercises 1 through 16, use the addition principle to solve the equations.

1. $x + 9 = 15$　　　**2.** $n + 7 = 11$　　　**3.** $y - 87 = 12$　　　**4.** $y - 28 = 18$

5. $15 = t + 4$　　　**6.** $13 = y + 6$　　　**7.** $18 = n - 92$　　　**8.** $22 = n - 24$

9. $348 + x = -132$　　**10.** $287 + t = -125$　　**11.** $x + 4 = -2.2$　　**12.** $t - 37 = -0.15$

13. $-\frac{3}{8} = \frac{1}{8} + n$　　**14.** $\frac{1}{7} + y = -\frac{5}{7}$　　**15.** $-1.7 = t + 9$　　**16.** $-36 = x - 0.38$

In Exercises 17 through 32, use the multiplication principle to solve the equations.

17. $19t = 57$　　　**18.** $85 = 15x$　　　**19.** $-9t = -12$　　　**20.** $-6x = -132$

21. $1.4n = 56$　　　**22.** $96 = 0.16x$　　　**23.** $\frac{3}{4}t = 24$　　　**24.** $\frac{2}{7}x = 36$

25. $0.2t = 14$　　　**26.** $27 = 0.03x$　　　**27.** $\frac{t}{66} = -11$　　　**28.** $\frac{x}{24} = -12$

29. $\frac{x}{12} = 132$　　　**30.** $560 = -\frac{n}{14}$　　　**31.** $-\frac{r}{18} = -900$　　　**32.** $-\frac{y}{8} = -176$

In Exercises 33 through 84, solve each equation.

33. $6x + 7 = 13$　　　**34.** $5x + 8 = 33$　　　**35.** $10y - 3 = 27$　　　**36.** $9y - 12 = 6$

37. $5x + 1.9 = 1.6$　　**38.** $6y + 2.1 = 1.2$　　**39.** $5.1m + 1.3 = 6.4$　　**40.** $7r - 1.4 = 3.5$

41. $\frac{x}{3} + 8 = 11$　　**42.** $\frac{r}{3} + 9 = -16$　　**43.** $\frac{x}{2} - 15 = -23$　　**44.** $\frac{r}{9} - 26 = -44$

45. $-5 - x = 22$　　**46.** $-8 - m = 16$　　**47.** $-3 - 6x = -0.15$　　**48.** $-4 + 2.4r = 8$

49. $18 = 0.6m + 30$　　**50.** $57 = 5y + 42$　　**51.** $-29 = \frac{x}{7} + 11$　　**52.** $-12 = \frac{x}{4} + 3$

53. $-7 = -0.25 - 5r$　　**54.** $1.2 = 0.8 + 2.5r$　　**55.** $-1.4 - 9y = -5.0$　　**56.** $-8 - 6m = -5.6$

57. $\frac{y}{6} + 9 = 26$　　**58.** $8 = 32 + \frac{m}{5}$　　**59.** $2x + 9 = 11x$　　**60.** $5y - 6 = 18y$

61. $-6 + 7x = 5x$　　**62.** $-88 - 18t = 26t$　　**63.** $-8.4 + 6x = 5x$　　**64.** $-3.2n - 51 = 7n$

65. $\frac{3}{5}x + 12 = \frac{1}{5}x$　　**66.** $-\frac{1}{4}y - 16 = \frac{1}{4}y$　　**67.** $4r - 6 = 9r + 12$　　**68.** $3v + 9 = 11v - 18$

69. $15(7x + 5) = -9$　　**70.** $11(2x + 6) = 22$　　**71.** $89x = 7(13x - 8)$　　**72.** $4(2x - 12) = 96$

73. $-8(3x - 2) = 8$　　**74.** $5(16 - 3x) = 20$　　**75.** $-2 = 12\left(\frac{5}{6} + 3x\right)$　　**76.** $\frac{1}{2} = -\frac{2}{5}\left(8x - \frac{3}{4}\right)$

77. $-5(-4 - v) + 9 = v + 12$　　　　　　**78.** $-12(8 - 2t) + 5 = 8t - 2$

79. $11(2x + 4) + 6 = -3(7x + 3)$

80. $7(3x - 2) - 11 = -3(8x - 5)$

81. $-6(2x + 4) = 9(3x - 8) + 5$

82. $-12(5x + 3) - 2 = 11(2x + 4)$

83. $(3x + 2)7 + 8 = 2(2x + 9)$

84. $(8 + t)2 + 7 = 3(15 + t)$

EXCURSIONS

Posing Problems

1. The following table lists the salaries of the New York Rangers hockey team during the 1994 season when they won the Presidents' Trophy and the Stanley Cup. Ask and answer four questions based on this data.

Rangers' Salaries*

Player	Salary	Player	Salary	Player	Salary
Mark Messier	$2,533,000	Eddie Olczyk	850,000	Stephane Matteau	425,000
Brian Leetch	1,805,000	Jeff Beukeboom	725,000	Brian Noonan	400,000
Glenn Anderson	1,250,000	Doug Lidster	700,000	Jay Wells	400,000
Adam Graves	1,150,000	Sergei Nemchinov	600,000	Mike Hudson	375,000
Steve Larmer	1,100,000	Craig MacTavish	550,000	Mike Hartman	310,000
Mike Richter	1,000,000	Nick Kypreos	525,000	Alexander Karpovtsev	275,000
Esa Tikkanen	979,000	Alexei Kovalev	450,000	Sergei Zubov	250,000
Kevin Lowe	950,000	Greg Gilbert	425,000		
Glenn Healy	850,000	Joe Kocur	425,000		

*Excluding bonuses other than for signing or reporting

Data Analysis

2. ✏ Using the following data, analyze the nutrients and calories per ounce of energy bars A through G. Use your analysis to recommend an energy bar for your health-conscious friends. Justify your choice.

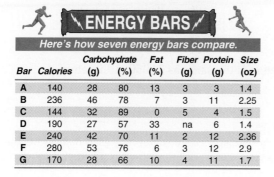

Bar	Calories	Carbohydrate (g)	Carbohydrate (%)	Fat (%)	Fiber (g)	Protein (g)	Size (oz)
A	140	28	80	13	3	3	1.4
B	236	46	78	7	3	11	2.25
C	144	32	89	0	5	4	1.5
D	190	27	57	33	na	6	1.4
E	240	42	70	11	2	12	2.36
F	280	53	76	6	3	12	2.9
G	170	28	66	10	4	11	1.7

CONNECTIONS TO *GEOMETRY*

Applying Ratio and Proportion

Sometimes information is given in one form but we want it in another form. We use measurements to describe length, area, volume, capacity, mass, time, and temperature. Two systems of measurement are used in the United States, the U.S. Customary system and the metric system. The metric system is used almost exclusively throughout the rest of the world.

In this Connection, we compare the two systems informally and **convert,** or rename, measurements within the systems and from system to system. The ideas of ratio and proportion are helpful in doing this.

The first measure we discuss is **length.** The most common units of length are shown in Table 1. (Note: All tables referred to in this Connection appear on the inside covers of this text.) Here are some examples of length:

> The wire in a paper clip is usually about 1 millimeter thick.
> An average-size man's long step is about 1 meter.
> (A meter is a little more than a yard.)

■■■
EXAMPLE

3245 centimeters is how many millimeters? (3245 cm = ? mm)

SOLUTION

To answer this question, we will set up a proportion involving ratios of centimeters to millimeters. We will use the fact that 10 millimeters is the same as 1 centimeter to help us convert this measurement. Then 3245 centimeters is to n millimeters as 1 centimeter is to 10 millimeters.

$$\frac{3245 \text{ cm}}{n \text{ mm}} = \frac{1 \text{ cm}}{10 \text{ mm}}$$

Once we check that the units are the same on both sides, we can drop them.

$$\frac{3245}{n} = \frac{1}{10}$$

$$3245 \times 10 = 1 \times n \quad \text{Cross-multiply}$$

$$32{,}450 = n$$

So 3245 centimeters is the same as 32,450 millimeters.

> In the metric system, spaces are sometimes used instead of commas to separate the periods (groups of three digits). Then 32,450 would be written 32 450.

Units of **area** are shown in Table 2. Area is a measure of the extent of a surface. One square unit of area can be thought of as the area of a square whose side is 1 unit long.

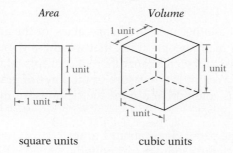

Area

1 unit

|← 1 unit →|

square units

Volume

1 unit

1 unit

1 unit

cubic units

Volume is the measure of the three-dimensional space in a container. Volume units are given in Table 3. The measure of liquid volume is usually called **capacity** (Table 4).

> One milliliter is the same as 1 cubic centimeter. A teaspoon holds about 5 mL. A liter is just a bit larger than a quart.

▪ ▪ ▪

EXAMPLE

1950 ounces is equal to how many quarts?

SOLUTION

We first must look up the relationship between ounces and quarts. Table 4 shows no direct relationship. However, it does show that 16 ounces is 1 pint and that 2 pints is 1 quart. We shall need two proportions.

Our first proportion is

$$\frac{1950 \text{ oz}}{n \text{ pt}} = \frac{16 \text{ oz}}{1 \text{ pt}} \quad \text{or} \quad \frac{1950}{n} = \frac{16}{1}$$

Then we get

$$1950 \times 1 = 16 \times n \quad \text{Cross-multiplying}$$

$$\frac{975}{8} = n \quad\quad\quad \text{Simplifying}$$

So 1950 ounces is the same as $\frac{975}{8}$ pints.

Now we want to determine the number of quarts equivalent to $\frac{975}{8}$ pints. We set up another proportion:

$$\frac{\frac{975}{8} \text{ pt}}{n \text{ qt}} = \frac{2 \text{ pt}}{1 \text{ qt}} \quad \text{or} \quad \frac{121.875}{n} = \frac{2}{1}$$

Then we have

$$121.875 \times 1 = 2 \times n \quad \text{Cross-multiplying}$$

$$60.9375 = n \quad\quad\quad \text{Simplifying}$$

So $\frac{975}{8}$ pints is the same as 60.9375 quarts and is also equivalent to 60 quarts 1 pint 14 ounces.

◢

Table 5 lists units of another measurement—**mass,** or the amount of matter in an object. Although many people think that weight and mass mean the same thing, in science, mass and weight are different concepts. Mass is the amount of matter in an object. Weight is a measure of the force of gravity exerted on an object. However, it is common to use the metric units for mass to describe the metric weight of an object.

> The mass of one paper clip is about 1 gram.
> A kilogram is slightly more than 2 pounds.
> The mass of a sub-compact car is about 1 ton.

Units of **time** are listed in Table 6. They are used in both the metric and the U.S. customary systems.

In the next conversion example, we will use ratios to change two units at the same time.

▪▪▪

EXAMPLE

Professor Bill runs a 9-minute mile (9 minutes per mile). How many feet per second does he run?

SOLUTION

He runs $\frac{9 \text{ minutes}}{\text{mile}}$. We want to write this as $\frac{? \text{ feet}}{\text{second}}$. We need to use these relationships:

$$1 \text{ minute} = 60 \text{ seconds} \longrightarrow \frac{60 \text{ sec}}{1 \text{ min}} \quad \text{or} \quad \frac{1 \text{ min}}{60 \text{ sec}}$$

$$1 \text{ mile} = 5280 \text{ feet} \longrightarrow \frac{5280 \text{ ft}}{1 \text{ mi}} \quad \text{or} \quad \frac{1 \text{ mi}}{5280 \text{ ft}}$$

We have to chose the units to rewrite

$$\frac{9 \text{ min}}{\text{mi}} \cdot \qquad = \frac{? \text{ ft}}{\text{sec}}$$

We choose

$$\frac{9 \,\cancel{\text{min}}}{\cancel{\text{mi}}} \left(\frac{60 \text{ sec}}{1 \,\cancel{\text{min}}} \right)\left(\frac{1 \,\cancel{\text{mi}}}{5280 \text{ ft}} \right) = \frac{5410 \text{ sec}}{5280 \text{ ft}}$$

Now we have the right units. However, they are in the wrong order. We invert the ratio to

$$\frac{5280 \text{ ft}}{540 \text{ sec}}$$

And then rewrite as a unit ratio.

$$9.8 \text{ ft/sec}$$

◢

PRACTICE

The heights of four South American volcanoes are given in Exercises 1 through 4. Use a calculator to convert them as requested. In all cases, round to the nearest hundredth of a unit.

1. Aconcagua, 22,834 feet. Convert to miles.

2. Chimborazo, 20,560 feet. Convert to yards.

3. Antisana, 18,713 feet. Convert to inches.

4. Cotopaxi, 19,344 feet. Convert to miles.

In modern rhythmic gymnastics, certain hand apparatus is used. The sizes of some of this equipment are given in Exercises 5 through 8. Rewrite these measurements in the given units.

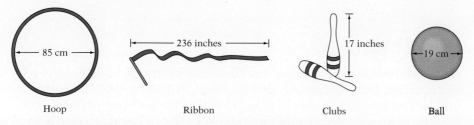

Hoop   Ribbon   Clubs   Ball

5. Hoop, 85 centimeters in diameter. Rewrite in millimeters.

6. Ribbon, 236 inches long. Rewrite in feet expressed as a fraction.

7. Clubs, 17 inches high. Rewrite in centimeters.

8. Ball, 19 centimeters in diameter. Rewrite in inches rounded to the nearest hundredth.

The weights of various playing balls are given in Exercises 9 through 12. Rewrite them in the requested units.

9. Table tennis ball, 0.09 ounce. Rewrite in grams. Give your answer to the nearest tenth of a gram.

10. Jai alai ball, 127 grams. Rewrite in ounces. Give your answer to the nearest ounce.

11. Badminton "bird," 5.5 grams. Rewrite in kilograms.

12. Racquetball, 1.4 ounces. Rewrite in pounds.

3.2 Solving Literal Equations and Formulas

SECTION LEAD-IN

Given

$$\frac{D}{F} = U$$

where F is the fuel used in gallons, D is the distance traveled, and U is the usage in miles per gallon, answer the following questions.

1. If U remains constant, how does F change as D gets larger?

2. If U remains constant and F becomes small, what happens to D?

3. I travel a distance D and my usage changes from 18 miles per gallon to 10 miles per gallon. In this case, does F become smaller or larger relative to what it was before?

WRITER'S BLOCK

What is the difference between a linear equation and a literal equation?

Literal Equations

A **literal equation** is an equation that contains several variables. We solve a literal equation, just as we solved other equations in the first part of Chapter 3, by using the addition and multiplication principles. The only thing that changes is the form of the solution. In most cases, the solution of a literal equation is an algebraic expression rather than a number.

■ ■ ■
EXAMPLE 1

Solve for n: $n + 5t = xy$

SOLUTION

We want to get the variable n alone on one side of the equation with a coefficient of 1. To remove $5t$ from the left side, we add $-5t$ to both sides of the equation.

$$
\begin{aligned}
n + 5t &= xy && \text{Original equation} \\
n + 5t + (-5t) &= xy + (-5t) && \text{Adding } -5t \text{ to both sides} \\
n &= xy - 5t && \text{Simplifying}
\end{aligned}
$$

So $n = xy - 5t$.

We can check by substituting in the original equation.

$$
\begin{aligned}
n + 5t &= xy \\
(xy - 5t) + 5t &= xy \\
xy - 5t + 5t &= xy \\
xy &= xy \quad \text{True}
\end{aligned}
$$

▶ *CHECK* **Warm-Up 1**

•••
EXAMPLE 2

Solve for r: $\frac{3r}{2} = t - 4$

SOLUTION

Our equation can be rewritten as $\left(\frac{3}{2}\right)r = (t - 4)$. Using the multiplication principle, we first multiply by $\frac{2}{3}$, the reciprocal of $\frac{3}{2}$.

$$\left(\frac{2}{3}\right)\left(\frac{3}{2}\right)r = \left(\frac{2}{3}\right)(t - 4)$$

$$r = \left(\frac{2}{3}\right)(t - 4)$$

Rewriting gives us

$$r = \frac{2(t - 4)}{3} \quad \text{or} \quad \frac{2t - 8}{3}$$

So $r = \frac{2t - 8}{3}$ in the equation $\frac{3r}{2} = t - 4$.

STUDY HINT

When multiplying one side of an equation by a term, be sure to multiply the entire side. To do this without mistakes, place parentheses around that entire side of the equation.

LOOK AGAIN

$$\frac{3r}{2} = t - 4$$

$$(2)\left(\frac{3r}{2}\right) = (2)(t - 4) \quad \text{Multiplying by 2}$$

$$3r = 2t - 8 \quad \text{Simplifying}$$

$$\frac{3r}{3} = \frac{2t - 8}{3} \quad \text{Dividing by 3}$$

$$r = \frac{2t - 8}{3} \quad \text{Simplifying}$$

To check the result $r = \frac{2t - 8}{3}$, substitute for r in the original equation

$$\frac{3\left(\frac{2t - 8}{3}\right)}{2} = t - 4 \quad \text{Substituting}$$

$$\frac{3\left(\frac{2t - 8}{\cancel{3}}\right)}{2} = t - 4 \quad \text{Cancelling}$$

$$\frac{2t - 8}{2} = t - 4 \quad \text{Simplifying}$$

$$\frac{2(t - 4)}{2} = t - 4 \quad \text{Distributive property}$$

$$t - 4 = t - 4 \quad \text{True}$$

▶ CHECK **Warm-Up 2**

WRITER'S BLOCK

Explain how "Dividing by 3" in the Look Again Solution is an application of the multiplication principle of equality.

Formulas

We are sometimes interested in how one variable changes in relation to others. We use formulas to examine these relationships. Formulas, such as Einstein's famous formula $E = mc^2$, are also literal equations.

▪▪▪
EXAMPLE 3

The distance formula is sometimes given as $t = \frac{d}{r}$, where t = time, d = distance, and r = rate. Solve this formula for the rate r.

SOLUTION

We want r alone on one side of the equation. Because it is now in a denominator, we multiply both sides by r, the multiplicative inverse of $\frac{1}{r}$.

$$t = \frac{d}{r}$$

$$(r)(t) = (r)\left(\frac{d}{r}\right) \quad \text{Multiplying by } r$$

$$rt = d$$

We now have r on the left, but we must still remove t. To do so, we multiply by the multiplicative inverse of t, which is $\frac{1}{t}$.

$$\left(\frac{1}{t}\right)rt = \left(\frac{1}{t}\right)d \quad \text{Multiplying by } \frac{1}{t}$$

$$r = \frac{d}{t}$$

So $r = \frac{d}{t}$ in the distance formula.

▶ *CHECK* **Warm-Up 3**

In Examples 4 and 5, some values for the variables are given. We solve for the final variable in two different ways. The first way involves substituting all known values and then solving for the unknown variable.

▪▪▪
EXAMPLE 4

Consider the formula $F = \frac{RU}{12}$, where F is the finance charge imposed on a buyer, R is the yearly rate of interest expressed in decimal form, and U is the unpaid balance. Find U when R is 0.18 and F is \$6.

SOLUTION

We first substitute in all known values of variables.

$$F = \frac{RU}{12}$$

$$6 = \frac{[0.18(U)]}{12}$$

Next we multiply both sides by 100 to clear the decimals.

$$(100)6 = \frac{0.18}{12}U(100) \quad \text{Multiplying by 100}$$

$$600 = \frac{18}{12}U \quad \text{Simplifying}$$

$$600 = \frac{3}{2}U \quad \text{Simplifying}$$

$$\left(\tfrac{2}{3}\right)600 = \left(\tfrac{2}{3}\right)\tfrac{3}{2}U \quad \text{Multiplying by } \tfrac{2}{3}$$

$$400 = U \qquad\qquad \text{Simplifying}$$

Thus U, the unpaid balance, is \$400.

▶ *CHECK* **Warm-Up 4**

In the second method of evaluating a variable in a formula, we solve for the unknown variable first. Then we substitute the given numerical values in the resulting equation and simplify.

▪▪▪
EXAMPLE 5

In the total-cost formula

$$T = F + QV$$

T is the total cost, F is the flat fee, Q is the amount per service, and V is the number of services.

Find Q when T is 78, F is 10, and V is 4.

SOLUTION

We want the variable Q alone, with a coefficient of 1. We remove F from the right side by adding $-F$ to both sides.

$$T = F + QV$$
$$T + (-F) = F + QV + (-F) \quad \text{Adding } -F$$
$$T - F = QV$$

Then we remove V on the right by multiplying both sides by $\tfrac{1}{V}$.

$$\left(\tfrac{1}{V}\right)(T - F) = \left(\tfrac{1}{V}\right)(QV) \qquad \text{Multiplying by } \tfrac{1}{V}$$
$$\tfrac{T - F}{V} = Q$$

So $Q = \tfrac{T - F}{V}$.

Now we substitute the given numerical values and simplify.

$$Q = \frac{T - F}{V}$$
$$= \frac{78 - 10}{4} \quad \text{Substituting}$$
$$= \frac{68}{4}$$
$$= 17$$

So Q is 17 when $T = 78$, $F = 10$, and $V = 4$.

To check this result, we can substitute all known values in the original formula. If the result is a true statement, the solution is correct.

Check:

$$T = F + QV$$
$$78 = 10 + (17)(4) \quad \text{Substituting}$$
$$= 10 + 68$$
$$78 = 78 \qquad \text{True}$$

 CHECK **Warm-Up 5**

It is often simpler to substitute first and then solve for the missing variable.

Both methods, as you can see, yield the same result. Choose whichever works best for you.

Practice what you learned.

SECTION FOLLOW-UP

Here are answers to the questions asked in the Section Lead-In using the equation

$$\frac{D}{F} = U$$

where F is the fuel used in gallons, D is the distance traveled, and U is the usage in miles per gallon.

1. F also gets larger.

2. D also becomes small.

3. F gets larger.

3.2 WARM-UPS

Work these problems before you attempt the exercises.

1.   Solve for t: $yx + t = j$ 2.   Solve for t: $\frac{2t}{3} = r - 5$

3.   Let S be the selling price, O be the original price, and D be the dollar amount of discount. The formula for the selling price of an item is $S = O - D$. Rewrite this formula in terms of D.

4.   Given $\frac{D}{F} = U$, where F is the fuel used in gallons, D is the distance traveled, and U is the usage in miles per gallon, find D when 87 gallons of fuel are used and your car averages 26 miles per gallon.

5.   Given $C = \frac{5}{9}(F - 32)$, where C is the temperature in degrees Celsius and F is the temperature in degrees Fahrenheit, find F when C is 78 degrees.

3.2 EXERCISES

Note: Use your graphing calculator to check your results whenever possible.

1. Solve for c: $c + h = e$

2. Solve for f: $y + f = e$

3. Solve for c: $c - h = e$

4. Solve for x: $h - x = e$

5. Solve for c: $\frac{c}{h} = e$

6. Solve for x: $\frac{x}{h} = e$

7. Solve for c: $\frac{h}{c} = e$

8. Solve for x: $\frac{h}{x} = e$

9. Solve for c: $2c - h = e$

10. Solve for x: $3x + h = e$

11. Solve for c: $6 + 3c = 5c - h$

12. Solve for x: $7 + 6x = 9x + h$

13. Solve for x: $\frac{x}{3} = e + 8$

14. Solve for c: $\frac{c}{3} = h - 8$

15. Solve for c: $c - \frac{x}{h} = t$

16. Solve for x: $\frac{x}{11} = h - 19$

17. Solve for c: $h - \frac{c}{k} = f$

18. Solve for x: $y - \frac{x}{f} = e$

19. Solve for c: $j + 2\left(\frac{c}{h}\right) = r$

20. Solve for x: $u - 3\left(\frac{x}{y}\right) = e$

21. Solve for c: $c(a + h) = e$

22. Solve for x: $\frac{x}{h} = \frac{e}{t}$

23. Solve for c: $j - \frac{h}{c} = r$

24. Solve for x: $u + \frac{y}{x} = e$

In Exercises 25 through 46, solve each problem by first rewriting the formula in terms of the variable indicated and then substituting the given values. Then check your answer by substituting first and then solving for the variable.

25. **Calories**   Given $C = \frac{A}{S}$, where A is the total number of calories, S is the number of servings, and C is the calories per serving, find the value of C when the total number of calories is 3060 and the number of servings is 25.

26. **Discount Price**   If S is the selling price, O is the original price, and D is the discount, find the value of D in the formula $S = O - D$ when the selling price is $99.95 and the original price is $400.

27. **Tax Rate**   Given $C = (P)(T) + P$, where C is the final cost of the item, P is the selling price, and T is the tax rate expressed as a decimal, find the value of C when the selling price is $50 and the tax rate is 0.0825.

28. Let $D = QM + R$, where D is the dividend, Q is the quotient, R is the remainder, and M is the divisor. Find the value of D when the divisor is 19, the quotient 3, and the remainder 9.

29. **Ohm's Law**   Ohm's law states that current (I amperes), voltage (E volts), and resistance (R ohms) in a simple electrical circuit are related by the equation

$$E = IR$$

If the resistance is 10 ohms and the voltage is 126 volts, what is the current?

30. **Salary** Given $A = BCH$, where A is the annual salary, B is the number of weeks worked, C is the average number of hours per week worked, and H is the hourly rate, find the value of H when a person works 52 weeks for an average of 35 hours per week and earns a salary of $43,500.

31. Given $S = \frac{n(n + 1)}{2}$, where S is the sum of the first n integers, what is the value of S for $n = 76$?

32. **Lifting Capacity** Given $L = V(D_a - D_h)$, where L is the lifting capacity of a blimp, V is the volume of the blimp, D_a is the density of air, and D_h is the density of helium, find the lifting capacity of a blimp 200 feet in diameter when V is 4,188,800 cubic feet, D_a is 0.0763 pound per cubic foot, and D_h is 0.0106 pound per cubic foot. (Round the answer to the nearest whole number.)

33. Let $d = m \div v$, where

> d is the density of an object in grams per cubic centimeter
> m is the mass of the object in grams
> v is the volume of the object in cubic centimeters

Find the value of d when the mass of a small plastic block is 44 grams and the volume is 11 cubic centimeters.

34. Let $V = at$, where

> V is the velocity of an object
> a is the rate of acceleration
> t is the elapsed time in seconds

Find the value of V for a sparrow when a is 16 feet per second^2 (that is, per second per second or per second squared) and the elapsed time is 4 seconds.

35. Let $S = O - D$, where

> S is the selling price
> O is the original price
> D is the discount

Find the value of S when the original price is $2420 and the discount is $50.

36. Let $Q = L \div M$, where

> Q is the cost per unit
> L is the price of an item
> M is the number of units in the item

Find the value of Q when the price of an item is $36 and there are 18 units in the item.

37. Let $P = 20s$, where P is the total number of pecks a woodpecker makes in s seconds. Find P when s is 60.

38. Let $S = C \div 2 + 4$, where S is the number of cups of sugar needed in a recipe and C is the number of cups of flour needed. What is the value of S when the number of cups of flour is 38?

39. Let $F = ma$, where

    F is the force on an object
    m is the mass of the object
    a is the rate of acceleration of the object

    Find F when a mass of 904 grams has an acceleration of 35 units.

40. Let $C = A \div S$, where

    A is the total number of calories
    S is the number of servings
    C is the number of calories per serving

    Find the value of C when there are 157 servings and a total of 18,055 calories.

41. Let $F = D \div U$, where

    F is the fuel in gallons used
    D is the distance traveled
    U is the rate of usage in miles per gallon

    Find the amount of fuel used when you get 25 miles per gallon and you travel 1025 miles.

42. Let $P = I^2R$, where

    P is the power loss in watts
    I is the current in amperes
    R is the resistance in ohms

    Find the power loss when a current of 6 amperes flows through a resistance of 62 ohms.

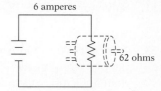

43. Let $rt = d$, where

    d is the distance in miles
    r is the rate or speed in miles per hour
    t is the time in hours

    How far will you travel if you move at 45 miles per hour for 17 hours?

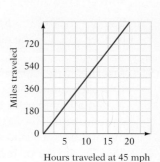

44. Let $A \div (B \times C) = H$, where

    A is the annual salary
    B is the number of weeks worked
    C is the average number of hours worked per week
    H is the hourly rate

    What is the hourly rate when a person works 52 weeks at an average of 40 hours per week and earns a salary of $52,000?

45. Let $S = NV + X$, where

> S is the score on an exam
> N is the number of questions correct
> V is the value for each question
> X is the number of extra credit points

Find the score on an exam when 27 questions are answered correctly at 3 points each, and an additional 5 points are given for extra credit.

46. Let $c + 40 = F$, where

> F is the Fahrenheit temperature
> c is the number of cricket chirps in 15 seconds

What is the temperature when the crickets are chirping 55 times each 15 seconds?

MIXED PRACTICE

By doing these exercises, you will practice the topics up to this point in the chapter.

47. Solve for e: $re + f = c$

48. Solve for x: $-186 + 4x = -127$

49. Solve for the unknown: $-4(2x + 3) = 7$

50. Solve for x: $-84 + x - 12(x - 4) = 62$

51. Solve for y: $-27 + 18y + 36 = 91 - 45y$

52. Solve for t: $\frac{t}{x} - h = r$

53. *Internet Site* On a mathematics education Web site, there were 1404 people who "visited" in November. This represents 30% more visitors than in October. October's number of visitors were 20% more than in September. How many people visited in September?

    Use $N = 0.3(0.2S + S) + 0.2S + S$, where S is the number visiting in September and N is the number visiting in November.

54. *Internet Site* On another Web site, the number of visitors increased 10% each month over the previous month's numbers. How many people visited altogether for three months? Is the equation to solve this problem equivalent to $T = 3.30V$, where T is the total number of visitors and V is the number of visitors in the first of the three months? If the answer is "No," write the correct equation.

55. Find x: $94 = 3x + 67$

56. Solve for x: $26 - x = -86$

57. Find t: $18 = \frac{1}{2}t + 26$

58. Solve for w: $4 = 11 - \frac{1}{7}w$

59. Solve for x: $-7x + 3 + 4x = -9x - 14 + 3x + 24$

60. Solve for t: $-9(2t - 4) = 18$

61. Solve for m: $22m - 7 = -15m + 5$

62. Solve for x: $\dfrac{-x}{15} = 126$

EXCURSIONS

Data Analysis

1. ✏ Compare and contrast the two telephone companies in the following table. Which company seems to offer the best value to you? Justify your answer.

Long-Distance Calls

City to City	Minutes	Telephone Company A	B	City to City	Minutes	Telephone Company A	B
New York City to Boston	17	$3.62	$1.86	Rockville Centre to Kansas City	5	1.34	0.68
Great Neck to Chicago	2	0.57	0.27	New Canaan to Cincinnati	26	6.19	3.32
Greenwich to Atlanta	20	4.79	2.66	Flushing to Milwaukee	3	0.81	0.40
White Plains to Philadelphia	6	1.26	0.63	Tarrytown to San Diego	15	4.01	2.21
Newark to Denver	14	3.56	1.96	New York City to Oklahoma City	59	14.68	8.13
Darien to New Orleans	42	10.48	5.74	Edison to Colorado Springs	23	5.79	3.23
Hackensack to Houston	21	5.29	2.91	Elmsford to Tucson	12	3.23	1.74
Jersey City to San Francisco	2	0.63	0.30	The Oranges to Baton Rouge	36	9.00	4.92
Forest Hills to Washington, D.C.	10	2.30	1.10	Stamford to Indianapolis	18	4.32	2.36
Garden City to Minneapolis	19	4.80	2.57	Farmingdale to Louisville	7	1.74	0.91
Elmsford to Salt Lake City	48	12.59	6.88	New York City to St. Louis	14	3.38	1.88
Paramus to Detroit	9	2.21	1.10	Brooklyn to Pittsburgh	5	1.26	0.58
New York City to Los Angeles	16	4.27	2.36				

2. Choose the car of your dreams and research the price. Using the information in the following table, finance your car and give the amount you would owe down and your cost per month (don't forget the interest rate). How would you find your *real* total cost by the end of your payments. Justify your bank choice.

CAR LOAN RATES -------------------

| | New Car | | | Used Car | | |
Bank	Term (months)	Interest rate (%)	% down	Term (months)	Interest rate (%)	% down
A	48	6.00	15	36	12.50	15
B	48	10.90	20	48	13.40	20
C	48	13.00	20	36	14.75	20
D	60	10.50	10	48	13.40	10
E	60	11.00	10	60	13.00	10
F	48	12.50	0	48	13.50	0
G	48	11.90	20	60	12.90	20
H	60	10.95	10	60	13.25	10
I	24	6.75	25	36	14.00	25
J	48	8.25	20	36	9.50	20
K	36	8.25	25	48	14.25	30

CONNECTIONS TO *STATISTICS*

Mean, Median, and Mode

Finding the Mean

An important concept that is used in calculating grades, in controlling the quality of products, and in many other situations is the "average" of a set of measurements. The average of a set of numbers is one of three measures that indicate where the "center" of the set is. The **mean** is one of those measures. The others are *median* and *mode*. We will discuss all three in this Connection. The mean can be thought of as the balance point in a set of measurements.

> To find the mean of a set of *n* numbers
>
> 1. Add the numbers together.
> 2. Divide by *n*.

Let's look at an example.

▪ ▪ ▪

EXAMPLE

A woman measures the heights of some shrubs in front of her house. They measure 33.4 inches, 35.2 inches, 42.5 inches, and 38.1 inches. What is the average height?

SOLUTION

Here $n = 4$, because there are four measurements. To find the mean, we add all the measurements and then divide that total by 4.

$$
\begin{array}{r}
37.3 \\
4\,\overline{)149.2} \\
\underline{12} \\
29 \\
\underline{28} \\
12 \\
\underline{12} \\
0
\end{array}
$$

Add: 33.4   Divide by 4:
      35.2
      42.5
      <u>38.1</u>
    149.2

The average of the four heights is 37.3 inches.

◢

Finding the Median

The **median** of a set of numbers is the "central number" of that set.

To find the median of a set of n numbers

1. Arrange the numbers in order.
2. Find the "central number" of those numbers. The "central number" is the number such that there are as many values greater than it as there are values less than it.

▪▪▪

EXAMPLE

Find the median of 43, 39, 47, 26, and 28.

SOLUTION

Arrange the numbers in order.

26   28   39   43   47

The central number is 39, because there are 2 numbers on either side of it. So 39 is the median.

◢

Finding the median involves more computation when we have an even number of values.

▪▪▪

EXAMPLE

A child cuts some string in the following lengths: 43.1 inches, 47.2 inches, 42 inches, and 48.5 inches. What is the median length?

SOLUTION

To find the median, we first write the numbers in order:

$$42 \quad 43.1 \quad 47.2 \quad 48.5$$

There is no "central number" because there are four numbers. To find the median in such a case, we must find the average of the two middle numbers— here 43.1 and 47.2. We add them and divide their sum by 2.

$$\frac{43.1 + 47.2}{2} = \frac{90.3}{2} = 45.15$$

So 45.15 inches is the median length.

◢

Finding the Mode

The **mode** is also relatively easy to find. It is the value that occurs most frequently in a set of numbers. A given set of numbers may have one mode, no mode, or many modes.

> **To find the mode of a set of *n* numbers**
>
> **1.** Determine how many times each number appears in the set (by tallying).
> **2.** The mode is the number (or numbers) that appears (or appear) most.

▪▪▪

EXAMPLE

Find the mode of the following numbers:

$$1, 3, 2, 4, 5, 3, 7, 2, 3, 7, 3$$

SOLUTION

First we write down all the numbers in order. Then we tally how many of each there are.

1	/
2	//
→ 3	////
4	/
5	/
7	//

Because the number 3 occurs four times, and no other number occurs as many times, 3 is the mode of this set of numbers.

◢

PRACTICE

1. The lengths of the spans of the five longest steel arch bridges in the world are given in the following table.

New River Gorge	1699.58 feet
Bayonne	1625.4 feet
Sydney Harbour	1649.66 feet
Fremont	1254.9 feet
Port Mann	1199.95 feet

   What are the mean and median lengths for these five steel arch bridges?

2. The lengths of three of the longest canals in the world are as follows: White Sea, 141.3 miles; Suez, 100.25 miles; and Volga, 61.75 miles. What is the mean for these canal lengths?

3. Find the mean land area of these five cities in the United States.

Columbus, Ohio	189.272	square miles
Honolulu, Hawaii	617	square miles
Washington, D.C.	68.25	square miles
Seattle, Washington	144.6	square miles
Cleveland, Ohio	79	square miles

4. ✏ Ask and answer four questions using data from Exercises 1 through 3.

5. The last six Apollo missions had durations of $245\frac{1}{3}$ hours, 143 hours, $215\frac{3}{4}$ hours, $295\frac{1}{4}$ hours, $266\frac{1}{3}$ hours, and $301\frac{1}{2}$ hours. What was the median duration of these six missions?

6. In a certain six-year period, the following numbers of people have finished the New York City marathon: 13,599; 14,546; 14,492; 15,887; 19,689; and 21,244. What was the median number of finishers during these six years? What was the mean?

7. On six of the entries in the Tall Ship race in New York City on July 4, 1976, the numbers of crew members were as follows: 104, 189, 99, 162, 236, and 16. What was the median number of crew members on these six ships?

8. Organize the following data in a table, and determine the mean number of people per year who immigrated to the United States from France in the period 1931–1980. The number of people who immigrated from France to the United States during the period 1931–1940 was 12,623. During 1941–1950, the number rose 26,186; during 1951–1960, the number increased 12,312. From 1961–1970, the number dropped by 5884, and during the period 1971–1980, the number dropped 20,168.

9. The following heights above and below sea level were recorded by a scientific team: 35 feet below, 46.3 feet above, 19 feet below, 25.17 feet above, 0.3 feet above, 135.4 feet below, 35 feet below, and 6.24 feet above. What is the range of these readings? What was the mean height recorded?

10. The estimated numbers of people who consider themselves Muslim, Buddhist, Protestant, Catholic, or Jewish are as follows:

Muslim	935 million
Buddhist	303 million
Protestant	73.5 million
Catholic	50.5 million
Jewish	5.9 million

What is the mean number of people (to the nearest tenth of a billion) who consider themselves part of the five major religions?

3.3 Further Applications of Linear Equations

SECTION LEAD-IN

Given the total-cost formula

$$T = F + QV$$

where T is the total cost, F is the flat fee, Q is the amount per service, and V is the number of services, answer the following questions.

1. If F remains constant and Q remains constant, how does V change as T gets larger? As T gets smaller?

2. If F and T remain constant and V gets larger, what happens to Q?

3. Think of another real-life situation that can be modeled by an equation in this form?

In Chapter 2, we used a four-step problem-solving procedure to solve word problems.

To solve a word problem

1. Interpret the problem. What is given? What is needed?
2. Decide on a method for solving the problem.
3. Apply the method. Do the necessary computations.
4. Reason. Check your result, making sure the answer fits the problem statement.

In this chapter, the "method" of steps 2 and 3 will be to write and solve a linear equation.

Recall that an equation is simply two expressions connected by an equals sign. You have already learned how to translate English phrases into algebraic expressions. The English words and phrases that we translate into an equals sign are "is equal to," "is," "is equivalent to," "is the same as," and others like them, including the plain word "equals." Accordingly,

---

To translate an English description into an equation

1. Find the word or phrase that means "is equal to."
2. Write algebraic expressions for the two quantities that are said to be equal, using a variable to represent the unknown quantity.
3. Connect the two expressions with an equals sign.

In the first example, we will solve a simple number problem.

▪▪▪

EXAMPLE 1

The sum of a number and 15 is 41. What is the number?

SOLUTION

INTERPRET What does the problem tell us? It says that if we add 15 and some number, we get a sum of 41. We are looking for that number.

DECIDE If we let n stand for a number, the statement "the sum of a number and 15 is 41" can be written as

<u>The sum of a number and 15</u> <u>is</u> <u>41</u>.
$$n + 15 \qquad\qquad = 41$$

APPLY We solve this equation by adding -15 to both sides.

$$n + 15 = 41$$
$$n + 15 + (-15) = 41 + (-15) \qquad \text{Addition principle}$$
$$n = 26 \qquad\qquad \text{Simplifying}$$

REASON To check, go back to the original statement. So the unknown number is 26.

▶ CHECK **Warm-Up 1**

> **WRITER'S BLOCK**
>
> In Example 1, why did we "let n stand for a number"?

> **STUDY HINT**
>
> *To check, always substitute the answer in the original statement, not in the translated equation. In that way, you will check both the answer and the translation.*

In the previous example, we could have written the problem's information in a little table:

$$\text{Let a number} = n$$
$$\text{a number plus 15} = n + 15$$
$$\text{total} = 41$$

So

$$n + 15 = 41$$

In the next example, we have an age problem with a more complex equation.

■ ■ ■
EXAMPLE 2

Mrs. Williams' age is twice Mark's age, plus 12. She is 84. How old is Mark?

SOLUTION

INTERPRET We want Mark's age. We know Mrs. Williams' age both as "twice Mark's age, plus 12," and as "84." Obviously, these quantities are equal.

DECIDE Let m represent Mark's age. Then

Twice Mark's age plus 12 equals 84.
$$2m \qquad\qquad + 12 \quad = \quad 84$$

Let
$$\text{Mark's age} = m$$
$$\text{Mrs. Williams' age} = 2m + 12$$

and
$$\text{Mrs. Williams' age} = 84$$

So
$$2m + 12 = 84$$

APPLY We then solve the equation.

$$2m + 12 + (-12) = 84 + (-12) \qquad \text{Addition principle}$$
$$2m = 72 \qquad\qquad\qquad \text{Simplifying}$$
$$\left(\tfrac{1}{2}\right)(2m) = \left(\tfrac{1}{2}\right)(72) \qquad \text{Multiplication principle}$$
$$m = 36 \qquad\qquad\qquad \text{Simplifying}$$

REASON So Mark is 36.

▶ CHECK Warn-Up 2

WRITER'S BLOCK

Explain why the same operation has to be performed on both sides of the equation.

The next example involves **consecutive integers,** or integers that follow each other in order. The numbers 1, 2, and 3, for example, are consecutive integers. So are -210 and -209. If n represents the first of a group of consecutive integers, the next is $n + 1$, then $n + 2$, then $n + 3$, and so on.

The numbers 1, 3, 5, . . . are **consecutive odd integers.** If m represents an odd integer, then the next consecutive odd integers are $m + 2$, $m + 4$, and so on. The same is true for **consecutive even integers:** If m represents an even integer, the next consecutive even integers are $m + 2$, $m + 4$, and so on.

ERROR ALERT

Identify the error and give a correct answer.

Translate into algebra and solve: The difference of 5 and y is 6.

Incorrect Solution:

$$y - 5 = 6$$
$$y - 5 + 5 = 6 + 5$$
$$y = 11$$

■ ■ ■
EXAMPLE 3

The sum of three consecutive integers is 36. Find the integers.

SOLUTION

INTERPRET Here we are asked to find three integers. Because they are consecutive integers, we can represent them as follows:

Let

$$\text{first integer} = x$$

Then

$$\text{second integer} = x + 1$$
$$\text{third integer} = x + 2$$

DECIDE The word "is" suggests that we write and solve an equation. To write the equation, we translate as follows:

The sum of three consecutive integers is 36.
$$x + (x + 1) + (x + 2) \qquad = 36$$

APPLY To solve this equation, we first collect like terms and then use the addition and multiplication principles.

$(x + x + x) + (1 + 2) = 36$	Associative property
$3x + 3 = 36$	Simplifying
$3x + 3 + (-3) = 36 + (-3)$	Addition pinciple
$3x = 33$	Simplifying
$\left(\frac{1}{3}\right)3x = \left(\frac{1}{3}\right)33$	Multiplication principle
$x = 11$	Simplifying

REASON If $x = 11$ (the first integer), then the second integer is $x + 1 = 12$ and the third integer is $x + 2 = 13$.

We leave the check to you.

▶ CHECK **Warm-Up 3**

In the last example, we work a mixture problem involving coins. When dealing with things that have monetary values, such as coins or stamps, we have to take into account those values. Follow along.

■ ■ ■
EXAMPLE 4

A pocketful of coins totaling $6.79 has been dropped on the floor. There are twice as many pennies as nickels and three times as many dimes as the number of nickels and pennies together. How many of each coin are there?

SOLUTION

INTERPRET We want to find the numbers of pennies, nickels, and dimes that were dropped. We are told their value in two ways, which we can equate. One is $6.79. To write their value the other way, we must first represent the number of one of the coins with a variable.

DECIDE The number of nickels does not depend on the number of dimes or pennies, so we choose it.

$$\text{Number of nickels} = n$$

Then, because there are twice as many pennies as nickels,

$$\text{Number of pennies} = 2n$$

There are three times as many dimes as *nickels and pennies together*. The total number of nickels and pennies is $n + 2n$, so

$$\text{Number of dimes} = 3(n + 2n) = 3(3n) = 9n$$

Now we need to write the value of all coins. A nickel is worth \$0.05, a dime \$0.10, and a penny \$0.01, so

$$n \text{ nickels are worth } (0.05)n = 0.05n$$
$$2n \text{ pennies are worth } (0.01)(2n) = 0.02n$$
$$9n \text{ dimes are worth } (0.10)(9n) = 0.90n$$
$$\text{Total} = \$6.79$$

APPLY So $\quad 0.05n + 0.02n + 0.90n = 6.79$

$\qquad\qquad\qquad\qquad 0.97n = 6.79 \quad$ Combining like terms

Multiplying both sides by 100 and simplifying gives us

$$(100)0.97n = (100)6.79$$
$$97n = 679$$
$$n = \frac{679}{97}$$
$$n = 7$$

REASON Now that we have found the value of n, we can go back and find that

$$\text{Number of nickels} = n = 7$$
$$\text{Number of pennies} = 2n = 2(7) = 14$$
$$\text{Number of dimes} = 9n = 9(7) = 63$$

We check by adding the value of 7 nickels and 2(7) pennies and 3(7 + 14) dimes.

Check: $\qquad\qquad 0.35 + 0.14 + \$6.30 = \$6.79$

So 7 nickels, 14 pennies, and 63 dimes fell on the floor.

▶ *CHECK* **Warm-Up 4**

> ■■■
> ### WRITER'S BLOCK
> In Example 4, how did we know to multiply by 100 instead of 10?

Practice what you learned.

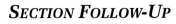

SECTION FOLLOW-UP

Here are answers to the questions asked in the Section Lead-In using the total-cost formula

$$T = F + QV$$

where T is the total cost, F is the flat fee, Q is the amount per service, and V is the number of services.

1. *V* becomes larger; *V* becomes smaller.

2. *Q* becomes smaller.

3. Answers may vary.

3.3 WARM-UPS

Work these problems before you attempt the exercises.

1. The difference between a number and 18 is 23. Find the number.

2. Nancy's age is 4 times Susan's age. If the sum of their ages is 80, find their ages.

3. The sum of three consecutive odd integers is 15. Find the integers.

4. I have in my pocket $3.75 in nickels, dimes, and quarters. If I have twice as many quarters as dimes and twice as many dimes as nickels, how many do I have of each coin?

3.3 EXERCISES

Note: Use your graphing calculator to check your results whenever possible.

For Exercises 1 through 48, solve each word problem by translating it into algebra.

1. Eleven decreased by four times a number is 39. What is the number?

2. The quotient of the sum of 17 and a number and 24 is equal to 2. Find the number.

3. Three more than five times an unknown number is twenty-eight. Find the unknown number.

4. Seven less than an unknown number is twelve and three-fourths. Find the unknown number.

5. Joe has three times as much money as John. Joe has $2.67. How much money does John have?

6. The sum of two consecutive even integers is equal to -10. Find the numbers.

7. The product of one-fifth and an unknown number is $63\frac{3}{5}$. Find the unknown number.

8. The difference between five-halves of an unknown number and thirteen is twelve. Find the unknown number.

9. The quotient of a number and -38 is -27. What is the number?

10. The difference of a number and -192 is -76. Find the number.

11. x less -517 is equal to 95. Find x.

12. t less than 74 is 81. Find t.

13. Find three consecutive odd integers such that the sum of the first and second is 27 less than three times the third.

14. The sum of the first and third in a series of three consecutive integers is 170. Find the integers.

15. The product of the difference between a number and 93 and 5 is 500. Find the number.

16. Three-quarters of the sum of 90 and a number is 72. Find the number.

17. The sum of a number and $\frac{1}{5}$ of that number is equal to 30. What is that number?

18. One-seventh of a number minus that number is 36. Find the number.

19. *Women's Javelin* The weight of a woman's javelin is $\frac{3}{4}$ that of a man's javelin. A javelin used by women weighs about 0.6 kilogram. What is the weight of a javelin used by men?

Weight = 0.6 kilograms   Weight = ?

20. *Grand Prix Racing* In Formula I Grand Prix racing, four times the length of the Monte Carlo circuit plus 0.235 mile is equal to the length of the Le Mans circuit. If the Le Mans circuit is 8.467 miles, how long is the Monte Carlo circuit?

21. *Competition Skateboards* Competition skateboards vary in length from less than a yard long to much longer. Twice the length of the smallest board, plus one foot, is equal to the length of the longest boards. If the longest boards are 72 inches, how long are the shortest boards?

22. *Long-Distance Racing* In 1929, a 5898-kilometer race was held from New York to Los Angeles. This is about 140 times the length of a marathon. About how many kilometers long is a marathon? (Round your answer to the nearest kilometer.)

23. *Dog Sled Team* The speed of a twelve-dog sled team is about one-fifth the speed of a pigeon. The dog team travels at an average speed of 15 miles per hour. About how fast can a pigeon fly?

24. *Pottery Clay* You are making pottery. Out of a sack of clay, $\frac{1}{3}$ will be used for mugs and the rest for planters. You need 25 pounds of clay for mugs. How much does the sack weigh?

25. *Cruise Ships* You are scheduling passengers for cruise ships. One ship holds twice as many passengers as a second ship. Together they hold 225 passengers. How many can you schedule into the smaller ship?

26. *Taxi Fare* A taxi company charges $1.15 for the first mile and $0.15 for each additional $\frac{1}{5}$ of a mile. If a taxi ride costs you $18, including a tip of $1.70, how far did you travel?

27. *Snake Length* For a certain snake species, the female's total length x and tail length y in millimeters are related as follows (approximately):

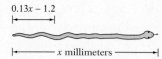

$$y = 0.13x - 1.2$$

A snake has a tail length of 154.8 millimeters. What is its total length?

28. *Rental Fees* In the early 1980s in some large cities, the relationship between the amount spent on a finder's fee x and the amount of monthly rent y one had to pay for an apartment in some desirable areas could be represented by the equation

$$y = -0.12x + \$1000$$

A person who rented through this procedure pays $520 a month rent. How much did he spend to find the apartment?

29. The sum of 4 consecutive integers is -2. What are the integers?

30. One-third of the sum of three consecutive integers is 0. What are the integers?

31. *Travel Time* At 3 P.M. two cars start traveling in opposite directions from the same point. One travels north at 40 mph, and the other travels south at 50 mph. After how many hours are they 315 miles apart?

32. *Travel Time* Car A travels 55 mph for 6 hours. Car B travels the same distance at 40 mph. How many hours does car B travel?

33. *Carpentry* Two-thirds of the length of a piece of teak will be used for a cutting board and the rest for bookends. A length of eighteen inches is needed for a cutting board. How long was the original piece of teak?

34. *Food* You need two servings of hamburger, one serving to be one-third the weight of the other. The smaller serving is to weigh B ounces. How much will the larger serving weigh?

35. Find three consecutive odd integers such that their sum is 43 more than -1000.

36. Twice the sum of three consecutive odd integers is -894. Find the integers.

37. *Part-time Work* Sherly and Sophia both work at a Learning Center. Sherly works 3 hours less than twice the number of hours that Sophia works. If together they work a total of 40 hours, how many hours does each work?

38. *Money* Marvin has $2.00 in dimes and nickels. The number of dimes exceeds the number of nickels by five. Find the number of dimes and the number of nickels he has.

39. **Age**   Three sisters' ages total 63 years. The first sister is 5 years older than the second, and the second is $\frac{1}{4}$ of the age of the third. Find the ages of the three sisters.

40. **Lending Library**   Special books can be borrowed from the lending library for $3.50 for the first day and a $0.30 charge for each additional day. If you keep the book for two weeks, what is the total charge?

41. **Money**   You owe $319 to your sister. This week you returned $69 and agreed to pay her back in weekly installments of $75. How many more weeks will it take you to eliminate the debt?

42. **Money**   I have twice as many nickels as I have dimes and quarters together. I have 26 nickels and 4 dimes. How many quarters do I have?

43. **Calorie Intake**   You are on a diet and would like to limit your caloric intake to 2000 calories each day. You have consumed a total of 760 calories for breakfast and $\frac{1}{5}$ of the total caloric limit for lunch. How many calories can you consume for dinner?

44. **Money**   Sean paid a bill of $3.70 with quarters, nickels, and dimes. The number of dimes was one less than the number of nickels, and the number of quarters was three less than the number of dimes. How many of each coin did he pay with?

45. ✏   If m is any integer, then $2m$ is an even integer. Explain this.

46. ✏   If m is any integer, then $2m + 1$ is an odd integer. Explain this.

47. ✏   If m and n are any two integers and $m > n$, then $2m - 2n$ is an even number. Explain this.

48. ✏   If m and n are any two odd integers and $m > n$, then $m - n$ is an even number. Explain this.

MIXED PRACTICE

By doing these exercises, you will practice the topics up to this point in the chapter.

49. Solve for the unknown: $-9x - 6 = 5x + 3$

50. Solve for c: $\frac{c}{h} - w = y$

51. Solve for t: $-\frac{1}{2}(t + 4) - \frac{3}{4}t = 8t - 15$

52. Solve for y: $-9(12y - 5) = 9$

53. Solve for x: $x - (-18) = -65$

54. Solve for t: $2t + 5 = 8t - 2$

55. Solve fot t: $\frac{t}{3} + 276 = 387$

56. Solve for x: $-4(-3x + 5) = 32$

57. *Rental Fees*   Boats can be rented at the shore for $34.50 for the first three hours plus $4.50 for each additional hour. How much will it cost you to rent the boat for *n* hours? Write an expression to represent the total cost.

58. *Geometry*   A circular mirror has a circular frame. The diameter of the mirror is 18 inches. The width of the circular frame is 1 inch. What area do the mirror and its frame cover? Use $A = \pi r^2$ and let $\pi \approx 3.14$.

EXCURSIONS

Data Analysis

1. Here is a list of men and women who completed the 1994 New York City marathon between 3 hours and 59 minutes and 3 hours 59 minutes 30 seconds. A second table on the following page shows the distribution of ages of men and women who ran the marathon. Use this information to answer the following questions.

1994 New York City Marathon Finishing Times

Men

9197	3745	Troy Bush, 31, NC	3:59:00
9198	3746	Harold Axelrod, 39, NY	3:59:00
9199	956	Dave Dutton, 58, IA	3:59:00
9200	3747	James Weant, 36, NC	3:59:00
9201	2932	Jose Vazquez Vera, 49, MEX	3:59:00
9202	3748	James Sterns, 31, MI	3:59:01
9203	2933	Jerry Smith, 41, MI	3:59:01
9204	957	Christian Rathey, 55, GER	3:59:02
9205	3749	Eusebio Hiraldo, 38, NY	3:59:02
9206	2934	Esko Polvi, 47, SWE	3:59:02
9207	2935	Goody Tyler, 42, VA	3:59:02
9208	3750	Bryan Touhey, 38, NY	3:59:02
9209	3751	Kurt Apostol, 35, UT	3:59:02
9210	2936	Peter Reynolds, 46, OR	3:59:02
9211	2937	Moreno Scanzi, 40, ITA	3:59:03
9212	1441	Keiji Oizumi, 28, JAP	3:59:03
9213	2938	Jacques Van Sevencoten, 43, BEL	3:59:03
9214	3752	Philip Vasquez, 39, TX	3:59:04
9215	3753	Tod Benton, 38, TX	3:59:04
9216	3754	Dominique Beauvir, 30, FRA	3:59:04
9217	1442	Sid Evans, 25, NY	3:59:04
9218	2939	Henry Allary, 43, FRA	3:59:05
9219	3755	Eduard Artner, 30, AUT	3:59:05
9220	3756	Patrick Vandenberghe, 35, FRA	3:59:05
9221	2940	Gary Null, 49, NY	3:59:05
9222	3757	Tad Smith, 35, CT	3:59:06
9223	3758	Harold Tinsley, 37, NY	3:59:06
9224	2941	Lance Goodwin, 40, NY	3:59:06
9225	3759	Andrew Oshrin, 30, NY	3:59:06
9226	2942	Paul Farinacci, 43, CA	3:59:06
9227	3760	Henry Menusan, 37, NY	3:59:06
9228	3761	Robert Van De Voort, 38, HOL	3:59:06
9229	3762	Kurt Kehl, 38, DC	3:59:06
9230	2943	Stephen Allen, 40, NY	3:59:07
9231	2944	Soeren Bjerrine, 45, DEN	3:59:07
9232	3763	Francis Pope, 39, CAN	3:59:07
9233	1443	John Ferris, 28, CAN	3:59:07
9234	3764	David Farmer, 30, GA	3:59:07
9235	3765	Howard Scruggs, 36, WV	3:59:07
9236	3766	Richard Bijenveld, 38, HOL	3:59:07

9237	2945	Alexander Scheffel, 44, HOL	3:59:08
9238	2946	Charles Pritchard, 43, NJ	3:59:09
9239	2947	John Lawton, 43, FL	3:59:09
9240	2948	Bernard Bourdette, 44, FRA	3:59:09
9241	2949	Rafael Velazquez, 42, MEX	3:59:09
9242	3767	Robert Somer, 32, VA	3:59:10
9243	1444	Darrin Zisk, 29, PA	3:59:10
9244	958	Masao Sugiura, 51, JAP	3:59:10
9245	3768	C Phillips, 38, NY	3:59:10
9246	2950	John Keane, 43, IRL	3:59:10
9247	3769	Robert Greene, 33, NY	3:59:11
9248	3770	John Urda, 31, NY	3:59:11
9249	2951	Carlo La Pasta, 45, ITA	3:59:11
9250	1445	Garrett Smith, 29, MD	3:59:11
9251	3771	Guido Konrad, 34, GER	3:59:11
9252	2952	Francisco Silva, 48, BRA	3:59:11
9253	2953	Werner Duerst, 45, SUI	3:59:11
9254	1446	Daniel Hendriksen, 24, GBR	3:59:12
9255	959	Eugene Gramzow, 53, OR	3:59:12
9256	3772	Gary Geurts, 34, GA	3:59:12
9257	960	Rene Skjold, 50, NOR	3:59:12
9258	2954	Daniel Sweeney, 45, MA	3:59:12
9259	961	Alain Saint-Marc, 55, FRA	3:59:12
9260	962	Ignacio Carrasquero, 51, VEN	3:59:12
9261	2955	Youn Kim, 44, NJ	3:59:13
9262	3773	Stephen Patti, 30, TX	3:59:13
9263	3774	Alexandre Volpi, 37, FRA	3:59:13
9264	963	Gaston Plante, 57, CAN	3:59:13
9265	3775	William Santoro, 37, PA	3:59:14
9266	1447	Frederic Massy, 28, FRA	3:59:14
9267	964	Edward Hynes, 51, NY	3:59:14
9268	2956	Mauricio Szkolnik, 46, VEN	3:59:15
9269	2957	Pedro Zoppi, 42, VEN	3:59:15
9270	3776	Benjamin Spang, 32, CT	3:59:15
9271	2958	John Markussen, 41, NOR	3:59:15
9272	1448	Carlos Salem, 28, BRA	3:59:16
9273	2959	Rommel Diawatan, 41, CA	3:59:16
9274	3777	Richard Van Den Boogaart, 36, HOL	3:59:16
9275	1449	Paolo Bolzon, 25, ITA	3:59:16
9276	1450	Christopher James, 26, NJ	3:59:16
9277	965	Rolando Vizhnay, 52, NY	3:59:16
9278	3778	Robert Kollar, 32, IN	3:59:16
9279	966	Kurt Sandig, 53, GER	3:59:16

9280	2960	Zlatko Jurisic, 40, FRA	3:59:17
9281	97	Klaus Kammerer, 61, GER	3:59:18
9282	3779	Scott Goodman, 35, GA	3:59:18
9283	2961	Roger Bohn, 40, GER	3:59:18
9284	2962	Peter Moeller, 43, DEN	3:59:18
9285	967	Lawrence Weis, 52, MI	3:59:18
9286	1451	Guy Prihar, 24, FL	3:59:18
9287	3780	Robert Boisvert, 34, MN	3:59:19
9288	2963	Arne Oiestad, 43, NOR	3:59:19
9289	3781	Scott Harper, 30, CA	3:59:19
9290	1452	Anders Gronli, 20, NOR	3:59:19
9291	1453	Jack Phillips, 29, OK	3:59:19
9292	1454	Jeffrey Campbell, 29, NY	3:59:20
9293	2964	Paul Moorefield, 40, NY	3:59:20
9294	968	James Muldoon, 53, CA	3:59:21
9295	3782	William-Paul Thomas, 36, TX	3:59:21
9296	3783	Thomas Lobig, 37, GER	3:59:21
9297	2965	Walter MacGowan, 47, NJ	3:59:21
9298	2966	Werner Keppler, 48, GER	3:59:21
9299	3784	Louis Perrelli, 39, CA	3:59:21
9300	98	Harry Lindell, 60, SWE	3:59:22
9301	2967	Lawrence Holen, 46, MI	3:59:22
9302	1455	Joshua Gross*, 24 NJ	3:59:23
9303	3785	Benjamin Steinmetz, 38, BEL	3:59:23
9304	2968	Jim Growney, 43, TX	3:59:23
9305	1456	James Turek, 26, WI	3:59:23
9306	2969	Noel Butler, 45, GBR	3:59:24
9307	3786	Enrique Lorente, 35, MEX	3:59:24
9308	2970	Ib Larsen, 43, DEN	3:59:24
9309	3787	Paul Swanenburg, 31, HOL	3:59:25
9310	2971	Sandy Engel, 47, PA	3:59:25
9311	2972	Yoshitaka Kaneko, 49, JAP	3:59:25
9312	2973	Michael Last, 42, PA	3:59:26
9313	3788	Victor Malu, 39, NJ	3:59:26
9314	1457	Heriberto Gonzalez, 28, MEX	3:59:26
9315	2974	Mike McCune, 40, CA	3:59:26
9316	3789	Gerrit-Jan Lagendijk, 33, HOL	3:59:27
9317	2975	Hans Bauer, 43, GER	3:59:27
9318	1458	Zev Zaidman Goodman, 27, MEX	3:59:27
9319	2976	Michael Bender, 42, GER	3:59:27
9320	3790	Patrice Gauthier, 35, CAN	3:59:27
9321	3791	Kevin Dougan, 31, NY	3:59:28
9322	2977	Matthias Rilling, 43, GER	3:59:28

9323	2978	Adolfo Kalach, 40, MEX	3:59:28
9324	3792	Arie Breederland, 36, HOL	3:59:28
9325	3793	Jeff Stein, 33, CAN	3:59:29
9326	1459	Yann Perrot, 22, FRA	3:59:29
9327	3794	Hans Schreyer, 38, OH	3:59:29
9328	2979	Michael Murtagh, 44, NY	3:59:29
9329	3795	Laurent Lefebvre, 30, FRA	3:59:29
9330	3796	Peter Waumans, 30, BEL	3:59:29

Women

1040	333	Claire Caussanel, 26, FRA	3:59:02
1041	246	Myriam Christiaen, 42, BEL	3:59:02
1042	247	Shirley Hume, 43, GBR	3:59:04
1043	248	Birthe Jensen, 45, DEN	3:59:06
1044	249	Ritsuko Hayashi, 42, JAP	3:59:11
1045	420	Montse Cot Rosell, 36, ESP	3:59:11
1046	421	Regina Hartikka, 32, FIN	3:59:12
1047	422	Carol Low, 34, NY	3:59:13
1048	250	Nicholette Sayer, 47, GBR	3:59:13
1049	334	Christine Fontes, 28, FRA	3:59:13
1050	335	Anita Beekers, 29, HOL	3:59:14
1051	423	Terri Guarino, 35, NY	3:59:14
1052	251	Dena Berlin, 40, NY	3:59:15
1053	424	Dana Hall, 32, BER	3:59:15
1054	252	Karen Martin, 45, SC	3:59:16
1055	253	Berendina Van De Pol, 46, HOL	3:59:17
1056	336	Jacqueline Allen, 28, CA	3:59:18
1057	254	Edith Schaffert, 49, SUI	3:59:18
1058	425	Christine Kaliardos, 33, CA	3:59:19
1059	426	Linda Haller, 37, DC	3:59:20
1060	427	Lisa Haller, 37, CA	3:59:20
1061	428	Jayne Blake, 31, GBR	3:59:21
1062	337	Angelica Ciatti, 28, ITA	3:59:21
1063	429	Guadalupe G Rodrigue, 39, MEX	3:59:22
1064	430	Heidi Kriel, 33, RSA	3:59:24
1065	431	Kathleen D'Alessandro, 39, NY	3:59:24
1066	338	Andrea Arena, 26, NY	3:59:27

Source: Running News, Fall 1994.

Entrants by Age Group

	Men	Women	Total
18–19	80	35	115
20–29	3,116	1,643	4,759
30–39	7,920	2,594	10,514
40–49	7,461	1,973	9,434
50–59	3,438	636	4,074
60–69	650	84	734
70–79	87	11	98
80–89	6	1	7
90 and older	0	0	0
Totals	**22,758**	**6,977**	**29,735**
Oldest runner (yrs)	**89**	**87**	

Source: Running News, Fall 1994.

a. How do the percent of men and the percent of women completing the marathon during this thirty-second time period compare to the percent of men and the percent of women entering the marathon?

b. Did you count the total number of men and the total number of women that completed the race during this time period? Look carefully at the information given. What other method could you have used?

c. What can you say about the age group spread of men and women entrants compared with the men and women completing the marathon during this thirty-second time period? Include as many observations as you can.

d. What is the mode (give this in terms of the age groups) completing during this thirty-second time period? How does this compare with the mode of age groups entering the marathon? Do the age group modes by gender (male or female) of the entrants also reflect the modes by gender completing during this thirty-second time period?

e. What percent of the entire population entering the marathon is 50 years old or older? What percent of the finishers during this thirty-second time period is 50 years old or older?

f. What percent of the entire population entering the marathon is 29 years old or younger? What percent of the finishers during this thirty-second time period is 29 years old or younger?

g. Ask and answer four additional questions about this data.

2. With only three games remaining in the 1995 shortened hockey season, the 1994 Stanley Cup winning New York Rangers had 45 points. The Eastern Conference NHL standings at that time are given in the following table:

NHL Standings, 1995

NHL Team	Points	NHL Team	Points
Philadelphia	58	Tampa Bay	37
Ottawa	19	NY Islanders	35
Québec	61	Florida	41
Boston	51	Montréal	42
Washington	49	NJ Devils	50
Buffalo	46	Hartford	43
NY Rangers	45	Pittsburgh	61

a. What is the mode?

b. What is the median?

c. What is the mean?

d. Which of these measures, in your opinion, best describes the "average" of the data? Justify your answer.

Posing Problems

3. What would you like to know about the world's water supply? For example, what is the approximate average depth of the oceans? the inland seas and saline lakes? Ask and answer four questions about these waters.

Water Supply of the World[1]

The Antarctic Icecap is the largest supply of fresh water, nearly 2 percent of the world's total of fresh and salt water. As can be seen from the table below, the amount of water in our atmosphere is over ten times as large as the water in all the rivers taken together. The fresh water actually available for human use in lakes and rivers and the accessible ground water amounts to only about one-third of one percent of the world's total water supply.

	Surface area (square miles)	Volume (cubic miles)	% of total
Salt water			
the oceans	139,500,000	317,000,000	97.2
inland seas and saline lakes	270,000	25,000	0.008
Fresh water			
freshwater lakes	330,000	30,000	0.009
all rivers (average level)	—	300	0.0001
Antarctic Icecap	6,000,000	6,300,000	1.9
Arctic Icecap and glaciers	900,000	680,000	0.21
water in the atmosphere	197,000,000	3,100	0.001
ground water within half a mile from surface	—	1,000,000	0.31
deep-lying ground water	—	1,000,000	0.31
Total (rounded)	—	**326,000,000**	**100.00**

1. All figures are estimated.

Source: Department of the Interior, Geological Survey, as cited in the *1996 Information Please*® *Almanac* (©1995 Houghton Mifflin Co.), p. 578. All rights reserved. Used with permission by Information Please LLC.

3.4 Linear Inequalities

SECTION GOALS

- *To solve linear inequalities using the addition and multiplication principles*
- *To graph linear inequalities*

SECTION LEAD-IN

1. Answer each question with one of the following: "remains true;" "is false;" or "cannot predict."

   What happens to the inequality

   $$ax \le b$$

   if

   a. both sides of the inequality are multiplied by a positive integer?

   b. both sides of the inequality are multiplied by zero?

   c. both sides of the inequality are multiplied by a positive number between zero and one?

   d. a negative number is added to both sides of the inequality?

2. Give a real-life situation that can be modeled by an inequality.

Solving Simple Linear Inequalities

> **Inequalities**
>
> An **inequality** is a mathematical statement in which two expressions are related by one of the symbols $<$ ("is less than"), $>$ ("is greater than"), $\le$ ("is less than or equal to"), or $\ge$ ("is greater than or equal to").

Here are three examples:

$$x > 5 \text{ (read "}x \text{ is greater than 5")}$$
$$2x - 17 \le 23 \text{ (read } 2x \text{ minus 17 is less than or equal to 23)}$$
$$13x + 5 \ge 21 \text{ (read } 13x \text{ plus 5 is greater than or equal to 21)}$$

An inequality that contains only constants and a single variable with an exponent of 1 is a **linear inequality**. (The inequalities $x > 5$, $2x - 17 \le 23$, and $13x + 5 \ge 21$ are linear.)

A **solution of an inequality** is any number that makes the inequality true when it replaces the variable. Most inequalities have many solutions; the first inequality in our example has all real numbers greater than 5 as solutions.

> **Solution Set**
>
> The **solution set for an inequality** consists of all the solutions to the inequality.

Thus the numbers 6, 7, 8.35, and many more are part of the solution set for the inequality $x > 5$.

Graphing Solutions to Inequalities

In Chapter 1, we graphed real numbers on a number line. The solution of a linear equation is a real number, so we can also graph such solutions on a number line.

For example, the solution to the equation $x + 3 = 5$ is $x = 2$. We represent this graphically by showing a filled-in circle at the point $x = 2$.

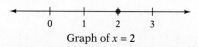

Graph of $x = 2$

Graphing the solution set for an inequality takes a little more work. For the inequality $x + 3 < 5$, the solution set is all real numbers less than 2. To show this on the graph, we place an open circle at the point 2, to indicate that the solution *does not include* the number 2 itself. Then we draw a thick arrow to the left along the number line, and this indicates that all real numbers less than 2 *are included.*

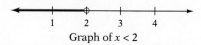

Graph of $x < 2$

To graph an inequality containing $\geq$ or $\leq$, we combine the techniques shown in the previous two figures. Such inequalities include the number itself, so we fill in the circle. To graph $x \geq 4$, for example, we draw a filled-in circle at the point 4, and we draw a solid arrow to the right of 4, along the number line.

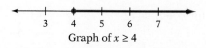

Graph of $x \geq 4$

How can we determine the solutions to an inequality such as $x + 2 < 6$? One way is to substitute real numbers for x and see whether the resulting statement is true or false. To test the number 3, we would write

$$3 + 2 < 6 \quad \text{Substituting 3 for } x$$
$$5 < 6 \quad \text{Simplifying}$$

Because the last inequality is true, 3 is a solution of $x + 2 < 6$.

To test the number 4, we would write

$$4 + 2 < 6 \quad \text{Substituting 4 for } x$$
$$6 < 6 \quad \text{Simplifying}$$

This statement is *false*, because 6 cannot be less than itself. So 4 is not part of the solution set for $x + 2 < 6$.

This method of solution is tedious because it requires many such tests. Instead, we can solve simple inequalities much as we solved linear equations. We apply two principles that enable us to rewrite inequalities in the very simple form $x <$ a number of $x >$ a number.

The Addition Principle of Inequality

The **addition principle of inequality** is much like the addition principle of equality.

Addition Principle of Inequality

For real numbers a, b, and c,

$$\text{If } a < b, \text{ then } a + c < b + c$$
$$\text{If } a > b, \text{ then } a + c > b + c$$

In words, if we add the same real number to both sides of an inequality, the solution set for the inequality remains unchanged. We can also substitute the symbol $\leq$ for $<$ and the symbol $\geq$ for $>$ in the addition principle.

We use the addition principle to get an inequality of the form

$$x + b > c$$

into the form

$$x > \text{some number}$$

That is, we add the same number to both sides of the inequality to isolate the variable on one side of the inequality.

▪▪▪
EXAMPLE 1

Find the solution set for $k + 5 > 12$.

SOLUTION

To isolate k, we add -5 (the additive inverse of 5) to both sides of the inequality and simplify.

$$
\begin{array}{ll}
k + 5 > 12 & \text{Original equation} \\
k + 5 + (-5) > 12 + (-5) & \text{Adding } -5 \text{ to both sides} \\
k > 7 & \text{Simplifying}
\end{array}
$$

To check the solution, substitute a number greater than 7 for k in the original inequality, and then substitute a number less than 7. If the first substitution gives a true statement and the second a false statement, then we know the solution is correct.

Check:	Let $k = 8$	Let $k = 6$
	$k + 5 > 12$	$k + 5 > 12$
	$8 + 5 > 12$	$6 + 5 > 12$
	$13 > 12$ True	$11 > 12$ False

So the solution set for $k + 5 > 12$ consists of all real numbers greater than 7.

From this point on in the text, we will always show the graph when we solve an inequality.

▶ *CHECK* **Warm-Up 1**

The next example involves combining like terms.

...

EXAMPLE 2

Find the solution set for $-3 + x + 5 < 7$.

SOLUTION

First we apply the commutative law and rewrite the inequality so that we can combine the constants on the left side.

$$-3 + x + 5 < 7 \quad \text{Original equation}$$
$$-3 + 5 + x < 7 \quad \text{Rewriting using commutative property}$$
$$2 + x < 7 \quad \text{Simplifying the left side}$$

Then we add -2 to both sides of the inequality and simplify.

$$2 + x < 7$$
$$2 + x + (-2) < 7 + (-2) \quad \text{Adding } -2 \text{ to both sides}$$
$$x < 5 \quad \text{Simplifying}$$

Check:

Let $x = 3$	Let $x = 6$
$-3 + x + 5 < 7$	$-3 + x + 5 < 7$
$-3 + 3 + 5 < 7$	$-3 + 6 + 5 < 7$
$5 < 7$ True	$8 < 7$ False

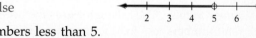

So the solution set for $-3 + x + 5 < 7$ contains all real numbers less than 5. See the accompanying figure.

▶ *CHECK* **Warm-Up 2**

> **WRITER'S BLOCK**
>
> How are equations and inequalities alike? Compare
>
> $3 + x > 5$ and
>
> $3 + x = 5$

Calculator Corner

The inequality in Example 2 can be worked on your graphing calculator in two ways. One way would be to have the calculator give the solution on a number line. Enter the inequality into Y_1 exactly as it appears in the text. You might need to consult your graphing calculator manual to learn how to insert inequality signs.

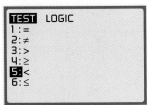

Press **2nd MATH.**

Press **ENTER.**

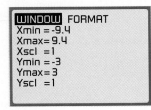

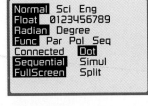

You may want to change the calculator to **DOT MODE.**

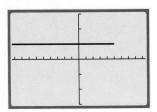

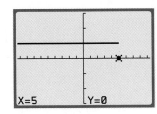

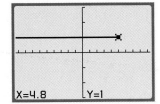

Press **GRAPH.**   Use **TRACE.**

For all x-values shown, when $y = 1$, the statement is true; when $y = 0$, the statement is false. The calculator uses the horizontal line $y = 1$ to show that for all real numbers less than 5 the inequality is true. However, for all real numbers greater than or equal to 5 the inequality is false.

A second way is to examine the inequality as two separate equations.

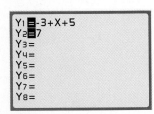

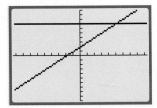

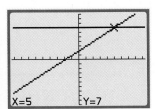

The graph now shows where the line $y = -3 + x + 5$ (the slanted line) is less than (in this case *below*) the line $y = 7$.

The Multiplication Principle of Inequality

Just as in solving equations, we need to rewrite the unknown term with a coefficient of one. To do this, we use the **multiplication principle of inequality.**

> **Multiplication Principle of Inequality, Positive Multipliers**
>
> For all real numbers a, b, and c, where $c > 0$,
>
> $$\text{If } a < b, \text{ then } ac < bc$$
> $$\text{If } a > b, \text{ then } ac > bc$$

In words, if we multiply both sides of an inequality by the same *positive* real number, the solution set for the inequality is not changed.

▪▪▪
EXAMPLE 3

Find the solution set for $3.2x < -7$.

SOLUTION

Using the multiplication principle, we multiply both sides of the inequality by $\frac{1}{3.2}$, the multiplicative inverse of 3.2.

$$3.2x < -7 \qquad \text{Original equation}$$
$$\left(\tfrac{1}{3.2}\right)3.2x < \left(\tfrac{1}{3.2}\right)(-7) \quad \text{Multiplying both sides}$$
$$x < -2.1875 \qquad \text{Simplifying}$$

The check is done as before. We leave this for you to do.

So the solution set for $3.2x < -7$ consists of all real numbers less than -2.1875. See the accompanying figure.

▶ *CHECK* **Warm-Up 3**

▪▪▪
EXAMPLE 4

Find the solution for $12m - 4m \le 18$.

SOLUTION

We first use the distributive property to combine like terms.

$$12m - 4m \le 18$$
$$(12 - 4)m \le 18 \qquad \text{Distributive property}$$
$$8m \le 18 \qquad \text{Combining like terms}$$

▪▪▪
WRITER'S BLOCK

How do equations and inequalities differ? Contrast

$3x > 15$ and

$3x = 15$

Then we use the multiplication principle.

$$\left(\frac{1}{8}\right)8m \le \left(\frac{1}{8}\right)18 \quad \text{Multiplying both sides by } \frac{1}{8}$$

$$m \le \frac{9}{4} \qquad \text{Simplifying}$$

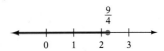

So the solution for the inequality $12m - 4m \le 18$ is $m \le \frac{9}{4}$. See the accompanying figure.

In the previous two examples, we multiplied both sides by a positive number, and the inequality symbol was not changed. If we multiply an inequality by a negative number, however, the symbol is reversed. For example, $7 > 5$ but $-7 < -5$ $(-1)(7) < (-1)(5)$. See the accompanying figure.

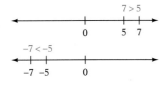

Multiplication Principle of Inequality, Negative Multipliers

For all real numbers a, b, and c, where $c < 0$,

$$\text{If } a < b, \text{ then } ac > bc$$
$$\text{If } a > b, \text{ then } ac < bc$$

In words, if we multiply both sides of an inequality by the same *negative* real number and *reverse* the inequality symbol, the solution set of the inequality is not changed.

We use this principle when the coefficient of the variable term is a negative number. We multiply both sides of the inequality by the multiplicative inverse of that coefficient and reverse the inequality symbol.

▪ ▪ ▪

EXAMPLE 5

Find the solution set for $-\frac{3}{4}r \ge 15$.

SOLUTION

To isolate r, we must multiply both sides of the inequality by $-\frac{4}{3}$, the multiplicative inverse of $-\frac{3}{4}$. Because $-\frac{3}{4}$ is negative, we must reverse the inequality symbol as well.

$$-\frac{3}{4}r \ge 15$$

$$\left(-\frac{4}{3}\right)\left(-\frac{3}{4}\right)r \le \left(-\frac{4}{3}\right)15 \quad \text{Multiplication principle}$$

Then we simplify.

$$r \le -20$$

To check the solution, we substitute a real number less than -20 and a real number greater than -20 for r in the original inequality.

WRITER'S BLOCK

Discuss how to solve

$$2x > 6 \quad \text{and}$$
$$-2x > 6$$

Explain how the results "make sense."

Check: Let $r = -24$ Let $r = -12$

$$-\frac{3}{4}r \geq 15$$ $$-\frac{3}{4}r \geq 15$$

$$-\frac{3}{4}(-24) \geq 15$$ $$-\frac{3}{4}(-12) \geq 15$$

$$18 \geq 15 \quad \text{True}$$ $$9 \geq 15 \quad \text{False}$$

So the solution set for $-\frac{3}{4}r \geq 15$ contains -20 and all **real numbers less than** -20. See the accompanying figure.

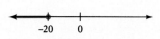

▶ *CHECK* **Warm-Up 5**

Calculator Corner

Look again at Example 5. One way to solve the inequality graphically would be to graph each side of the inequality separately.

 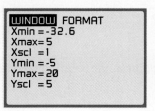 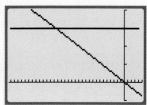 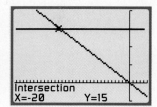

Remember that the inequality is asking, "Where is the line $y = \left(-\frac{3}{4}\right)x$ greater than or equal to the horizontal line $y = 15$." The fourth screen shows that the point of intersection for the two lines is at $x = -20$. So the answer is all real numbers less than or equal to -20.

Solving More Complex Linear Inequalities

Solving Inequalities

To solve inequalities such as $ax + b > c$, we must use both the addition principle and the multiplication principle. We always apply the addition principle first, to isolate the term that contains the variable. Then we use the multiplication principle to isolate the variable—obtain a coefficient of 1 on the variable term.

> **To solve a linear inequality**
> 1. Use the properties of real numbers to simplify both sides of the inequality.
> 2. Use the addition principle to isolate the term containing the variable.
> 3. Use the multiplication principle to give the variable a coefficient of 1.

■ ■ ■
EXAMPLE 6

Find the solution set for $3x + 4 \geq 9$.

SOLUTION

To isolate the term $3x$, we add -4 to both sides of the inequality.

$$3x + 4 \geq 9 \qquad \text{Original inequality}$$
$$3x + 4 + (-4) \geq 9 + (-4) \qquad \text{Addition principle}$$
$$3x \geq 5 \qquad \text{Simplifying}$$

Next, to obtain a coefficient of 1 for x, we multiply both sides of the inequality by $\frac{1}{3}$ (the multiplicative inverse of 3).

$$3x \geq 5$$
$$\left(\frac{1}{3}\right)3x \geq \left(\frac{1}{3}\right)(5) \qquad \text{Multiplication principle}$$
$$x \geq \frac{5}{3} \qquad \text{Simplifying}$$

We check by choosing a point in the solution set, such as 2, and a point outside the solution set, such as 1.

Check:	Let $x = 2$	Let $x = 1$
	$3x + 4 \geq 9$	$3x + 4 \geq 9$
	$3(2) + 4 \geq 9$	$3(1) + 4 \geq 9$
	$6 + 4 \geq 9$	$3 + 4 \geq 9$
	$10 \geq 9$ True	$7 \geq 9$ False

So the solution set for the inequality $3x + 4 \geq 9$ contains all real numbers x such that $x \geq \frac{5}{3}$. See the accompanying figure.

▶ *CHECK* **Warm-Up 6**

■ ■ ■
EXAMPLE 7

Find the solution set for $-14r - 3(r + 9) > 7 + 2r$.

SOLUTION

First we simplify the left side by using the distributive property and combining like terms. We obtain

$$-14r - 3r - 27 > 7 + 2r \qquad \text{Distributive property}$$
$$-17r - 27 > 7 + 2r \qquad \text{Combining like terms}$$

Then we use the addition principle twice, adding first $-2r$ and then 27 to both sides.

$$-17r - 27 + (-2r) > 7 + 2r + (-2r) \quad \text{Addition principle}$$
$$-19r - 27 > 7 \quad \text{Simplifying}$$
$$-19r - 27 + (27) > 7 + (27) \quad \text{Addition principle}$$
$$-19r > 34 \quad \text{Simplifying}$$

Finally, we multiply both sides by $-\frac{1}{19}$ to give the variable a coefficient of 1. Remember that when we multiply an inequality by a negative number, we must reverse the direction of the inequality.

$$-19r > 34$$
$$\left(-\frac{1}{19}\right)(-19r) < \left(-\frac{1}{19}\right)(34) \quad \begin{array}{l}\text{Multiplying both sides by } -\frac{1}{19} \text{ and} \\ \text{changing direction of inequality}\end{array}$$
$$r < -\frac{34}{19} \quad \text{Simplifying}$$

So the solution set for $-14r - 3(r + 9) > 7 + 2r$ contains all real numbers less than $-\frac{34}{19}$. See the accompanying figure. We leave the check to you to do.

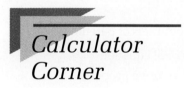

▶*CHECK* **Warm-Up 7**

Calculator Corner

Look again at Example 7, an inequality involving two lines. This type of inequality can also be solved easily using your graphing calculator. Put your calculator in DOT mode.

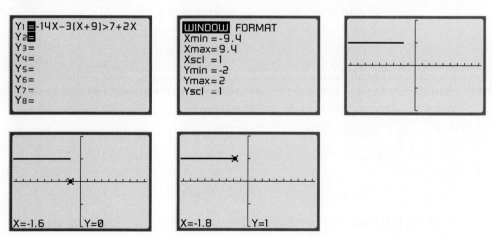

By using **TRACE** the calculator indicates that the inequality is *true* for x's less than -1.8. If you converted the algebraic answer $r < -\frac{34}{19}$ to a decimal answer, the result would be -1.789473684, which would round off to $r < -1.8$.

The same result can also be found by graphing the two sides of the inequality separately. You are then asking where is Y1 > Y2. This would be the same as asking where Y1 is above Y2. The point of intersection of the two lines is −1.789474, which is −1.8 if rounded to one decimal position.

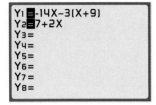 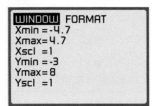 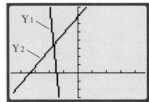 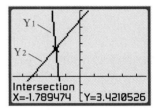

Graphing More Complex Inequalities

▪ ▪ ▪

EXAMPLE 8

Solve and graph the solution set for $-4(y + 6) \leq 2y - 5 + 4(y - 7)$.

SOLUTION

First we use the distributive property to remove both sets of parentheses.

$$-4(y + 6) \leq 2y - 5 + 4(y - 7) \quad \text{Original equation}$$
$$-4y - 24 \leq 2y - 5 + 4y - 28 \quad \text{Distributive property}$$
$$-4y - 24 \leq 6y - 33 \quad \text{Combining like terms}$$

Next we use the addition principle twice.

$$-4y - 24 + 24 \leq 6y - 33 + 24 \quad \text{Adding 24 to both sides}$$
$$-4y \leq 6y - 9 \quad \text{Simplifying}$$
$$-4y + (-6y) \leq 6y - 9 + (-6y) \quad \text{Adding } -6y \text{ to both sides}$$
$$-10y \leq -9 \quad \text{Simplifying}$$

This time we must change the direction of the inequality when we use the multiplication principle, because we have to multiply by a negative number to obtain a positive coefficient for y.

$$-10y \leq -9$$
$$\left(-\frac{1}{10}\right)(-10)y \geq \left(-\frac{1}{10}\right)(-9) \quad \begin{array}{l}\text{Multiplying both sides by } \left(-\frac{1}{10}\right) \\ \text{and reversing the inequality sign}\end{array}$$
$$y \geq \frac{9}{10} \quad \text{Simplifying}$$

To graph this inequality, we mark a filled-in circle at $\frac{9}{10}$ and draw an arrow extending to the right.

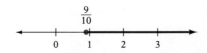

So the solution set of $-4(y + 6) \leq 2y - 5 + 4(y - 7)$ contains all real numbers greater than or equal to $\frac{9}{10}$.

▶ CHECK **Warm-Up 8**

Practice what you learned.

SECTION FOLLOW-UP

Here are answers to the questions asked in the Section Lead-In.

1.   a. The inequality remains true.

     b. The inequality remains true.

     c. The inequality remains true.

     d. The inequality remains true.

2.   Answers may vary.

3.4 WARM-UPS

Work these problems before you attempt the exercises.

1.   Find and graph the solution set for $\frac{1}{3} + x > 1\frac{8}{9}$.

2.   Find and graph the solution set for $-7.8 + x + 1.9 < 8.2$.

3.   Find and graph the solution set for $9x < 18$.

4.   Find and graph the solution set for $55x + 9x \leq 16$.

5.   Find and graph the solution set for $-12x < 156$.

6.   Find and graph the solution set for $1.2x + 8.4 < 12$.

7.   Find and graph the solution set for $2x - 3 - 9x > 10$.

8.   Find and graph the solution set for
     $-5(x - 2) \geq 6x - 4 + 3(x + 5)$.

3.4 EXERCISES

Note: Use your graphing calculator to check your results whenever possible.

In Exercises 1 through 8, determine whether the given number is in the solution set for the given equality.

1. $3: m + 4 > 7$

2. $2: 6 + x \leq 9$

3. $4: -12 < x - 8$

4. $-10: 17 \geq -x + 5$

5. $-1: 6m \geq 8$

6. $-3: 45 > 15y$

7. $-26: -125 < 5y$

8. $7: 15r \leq 120$

In Exercises 9 through 16, solve the inequality using the addition principle. Check your results.

9. $-24 + x > 9$

10. $x + 32 < 16$

11. $-26 \leq y - 18$

12. $15 \geq t - 34$

13. $11 \geq 4 - r$

14. $-15 - y < 12$

15. $-t + 3 > -17$

16. $26 \leq 18 - n$

In Exercises 17 through 24, solve the inequality using the multiplication principle. Check your results.

17. $12 > -4m$

18. $104 \geq -8r$

19. $-12x \leq -204$

20. $-11t < -1144$

21. $16x \geq -124$

22. $18r > -168$

23. $230 < -20y$

24. $-225 \leq 15m$

In Exercises 25 through 60, solve and check.

25. $16 < 3.2r + 3.6r$

26. $11.2y - 0.5y \geq 175$

27. $\frac{4}{5}t > 36$

28. $\frac{6}{7}x + \frac{1}{3}x \leq 27$

29. $\frac{1}{2} + y + \frac{3}{4} \leq \frac{5}{6}$

30. $4.6 + n > -7.3 + 5.3$

31. $7.8 \geq 11.4 + x - 9.3$

32. $\frac{-3}{8} < \frac{-3}{4} + y$

33. $-6.5 < 0.5y + 3.5y$

34. $-11.6x - 2.4x > 31.2$

35. $15.6t - 3.1t \geq 55.8$

36. $24 \leq \frac{-4}{7}x$

37. $-0.3t + -4.5t < -1.8$

38. $85 - 64 \geq \frac{-2}{3}y$

39. $\frac{3}{7} + x + \frac{3}{5} \leq \frac{-2}{5}$

40. $\frac{-6}{11} + t \geq \frac{-1}{5}$

41. $7.4 < 2.8 + y - 9.8$

42. $v - \frac{2}{5} > \frac{-2}{3}$

43. $3r + 5 < 12$

44. $13 > 4t + 9$

45. $18 \leq 12 + 7y$

46. $3y + 27 \geq 23$

47. $1.5r - 3 \geq 6$

48. $2.4 < 6t - 1.8$

49. $7.7 > 1.1r - 4.4$

50. $0.5y - 7.1 \leq 9.3$

51. $\frac{-1}{6} - \left(\frac{2}{3}\right)r \leq \frac{1}{2}$

52. $\frac{7}{12} \geq \frac{-1}{2}t - \frac{1}{3}$

53. $\frac{4}{15} < \frac{-3}{5}x - \frac{2}{3}$

54. $-3\frac{3}{4}y - \frac{3}{4} > \frac{1}{8}$

55. $-2r + 3r > 9$

56. $32 \leq 27t - 9t$

57. $45 \geq -5n + 4n$

58. $-8y + 6y < 42$

59. $12 - 8t > -11t + 18t$

60. $-3t + 7 \leq -4t + 5t$

In Exercises 61 through 72, solve each inequality and graph the results.

61. $-2r + 6 < 27$

62. $14 \geq -13n + 25$

63. $12r + 6 < 9r + 8$

64. $12t + 3 \geq 10t + 5$

65. $8 + 21y \leq 11y + 5$

66. $11(7x - 5) > 2(x - 3)$

67. $16(y - 4) \leq 2(y - 7)$

68. $4(y - 4) + 3 > 4(y + 5)$

69. $-3(t + 4) > -1(t - 3) + 5t$

70. $6 + 5(x - 5) < 5(x - 2)$

71. $2x - 3x + 5 - 7 < 3(x - 6)$

72. $-9 + 3y + 5y - 7 \leq 6(y - 12)$

In Exercises 73 through 82, solve each inequality.

73. $-27y + 11y < 15 - 3y$

74. $(4r + 5)2 \geq 6r$

75. $3(t + 2) \leq 8t$

76. $8x > 4(8x + 3)$

77. $14y < 7(3y + 5)$

78. $0.5(6r + 8) < 0.9r - 7$

79. $4(2t + 3.4) \geq -2.2 + 2t$

80. $3r - 4.2 \leq 8(3r + 2.2)$

81. $-8(x - 2) \geq 6x - 4 + 8(x + 9)$

82. $0.6y - 8 > 0.3(4y + 9)$

MIXED PRACTICE

By doing these exercises, you will practice the topics up to this point in the chapter.

83. Solve for x: $5x + y = -3x + v$

84. Solve: $8 - \frac{x}{6} < 234$

85. Solve for v: $5v - x = -9v + x$

86. The difference between a number and -212 is 113. What is the number?

87. Solve: $\frac{1}{5} + y > \frac{23}{2}$

88. Solve for the variable: $6x = 36 + 23(x - 6)$

89. Solve for y: $-14y + 11 = 21y - 4 + 5(y - 5)$

90. Solve: $-7y + 5 < 12$

91. Solve for t: $\frac{3t}{2} + 26 = -37$

92. Solve for z: $3(z + x) = x$

93. Solve for w: $-5(w + 5) = -180$

94. Solve for x: $7x - 8 = 8x - 9$

95. Solve for c: $ce - x = f$

96. Solve for t: $\frac{-5}{6} + t = \frac{1}{6}$

97. Solve for m: $-\frac{m}{84} = 26$

98. Solve for y: $11 - 6y = 18$

99. You need two pieces of wood, one three times the length of the other. The shorter piece is 5.2 feet long. How long is the longer piece?

100. How many pounds of hard candy worth $1.80 per pound must be mixed with 15 pounds of caramels worth $1.10 per pound to produce a mixture worth $1.50 per pound?

EXCURSIONS

Class Act

1. Seven critics have graded ten current movies in the following table.

Movie	C_1	C_2	C_3	C_4	C_5	C_6	C_7
I	A–	B	B	C–	C	B–	B
II	–	B+	B	B–	–	B–	B+
III	A	A	B–	A	B	B+	A
IV	–	C	C	C	B	B+	B+
V	–	A	B+	A	A–	B–	B+
VI	–	B+	C+	B	–	–	C
VII	A–	B	B+	B+	C+	B	B+
VIII	–	B	B–	–	B	C–	C–
IX	A	B+	C+	B	B–	C+	B–
X	B–	C	C	F	C+	C–	C–

a. Use the plus-minus scale to calculate the grade point average (GPA) of each movie and order the movies based on this.

b. How would the final average change if we eliminate + and – and go to a four-point scale?

Plus/minus Scale		Four-Point Scale	
A+	4.5	A	4.0
A	4.2	B	3.0
A–	4.0	C	2.0
B+	3.5	D	1.0
B	3.2	F	0.0
B–	3.0		
C+	2.5		
C	2.2		
C–	2.0		
D+	1.5		
D	1		
F	0		

Data Analysis

2. For the 1994 New York City Marathon, the ages of the first 15 male finishers were

26	33	31	30	30
25	28	33	32	25
29	28	32	24	33

a. What is the mode?

b. What is the median?

c. What is the mean?

d. Which of these measures, in your opinion, best describes the "average" of the data. Justify your answer.

3. For the 1994 New York City Marathon, the ages of the last 10 men to complete the race were

23	58	60	49	71
48	30	33	48	56

a. What is the mode?

b. What is the median?

c. What is the mean?

d. Which of these measures, in your opinion, best describes the "average" of the data. Justify your answer.

3.5 Linear Equations in Two Variables

SECTION LEAD-IN

1. Use the equation

$$\text{Distance} = \text{rate} \times \text{time}$$

$$d = rt$$

Does the rate increase or decrease when

a. the distance increases and the time remains the same?

b. the time increases and the distance remains the same?

c. both the distance and the time increase?

d. both the distance and the time decrease?

2. Ask and answer questions (as a group) about variables in another literal equation. Choose your own literal equation. Use numbers to check your decisions.

Solving Linear Equations in Two Variables

You learned earlier to solve linear equations and inequalities in one variable. A much larger and more useful group of linear equations contains two variables.

Linear equation in two variables

A **linear equation in two variables** is an equation that can be put in the *general form* $Ax + By = C$. Here, A, B, and C are constants, and A and B are not both equal to zero. The exponent on each variable must be 1, and each term includes at most one variable.

All of these are linear equations:

$$2x + 3y = -6.7 \qquad y = 2x + 8$$
$$x = 6 \qquad 3x = 2y$$

$xy = 7$ is not a linear equation, and neither is $x^2 + 3 = y$. Do you see why not?

▪▪▪

EXAMPLE 1

Which of the following are linear equations?

a. $2x = 6$

b. $-12x - 7y = 4$

c. $-2x = 3.5 + 4y$

d. $x^2 - 3y = 12$

e. $xy = 9$

SOLUTION

a. $2x = 6$ is a linear equation. The coefficient of y is 0. That is, $2x + 0y = 6$.

b. $-12x - 7y = 4$ is a linear equation.

c. $-2x = 3.5 + 4y$ is a linear equation.

d. $x^2 - 3y = 12$ is not a linear equation because the exponent on the variable x is greater than 1.

e. Because the term xy includes more than one variable, the equation is not linear.

▶ *CHECK* **Warm-Up 1**

> ## ! ! !
> ### *ERROR ALERT*
> Identify the error and give a correct answer.
>
> $2x + 3y = z^2$ is a linear equation.

Solutions

A **solution of an equation in two variables,** such as x and y, is a pair of numbers that, when substituted for the variables, makes the equation true. The pair of numbers is usually written in parentheses in the form (x-value, y-value) or simply (x, y). In this form it is called an **ordered pair** of numbers. The pair is *ordered* because the x-value is always the first one listed.

An equation can have many solutions. To determine whether a particular ordered pair is a solution of an equation, we substitute the values into the equation. If the result is a true statement, the ordered pair is a solution.

▪▪▪

EXAMPLE 2

Which of the following ordered pairs is a solution of the equation $2x + 4y = 8$?

a. $(1,0)$ **b.** $(4,0)$ **c.** $(-2,3)$

> ## ! ! !
> ### *ERROR ALERT*
> Identify the error and give a correct answer.
>
> *Incorrect Solution:*
>
> $(-2, -3)$ is a solution to $-2x + 3y = -13$ because
> $(-2)2 + 3(-3) = -4 - 9 = -13.$

SOLUTION

We must substitute each ordered pair into the equation and evaluate the result.

a. To test the ordered pair $(1,0)$, we substitute 1 for x and 0 for y.

$$2x + 4y = 8$$
$$2(1) + 4(0) = 8$$
$$2 + 0 = 8$$
$$2 = 8 \quad \text{False}$$

So $(1,0)$ is *not* a solution of $2x + 4y = 8$.

b. We substitute $(4,0)$ for x and y and evaluate.

$$2x + 4y = 8$$
$$2(4) + 4(0) = 8$$
$$8 + 0 = 8$$
$$8 = 8 \quad \text{True}$$

So $(4,0)$ is a solution of $2x + 4y = 8$.

c. We substitute $(-2,3)$ and evaluate.

$$2x + 4y = 8$$
$$2(-2) + 4(3) = 8$$
$$-4 + 12 = 8$$
$$8 = 8 \quad \text{True}$$

So $(-2,3)$ is a solution of $2x + 4y = 8$.

▶ *CHECK* **Warm-Up 2**

Calculator Corner

You can also test each ordered pair using your graphing calculator's Home Screen. The work for Example 2 part (a) using the ordered pair $(1,0)$ should look like the screen at the right. The result is 2, not 8. Therefore, the ordered pair $(1,0)$ is not a solution to the equation.

Some graphing calculators have the capability to recall the previous line so that you can substitute new numbers for x and y. For instance, pressing the **2nd EN-TRY** keys on the *TI-82* graphing calculator would result in the screen at the left on page 186. Replace the ordered pair $(1,0)$ with the ordered pair $(4,0)$ and press **ENTER** to get the screen at the right on page 186. The result is 8, so the ordered pair $(4,0)$ is a solution to the equation.

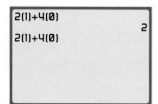
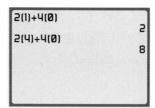

Now try the ordered pair $(-2, 3)$ on your own.

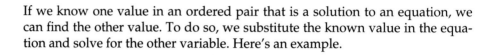

If we know one value in an ordered pair that is a solution to an equation, we can find the other value. To do so, we substitute the known value in the equation and solve for the other variable. Here's an example.

▪▪▪
EXAMPLE 3

Find a solution of the equation $4x + 72 = -8y$ when $y = -9$.

SOLUTION

We want to find x in the ordered pair $(x, -9)$. We substitute -9 for y in the equation and solve for x.

$$4x + 72 = (-8)(-9) \qquad \text{Substitution}$$
$$4x + 72 = 72$$
$$4x + 72 + (-72) = 72 + (-72) \qquad \text{Addition principle}$$
$$4x = 0$$
$$\left(\tfrac{1}{4}\right)4x = \left(\tfrac{1}{4}\right)0 \qquad \text{Multiplication principle}$$
$$x = 0$$

So $(0, -9)$ is a solution of the equation $4x + 72 = -8y$. You can check by substituting both values in the equation.

▶ *CHECK* **Warm-Up 3**

Constructing a Table of Values

The method in Example 3 can be used to make a **table of values** for an equation in two variables. Such a table lists the values for several solution pairs. We will use such tables in Chapter 4 to graph linear equations.

▪▪▪
EXAMPLE 4

Construct a table of values for the equation

$$2y + x = 4$$

SOLUTION

We choose a value for one variable and solve for the other variable. We will start by choosing $x = 0$, $y = 0$, and $x =$ any other number—we are choosing $x = 2$.

x	y
0	
	0
2	

Next we substitute the chosen values, one at a time, and solve.

For $x = 0$,

$$2y + x = 4$$
$$2y + 0 = 4 \qquad \text{Substitution}$$
$$2y = 4 \qquad \text{Addition of zero}$$
$$\left(\tfrac{1}{2}\right)2y = \left(\tfrac{1}{2}\right) \times 4 \quad \text{Multiplication principle}$$
$$y = 2$$

For $y = 0$,

$$2y + x = 4$$
$$2(0) + x = 4 \quad \text{Substitution}$$
$$0 + x = 4 \quad \text{Addition of zero}$$
$$x = 4$$

For $x = 2$,

$$2y + x = 4$$
$$2y + 2 = 4 \qquad \text{Substitution}$$
$$2y + 2 + (-2) = 4 + (-2) \quad \text{Addition principle}$$
$$2y = 2$$
$$\left(\tfrac{1}{2}\right)2y = \left(\tfrac{1}{2}\right)2 \qquad \text{Multiplication principle}$$
$$y = 1$$

Finally, we add these values to the table

x	y
0	2
4	0
2	1

This is our table of values for the equation $2y + x = 4$.

▶ *CHECK* **Warm-Up 4**

There are many ways to find values for a table.

▪▪▪

EXAMPLE 5

Use a table of values to find ordered pairs that are solutions of $3x + y = -12$ corresponding to $x = -1$, $x = 0$, and $x = 1$.

SOLUTION

We will solve the equation for y.

$$3x + y = -12$$
$$3x + y + (-3x) = -12 + (-3x) \quad \text{Addition principle}$$
$$y = -12 - 3x \quad \text{Simplifying}$$

Then we prepare a table of values, entering the given x values.

x	y
-1	
0	
1	

We next substitute the given values into the equation, in turn, and solve for y.

For $x = -1$,

$$y = -12 - 3x = -12 - 3(-1)$$
$$= -12 + 3$$
$$= -9$$

For $x = 0$,

$$y = -12 - 3x = -12 - 3(0)$$
$$= -12$$

For $x = 1$,

$$y = -12 - 3x = -12 - 3(1)$$
$$= -15$$

Then we complete the table and write the ordered pairs.

x	y
-1	-9
0	-12
1	-15

Our solutions are $(-1, -9)$, $(0, -12)$, and $(1, -15)$.

▶ *CHECK* **Warm-Up 5**

Calculator Corner

A table of values can also be constructed on some graphing calculators using the **2nd TblSet** and the **2nd TABLE** utilities. For example, using the equation given in Example 5, first enter the equation into the calculator. Remember to use the negative key $\boxed{(-)}$ for -12 and the subtract key $\boxed{-}$ $-3x$.

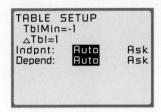

Press the **2nd TblSet** keys and let TblMin equal -1 and the increment ΔTbl equal 1, as shown at the left below. The increment is how much the x-value in the table will change. (Note: The *TI-83* graphing calculator uses TblStart instead of TblMin.) Press the **2nd TABLE** key to obtain the table shown at the right below. You can see that you get the same values for y that you obtained when constructing the table by the paper-and-pencil method.

```
TABLE SETUP
  TblMin=-1
  △Tbl=1
Indpnt:  Auto   Ask
Depend:  Auto   Ask
```

X	Y₁
-1	-9
0	-12
1	-15
2	-18
3	-21
4	-24
5	-27

X=-1

Practice what you learned.

SECTION FOLLOW-UP

Here are answers to the questions asked in the Section Lead-In using the equation

$$\text{Distance} = \text{rate} \times \text{time}$$
$$d = rt$$

1. **a.** increases

    b. decreases

    c. either is possible

    d. either is possible

2. Answers may vary.

3.5 WARM-UPS

Work these problems before you attempt the exercises.

1. Tell whether each of the following is a linear equation.

    a. $-13 - 4g = h$　　　　　　b. $2m + 6 = 48$

    c. $2y - 3 = x^2$　　　　　　d. $2c + 4k = 65c$

2. Which of the following ordered pairs is a solution of the equation $-3x + 7 = 4y$?

    a. $(1, -1)$　　　　　　b. $(-1, 1)$

    c. $(3, 0)$　　　　　　d. $\left(0, \frac{7}{4}\right)$

3. In the equation $x - y = 0$, what solution for x corresponds to $y = -4.5$?

4. Complete this table of values for $x - 9y = 7$.

x	y
	-1
	0
	1

5. Find the ordered pairs whose y values are 3, 0, and -4 that are solutions to $-3y = 3x + 24$.

3.5 EXERCISES

Note: Use your graphing calculator to check your results whenever possible.

In Exercises 1 through 8, determine whether each equation is a linear equation. Justify your answers.

1. $-10 + 2m = n$　　　　2. $3x = 7$　　　　3. $3x + 4 = 16y$　　　　4. $-6x + 5w = 11$

5. $x^2 - 4y = 6$　　　　6. $-7 + 12y = 4x$　　　　7. $3r + 4nz = 18$　　　　8. $x^2 - xy = -17$

In Exercises 9 through 20, determine whether the ordered pairs are solutions of the given equation.

9. $(-4, 3), (5, 3)$: $y = 3$　　　　10. $(-5, 21), (0, 9)$: $x + y = 9$　　　　11. $(2, 4), (-6, 7)$: $y = 2x + 2$

12. $\left(-3, \frac{1}{3}\right), (-2, -6)$: $y = 3x - 2$　　　　13. $(3, 3), (-2, 0)$: $y = x$　　　　14. $\left(0, 4\frac{1}{2}\right), (4, 11)$: $x - y = -7$

15. $(5, 4), (-4, 9)$: $x + y = 5$　　　　16. $(1, 3), (-7, -5)$: $y - 2x = 1$　　　　17. $(7, 1), (8, -6)$: $x = y + 6$

18. $(-6, -14), (-9, 2)$: $3x - 2y = -31$

19. $\left(\frac{3}{20}, 3\frac{3}{4}\right), \left(-\frac{7}{20}, 1\frac{1}{4}\right)$: $5x + 3 = y$

20. $\left(1, \frac{1}{2}\right), \left(-1, -\frac{1}{2}\right)$: $2x - 4y = 0$

In Exercises 21 through 28, find a solution when:

21. $x = 5$ in $5x - y = 8$

22. $x = -3$ in $7x = y - 11$

23. $x = 2$ in $6x - 2y = 5$

24. $x = 11$ in $-8x + 3y = 9$

25. $y = -2$ in $3x - 7 = y$

26. $y = 9$ in $2y + 7 = 4x$

27. $y = -5$ in $y + 3x = 10$

28. $y = 0$ in $3x + 5y = 12$

In Exercises 29 through 36, find a value for A such that the given ordered pair is a solution of $Ax + 2y = 1$.

29. $(0, 1)$

30. $(2, 3)$

31. $(1, -1)$

32. $(-4, 0)$

33. $(2.5, 1)$

34. $(-5, -5)$

35. $(2, 1)$

36. $(0, 0)$

Calculator
Corner

Exercises 37 through 46 can be checked with your graphing calculator's **TABLE** utility by using **ASK** on the **TblSet** menu as follows. Note that it is then possible to input various values of x, but it is not possible to input values for y (the calculator automatically computes y in this setting). Consider the following for Exercise 37.

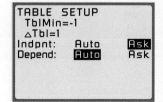

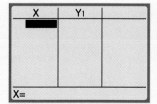

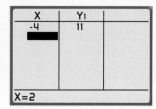

In Exercises 37 through 44, complete the table of values to determine three points that are solutions of the given equation.

37. $y = -\frac{3}{2}x + 5$

x	y
-4	
2	
	1

38. $y = \frac{5}{3}x + \frac{1}{3}$

x	y
1	
-2	
	0

39. $2x + 5y = -10$

x	y
0	
	0
	2

40. $15y = 10x$

x	y
1	
	0
	-2

41. $3x = 6$

x	y
	5
	3
	$-\frac{15}{16}$

42. $-16y = 32$

x	y
-16.3	
-1	
-19	

43. $0.1x = 3y - 4$

x	y
0	
10	
	$\frac{1}{3}$

44. $2y = 2.5x + 6$

x	y
0	
-4	
	0

In Exercises 45 through 60, set up a table of values to determine three points that are solutions of the given equation. Tables will vary.

45. $x - y = 4$

46. $2x + y = 5$

47. $y - 4x = 10$

48. $y + x = 2$

49. $5x = 3 - y$

50. $2y + 5 = x$

51. $2y = 4x$

52. $y = x - 6$

53. $2x + 2 = y$

54. $3y + 2 = 4x$

55. $y - 3x = 1$

56. $-2y + x = 2$

57. $5x = 3 - 5y$

58. $2y + 4 = 10x$

59. $2x + 2 = 4y$

60. $y + 12 = 4x$

In Exercises 61 through 64, find the solution of each equation given the information indicated.

61. *Age* Let y be Marcy's age and x be Victor's age. The relationship between their ages is $y = x + 4$. What age was Marcy when Victor was 2 years old and when he was 15 years old?

62. *Ice Trays* The relationship between the number of ice trays (T) and the number of ice cubes (I) is $I = 12T$. Find the number of ice cubes in 6 trays and the number in 12 trays. How many trays are needed for 264 cubes?

63. ***Travel Distance*** The relationship between how far a certain car travels (y) and how far a bus travels (x) is $y = \left(1\frac{1}{2}\right)x - 10$ (the bus has a head start). Find the distance the car has traveled when the bus has traveled 50 miles and when the bus has traveled 80 miles. How far has the bus traveled when the car has gone 65 miles?

64. ***d = rt*** At a certain speed, the relationship between the distance traveled (d) and the time traveled (t) is $d = 60t$.

   a. Find d when the time is 3 hours.

   b. Find d when the time is 4 hours.

   c. Find t when the distance is 1800 miles.

MIXED PRACTICE

By doing these exercises, you will practice the topics up to this point in the chapter.

65. Solve for y: $3y - 4 = 7$

66. Solve for x: $-2(x + 5) = 6$

67. Solve for t: $-5t - 3t - 6(t - 4) = 11 - t$

68. Solve for r: $6 = \frac{3r}{19}$

69. The width of a picture frame is twice its length. The perimeter is 4.8 feet. What are the length and width?

70. The sum of two numbers is -5. One number is 12. What is the other?

71. The sum of three consecutive odd integers is -117. Find the integers.

72. My age is equal to twice Ballow's age plus 15. If I am 49 years old, how old is Ballow?

73. The base of a rectangle is three times its height. Write an expression for the area of the rectangle in one variable, x.

74. Solve for e: $4e + 37 = 9e - 87$

EXCURSIONS

Data Analysis

1. Study Table 1 on page 194. It contains information about various automobiles that cost under $10,000.

   a. Use the data in this table to identify your three "best buys." Justify your choice.

   b. Ask and answer four questions using this data.

   c. Describe the "average" car. Then discuss your definition of "average" and justify it. Use the data in the table for this problem.

TABLE 1

Car	Suggested retail price	Estimated dealer's cost	Your target price	Dealer's cost for options (1% of retail)	Five-year resale value (% of list price)	Miles per gallon (city/ hwy.)	Five-year totals		Total ownership costs
							Mainte-nance	Repairs	
A	$9,836	$9,387	$9,669	86	53	32/40	$3,559	$770	$22,489
B	9,680	8,958	9,227	89	62	26/34	3,932	780	23,986
C	8,085	7,616	7,844	89	66	44/49	3,527	630	19,680
D	9,085	8,467	8,721	89	67	39/43	3,495	630	20,859
E	8,385	7,815	8,049	89	69	39/43	3,553	630	20,467
F	9,485	8,840	9,105	89	67	39/43	3,495	630	21,109
G	9,750	9,146	9,420	85	68	42/46	3,380	385	19,778
H	9,894	9,099	9,372	76	58	32/39	3,962	410	21,699
I	9,995	8,996	9,995	90	82	27/36	3,317	674	20,650
J	9,998	9,329	9,609	84	68	33/39	3,894	615	21,088

Source: Adapted from *Money,* March 1995, p. 149.

2. The production of "trash" is a continuing problem for the world. The top twenty "trash" producers are listed below (not in any particular order).

Country	Waste (lbs. per capita per day)	Country	Waste (lbs. per capita per day)
Australia	4.2	Denmark	2.6
Saudi Arabia	2.4	Israel	2.4
Switzerland	2.2	Qatar	2.4
United Arab Emirates	2.4	United Kingdom	2.2
New Zealand	4	Oman	2.4
Finland	2.4	Canada	3.7
Iraq	2.4	United States	3.3
Bahrain	2.4	Netherlands	2.6
France	4	Kuwait	2.4
Norway	2.9	Luxembourg	2.2

Source: 1993 Information Please® Environmental Almanac (©1995 Houghton Mifflin Co.), p. 339. All rights reserved. Used by permission from Information Please LLC.

a. What is the mode?

b. What is the median?

c. What is the mean?

d. Which of these measures, in your opinion, best describes the "aver-age" of the data? Justify your answer.

CHAPTER LOOK-BACK

Refer to the Chapter Lead-In on page 122. Using the formulas we are given relating bone length to height, we asked the following questions:

1. If I am 5 feet 5 inches tall, how long are my femur, tibia, humerus, and radius?

    Because the formulas are given in terms of centimeters, we first write my height in centimeters.

    $$5 \text{ feet } 5 \text{ inches} = 5(12) + 5 \text{ inches} \quad \text{Why?}$$
    $$= 65 \text{ inches}$$

    One inch is approximately 2.54 centimeters. Using this relationship, we solve the following problem:

    $$= \frac{65 \text{ in.}}{1} \times \frac{2.54 \text{ cm}}{1 \text{ in.}}$$
    $$= (65)(2.54 \text{ cm})$$
    $$= 165.1 \text{ cm}$$

    Because I am female, I use the second set of equations.

    Female

    $$h = 61.41 + 2.32F$$
    $$h = 72.57 + 2.53T$$
    $$h = 64.98 + 3.14H$$
    $$h = 73.50 + 3.88R$$

    I substitute my height for h and solve for each variable to find F, T, H, and R.

    Female

    $$165.1 = 61.41 + 2.32F$$
    $$165.1 - 61.41 = 2.32F$$
    $$103.69 = 2.32F$$
    $$103.69 \div 2.32 = F$$
    $$44.69 = F$$

    You should do the work for the remaining three equations.

    $$165.1 = 72.57 + 2.53T$$
    $$36.57 = T$$
    $$165.1 = 64.98 + 3.14H$$
    $$31.89 = H$$
    $$165.1 = 73.50 + 3.88R$$
    $$23.61 = R$$

2. My good friend Pocahontas Fox is 76 years old (2/18/20). How many inches taller was she when she was 30 years old?

We know that the height of a person begins to decrease at the rate of approximately 0.06 centimeters per year. We also know that Pocahontas was 30 years old 46 years ago. To find how much taller she was then in centimeters, we will multiply:

$$46(0.06) = 2.76 \text{ cm}$$

To change 2.76 centimeters to inches, we set up and solve this equation:

$$= \frac{2.76 \text{ cm}}{1} \times \frac{1 \text{ in.}}{2.54 \text{ cm}}$$
$$= \frac{2.76 \text{ in.}}{2.54}$$
$$= 1.09 \text{ in.}$$

3. A coroner reports that a bone from a 57-year-old man has been found. It has been identified as a femur. It measures 58.5 centimeters. How tall was this man when he was 30 years old? Give your answer in inches.

We use this formula:

$$h = 69.09 + 2.24F$$
$$h = 69.09 + 2.24(58.5)$$
$$h = 200.13 \text{ cm}$$

This is his height now. We want his height at 30 years of age.

$$57 - 30 = 27 \text{ yrs}$$

He is $(27)(0.06)$ centimeters shorter now than he was 27 years ago. So his height is $200.13 + 27(0.06) = 201.75$ cm.

In inches, 201.75 cm is 201.75 cm × 1 in. ÷ 2.54 cm = 201.75 ÷ 2.54 = 79.4 in.

The man was 6 feet 7 inches tall.

What questions did you ask?

CHAPTER **3**
REVIEW PROBLEMS

The following exercises will give you a good review of the material presented in this chapter.

SECTION 3.1

1. Solve: $2x = -84$

2. Solve: $x + 3.6 = -8.9$

3. Solve: $3x + 7 = 14$

4. Solve: $-46 + \frac{7}{9}x = -63$

5. Solve: $4(2x + 5) = 21$

6. Solve: $-9 + 8(x + 3) \leq 14$

7. Solve: $3x - 4x + 8 - 3x = 9 - 2x$

8. Find the solution: $-7 + \frac{3}{8}x + 12 = 19 - 4x$

SECTION 3.2

9. Solve for r: $\frac{r}{v} = m - x$

10. Solve for p: $mp + tr = st$

11. Solve for x: $rx + e = fx$

12. Solve for t: $ts - 4t = 11r$

13. ***Electric Circuit*** Ohm's law states that current (I amperes), voltage (E volts), and resistance (R ohms) in a simple electrical circuit are related by the equation

$$E = IR$$

If the resistance is 20 ohms and the voltage is 160 volts, what is the current?

14. Given $S = \frac{n(n + 1)}{2}$, where S is the sum of the first n integers, what is the value of S for $n = 26$?

SECTION 3.3

15. The sum of two consecutive even integers is -86. Find the integers.

16. A bag full of 70 coins contains dimes and nickels. If the value of the coins is $5.65, find how many of each coin the bag contains.

17. Bimla is ten years older than her sister. If the sum of their ages is 54, find the age of each.

18. The difference of a number and 25 is 34. Find the number.

19. The sum of a number and 12 is -24. Find the number.

20. The sum of three consecutive integers is -36. Find the integers.

SECTION 3.4

21. Solve: $-3(5 + 3x) > 9$

22. Solve: $6 - 3x > 24$

23. Find and graph the solution for $-3(12x - 3) < 12$.

24. Find and graph the solution for $3x - 4x + 8 - 3x > 9 - 2x$.

25. Solve: $3x < 1.5$

26. Solve: $x - 1.2 \geq 8$

27. Solve: $-4.5 + x \leq 9.6$

28. Solve: $-9x > 27.9$

SECTION 3.5

29. Construct a table of values for the equation $2x = y + 3$, using $x = -1, -2$, and 3.

30. Find a value for A such that $(-2, 3)$ is a solution of $Ax + 3y = 1$.

31. Find the coordinates of the point whose y-coordinate is three times its x-coordinate, if the x-coordinate is $-\frac{2}{5}$.

32. Find a solution for

$$2x = 8y - 12$$

when $x = 10$.

MIXED REVIEW

33. Solve: $2x - 7 < 15$

34. Solve: $-3(4 - 2x) < 19$

35. Solve for r: $rt + v = wx$

36. n less than 18 is -84. Find n.

37. Solve for c: $2c + w = 3c - t$

38. Solve and graph: $8x \leq 1.6$

39. Three times the sum of a number and 12 is 81. Find the number.

40. You need two pieces of ribbon. One must be twice the length of the other. You have 36 inches of ribbon. How long will each piece be if you use the entire 36 inches?

CHAPTER 3 TEST

This exam tests your knowledge of the material in Chapter 3.

1. Solve for x:

    a. $\frac{x}{6} = 12$

    b. $\frac{5}{7} + x = \frac{2}{3}$

    c. $17 + 5x = 22x$

2. Solve.

   a. A rectangle has a length that is three times its width. If the perimeter of the rectangle is 56 inches, find the dimensions of the rectangle.

   b. $5(-4y - 3) = 6(8y + 3)$ **c.** $-2r + 8 - 7r = -9 - 15 + 9r$

3. Solve for the indicated variable.

   a. Solve for e: $ef - g = 4$ **b.** Solve for f: $\frac{e}{f} + 4 = g$ **c.** Solve for m: $mt + r = x$

4. Translate each problem into an equation and solve the equation.

   a. The difference of a number and 12 is seventeen. Find the number.

   b. A number less 75 is 26. Find the number.

   c. A man has both 20-cent stamps and 23-cent stamps. If he has 3 more 23-cent stamps than 20-cent stamps and the value of the stamps is 542 cents, how many of each type does he have?

5. Find and graph the solution set.

   a. $13 + x > 12$ **b.** $-7x - 4 \geq 10$ **c.** $6r \leq 24$

6. Solve the following problems.

   a. Find A so that the equation
   $$Ax - 2 = 3y$$
   has $(8, -1)$ as a solution.

   b. Complete this table of values for $2y - x = -3$.

x	y
	$-\frac{1}{2}$
1	
	$-\frac{7}{2}$

   c. Find y when $x = -5$ in $-x - y = 11$.

CUMULATIVE REVIEW

CHAPTERS 1–3

The following exercises will help you maintain the skills you have learned in this and previous chapters.

1. Solve for x: $14 - 3x = 7(x - 3)$

2. Solve for t: $6 - 2t \geq 9 + t$

3. Evaluate $c^2e^2 - 4(c - et^2)$ when $c = 4$, $e = -5$, and $t = -2$.

4. Solve: $\frac{x}{9} + 3 = 5$

5. Are $y - 5$ and $-(5 - y)$ equal? Justify your answer.

6. Write -6^2 without exponents.

7. Solve: $-3(n - 5) \leq 2(n + 6) - 3$

8. Find the sum of 3.4^2 and -8^2.

9. Simplify: $2^3 + (-5 + 6)^3 - 3^2$

10. Divide: $(-4.602) \div (-0.0002)$

11. Simplify: $3r^2t + 3.7tr^2 - 2.6rt + 7.5rt - 4.6r - 1.2r$

12. Simplify: $\left(-18\frac{2}{3}\right) + \left(15\frac{1}{4}\right) - \left(27\frac{1}{5}\right)$

13. What is the reciprocal of $-2\frac{1}{3}$?

14. Which is greater, $-|18.6|$ or $-(18.8)$?

15. Simplify: $(-5)^3$

16. Simplify: $\{-[(-1^3)^3]^3\}^3$

17. Evaluate $xy^2 - 3xy + y^2$ when $x = -1$ and $y = -2$.

18. Evaluate $-(r^3t^2) + (rt - 5rt) - 5t + 3r$ when $t = 0$ and $r = -2$.

19. Find the difference of $15\frac{4}{5}$ and $17\frac{3}{8}$.

20. Write in symbols: The product of -85 and a number is less than 42.5.

GRAPHS OF LINEAR EQUATIONS

A brilliant-cut diamond has a round top and a conical base that meet at the girdle. Viewed from the side, the top is a flat "table." A diamond's brilliance—and its price—depends in part on the slopes of the lines of facets that run from girdle to table and girdle to the base's apex. Plotting the lines on a graph where the girdle is the *x*-axis, the slope from the top to the girdle might be about $\frac{-7}{6}$; the slope from the base to the girdle about $\frac{+8}{9}$. A gemcutter's error can mean the stone reflects less light, resulting in the loss of thousands of dollars in value.

■ *Think of a sport in which slope can affect performance or enjoyment.*

SKILLS CHECK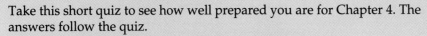

Take this short quiz to see how well prepared you are for Chapter 4. The answers follow the quiz.

1. Add: $-27 + 18$

2. Solve: $3x = 2x - 4$

3. Solve for q: $x + 3q = r$

4. Subtract: $18 - (-36)$

5. Multiply: $\left(\frac{1}{2}\right)(5)$

6. Simplify: $-7y + 3y - (-8y)$

7. Find the value of x when $y = -3$ in $y = -\frac{1}{6}x - \frac{2}{3}$.

8. Solve: $3y - 8 > 9$

CHAPTER LEAD-IN

A teenager is rollerblading in the park. He is stopped by a park ranger for speeding and not wearing a helmet. He puts on a helmet and starts skating again, this time determined to stay within the speed limit. The graph shown here is the teen's speed graphed against time.

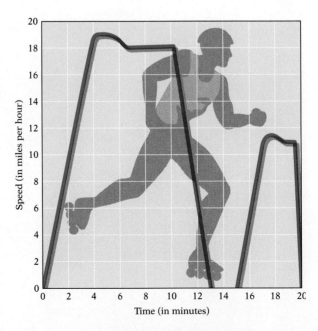

What information can you get from this graph? Ask and answer four questions about the graph. See what we asked at the end of the chapter.

4.1 Graphs of Linear Equations

SECTION LEAD-IN

Travel takes many forms—we can walk, blade, fly, or even take a taxi.

When I take a taxi from the airport to my home, it costs $2.00 plus $0.25 for every $\frac{1}{5}$ mile.

So for a trip of one mile, I will owe

$$\$2.00 + 5(\$0.25) \quad \text{A mile is } 5\left(\frac{1}{5}\right)$$

My total cost will be $3.25.

A table of costs for 1, 2, 3, . . . miles in this situation is

miles	1	2	3	. . .	10
cost	3.25	4.50	5.75		14.50

A "rule" that gives me the total cost for any number of miles is

$$\text{cost} = \$2.00 + (\$1.25 \times \text{the number of miles}) \quad \text{Why?}$$

I can graph the costs in several ways. One way is shown in the following figure.

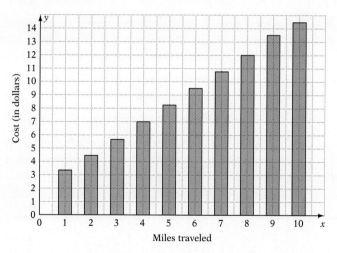

Does this graph accurately represent the taxi information?

Justify your answer.

In this chapter we will explore other ways to represent this information.

Graphing Ordered Pairs of Numbers

As you saw in Chapter 1, we graph a real number on a number line. We can also graph an ordered pair of numbers. The graph is a point on a **Cartesian** (or rectangular) **coordinate system.** To graph a point we use a horizontal number line and a vertical number line that intersect (cross) at their zero points. When we are graphing ordered pairs (x, y), the horizontal number line is called the **x-axis** and the vertical number line is called the **y-axis.**

The axes (plural of axis) divide the **coordinate system** (or **coordinate plane**) into four regions called **quadrants.** These are numbered in counterclockwise order, starting at the upper right. The point where the axes intersect is called the **origin.**

To graph an ordered pair of numbers *(x, y)*

1. Starting from the origin, move right (for positive x) or left (for negative x) along the x-axis to x.
2. Then move vertically up (for positive y) or down (for negative y) exactly $|y|$ units.
3. Mark a dot there on the coordinate plane, and label it (x, y). The dot represents the ordered pair (x, y).

WRITER'S BLOCK

Define *axes, origin,* and *quadrant* in your own words.

▪▪▪

EXAMPLE 1

Graph the following ordered pairs.

a. $(1, 4)$　**b.** $(0, 0)$　**c.** $(-2, 2)$　**d.** $(2, 0)$

SOLUTION

a. $(1, 4)$: The first coordinate, 1, tells us the distance (1 unit) and direction (right, positive) from the origin along the x-axis. The second coordinate, 4, tells us the distance (4 units) and the direction (up) parallel to the y-axis.

b. The point $(0, 0)$ is zero units left or right from the origin and zero units up or down. Thus the point $(0, 0)$ is the origin.

c. The point $(-2, 2)$ is 2 units left and 2 units up from the origin.

d. The point $(2, 0)$ is 2 units right and zero units up. It is on the x-axis, as you can see in the accompanying figure, in which the points $(0, 0)$ $(-2, 2)$, and $(1, 4)$ are also graphed.

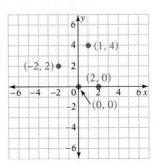

▶ CHECK **Warm-Up 1**

∎∎∎
EXAMPLE 2

Determine the coordinates of points A, B, and C in this figure.

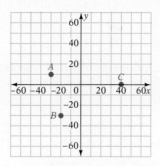

STUDY HINT

We can use any scale on the graph. We usually choose units that make it easy for us to graph.

SOLUTION

a. Point A is 30 units to the left of the origin (in the negative x direction), so its x-coordinate is -30. Point A is also 10 units above the x-axis (in the positive y direction), so its y-coordinate is 10. Its cordinates are $(-30, 10)$.

b. Point B has an x-coordinate of -20 and a y-coordinate of -30, so its coordinates are $(-20, -30)$.

c. Point C has coordinates $(40, 0)$.

▶ *CHECK* **Warm-Up 2**

Graphing a Linear Equation in Two Variables

All points on a graph of a line represent *solutions* of the equation of that line.

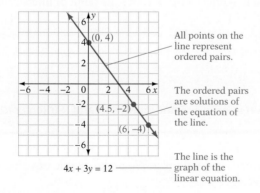

All points on the line represent ordered pairs.

The ordered pairs are solutions of the equation of the line.

$4x + 3y = 12$ — The line is the graph of the linear equation.

In the graph of $4x + 3y = 12$, the points $(0, 4)$ and $(6, -4)$ are solutions of this equation and determine its graph.

Exactly two points determine the graph of a linear equation.

STUDY HINT

In all cases so far, we have used integer values in graphing. This was simply to make graphing easier. In fact, non-integer values may also be used, as we will see later in this book.

However, the line consists of an infinite number of points. One such point is the ordered pair (4.5, −2).

$$4x + 3y = 12 \quad \text{Original equation}$$
$$4(4.5) + 3(-2) = 12 \quad \text{Substituting}$$
$$18 + (-6) = 12 \quad \text{Simplifying}$$
$$12 = 12 \quad \text{True}$$

Because substituting these values results in a true statement, this ordered pair is a solution of the equation. Inspection will show also that the point (4.5, −2) lies on the graph of the line.

To graph a linear equation in two variables

1. Find two or more solutions of the equation.
2. Graph the solutions (ordered pairs) as points.
3. Draw a straight line through these points.

We will use a **table of values** to list solutions of an equation so that we can graph it.

Finding x- and y-Intercepts

When a graph intersects the x- or y-axis, we call the points of intersection **intercepts**.

When a line intersects the x-axis, its point of intersection (the x-intercept) is $(x, 0)$. When a line intersects the y-axis, its point of intersection (the y-intercept) is $(0, y)$.

■ ■ ■
EXAMPLE 3

Graph the equation $y = 2x + 2$ and find its x- and y-intercepts.

SOLUTION

The equation is already solved for y. We make a table of values using the x-values −2, 1, and 2. We get

x	y	
−2	−2	$y = 2(-2) + 2 = -2$
1	4	$y + 2(1) + 2 = 4$
2	6	$y = 2(2) + 2 = 6$

Next, we graph these three ordered pairs and draw a line through them, obtaining

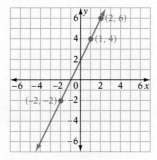

The line appears to intersect the x-axis at $(-1, 0)$, and it appears to intersect the y-axis at $(0, 2)$. We substitute these points into the original equation to verify that they are on the line.

$$y = 2x + 2$$

Substituting $(-1, 0)$:

$$0 = 2(-1) + 2$$
$$0 = -2 + 2$$
$$0 = 0 \qquad \text{True}$$

So $(-1, 0)$ is the x-intercept.

Substituting $(0, 2)$:

$$2 = 2(0) + 2$$
$$2 = 0 + 2$$
$$2 = 2 \qquad \text{True}$$

So $(0, 2)$ is the y-intercept.

▶ *CHECK* **Warm-Up 3**

We also have a way to find the intercepts algebraically, without having to graph the equation:

To find the x-intercept of the graph of an equation

Set $y = 0$ and solve the equation for x. The x-intercept is of the form $(a, 0)$.

To find the y-intercept of the graph of an equation

Set $x = 0$ and solve the equation for y. The y-intercept is of the form $(0, b)$.

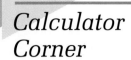

Calculator Corner

Example 3 can incorporate the graphing calculator's **TABLE, GRAPH**ing, and **Home Screen** utilities as follows. (Note: See the Calculator Corners on pages 189 and 191 to review how to create a TABLE.)

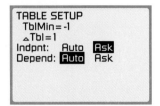

Many graphing calculators have a special feature that enables you to have **DECIMAL** numbers when you graph a function and then want to **TRACE** on that function. Consult your calculator's manual to see how to do this on your calculator. On the *TI-82* the steps are as follows.

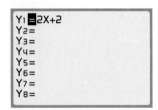
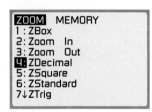
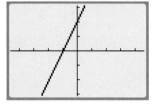

Now **TRACE** on the graph to find the x- and y-intercepts. The next two screens show that the y-intercept is $(0, 2)$ and the x-intercept is $(-1, 0)$.

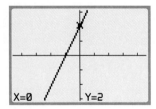
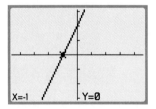

You can use the Home Screen to check to see if your answers are correct. Press **2nd QUIT** to go from the **GRAPH** to the **Home Screen.**

First, let $x = -1$ and see if $y = 0$.   Now let $x = 0$ and see if $y = 2$.

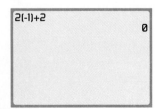
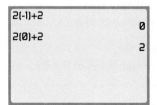

···

EXAMPLE 4

Find the x- and y-intercepts of the graph of the equation $2x + 5y = 9$ algebraically. Then use them to graph the equation.

SOLUTION

To find the x-intercept, we substitute 0 for y and solve.

$$2x + 5y = 9$$
$$2x + 5(0) = 9$$
$$2x = 9 \qquad \text{Because } 5(0) = 0$$
$$x = 4\tfrac{1}{2} \qquad \text{Because } \left(\tfrac{1}{2}\right)(9) = 4\tfrac{1}{2}$$

To find the y-intercept, we substitute 0 for x and solve.

$$2x + 5y = 9$$
$$2(0) + 5y = 9$$
$$5y = 9 \qquad \text{Because } 2(0) = 0$$
$$y = \tfrac{9}{5} = 1\tfrac{4}{5} \qquad \text{Because } \left(\tfrac{1}{5}\right)9 = 1\tfrac{4}{5}$$

So the x- and y-intercepts are $\left(4\tfrac{1}{2}, 0\right)$ and $\left(0, 1\tfrac{4}{5}\right)$, respectively.

We check our work by finding a third point on the line. We can substitute any value for x or y. Say we let $x = 2$. We get

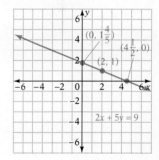

$$2x + 5y = 9$$
$$2(2) + 5y = 9$$
$$5y = 5 \qquad \text{Adding } -4 \text{ to both sides.}$$
$$y = 1 \qquad \text{Multiplying by } \tfrac{1}{5}$$

So $(2, 1)$ is a point on the line. We show the graph of $(2, 1)$ and the two intercepts in the accompanying figure. The line representing $2x + 5y = 9$ passes through all three points.

▶ *CHECK* **Warm-Up 4**

Vertical and Horizontal Lines

If a linear equation contains only one variable, its graph on the Cartesian coordinate system is either a vertical line or a horizontal line.

> Let a and b be constants. Then the graph of the equation $x = a$ is a **vertical line** that crosses the x-axis at the point $(a, 0)$.
>
> The graph of the equation $y = b$ is a **horizontal line** that crosses the y-axis at the point $(0, b)$.

...

EXAMPLE 5

Graph $x = 5$ and $y + 3 = 0$.

SOLUTION

a. The equation $x = 5$ tells us that x is 5 no matter what y is. Here is a possible table of values, along with the graph of the equation.

x	y
5	−2
5	0
5	1

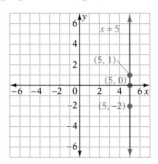

The graph is a vertical line 5 units to the right of the origin. Its x-intercept is (5, 0); it has no y-intercept.

b. By adding −3 to both sides of the equation $y + 3 = 0$, we get $y = -3$. This equation tells us that $y = -3$ no matter what x is. Its graph is a horizontal line 3 units below the origin. Its y-intercept is (0, −3); it has no x-intercept.

x	y
−6	−3
0	−3
2	−3

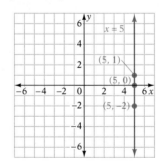

► CHECK Warm-Up 5

Practice what you learned.

SECTION FOLLOW-UP

A good way to represent the taxi data presented in the Section Lead-In is to graph it as a line.

miles	1	2	3	...	10
cost	3.25	4.50	5.75		14.50

a. Plot the points.

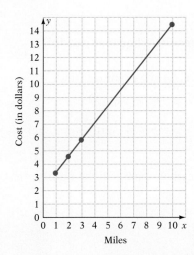

b. Comment on the information shown on this graph. Does this accurately portray the taxi data? Justify your answer.

4.1 WARM-UPS

Work these problems before you attempt the exercises.

1. Graph the points $(6, 3)$, $(2, -3)$, $(-4, 2)$, and $(-3, -2)$.

2. Find the coordinates of points A, B, C, and D.

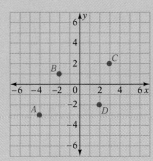

3. Graph $y - 3x = 6$ and find its x- and y-intercepts.

4. Find the x- and y-intercepts of $3x + 4y = 9$ algebraically.

5. Graph $y = 4$.

4.1 EXERCISES

Note: Use your graphing calculator to check your results whenever possible.

In Exercises 1 through 8, identify the quadrant in which each point is found.

1. $(-3, -4)$ **2.** $(7, -2)$ **3.** $(-5, 12)$ **4.** $(1, 8)$

5. $(-8, -5)$ **6.** $(-1, 2)$ **7.** $(1, -7)$ **8.** $(-2, -6)$

In Exercises 9 through 12, find the coordinates of the given points in the figure.

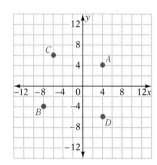

9. Point A **10.** Point B

11. Point C **12.** Point D

In Exercises 13 through 16, find a point indicated on the line in the figure that meets the conditions given. Write your answer as an ordered pair.

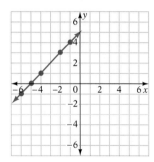

13. The x-coordinate is -5.

14. The y-coordinate is 4.

15. The x-coordinate is -6.

16. The y-coordinate is 5 more than the x-coordinate.

In Exercises 17 through 24, find a point that will complete the figure named.

17. Right triangle

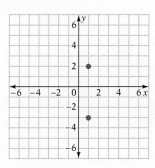

18. Rectangle

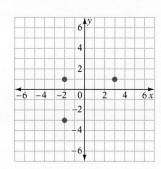

19. Square

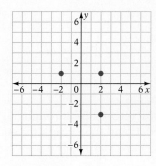

20. Parallelogram

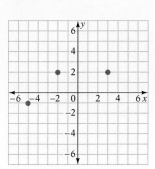

21. Rhombus

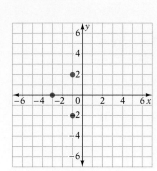

22. Hexagon

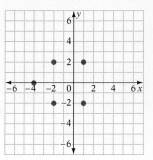

23. Octagon

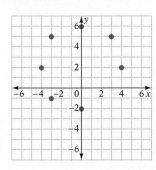

24. Pentagon

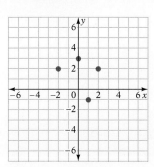

In Exercises 25 through 36, graph the equations by constructing a table of values with the *x*-coordinates that are indicated.

25. $5x - y = 4$
Let $x = 1, 0, 2$

26. $5x + y = 5$
Let $x = 1, 0, 2$

27. $y - 4x = 4$
Let $x = 0, -1, -2$

28. $7y + 7x = 0$
Let $x = 1, 0, -1$

29. $5x = 3 - 5y$
Let $x = 1, 0, -1$

30. $y + 5 = x$
Let $x = 1, 0, 5$

31. $2y = 6x$
Let $x = 0, 1, -1$

32. $6y = 6x - 6$
Let $x = 0, 1, -1$

33. $2x + 2 = 2y$
Let $x = 0, 1, -1$

34. $6y + 2 = 4x$
Let $x = 0, 1, -1$

35. $-5y + 10 = x$
Let $x = 0, 5, -5$

36. $8 = 3y - x$
Let $x = 0, 1, -2$

In Exercises 37 through 56, find the x- and y-intercepts algebraically (if they exist) and then graph the equation.

37. $x - 5y = 4$

38. $2x + 2y = 5$

39. $4y - 4x = 10$

40. $6y + x = 2$

41. $-5x = 3 - 5y$

42. $12y + 15 = 12x$

43. $3y = 12x$

44. $y = 3x - 6$

45. $2x + 6 = 2y$

46. $8y + 12 = 4x$

47. $x = 3$

48. $y = -2$

49. $x - 2y = 1$

50. $y + x = 1$

51. $x = -1$

52. $3y = 15$

53. $y = -4$

54. $x = -6$

55. $2x + 3y = 7$

56. $-x + 4y = 8$

57. *Weather* The distance d in miles between you and a lightning strike can be estimated by the relationship $d = \frac{1}{5}t$, where t is the number of seconds it takes for you to hear thunder after seeing the lightning.

a. Graph this equation, and use your graph to estimate how long it will take before you hear the thunder when lightning strikes three miles away.

b. If the time between lightning and thunder is 30 seconds, how far away did the lightning strike?

58. *Commuting* A commuter train ride costs $2.00 plus 10 cents for each mile. The relationship between cost and distance is thus $y = \frac{1}{10}x + 2$, where y is the total cost in dollars and x is the distance in miles.

a. Graph this equation, and use the graph to find the cost of a 10-mile trip.

b. If a trip costs $4, what distance did you go?

EXCURSIONS

Data Analysis

1. How have Olympic Marathon winning times changed from 1896 to 1996? Graph the data from the following table. Ask four questions and answer them using your data.

Olympic Marathon

Year	Winner	Time	Year	Winner	Time
1896	Spiridon Loues, Greece	2h58m50s	1952	Emil Zatopek, Czechoslovakia	2h23m3.2s
1900	Michel Teato, France	2h59m45s	1956	Alain Mimoun, France	2h25m
1904	Thomas Hicks, United States	3h28m53s	1960	Abebe Bikila, Ethiopia	2h15m16.2s
1906	William J. Sherring, Canada	2h51m23.65s	1964	Abebe Bikila, Ethiopia	2h12m11.2s
1908	John J. Hayes, United States	2h55m18.4s	1968	Mamo Wold, Ethiopia	2h20m26.4s
1912	Kenneth McArthur, South Africa	2h36m54.8s	1972	Frank Shorter, United States	2h12m19.8s
1920	Hannes Kolehmainen, Finland	2h32m35.8s	1976	Walter Cierpinski, East Germany	2h9m55s
1924	Albin Stenroos, Finland	2h41m22.6s	1980	Walter Cierpinski, East Germany	2h11m3s
1928	A.B. El Quafi, France	2h32m57s	1984	Carlos Lopes, Portugal	2h9m55s
1932	Juan Zabala, Argentina	2h31m36s	1988	Gelindo Bordin, Italy	2h10m47s
1936	Kitei Son, Japan	2h29m19.2s	1992	Hwang Young-Cho, South Korea	2h13m23s
1948	Delfo Cabrera, Argentina	2h34m51.6s	1996	Josia Thugwane, South Africa	2h12m36s

Source: 1997 Information Please® Almanac (©1995 Houghton Mifflin Co.), pp. 893–894. All rights reserved. Used with permission by Information Please LLC.

SECTION GOALS

▪ *To find the slope of a graph of a line*

▪ *To determine whether points are collinear*

4.2   Finding the Slope of a Line

SECTION LEAD-IN

In the taxi problem presented in the Section 4.1 Lead-In, our table of taxi costs can be thought of as a table of values.

x	y
1	3.25
2	4.5
3	5.75
10	14.5

The graph we made is a broken-line graph—we simply connected the four points with straight line segments, or pieces. The graph is also the graph of one line—a linear graph.

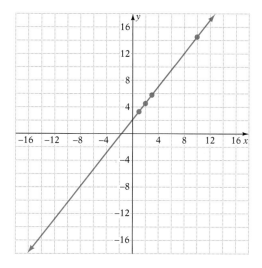

Note that when we graph this as a line, we use all four quadrants and place arrows at each end. This indicates that we used the four points we were given to determine the line but many other points are on this line—an infinite number. Does this graph correctly represent our taxi data? Justify your answer.

WRITER'S BLOCK

Sometimes we refer to the formula of the slope as "rise over run." Investigate the graphs and explain this interpretation.

Introduction

The following figure on the left shows a line and two points on the line, (0, 0) and (4, 2). To move from (0, 0) to (4, 2), we must move vertically up 2 units on the graph and then horizontally to the right 4 units. We say that the *slope* of this line is

$$\text{Slope} = \frac{\text{change in } y}{\text{change in } x} = \frac{2 \text{ units (up)}}{4 \text{ units (right)}} = \frac{2}{4} = \frac{1}{2}$$

The **slope** of a line is a measure of its steepness.

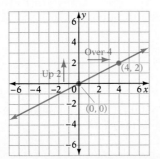
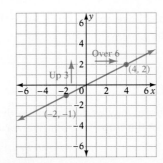

The figure at the right above shows the same line with another point, $(-2, -1)$, marked on it. To move from $(-2, -1)$ to $(4, 2)$, we would have to move up 3 units and to the right 6 units. The slope of the line is

$$\text{Slope} = \frac{\text{change in } y}{\text{change in } x} = \frac{3 \text{ units (up)}}{6 \text{ units (right)}} = \frac{3}{6} = \frac{1}{2}$$

which is the same slope we calculated before. A line has only one slope, and any two points on the line can be used to find it.

The Slope Formula

If we know the coordinates of any two points on a line, we can find the slope of the line with the following formula:

> **Slope of a Line**
>
> Let (x_1, y_1) and (x_2, y_2) be any two points on a line. Then the slope m of the line is
>
> $$m = \frac{\text{change in } y}{\text{change in } x} = \frac{y_2 - y_1}{x_2 - x_1}$$
>
> provided that x_2 is not equal to x_1. If $x_2 = x_1$, the slope is undefined.

■■■

EXAMPLE 1

Find the slope of the line that contains the points $(-1, -3)$ and $(2, -1)$.

SOLUTION

We let $(-1, -3)$ be the point (x_1, y_1), and we let $(2, -1)$ be the point (x_2, y_2). Then $x_1 = -1$, $y_1 = -3$, $x_2 = 2$, and $y_2 = -1$. So

$$m = \frac{y_2 - y_1}{x_2 - x_1} = \frac{-1 - (-3)}{2 - (-1)} = \frac{-1 + 3}{2 + 1} = \frac{2}{3}$$

So the slope is $\frac{2}{3}$.

 CHECK **Warm-Up 1**

In Example 1 we might have let $(-1, -3)$ be the point (x_2, y_2) and $(2, -1)$ be (x_1, y_1). Then we would have had

$$m = \frac{y_2 - y_1}{x_2 - x_1} = \frac{-3 - (-1)}{-1 - 2} = \frac{-3 + 1}{-3} = \frac{-2}{-3} = \frac{2}{3}$$

Either point may be designated as the first point (x_1, y_1) or the second point (x_2, y_2) in the slope formula. The result will be the same.

Calculator Corner

Use the Home Screen to confirm the numerical slope of the line containing the points $(-1, -3)$ and $(2, -1)$ from Example 1. Many graphing calculators can give results in fractional form, as shown at the right. (Note: See the Calculator Corner on pages 6 and 7 to review how to get an answer in fractional form.)

It is also possible to enter a formula on the Home Screen and use it repeatedly without having to enter it on the Home Screen each time. For example, rewrite the formula for slope as $m = \frac{D - B}{C - A}$ and the two points as $(A, B) = (-1, -3)$ and $(C, D) = (2, -1)$.

STORE -1 into A, -3 into B, 2 into C, and -1 into D. (Note: See the Calculator Corner on page 85 to review how to store values for variables.)

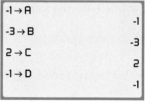

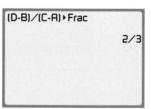

Then type in the new equation for slope: $\frac{D - B}{C - A}$.

Now check your numerical work for Example 1 using your graphing calculator. You can press **2nd ENTRY** repeatedly and use your graphing calculator's capability to scroll backwards to recall lines you have entered on the Home Screen.

```
3→A
              3
4→B
              4
6→C
              6
-2→D
             -2
```

Now press **2nd ENTRY** five times to recall the equation for slope.

```
6→C
              6
-2→D
             -2
(D-B)/(C-A)▸Frac
             -2
```

Positive and Negative Slope

When we describe the graph of a line, we often give its slope.

■ ■ ■
EXAMPLE 2

Find the slope of the line that passes through the points (3, 4) and (6, 2).

SOLUTION

Let (3, 4) be the point (x_1, y_1) and let (6, −2) be the point (x_2, y_2).

Then

$$m = \frac{y_2 - y_1}{x_2 - x_1} = \frac{(-2) - 4}{6 - 3} = \frac{-6}{3} = -2$$

▶ *CHECK* **Warm-Up 2**

The line in Example 2 slants down from left to right and has a negative slope. Sketch it to verify this statement.

A line that slants downward from left to right has a **negative slope.**

Now sketch the line in Example 1. It slants upward from left to right. Its slope is positive.

A line that slants upward from left to right has a **positive slope.**

Slopes of Horizontal and Vertical Lines

The slopes of horizontal and vertical lines can be found with the same formula we use for all other lines.

■ ■ ■
EXAMPLE 3

Find the slope of the line that passes through the points (−3, 2) and (0, 2).

SOLUTION

The points and the line that passes through them are shown here. The line is horizontal.

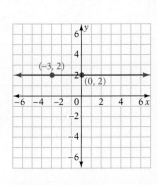

Let (−3, 2) be the point (x_1, y_1) and let (0, 2) be the point (x_2, y_2). Then

$$m = \frac{y_2 - y_1}{x_2 - x_1} = \frac{2 - 2}{0 - (-3)} = \frac{0}{3} = 0$$

So the slope is 0.

Note that $y_1 = y_2$ for all points (x_1, y_1) and (x_2, y_2) on a horizontal line.

▶ CHECK **Warm-Up 3**

Example 3 shows us that

> Horizontal lines have a slope of zero.

■ ■ ■
EXAMPLE 4

Find the slope of the line that contains the points $(5, 2)$ and $(5\ {-}3)$.

SOLUTION

The points and the line are shown here; the line is vertical.

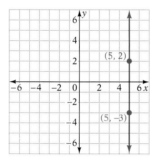

Let $(5, 2)$ be (x_1, y_1) and let $(5, -3)$ be (x_2, y_2). Then

$$m = \frac{y_2 - y_1}{x_2 - x_1} = \frac{-3 - 2}{5 - 5} = \frac{-5}{0}$$

Because division by zero is not defined for the real numbers, we say that the slope of this line is undefined.

Note that $x_1 = x_2$ for all points (x_1, y_1) and (x_2, y_2) on a vertical line.

▶ CHECK **Warm-Up 4**

> Vertical lines have a slope that is undefined.

Collinear Points

Collinear points are points that lie on the same line. Given three or more points, you can determine whether these points lie on the same line without drawing the graph. This is done by finding the slope of the line between any two pairs of these points. If the slopes are the same, the points are collinear. Look at the next example.

▪▪▪
EXAMPLE 5

Given the points (1, 2), (3, 5), and (5, 6), determine whether these three points lie on the same line.

SOLUTION

We first find the slope of the line that passes through (1, 2) and (3, 5).

$$m = \frac{y_2 - y_1}{x_2 - x_1} = \frac{5 - 2}{3 - 1} = \frac{3}{2}$$

Then we find the slope of the line that passes through (3, 5) and (5, 6).

$$m = \frac{y_2 - y_1}{x_2 - x_1} = \frac{6 - 5}{5 - 3} = \frac{1}{2}$$

Because the slopes are *not* the same, the three points do *not* lie on the same line. We can check this by graphing the three points.

▶CHECK **Warm-Up 5**

Often, we can use the slope of a line to work with applications.

▪▪▪
EXAMPLE 6

Can (0, 0), (1, 1), and (−1, −1) be the vertices of a triangle?

SOLUTION

Either three points are collinear or they could be the vertices of a triangle. We can tell which by finding the slopes of the lines between any two pairs. If the slopes are different, the points form a triangle.

For (1, 1) and (0, 0):   For (1, 1) and (−1, −1):

$$m = \frac{0 - 1}{0 - 1} = 1 \qquad\qquad m = \frac{1 - (-1)}{1 - (-1)} = 1$$

The slopes are the same. The lines do not form a triangle.

▶CHECK **Warm-Up 6**

You can use a graphing calculator to graph lines and to check your work.

Practice what you learned.

SECTION FOLLOW-UP

a. What is the slope of the line representing the taxi data shown in the Section Lead-In?

b. Use the slope to prove that this "line" is, in fact, a straight line?*

Hint: Check that the slopes of the lines between each pair of points are the same.

4.2 WARM-UPS

Work these problems before you attempt the exercises.

1. Draw the line containing the points $(-2, 1)$ and $(1, -2)$ and find its slope.

2. Find the slope of the line shown in the figure.

3. Find the slope of the line that contains the points $(-1, -2)$ and $(-3, -2)$.

4. Find the slope of the line that contains the points $(3, 6)$ and $(3, -2)$.

5. Determine whether the points $(3, 5)$, $(2, 6)$, and $(-6, 2)$ lie on the same line by finding the slopes of the lines they determine.

6. Is the figure formed by the points $(1, 0)$, $(-1, 2)$, $(0, 1)$, and $(-5, -3)$ a triangle or a rectangle?

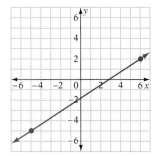

Figure for 2

4.2 EXERCISES

Note: Use your graphing calculator to check your results whenever possible.

In Exercises 1 through 16, find the slope of the line that passes through the given points.

1. $(2, 0)$ and $(-3, 5)$
2. $(3, 4)$ and $(4, 3)$
3. $(-2, 1)$ and $(2, 5)$
4. $(-1, 5)$ and $(-6, 3)$

5. $(0, 0)$ and $(5, 3)$
6. $(0, 2)$ and $(5, 0)$
7. $(6, -2)$ and $(16, -2)$
8. $(-1, 1)$ and $(0, 1)$

9. $(0, 4)$ and $(2, 6)$
10. $(5, 1)$ and $(6, -4)$
11. $(-8, -3)$ and $(-7, 2)$
12. $(2, 5)$ and $(2, -1)$

13. $(2, 9)$ and $(2, 5)$
14. $(-3, 5)$ and $(5, 1)$
15. $(1, -3)$ and $(-2, -1)$
16. $(7, -1)$ and $(-6, 2)$

In Exercises 17 through 24, you are given the slope of a line and a point on the line. Find a second point on the line. Answers may may.

17. slope $= \frac{1}{3}$ point: $(1, 2)$
18. slope $= \frac{1}{4}$ point: $(-6, 4)$
19. slope $= -3$ point: $(5, -2)$

20. slope $= -\frac{3}{8}$ point: $(1, -1)$
21. slope $= -2$ point: $(0, -1)$
22. slope $= -6$ point: $(0, 1)$

23. slope $= \frac{1}{2}$ point: $(-1, 3)$
24. slope $= -1$ point: $(5, 9)$

In Exercises 25 through 28, find the slope of the line described.

25. Find the slope of the *x*-axis.

26. Find the slope of the *y*-axis.

27. Find the slope of the line that passes through the origin and the point $(-2, 3)$.

28. Find the slope of the line that passes through the point $(-1, -6)$ and the origin.

In Exercises 29 through 32, determine whether the points are collinear.

29. $(3, 2), (5, 1), (1, 3)$

30. $(0, 0), (5, 5), (-1, -3)$

31. $(-1, -2), (-3, -4), (-6, -8)$

32. $(-2, 4), (-6, 2), (-12, 4)$

33. *Forestry*   Trees are to be planted in a row. Their positions are recorded on a grid, and two of them are located at $(0, -3)$ and $(2, -1)$. If the third tree is planted at $(7, 6)$, will this tree be aligned (on the same line) with the other two?

34. *Geometry*   Verify that the figure formed with the coordinates $(2, 5)$, $(4, 8), (7, 6)$, and $(5, 3)$ is a rectangle.

35. *Geometry*   Show that $(0, -2), (4, 5), (3, 1)$, and $(1, 2)$ are not the vertices of a rectangle.

36. *Geometry*   Are $(-2, -5), (1, -2)$, and $(11, 8)$ the vertices of a triangle? How do you know?

37. Use your graphing calculator to determine whether the graph of each of the following equations has a negative or a positive slope.

   a. $y = -5x + 4$ **b.** $y = 3x$ **c.** $3y = 4$ **d.** $2y - 8x = 5$

38. Use your graphing calculator to determine whether the graph of each of the following equations has a negative or a positive slope.

   a. $2y = 3 - 5x$ **b.** $2y = -2x$ **c.** $6 = 2y$ **d.** $0 = 4x + 2y - 12$

39. Using the **TRACE** function on your graphing calculator, find five points on the graph of $y = 3x + 2$.

40. Using the **TRACE** function on your graphing calculator, find three points on the graph of $2y = x + 3$.

MIXED PRACTICE

By doing these exercises, you will practice the topics up to this point in the chapter.

41. Find the missing coordinate of the point $(x, -3)$ that lies on the graph of $2y = 3x - 4$.

42. Find *x* when $y = 1.2$ in $2x - 5y = 9$.

43. Find the *x*- and *y*-intercepts of $6x - 4y = 11$ algebraically.

44. For $3x = 5$, a linear equation, what is the value of *x* when $y = 29$? Explain.

45. Is $(-3, -4)$ a solution of the equation $4x - 3y = 0$?

46. Find the slope of the line that passes through the points $(3, -2)$ and $(-4, 1)$.

47. Graph $3x - 2y = 18$ and determine the slope.

48. Find the x- and y-intercepts of $-3x - 2y = -6$ algebraically, and then graph the line determined by these two intercepts.

49. Graph $3x - 4y = 2$ using a table of values with $x = -4, 2$, and -6.

50. Graph $10x - 5y = 20$ using a table of values.

51. Is $(-3, -2)$ a solution of the equation $4x - 6y = 8$?

52. When $3y = 8$, what does x equal?

53. What does x equal in the equation $x - y = 8$ when $y = 2\frac{1}{3}$?

54. Find the x- and y-intercepts of $-13y - 7x = 5$ algebraically.

EXCURSIONS

Data Analysis

1. An equation often given for your optimum target heart rate is

$$\text{target heart rate} = 70\%(220 - \text{your age})$$

Other people have argued that any level from 70% to 85% is acceptable.

Draw a graph showing the target heart rate for exercising adults for maximum health benefits.

a. Use what you know about linear equations to write an equation that describes the upper level of the target zone (at the 85% level).

b. Estimate the equation for an 80% level.

c. Using your graph, discuss the information presented. That is, analyze what data is given and describe it to a friend.

d. Ask and answer four questions about this data.

2. ✏ Write a newspaper article discussing the data presented in the following two graphs.

Chart 1. Civilian employment and unemployment, January-December 1993

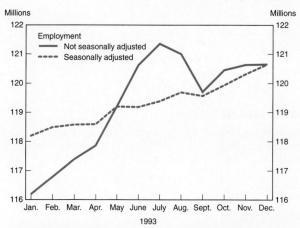
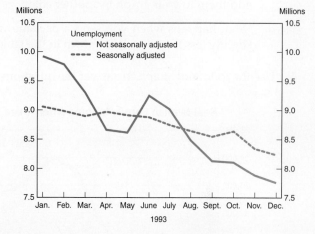

Source: Bureau of Labor Statistics, Report 864, March 1994.

3. Study the following postage tables and then answer the questions.

Airmail Rates
All Countries except Canada & Mexico

Weight not over (oz)	Cost	Weight not over (oz)	Cost
0.5	$ 0.60	9.0	$ 7.40
1.0	1.00	9.5	7.80
1.5	1.40	10.0	8.20
2.0	1.80	10.5	8.60
2.5	2.20	11.0	9.00
3.0	2.60	11.5	9.40
3.5	3.00	12.0	9.80
4.0	3.40	12.5	10.20
4.5	3.80	13.0	10.60
5.0	4.20	13.5	11.00
5.5	4.60	14.0	11.40
6.0	5.00	14.5	11.80
6.5	5.40	15.0	12.20
7.0	5.80	15.5	12.60
7.5	6.20	16.0	13.00
8.0	6.60	16.5	13.40
8.5	7.00		

Airmail Rates
Canada and Mexico

Weight not over (lbs)	(oz)	Cost Canada	Mexico
0	0.5	$.48	$.40
0	1	.52	.48
0	1.5	.64	.66
0	2	.72	.86
0	3	.95	1.26
0	4	1.14	1.66
0	5	1.33	2.06
0	6	1.52	2.46
0	7	1.71	2.86
0	8	1.90	3.26
0	9	2.09	3.66
0	10	2.28	4.06
0	11	2.47	4.46
0	12	2.66	4.86
1	0	3.42	6.46
1	8	4.30	9.66

Source: 1996 Information Please® Almanac (©1995 Houghton Mifflin Co.), p. 1024. All rights reserved. Used with permission by Information Please LLC.

a. On the same axes, graph the airmail rates for Canada, Mexico, and all other countries.

b. Ask and answer four questions about your graphs. Use line graphs.

c. **Research** Are these prices current? If not, find the new ones and add them to your graph in another color.

d. What happens when you send a package weighing slightly more or slightly less than a weight shown in the table?

4. Use the following graph to answer the questions.

Real Median Family Income After Taxes, 1981 to 1994 ($)

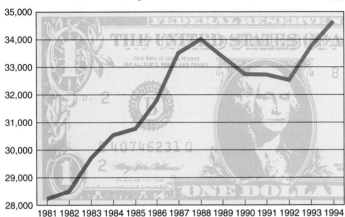

Source: *1996 Information Please® Business Almanac* (©1995 Houghton Mifflin Co.), p. 102. All rights reserved. Used with permission by Information Please LLC.

a. Calculate the slopes of the line segments for each year to find between what two years the income changed most rapidly.

b. ✏ Explain why we can use the method in part (a).

c. What information will you get if you calculate the change for every two years?

d. Who might be interested in information from this graph? Tell them four noteworthy facts about this data.

CONNECTIONS TO *STATISTICS*

Broken-Line Graphs

Reading Broken-Line Graphs

In some graphs, points are plotted and connected by lines to present information. The following figure is such a **broken-line graph.** The position of each dot represents the average price of tea, in cents per pound, for the year shown directly beneath it. The dots are connected by lines so that readers can see more clearly the changes, up and down, in the data.

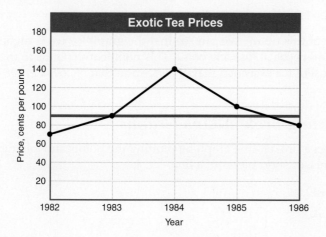

■ ■ ■

EXAMPLE

In what year did tea first average more than 90 cents per pound?

SOLUTION

We locate 90 cents on the price scale at the left. A horizontal line at 90 cents crosses the graph just about at the 1983 mark. Between 1983 and 1984, the graph rises higher than 90 cents.

So tea first cost more than 90 cents a pound in 1983.

We can sometimes use a broken-line graph to represent two or more sets of data. The next example shows such a use.

■ ■ ■

EXAMPLE

This next graph shows the average prices in cents per pound of imported coffee, cocoa, and tea from 1982 to 1986.

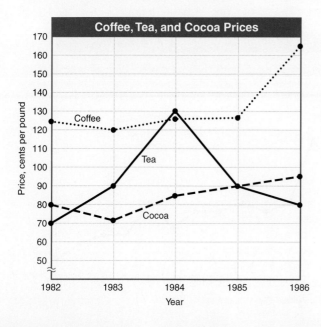

In what year did the price of tea exceed that of coffee and cocoa?

SOLUTION

As we look at this figure, we can see that the graph for tea is higher than that for coffee and cocoa at just one point. This point corresponds to the year 1984. Thus, the cost of tea exceeded that of coffee and that of cocoa in 1984.

◢

Constructing Broken-Line Graphs

We construct a broken-line graph by plotting the points and joining them with line segments, or short pieces of a line.

PRACTICE

Use the Coffee, Tea, and Cocoa Prices graph to answer Exercises 1 through 6.

1. What is the difference in cost between the the most expensive and the least expensive of these three goods in 1986?

2. In what year were the costs of cocoa and tea the same?

3. In 1984 what was the difference in the prices of coffee and cocoa?

4. In 1985 what would you have paid altogether for a pound of cocoa and a pound of tea?

5. If you had bought a pound each of coffee, cocoa, and tea in 1986, about what would you have paid altogether (to the nearest 10 cents)?

6. Ask and answer a question about this graph.

The following graph shows the winning times for the men's Olympic 800-meter run for the years 1896 through 1960. Use this graph to answer Exercises 7 through 10.

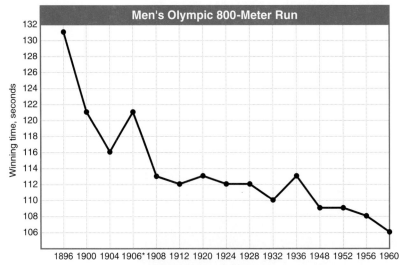

* Interim games in Athens but *not* official Olympics games.

7. What is the difference between the longest and the shortest record times shown?

8. During which two pairs of succeeding Olympics did the winning time remain the same?

9. Make up three additional questions about the graph and answer them.

10. **Research** Find the winning times for the men's Olympic 800-meter run for the years 1964–1996. Redraw the graph, adding those points.

4.3 Linear Graphs and Their Equations

SECTION LEAD-IN

The line in the taxi problem was graphed using the four points in the table (see the Section 4.2 Lead-In). We could also have used an equation.

The equation

$$\text{cost} = \$2.00 + (\$1.25 \times \text{the number of miles})$$

can be written in terms of an independent variable—a variable we choose (the number of miles we travel)—and a dependent variable—one that depends on the other (the cost). We let

$$x = \text{number of miles traveled}$$
$$y = \text{cost in dollars}$$

The equation is

$$y = 2 + 1.25x$$

We can also write it as

$$y = 1.25x + 2$$

SECTION GOALS

- *To rewrite linear equations to identify the slope and y-intercept*

- *To graph lines using the slope and y-intercept*

- *To determine when two lines are parallel or perpendicular*

- *To write the equation of a line, given the slope and y-intercept*

- *To write the equation of a line, given the slope and a point on the line*

- *To write the equation of a line, given two points on the line*

Slope-Intercept Form of a Linear Equation

The *slope-intercept form of a linear equation* enables us to obtain information about its graph quickly.

Slope-Intercept Form

The **slope-intercept form of a linear equation** is

$$y = mx + b$$

In that form, m is the slope of the line, and $(0, b)$ is the y-intercept.

Finding the Slope and y-Intercept

Any linear equation in two variables can be rewritten in slope-intercept form by solving the equation for y. Then the slope-intercept form immediately tells us the slope and y-intercept of the graph of the equation.

▪ ▪ ▪
EXAMPLE 1

Find the slope and y-intercept of the graph of $3x - 4y = 5$.

SOLUTION

First we put the equation in slope-intercept form. We do this by solving for y.

$$3x - 4y = 5$$
$$3x - 4y + (-3x) = 5 + (-3x) \qquad \text{Adding } -3x \text{ to both sides}$$
$$-4y = -3x + 5 \qquad \text{Simplifying}$$
$$\left(-\tfrac{1}{4}\right)(-4y) = \left(-\tfrac{1}{4}\right)(-3x + 5) \qquad \text{Multiplying both sides by } \tfrac{-1}{4}$$
$$y = \tfrac{3}{4}x - \tfrac{5}{4} \qquad \text{Simplifying}$$

By comparing this result with the general form, we easily read off the slope and y-intercept.

$$y = \tfrac{3}{4}x - \tfrac{5}{4}$$
$$y = mx + b$$

So the slope m is $\tfrac{3}{4}$, and the y-intercept $(0, b)$ is $\left(0, -\tfrac{5}{4}\right)$.

▶ CHECK **Warm-Up 1**

❗❗❗
ERROR ALERT

Identify the error and give a correct answer.

Find the slope of the line whose equation is $-y + 2x = 5$.

Incorrect Solution:

$$-y + 2x = 5$$
$$-y = -2x + 5$$
$$\text{Slope} = -2$$

Graphing with the Slope and y-Intercept

Once we know the slope and y-intercept, we can graph a linear equation easily.

▪ ▪ ▪
EXAMPLE 2

Graph the line that has a slope of -2 and y-intercept $(0, 4)$.

SOLUTION

The y-intercept is a point on the line, so we graph it first. We are told that the slope is -2, but to use it in graphing, we first write it as a fraction, $\tfrac{-2}{1}$. This tells us that for a change in x of $+1$, the change in y is -2 or, for a change in x of -1, the change in y is $+2$. Thus we start at the graphed point $(0, 4)$, and, to find another point on the line, we move

> 1 unit right (in the *positive x* direction), and
>
> 2 units down (in the *negative y* direction)

This gives us the point $(1, 2)$. Finally, we draw the line through the two points.

❗❗❗
ERROR ALERT

Identify the error and give a correct answer.

Find the y-intercept of the graph of the equation $-3y + x = 6$.

Incorrect Solution:

$$-3y + x = 6$$
$$-3(0) + x = 6$$
$$0 + x = 6$$
$$x = 6$$

So the y-intercept is $(6, 0)$

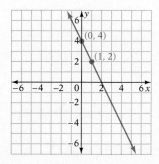

▶ *CHECK* **Warm-Up 2**

▪▪▪

EXAMPLE 3

Graph $-5x + 2y - 3 = 0$ by finding the slope and y-intercept.

SOLUTION

First we must put this equation in slope-intercept form.

$$-5x + 2y - 3 = 0$$
$$-5x + 2y - 3 + (5x) = 0 + (5x) \qquad \text{Adding } 5x \text{ to both sides}$$
$$2y - 3 = 0 + 5x \qquad \text{Simplifying}$$
$$2y - 3 + 3 = 0 + 5x + 3 \qquad \text{Adding 3 to both sides}$$
$$2y = 5x + 3 \qquad \text{Simplifying}$$
$$\left(\tfrac{1}{2}\right)(2y) = \left(\tfrac{1}{2}\right)(5x + 3) \qquad \text{Multiplying both sides by } \tfrac{1}{2}$$
$$y = \tfrac{5}{2}x + \tfrac{3}{2} \qquad \text{Simplifying}$$

By comparing this result with the equation $y = mx + b$, we see that the slope is $\frac{5}{2}$. So for a change in x of 2, the change in y is 5.

We first plot the y-intercept, which is $\left(0, \frac{3}{2}\right)$, or $\left(0, 1\frac{1}{2}\right)$. Next, starting at $\left(0, 1\frac{1}{2}\right)$, we move 2 units to the right and 5 units up to locate a second point on the line. Then we draw the line, as shown in the accompanying figure.

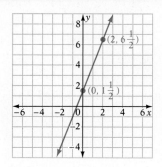

▶ *CHECK* **Warm-Up 3**

Calculator Corner

GRAPH each of the following equations on your graphing calculator. What do these four lines have in common?

GRAPH each of the following equations on your graphing calculator. What do these four lines have in common?

State a conjecture about how the value of m affects the graph of a straight line.

GRAPH each of the following groups of equations on your graphing calculator.

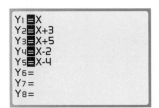   

State a conjecture about how the value of the constant b affects the graph of a straight line.

Parallel and Perpendicular Lines

The slope-intercept form also enables us to determine quickly when two lines are *parallel* or *perpendicular* to each other.

▪▪▪

EXAMPLE 4

Graph $y = 2x - 7$ and $3y = 6x - 9$ on the same coordinate axes.

SOLUTION

We need both equations in slope-intercept form. The first is already written in that form. And multiplying both sides of $3y = 6x - 9$ by $\frac{1}{3}$ yields

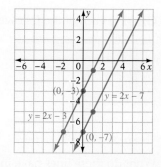

$$y = 2x - 3$$

Placing the two equations together,

$$y = 2x - 7$$
$$y = 2x - 3$$

we find that they have the *same* slope.

Using the common slope and the y-intercepts of the two equations, we graph them. The lines are parallel.

▶ *CHECK* **Warm-Up 4**

The two lines in Example 4 have exactly the same slope but intersect the y-axis at different points. They will *never* intersect each other.

> Two distinct lines are **parallel** if their slopes are equal.

▪▪▪

EXAMPLE 5

Graph $y = \frac{1}{2}x + 3$ and $y = -2x + 5$.

SOLUTION

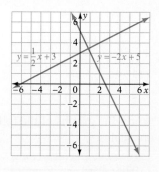

These equations are both already in slope-intercept form. The first has a slope of $\frac{1}{2}$ and a y-intercept of $(0, 3)$ and is graphed in the accompanying figure. The second equation, with a slope of -2 and a y-intercept of $(0, 5)$ is also graphed in the figure. The lines intersect in 90° angles, just as the x- and y-axes do.

▶ *CHECK* **Warm-Up 5**

When two lines intersect in a 90° angle, or right angle, they are called *perpendicular lines*. Their slopes (-2 and $\frac{1}{2}$ in Example 5) are negative reciprocals.

> Two lines are **perpendicular** if their slopes are negative reciprocals of each other—that is, if the product of their slopes is -1.

▪▪▪

WRITER'S BLOCK

The equations

$y = 5x + 6$ and

$y = -5x + 6$

are alike in what ways? What about their graphs?

Calculator Corner

GRAPH each of the following equations on your graphing calculator.

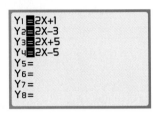

What do these lines have in common? State a conjecture about lines having the same slope.

Study the graphs of the following two equations on grids that have two different scales.

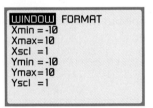

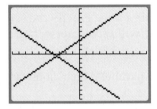

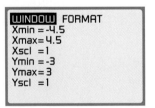

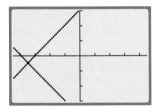

Study the graphs of the following two equations on grids that have two different scales.

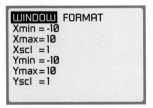

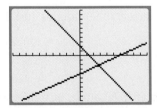

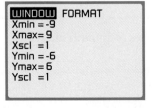

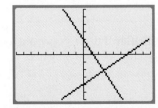

The second graph is the most "visually correct" representation of these perpendicular lines. Why is this true? State a conjecture about the slope of perpendicular lines.

Give the equations of two perpendicular lines and graph them on a grid that shows them to be perpendicular.

▪▪▪

EXAMPLE 6

Which of the following equations have graphs that are parallel to each other or are perpendicular to each other? Use only the equations themselves to answer. Do not graph them.

a. $-2y = 7 + 3x$ **b.** $y + 12 = \frac{2}{3}x$

c. $y = \frac{-3}{2}x + 4$ **d.** $y = \frac{-2}{3}x + 4$

SOLUTION

The four equations, in slope-intercept form, are

a. $y = \frac{-3}{2}x - \frac{7}{2}$ **b.** $y = \frac{2}{3}x - 12$

c. $y = \frac{-3}{2}x + 4$ **d.** $y = \frac{-2}{3}x + 4$

Those that have parallel graphs must have the same x-coefficient, and that is true only of equations (a) and (c). So the graphs of (a) and (c) are parallel.

Equations (a) and (b) and equations (c) and (b) have x-coefficients that are negative reciprocals because

$$\left(-\frac{3}{2}\right)\left(\frac{2}{3}\right) = -1$$

So equations (a) and (b) have graphs that are perpendicular to each other, as do equations (c) and (b).

▸ *CHECK* **Warm-Up 6**

▪▪▪

WRITER'S BLOCK

Give examples in your environment that have parallel lines and perpendicular lines.

Writing the Equation of a Line

"The equation of a line" really means "the equation whose graph is that line." We can write the equation of any line if we have certain information about the line. For example, if we know the slope and y-intercept of a line, we can use the slope-intercept form to write its equation.

Writing the Slope-Intercept Form

> **To write an equation of a line given the slope *(m)* and *y*-intercept (0, *b*)**
>
> **1.** Substitute the values for m and b into the slope-intercept form,
> $$y = mx + b$$
> **2.** Simplify the resulting equation.

▪▪▪

EXAMPLE 7

Write the equation of the line that has slope 5 and y-intercept $(0, -3)$.

SOLUTION

We have $m = 5$ and $b = -3$. Substituting into the slope-intercept form gives us

$$y = mx + b$$
$$y = (5)x + (-3)$$

Simplifying yields

$$y = 5x - 3$$

So the equation of the line is $y = 5x - 3$.

▶ *CHECK* **Warm-Up 7**

In Example 7 we knew the slope and the y-intercept. But we can also write the equation of a line if we know the slope and *any* point on the line.

To write an equation of a line given the slope m and a point (x, y) on the line

1. Substitute the value for m and the coordinates of the point (x, y) into the slope-intercept form, $y = mx + b$.
2. Solve for b.
3. Substitute the values for m and b in the slope-intercept form.
4. Simplify the resulting equation.

▪▪▪

EXAMPLE 8

Find the equation of the line that has slope 4 and passes through the point $(2, 3)$.

SOLUTION

We have $m = 4$ and $(x, y) = (2, 3)$. Substituting these into the slope-intercept form gives us

$$y = mx + b$$
$$3 = 4(2) + b$$

Solving for b, we obtain

$3 = 8 + b$	Simplifying
$3 + (-8) = 8 + b + (-8)$	Adding -8 to both sides
$-5 = b$	Simplifying

Now we know that the slope is 4 and that the y-intercept is $(0, -5)$. Substituting m and b into the slope-intercept form yields

$y = mx + b$	
$y = (4)x + (-5)$	Substituting
$y = 4x - 5$	Simplifying

! ! !
ERROR ALERT

Identify the error and give a correct answer.

Find the equation of the line that has slope 4 and passes through the point $(2, 3)$.

Incorrect Solution:

$$y = mx + b$$
$$3 = 4(2) + b$$
$$3 = 8 + b$$
$$-5 = b$$

Equation: $3 = 4(2) - 5$

So the equation of the line is $y = 4x - 5$.

▶ *CHECK* **Warm-Up 8**

We can also use the slope-intercept form to write an equation when we know two points that are on the line. But first, we must use the two given points to find the slope.

To write an equation of a line given two points (x_1, y_1) and (x_2, y_2) on the line

1. Find the slope.

$$m = \frac{y_2 - y_1}{x_2 - x_1}$$

2. Substitute the slope and the coordinates of either point into the slope-intercept form, $y = mx + b$.
3. Solve for b.
4. Substitute the values for m and b into the slope-intercept form.
5. Simplify the resulting equation.

▪▪▪

EXAMPLE 9

Find the equation of the line that passes through the points $(4, 3)$ and $(-1, -2)$.

SOLUTION

Because we do not have the slope, we must calculate it. We let $(4, 3)$ be the point (x_1, y_1), and we let $(-1, -2)$ be the point (x_2, y_2). Then

$$m = \frac{y_2 - y_1}{x_2 - x_1} = \frac{(-2) - 3}{(-1) - 4} = \frac{-5}{-5} = 1$$

Now we need to find b. To do so, we substitute $m = 1$ and the coordinates of either point into the slope-intercept form. We'll use the first point because its coordinates are both positive so our calculations will be easier. Thus we have $x_1 = 4, y_1 = 3$, and $m = 1$. We get

$$
\begin{aligned}
y &= mx + b & & \\
3 &= (1)(4) + b & & \text{Substituting} \\
3 &= 4 + b & & \text{Simplifying} \\
3 + (-4) &= 4 + b + (-4) & & \text{Adding } -4 \text{ to both sides} \\
-1 &= b & & \text{Simplifying}
\end{aligned}
$$

So the slope m is 1, and b is -1. We substitute these values into $y = mx + b$ to get

$$
\begin{aligned}
y &= (1)x + (-1) \\
y &= x - 1
\end{aligned}
$$

▶ *CHECK* **Warm-Up 9**

Practice what you learned.

◢ **SECTION FOLLOW-UP**

1. What is the slope of the line in the taxi problem? What is the *y*-intercept?

2. Write an equation of a line parallel to the line in the taxi problem.

3. ✏ Write a real-life word problem that this parallel line might describe.

4. ✏ What other real-life situations might be described by straight lines?

5. ✏ What real-life situation might be described by a straight line that passes through the point $(0, 0)$?

4.3 WARM-UPS

Work these problems before you attempt the exercises.

1. Find the slope and *y*-intercept of the graph of $6x + 3y = 9$.

2. Graph the line that has slope 5 and *y*-intercept $(0, -4)$.

3. Graph $x + 2y - 3 = 0$ by finding the slope and *y*-intercept.

4. Graph $-2y = 7x + 12$ and $7x - 2 = -2y$.

5. Graph $5y = x + 5$ and $y + 5x + 3 = 0$.

6. Are the graphs of $3x + 4y = 7$ and $y = -\frac{3}{4}x - 8$ parallel, perpendicular, or neither?

7. Find an equation of the line with slope 4 and *y*-intercept $(0, 5)$.

8. Find an equation of the line that has slope -3 and passes through the point $(-3, 0)$.

9. Find an equation of the line that passes through the points $(2, 0)$ and $(0, -2)$.

4.3 EXERCISES

Note: Use your graphing calculator to check your results whenever possible.

In Exercises 1 through 8, rewrite each equation in slope-intercept form and find the slope and y-intercept of its graph.

1. $3y - 2x = 1$

2. $5(x + y) = 12$

3. $7x - 3y = 5$

4. $-7(y + 4x) = 5$

5. $-3(x + y) = 8y;$

6. $5x - 7 = y + 3$

7. $6x - y = 2x + 5$

8. $-8(x + 4y) = 12y$

In Exercises 9 through 14, use the slope and y-intercept to graph the line.

9. $y + 2x = 4$

10. $3x + y = 7$

11. $3x - y = 5$

12. $-x + y = 2$

13. $y - 3x = 9$

14. $2x + 6y = 9$

In Exercises 15 through 18, graph the line that is described.

15. Parallel to the line $-2x = y + 2$ and having the y-intercept $(0, 7)$

16. Parallel to the line $6x - y = 4$ and having the y-intercept $(0, -2)$

17. Perpendicular to the line $x + 5y = 27$ and passing through the point $(2, 5)$

18. Pependicular to the line $y = -x + 1$ and passing through the point $(5, -4)$

In Exercises 19 through 30, determine whether the given equations have graphs that are parallel, perpendicular, or neither. Use the equations only. Do not graph the lines.

19. $y = -2x + 1$ and $y = -\frac{1}{2}x + \frac{5}{2}$

20. $y = 2x + 5$ and $15 + 8y = x$

21. $y = 8x + 3$ and $\frac{1}{8}x + y = 7$

22. $y = 3x - 4$ and $-3y = x - 5$

23. $y = -3x + 4$ and $3y = x + 2$

24. $y = 2x - 7$ and $8y + 4x = 16$

25. $y = x + 7$ and $y = -x - 7$ **26.** $y = \frac{1}{4}x - 6$ and $-x + 4y = 9$ **27.** $y = 2x + 4$ and $2x - y = 8$

28. $y = -x + 1$ and $2x - y = 5$ **29.** $y = 8x - 5$ and $8x - y = 6$ **30.** $y = x - 1$ and $y = x + 1$

In Exercises 31 through 34, use your graphing calculator to graph the first equation. Then graph each of the other equations in turn and record its relationship to the first equation—parallel, perpendicular, or neither.

31. $3y - 7 = 4x$ and

   a. $2y + 8 = 6x$ **b.** $9y - 3 = 12x$ **c.** $12x - 8 = -y \cdot 16$

32. $y = 2x - 5$ and

   a. $-\frac{1}{2}x = y + 4$ **b.** $-3x = 6y + 3$ **c.** $16x = 8y - 4$

33. $-y + 5 = 8x$ and

   a. $-2y + 6 = x \cdot 2$ **b.** $\frac{1}{2}y + 8x = 4$ **c.** $y = -8x - 3$

34. $2x + 4y = 7$ and

   a. $3x + 6y = 14$ **b.** $8x - 4y = 12$ **c.** $y = 2x - 13$

In Exercises 35 through 42, write an equation of the line that has the given slope and y-intercept. Write your answers in slope-intercept form.

35. $m = -3; \left(0, \frac{1}{2}\right)$ **36.** $m = -5; (0, -8)$ **37.** $m = \frac{4}{5}; \left(0, \frac{2}{5}\right)$ **38.** $m = \frac{3}{5}; (0, -5)$

39. $m = \frac{3}{4}; (0, 8)$ **40.** $m = \frac{1}{2}; (0, 7)$ **41.** $m = 9; (0, -5)$ **42.** $m = \frac{5}{6}; (0, -2)$

In Exercises 43 through 58, write an equation of the line that has the given slope and passes through the given point. Use the slope-intercept form.

43. $m = 4; (-1, 2)$ **44.** $m = 6; (2, 4)$ **45.** $m = 2; (3, 5)$ **46.** $m = 9; (4, 6)$

47. $m = -5; (-2, -2)$ **48.** $m = -8; (0, 0)$ **49.** $m = -7; (1, -3)$ **50.** $m = -3; (3, -4)$

51. $m = 0; (5, -2)$ **52.** $m = 2; (1, 2)$ **53.** $m = -4; (-6, 3)$ **54.** $m = -1; (3, 2)$

55. $m = \frac{3}{4}; (0, 0)$ **56.** $m = \frac{2}{3}; (2, 0)$ **57.** $m = \frac{1}{5}; (5, -4)$ **58.** $m = \frac{8}{9}; (-9, -2)$

In Exercises 59 through 74, write an equation of the line that passes through the two given points. Use the slope-intercept form.

59. $(2, 5)$ and $(-2, 1)$ **60.** $(0, 12)$ and $(2, 2)$ **61.** $(8, 14)$ and $(1, 4)$ **62.** $(2, -1)$ and $(3, 5)$

63. $(4, 6)$ and $(-2, 0)$ **64.** $(0, 0)$ and $(7, -3)$ **65.** $(3, 2)$ and $(5, 6)$ **66.** $(2, -1)$ and $(-2, -1)$

67. $(-1, 3)$ and $(0, 5)$ **68.** $(-1, 1)$ and $(-2, 4)$ **69.** $(0, 1)$ and $(-1, 5)$ **70.** $(0, -3)$ and $(1, 3)$

71. $(6, 3)$ and $(1, 2)$ ⠀⠀**72.** $(6, -4)$ and $(-1, 2)$ ⠀⠀**73.** $(3, 2)$ and $(1, -1)$ ⠀⠀**74.** $(3, -2)$ and $(1, -1)$

In Exercises 75 through 82, write an equation of the line that passes through the given point and is parallel to the line represented by the given equation.

75. $(-1, 4); 3x + y = 2$ ⠀⠀**76.** $(1, 2); y = 4x + 5$ ⠀⠀**77.** $(1, -8); x = 2y - 1$ ⠀⠀**78.** $(0, 2); y = 2x - 4$

79. $(-2, 3); y = x + 3$ ⠀⠀**80.** $(2, 5); y + 7x = 5$ ⠀⠀**81.** $(0, -6); -x + y = 6$ ⠀⠀**82.** $(-11, 7); -y + x = 12$

In Exercises 83 through 90, write an equation of the line that passes through the given point and is perpendicular to the line represented by the given equation.

83. $(-2, -2); x + y = -3$ ⠀⠀⠀⠀**84.** $(6, -5); y - x = 2$

85. $(-3, -5); \frac{1}{2}y + 7 = x$ ⠀⠀⠀⠀**86.** $(3, 1); y = 2x - 4$

87. $(0, -1); y = 3x - 5$ ⠀⠀⠀⠀**88.** $(5, 0); 5x + y = 8$

89. $(-2, 1); -2x = 5 + y$ ⠀⠀⠀⠀**90.** $(-5, 2); y - 9 = 10x$

MIXED PRACTICE

By doing these exercises, you will practice the topics up to this point in the chapter.

91. Graph the line that has slope 2 and y-intercept 3.

92. Write an equation of the line that passes through the points $(-3, 2)$ and $(0, 0)$.

93. Is $(-2, -2)$ a solution of the equation $y - 8x = 14$?

94. Determine whether $2y + 5 = x$ and $2x + 6 = y$ are parallel, perpendicular, or neither.

95. Graph $-x + y = 2$ by finding the y-intercept and slope.

96. Write the equation of the line that passes through the point $(-9, 3)$ and has slope 4.

97. Graph $2x - 2y = 4$ using a table of values.

98. Is $\left(-\frac{3}{2}, -12\right)$ a solution of $2x - y = 9$?

99. Find the x-intercept of the graph of $y = 5x + 12$.

100. Determine which of these two equations, $3x - 4y = 8$ or $-2x - 5y = 8$, has a graph with negative slope.

EXCURSIONS

Class Act

1. **a.** Write two different verbal situations that could be shown by the following graph.

    b. Make a list that gives the "points" on this graph.

    c. Write equations for the lines whose "pieces" make up this graph. (There are 5.)

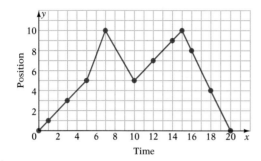

2.* **a.** Choose the graph that most closely represents the situation. Justify your choice by analyzing the graph and writing a paragraph about it.

    i. A boy drags a sled up a hill and then slides back down.

    ii. A car pulls up to the curb and a passenger gets in.

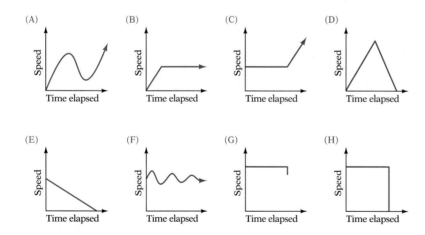

b. Write a verbal "scene" that could represent one of the remaining graphs. Give it to a classmate and see if he or she can identify the graph you described.

* Adapted with permission from "Relating to Graphs in Introductory Algebra," by Frances Van Dyke, *Mathematics Teacher*, copyright September 1994 by the National Council of Teachers of Mathematics.

4.4 Graphing Linear Inequalities in Two Variables

SECTION GOAL

■ *To graph linear inequalities*

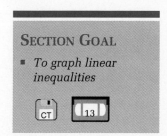

SECTION LEAD-IN

I have only $6. How far can I travel in the taxi? Let's go back to our equation from the Lead-In to Section 4.3.

$$y = 1.25x + 2.00, \text{ where } x = \text{number of miles}$$

We need to find possible x values when $y \leq 6$. Therefore, we want to know when

$$1.25x + 2.00 \leq 6$$
$$1.25x \leq 4$$
$$x \leq \frac{4}{1.25}$$
$$x \leq 3.2$$

Now we need to know how many fifths of a mile are represented by 0.2. [Why?]

$$0.2 = \frac{2}{10} = \frac{1}{5} \quad \text{How did we do that?}$$

So we can travel $3\frac{1}{5}$ miles on $6.00.

The graph of a linear equation in one variable is a point (figure on the left), but the graph of a linear inequality in one variable contains many points and is a portion of the number line (figure on the right).

Similarly, the graph of a linear equation in two variables is a line (figure below on the left), but the graph of a linear inequality in two variables is a portion of the coordinate plane (the shaded region in the figure on the right).

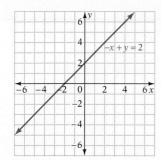
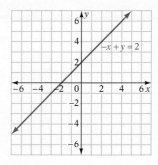

The line in this last figure, on the right, is the graph of an equation. Accordingly, it is part of the graph of the inequality only if the inequality includes an equality—that is, only for ≤ and ≥ inequalities. For < and > (the inequalities that don't include an equality), we draw the line *broken* to show that it is *not* part of the graph.

To graph a linear inequality in two variables

1. Replace the inequality sign with an equal sign and graph the resulting equation. (Its graph is a line.)
2. Draw the line dashed for < or > inequalities, and draw it solid for ≤ and ≥ inequalities.
3. Test a point on each side of the line by substituting it into the inequality. The point that results in a true statement belongs to the graph of the inequality. Shade *its* region of the plane.

▪ ▪ ▪

EXAMPLE 1

Graph $x + y < 6$.

SOLUTION

We begin by replacing the inequality sign with an equal sign.

$$x + y = 6$$

We can graph this equation easily with a table of values. Letting x be 1, 0, and 3 in turn we obtain the table

x	y	$x + y = 6$	
1	5	$1 + y = 6$	$y = 5$
0	6	$0 + y = 6$	$y = 6$
3	3	$3 + y = 6$	$y = 3$

We next graph these points and draw a line through them. The original inequality is of the "less than" type, so we draw a broken line to show that the equality $x + y = 6$ is not part of the graph.

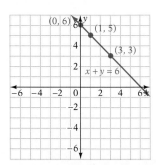

As the figure shows, the line divides the plane into two regions. We have to determine which region is part of the inequality. The point $(0, 0)$ is usually a good point to test. *Note:* If the line goes through the origin $(0, 0)$, we need to choose a

different test point. Substituting its coordinates into the original inequality yields

$$x + y < 6$$
$$0 + 0 < 6$$
$$0 < 6 \quad \text{True}$$

The point $(0, 0)$ leads to a true statement, so it must be part of the graph of $x + y < 6$. Because $(0, 0)$ is *below* the broken line, we shade *below* the line as in the figure that follows.

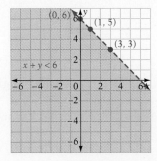

As a check, we test a point on the other side of the line. Let's try $(2, 5)$. It gives us

$$x + y < 6$$
$$2 + 5 < 6$$
$$7 < 6 \quad \text{False}$$

So we are correct in shading the lower region of the plane.

▶ *CHECK* **Warm-Up 1**

Every point on the graph of a linear inequality is a *solution* of that inequality. For example, every shaded point in the preceding figure has coordinates that, when substituted into the inequality, make the inequality true. Try one and see for yourself.

•••

EXAMPLE 2

Graph the inequality $2y - 4x \geq 6$.

SOLUTION

We must first graph the equation $2y - 4x = 6$. To start, we solve for y and construct a table of values.

x	y		
1	5	$y = 2x + 3$	
0	3	$y = 2(1) + 3$	$y = 5$
-1	1	$y = 2(0) + 3$	$y = 3$
		$y = 2(-1) + 3$	$y = 1$

We graph these points and draw a *solid line* through them, because the graph of a $\geq$ inequality *includes* the line.

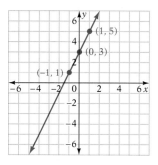

Again, we test the point (0, 0) to determine which region to shade.

Substituting these coordinates into the original inequality gives us

$$2y - 4x \geq 6$$

$$2(0) - 4(0) \geq 6$$

$$0 \geq 6 \quad \text{False}$$

Because (0, 0) is below (or to the right of) the line, and the resulting inequality is false, we shade above (to the left of) the line.

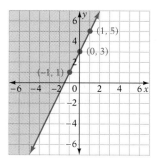

To check, you should test a point on the other side of the line, say $(-1, 3)$. You will find that we have shaded the correct region.

▶ *CHECK* **Warm-Up 2**

Practice what you learned.

SECTION FOLLOW-UP

In the taxi problem, what would be my cost if I travel $6\frac{1}{3}$ miles?
I substitute $x = 6\frac{1}{3}$ into the taxi cost equation:

$$y = 1.25 + 2.00$$
$$y = (1.25)\left(6\frac{1}{3}\right) + 2.00$$
$$y \approx (1.25)(6.33) + 2.00 \qquad \left[6\frac{1}{3} \approx 6.33\right]$$
$$y = 7.9125 + 2.00$$
$$y = 9.9125$$

The result of my work is $y \approx$ \$9.91. Is this what I must pay for my
taxi ride? No. Taxi meters charge 25 cents for each $\frac{1}{5}$ mile. I
traveled $6\frac{1}{3}$ miles. And the meter shows the charge for each distance
at the beginning of the fifth of a mile. We will compare $\frac{1}{3}$ and $\frac{1}{5}$ using
division.

$$\frac{1}{3} \div \frac{1}{5} = \frac{1}{3} \cdot \frac{5}{1} = \frac{5}{3} = 1\frac{2}{3}$$

Now what? This answer, $1\frac{2}{3}$, tells us that when we travel $\frac{1}{3}$ mile,
that is the same as $\frac{1}{5}$ plus a little more. But we pay for *two*-fifths.

Our cost will be

$$\$9.50 + 0.25 + 0.25 = \$10.00$$

cost for 6 miles $\frac{1}{5}$ mile $\frac{1}{5}$ mile

Consider the result $y =$ \$9.91. If we round this *up* to the next \$0.25,
we also obtain $y =$ \$10.00. We pay \$10.00 for our taxi ride.

4.4 WARM-UPS

Work these problems before you attempt the exercises.

1. Graph the inequality $x + y < 5$.

2. Graphically find the solution set for $-2x + 3y \leq -3$.

4.4 EXERCISES

Note: Use your graphing calculator to check your results whenever possible.

In Exercises 1 through 24, graph the inequality.

1. $x - 3 < y$

2. $x + y > 1$

3. $4x + y > -2$

4. $5x + y < 4$

5. $x - 4y \geq 4$

6. $4x + y < -3$

7. $y \geq x + 5$

8. $5x + y \leq 11$

9. $y < 5x + 5$

10. $x + y \geq 7$

11. $y + x < 10$

12. $4x - 5 \leq y$

13. $4x - 5 \geq y$

14. $5x + y \geq -3$

15. $8x - 2y \leq 14$

16. $y > 4x$

17. $x + y \geq 4$

18. $x + y \leq -1$

19. $x - 4y \leq 8$

20. $4x + 1 \leq y$

21. $4y \geq 12$

22. $y < 5$

23. $x > 2$

24. $-5x \leq 10$

In Exercises 25 through 42, graph the inequality and name three points that are in the solution set.

25. $y > 5$

26. $4x \geq 8$

27. $4x - 4y \geq -8$

28. $4x - 4y \geq 12$ $\qquad$ **29.** $5x + y < 15$ $\qquad$ **30.** $y + (-3) > x$

31. $-3x < y + 5$ $\qquad$ **32.** $4x + 1 \geq y$ $\qquad$ **33.** $x + y > -2$

34. $y - x \leq 1$ $\qquad$ **35.** $x - 4y > -1$ $\qquad$ **36.** $5y + 2x \geq -10$

37. $-7x + y \leq -3$ $\qquad$ **38.** $5x + y > 2$ $\qquad$ **39.** $y > -1$

40. $x < -5$ $\qquad$ **41.** $y - 4x \leq 4$ $\qquad$ **42.** $-2x + y > 4$

MIXED PRACTICE

By doing these exercises, you will practice the topics up to this point in the chapter.

43. Use the slope-intercept form to find an equation of the line that passes through the points $(6, 2)$ and $(-4, -5)$.

44. Find the slope of the line whose equation is $-3y + 2x = 12$.

45. Find an equation of the line whose graph has slope -4 and y-intercept $(0, 12)$.

46. Determine whether the graphs of $-3x + y = 7$ and $-y + 3x = 12$ are parallel, perpendicular, or neither.

47. $(6, n)$ is a point on the graph of the line $7 + y = x$. What is n?

48. Write $11 = -3y + x$ in slope-intercept form.

49. Find the x- and y-intercepts of the graph of $x - 8y = 9$.

50. Find an equation of the line that passes through the points $(-4, 0)$ and $(0, -5)$.

51. Find three points that lie on the graph of $-3x + y \geq -9$.

52. Graph $4y - 3x \geq 2$.

53. Graph $4x - 9 = 7$.

54. Complete this table of values to find three points that are on the graph of $y - x = 2$.

x	y
2	
	0
0	

EXCURSIONS

Class Act

1. Use the information from these two graphs to answer the questions.

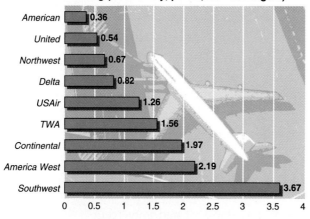

Denied Boarding (Involuntary, per 10,000 Passengers)

American 0.36
United 0.54
Northwest 0.67
Delta 0.82
USAir 1.26
TWA 1.56
Continental 1.97
America West 2.19
Southwest 3.67

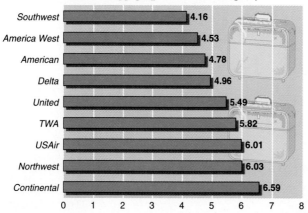

Mishandled Baggage (per 1000 Passengers)

Southwest 4.16
America West 4.53
American 4.78
Delta 4.96
United 5.49
TWA 5.82
USAir 6.01
Northwest 6.03
Continental 6.59

a. Complete the table.

	Rank Denied Boarding	Rank Mishandled Baggage
American	1	3
America West		
Continental		
Delta		
Northwest		
TWA		
Southwest		
United		
USAir		

b. Plot the points from part (a) on the graph.

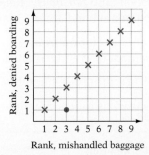

c. ✏ If the points line up perfectly with the ✗s, a perfect correlation would have occurred. Comment on how well correlated these two rankings are. If these two rankings had been perfectly correlated, what would that have meant in terms of how "denied" and "mishandled" were related?

d. Use the information from the Airlines Ranked by Quality table and graph those rankings against each of the other two rankings.

e. Which of the other two rankings is most closely correlated to the Quality ranking? Justify your decision.

Airlines Ranked by Quality

Rank	Airline
1	Southwest
2	American
3	United
4	Delta
5	USAir
6	Northwest
7	TWA
8	America West
9	Continental

Source: Wichita State University and University of Nebraska at Omaha, as cited in the *1996 Information Please® Business Almanac* (©1995 Houghton Mifflin Co.), p. 540. All rights reserved. Used with permission by Information Please LLC.

4.5 Optional Topic: Functions and Graphs

SECTION LEAD-IN

Think about the taxi problem that was first introduced in the Section 4.1 Lead-In. Remember that the taxi fare is $2.00 plus 25 cents for every $\frac{1}{5}$ mile. The taxi meter clicks at the beginning of the fifth of a mile. Use what you know about fractions, decimals, and the taxi fare to fill in the following table.

Distance in miles	Taxi fare
1.2	
3.3	
$16\frac{1}{2}$	
$\frac{3}{4}$	
5.25	
$2\frac{1}{3}$	

Draw a graph that shows how the taxi meter prices the cab ride. Justify your graph.

The graphs that we sketched in previous sections of this chapter serve as examples of the graphs of *relations* and *functions*. "Relation" and "function" are basic concepts. They form a common bond among many areas of mathematics.

Let A and B be sets. Then the **relation** from A to B consists of all ordered pairs (a, b) such that a is an element of A and b is an element of B. That is,

$$\{(a, b) \mid a \text{ is in } A \text{ and } b \text{ is in } B\}$$

We will confine our discussions to sets containing real numbers.

> **Domain and Range of a Relation**
>
> The set of all first elements of a relation is called the **domain** of the relation. The set of all second elements is called the **range.**

The relations that interest us the most are the relations that are predictable— the functions.

Functions and Graphs of Functions

A *function* is a special kind of relation. Let X and Y be sets. Then a **function** f from X to Y is a rule or other means for assigning to each element of X *exactly one element of Y*. The set X is the domain of the function, and the set Y is its range.

A function f is actually an *instruction* for making a set of ordered pairs (x, y) such that each x-value is used with only one y-value. The **graph of a function** is the graph of its ordered pairs.

▪▪▪
EXAMPLE 1

Let f = "add 3 to the first element to get the second element," and let $X = \{1, 2, 3, 5\}$.

a.   Find the ordered pairs produced by the function.

b.   Find its range.

SOLUTION

a.   The function assigns to each element x of the domain X the number $x + 3$. Taking the elements of the domain in order, we obtain the set of ordered pairs

$$\{(1, 4), (2, 5), (3, 6), (5, 8)\}$$

b.   The range is $\{4, 5, 6, 8\}$.

▶ *CHECK* **Warm-Up 1**

▪▪▪
EXAMPLE 2

Determine which of the following are functions. Identify the domain and range of each function.

a. $\{(0, 15), (2, 15), (3, 15), (4, 15)\}$ b. $\{(1, 2), (2, 3), (3, 4), (4, 5)\}$

c. $\{(1, 2), (3, 4), (4, 3), (1, 3)\}$ d. $\{(1, 2), (3, 4), (4, 3), (1, 2)\}$

SOLUTION

a.   This is a function. Each first element is assigned just one second element.

Domain: $\{0, 2, 3, 4\}$

Range: $\{15\}$

b.   This is a function. Again, each first element is associated with just one second element.

Domain: $\{1, 2, 3, 4\}$

Range: $\{2, 3, 4, 5\}$

c.   This is not a function, because the first and fourth pairs have the same first element and different second elements.

d.   This is a function. The first and fourth pairs are the same pair. One of them is not needed.

Domain: $\{1, 3, 4\}$

Range: $\{2, 3, 4\}$

▶ *CHECK* **Warm-Up 2**

Functional Notation

The most useful functions have all the real numbers as the domain; then, however, lists of ordered pairs become awkward as means for stating functions. Instead, mathematicians use what is called **functional notation.** In this notation, the function rule is written as a mathematical expression that includes a variable. The variable (usually x) represents domain elements. The corresponding range elements are represented by the symbol $f(x)$, which is read "f of x." The two are equated to show that each range value depends on the corresponding domain value.

As an example, the function "add 3 to the first element to get the second element" is written

$$f(x) = x + 3$$

Suppose the domain of f is all real numbers. Then this notation tells us that

when x is 2, then $f(2) = 2 + 3 = 5$

This gives us the pair (2, 5). Similarly,

when $x = -1.7$, then $f(-1.7) = -1.7 + 3 = 1.3$

This gives us the pair (−1.7, 1.3).

More generally, a function f generates the pairs $(x, f(x))$ for all x in the domain of f. Finding the value of $f(x)$ for a given value of x is called "evaluating the function at x."

▪▪▪

EXAMPLE 3

The function $f(x) = 2x - 2$ has the domain $X = \{x \mid x \text{ is a real number}\}$. Evaluate the function at $x = 0$, $x = 1$, and $x = 2$. Then graph the function.

SOLUTION

Substituting 0, 1, and then 2 for x yields

$$f(0) = 2(0) - 2 = -2$$
$$f(1) = 2(1) - 2 = 0$$
$$f(2) = 2(2) - 2 = 2$$

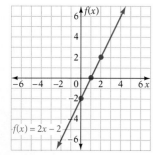

To graph the function, first note that $f(0)$, $f(1)$, and $f(2)$ gives us the function pairs (0, −2), (1, 0), and (2, 2). Now look at the function $f(x) = 2x - 2$.

It is not only a function; it is also a linear equation with independent variable x, dependent variable $f(x)$, and solutions (0, −2), (1, 0), and (2, 2). The graph of this equation is a straight line through the points that correspond to those solutions, as sketched in the accompanying figure.

Moreover, every solution of the equation $f(x) = 2x - 2$ is an ordered pair of the function $f(x) = 2x - 2$. Thus the graph in this figure is also the graph of the function.

▶ CHECK **Warm-Up 3**

A function of the form $f(x) = ax + b$ is called a **linear function.** Its graph is a straight line with slope a and vertical intercept $(0, b)$.

The Vertical-Line Test for Functions

We can determine whether any graph is the graph of a function by applying the following test:

Vertical-Line Test

A graph is the graph of a function if no vertical line intersects it at more than one point.

▪ ▪ ▪

EXAMPLE 4

Determine whether each of the following graphs is the graph of a function.

a.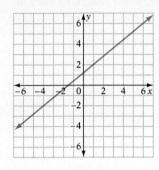

b.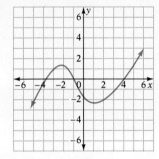

c.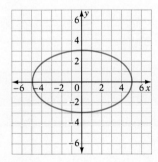

d.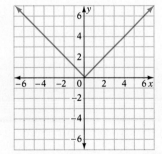

SOLUTION

a. This is the graph of a function. At no place along the graph does a vertical line cross more than one point. Every line except a vertical line is the graph of a function.

b. This is also the graph of a function, even though it is curved.

c. This is not the graph of a function, because some vertical lines intersect it in two points. It is actually an ellipse. Note that it has two *y*-intercepts.

d. This too is the graph of a function. It is actually the graph of a $f(x) = |x|$.

▶ *CHECK* **Warm-Up 4**

Practice what you learned.

SECTION FOLLOW-UP

Taxi drivers want a fare hike. They want the ride to cost $2.00 plus 50 cents for every $\frac{1}{3}$ mile. They also expect a tip of 15% per trip.

a. Make a table showing the cost for a ride of 1, 3, and 5 miles.

b. Graph this information.

c. Write the equation needed to compute the total cost for the ride. (*Hint:* Write the verbal *rule* first; then write an equation.)

d. Compute the fare a taxi driver would expect to receive for

10 miles

and for

$10\frac{1}{2}$ miles

4.5 WARM-UPS

Work these problems before you attempt the exercises.

1. Let f = "twice the first element is the second element," and let $X = \left\{\frac{1}{2}, 1, 1\frac{1}{2}\right\}$. Find the ordered pairs produced by the function and find the range.

2. Determine which are functions.

a. $\{(1, 3), (3, 1), (2, 1)\}$

b. $\{(0, 2), (2, 4), (4, 6), (6, 8), (6, 2)\}$

c. $\{(0, 3), (5, 3), (6, 3)\}$

d. $\{(0, 2), (0, 3), (0, 4)\}$

3. Evaluate $f(0)$, $f(2)$, and $f(a)$ when $f(x) = 3x^2 - 2$.

4. Use a vertical-line test to determine which are functions.

a.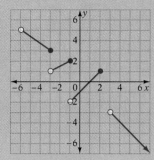

b.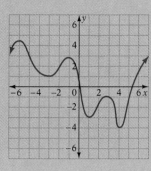

4.5 EXERCISES

Note: Use your graphing calculator to check your results whenever possible.

In Exercises 1 through 8, the functions are defined on the real numbers. Rewrite using functional notation, and list four ordered pairs in the function.

1. $f =$ "the second element is 3 times the first element."

2. $f =$ "twice the first element plus two is the second element."

3. $f =$ "the first element minus three is the second element."

4. $f =$ "the second element is one-third the first element."

5. $f =$ "the second element is four less than twice the first element."

6. $f =$ "one half the first element minus 5 is the second element."

7. $f =$ "the second element is the product of the first element and the first element minus one."

8. $f =$ "the second element is the product of the first element and 6."

STUDY HINT

In Exercises 1 through 8, the ordered pairs you find may differ from the ones we chose. As long as yours meet the requirements of the instruction given, the pairs are correct.

In Exercises 9 through 16, given the set of ordered pairs in the relation R, identify the domain and range. Determine whether the relation is a function.

9. $R = \{(0, 3), (0, 5), (0, 7)\}$

10. $R = \{(3, 0), (5, 0), (7, 0)\}$

11. $R = \{(3, 2), (6, 7), (3, 2)\}$

12. $R = \{(5, 7), (6, 8), (9, 10)\}$

13. $R = \{(0, 3), (5, 0), (6, 0)\}$

14. $R = \{(8, 2), (9, 2), (6, 3)\}$

15. $R = \{(5, 7), (6, 2), (8, 7)\}$

16. $R = \{(5, 7), (6, 2), (6, 7)\}$

In Exercises 17 through 24, evaluate $f(x)$ at the given values.

17. $f(x) = 2x + 7$; evaluate at $x = 0, 2, -1$

18. $f(x) = 3x - 8$; evaluate at $x = 5, 0, -3$

19. $f(x) = x^2 - 2$; evaluate at $x = 0, 4, -3$

20. $f(x) = 2x^2$; evaluate at $x = 0, 3, -2$

21. $f(x) = -2x + 1$; find $f(0), f(a), f(x + 2)$
(*Hint:* Treat "$x + 2$" as just another number.)

22. $f(x) = x - 2$; find $f(0), f(a), f(x + 3)$

23. $f(x) = 3 - 2x$; find $f(0), f(a), f(x - 1)$

24. $f(x) = 2 - 3x$; find $f(0), f(a), f(x - 2)$

In Exercises 25 through 32, use the vertical-line test to determine which graphs are the graphs of functions.

25.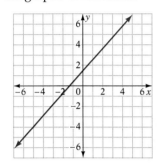

26.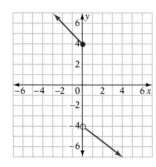

27.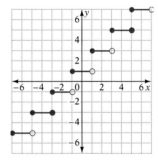

28.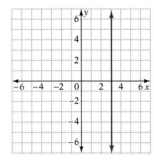

29.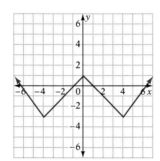

30.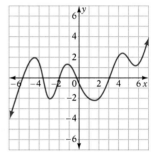

31.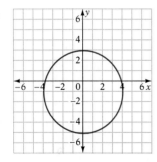

32.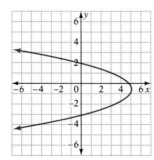

EXCURSIONS

Class Act

1. **a.** Graph both sets of this data and compare and contrast the information about the sites. They are both located 450 m above sea level. Their latitudes are 43° S and 45° N; their longitudes are 103° W and 147° E.

   b. See if you can determine what countries these sites are in.

Mean Monthly Temperature (°C)

Site 1		Site 2	
January	−17.9	January	12.3
February	−13.8	February	12.8
March	−7.8	March	11.0
April	3.1	April	8.6
May	10.6	May	5.9
June	15.7	June	4.3
July	18.6	July	3.5
August	17.5	August	4.0
September	11.5	September	5.4
October	5.3	October	7.4
November	−5.1	November	8.9
December	−12.9	December	10.5

Mean Monthly Precipitation Including Snow (mm)

Site 1		Site 2	
January	21	January	88
February	18	February	82
March	22	March	97
April	28	April	132
May	50	May	142
June	74	June	124
July	53	July	142
August	56	August	139
September	42	September	144
October	25	October	138
November	17	November	132
December	22	December	117

Data Analysis

2.* **a.** ✎ Choose the graph most closely representing the situation. Justify your choice by analyzing the graph and writing a paragraph about it.

    i. A monkey swings on a rope.   **ii.** Two children ride on a ferris wheel.

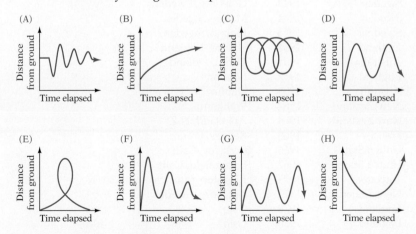

  b. ✎ Write a verbal "scene" that could represent one of the remaining graphs. Give it to a classmate and see if he or she can identify the graph you described.

* Adapted with permission from "Relating to Graphs in Introductory Algebra," by Frances Van Dyke, *Mathematics Teacher*, copyright September 1994 by the National Council of Teachers of Mathematics.

3. Study the following table and then answer the questions.

   a. Graph these data and comment on the shape of the graph. Ask and answer four questions about these data.

   b. *Research* Is 3:44.39 the fastest time ever for the mile run? Add any additional records to your graph.

History of the Record for the Mile Run

Time	Athlete	Country	Year	Location
4:36.5	Richard Webster	England	1865	England
4:29.0	William Chinnery	England	1868	England
4:28.8	Walter Gibbs	England	1868	England
4:26.0	Walter Slade	England	1874	England
4:24.5	Walter Slade	England	1875	London
4:23.2	Walter George	England	1880	London
4:21.4	Walter George	England	1882	London
4:18.4	Walter George	England	1884	Birmingham, England
4:18.2	Fred Bacon	Scotland	1894	Edinburgh, Scotland
4:17.0	Fred Bacon	Scotland	1895	London
4:15.6	Thomas Conneff	United States	1895	Travers Island, N.Y.
4:15.4	John Paul Jones	United States	1911	Cambridge, Mass.
4:14.4	John Paul Jones	United States	1913	Cambridge, Mass.
4:12.6	Norman Taber	United States	1915	Cambridge, Mass.
4:10.4	Paavo Nurmi	Finland	1923	Stockholm
4:09.2	Jules Ladoumegue	France	1931	Paris
4:07.6	Jack Lovelock	New Zealand	1933	Princeton, N.J.
4:06.8	Glenn Cunningham	United States	1934	Princeton, N.J.
4:06.4	Sydney Wooderson	England	1937	London
4:06.2	Gunder Hägg	Sweden	1942	Goteborg, Sweden
4:06.2	Arne Andersson	Sweden	1942	Stockholm
4:04.6	Gunder Hägg	Sweden	1942	Stockholm
4:02.6	Arne Andersson	Sweden	1943	Goteborg, Sweden
4:01.6	Arne Andersson	Sweden	1944	Malmo, Sweden
4:01.4	Gunder Hägg	Sweden	1945	Malmo, Sweden
3:59.4	Roger Bannister	England	1954	Oxford, England
3:58.0	John Landy	Australia	1954	Turku, Finland
3:57.2	Derek Ibbotson	England	1957	London
3:54.5	Herb Elliott	Australia	1958	Dublin
3:54.4	Peter Snell	New Zealand	1962	Wanganui, N.Z.
3:54.1	Peter Snell	New Zealand	1964	Auckland, N.Z.
3:53.6	Michel Jazy	France	1965	Rennes, France
3:51.3	Jim Ryun	United States	1966	Berkeley, Calif.
3:51.1	Jim Ryun	United States	1967	Bakersfield, Calif.
3:51.0	Filbert Bayi	Tanzania	1975	Kingston, Jamaica
3:49.4	John Walker	New Zealand	1975	Goteborg, Sweden
3:49.0	Sebastian Coe	England	1979	Oslo
3:48.8	Steve Ovett	England	1980	Oslo
3:48.53	Sebastian Coe	England	1981	Zurich, Switzerland
3:48.40	Steve Ovett	England	1981	Koblenz, W. Ger.
3:47.33	Sebastian Coe	England	1981	Brussels
3:46.31	Steve Cram	England	1985	Oslo
3:44.39	Noureddine Morceli	Algeria	1993	Rieti, Italy

Source: USA Track & Field, as cited in the *1996 Information Please® Almanac* (©1995 Houghton Miffin Co.), p. 957. All rights reserved. Used with permission by Information Please LLC.

CHAPTER LOOK-BACK

Refer to the Chapter Lead-In on page 202. Here are the questions we asked.

1. How fast was the teen skating when the park ranger stopped him?

He was skating 18 miles per hour.

2. How long was he stopped talking to the ranger?

He was stopped for about 2 minutes.

3. What was his top speed?

His top speed was 19 miles per hour.

4. How long did he skate?

He skated 18 minutes altogether.

5. What do you think the speed limit was?

Probably about 12 miles per hour, but we can't really tell.

What did you want to know?

CHAPTER 4
REVIEW PROBLEMS

The following exercises will give you a good review of the material presented in this chapter.

SECTION 4.1

1. Find the x- and y-intercepts of the graph of $-y + x = 1$.

2. Graph $-2y + 6x = 3$ by finding the x- and y-intercepts.

3. Graph $y = 4$. What are the x- and y-intercepts?

4. Graph $y = -7x$. What are the x- and y-intercepts?

SECTION 4.2

5. Write $12x - y = 6$ in slope-intercept form.

6. What is the slope of the graph of $-9x + 2y = 6$?

7. Graph $y = -5$ and $x = 3$ on the same set of axes. Where do they intersect?

8. Find the slope of the line that contains the points $(-1, 3)$ and $(5, 7)$.

9. Which pair of points, $(-1, -3)$ and $(5, -7)$ or $(-6, -2)$ and $(-1, 5)$, lie on a line with a negative slope?

10. Which pair of points, $(5, 4)$ and $(2, 3)$ or $(-1, -2)$ and $(3, -4)$, lie on a line with a positive slope?

11. Find the slope of the line that contains the points $(5, -4)$ and $(3, -4)$.

12. Find the slope of the line that contains the points $(-1, 1)$ and $(-1, 2)$.

13. Without graphing, determine whether $(-3, -2)$, $(-5, -4)$, and $(1, 2)$ are collinear.

SECTION 4.3

14. Find the slope of the line whose equation is $-1x + 2y = 13$.

15. Find the slope and y-intercept of the graph of $3x - 2y = 11$.

16. Determine whether the graphs of $2x - 7y = 8$ and $\frac{1}{2}x + 8y = 3$ are parallel, perpendicular, or neither. Do not graph them.

17. Graph the line whose slope is -2 and whose y-intercept is $(0, 4)$.

18. Write $-5 + y = -x$ in slope-intercept form.

19. Determine by graphing whether $3y + x = 7$ and $-3x + y = 1$ are parallel, perpendicular, or neither.

20. Write an equation of the line that has slope -5 and y-intercept $(0, -2)$.

21. Use the slope-intercept form to find an equation of the line that passes through the points $(0, 0)$ and $(-4, 5)$.

22. Find an equation of the line that passes through the points $(9, 8)$ and $(-2, -3)$.

23. Find an equation of the line that has slope -1 and passes through the point $(2, -4)$.

24. Find an equation of the line that passes through the points $(0, 0)$ and $(-3, -8)$.

SECTION 4.4

25. Graph $-11x + 2y \le 0$ and identify two points that are part of the solution.

26. Graph $\frac{1}{3}x + y > 8$.

27. Graph $-8x + 5 \ge y$.

28. Graph $-x < 4$ and $-x \ge -0.5$.

MIXED REVIEW

29. Find the x-intercept of the graph of $5y - 6 = 3x$.

30. Find the y-intercept of the graph of $y - 2x = 7$.

31. Determine whether the graphs of $3x + 3 = -y$ and $y = -3x + 5$ are parallel, perpendicular, or neither.

32. Find an equation of the line that has slope -6 and y-intercept $(0, -4)$.

33. Find an equation of the line that passes through the points $(-1, -1)$ and $(-3, -2)$.

34. Find an equation of the line that has slope 5 and y-intercept $(0, -12)$.

CHAPTER 4 TEST

This exam tests your knowledge of the material in Chapter 4.

1. For parts (a) and (b), verify whether the given ordered pair is a solution to the given linear equation.

   a. $(-2, 0); y = \frac{1}{2}x + 1$

   b. $(3, -5); x + y = 2$

   c. Graph $-2y + 1 = x$ using a table of values.

2. **a.** Find the x- and y-intercepts of the graph of $-2y - 1 = y$.

   b. Find the slope of the line that passes through the points $(0, 0)$ and $(-5, -4)$.

   c. Find the slope of the line that passes through the points $(-1, 2)$ and $(-3, -9)$.

3. For parts (a) and (b) find the slope and y-intercept of the graph of each equation.

   a. $3x - 7y = 9$

   b. $-1y - 2x = 8$

   c. Graph $y = -3x + 6$ using the slope-intercept form.

4. **a.** Write the equation of the line that has slope -6 and y-intercept $(0, 3)$.

   b. Write the equation of the line that passes through the points $(0, 0)$ and $(-2, -8)$.

   c. Determine whether the graphs of this pair of equations are parallel, perpendicular, or neither. Do not graph the equations.

   $$y = 4x + 5$$
   $$4x + 2y = 7$$

5. Graph each of the following inequalities.

   a. $x - y > 8$

   b. $y + 3x \geq 11$

   c. $-y \leq -\frac{1}{4}x - 3$

CUMULATIVE REVIEW

CHAPTERS 1–4

The following exercises will help you maintain the skills you have learned in this and previous chapters.

1. Solve for x: $-3x + 5 = -9(x - 7)$

2. Solve for x: $1.1 - 5x \leq 1.2 + x$

3. Symbolize the quotient of 3 divided by a number.

4. Compare $|-0.56|$, $-|-0.5|$, and $-\left|\frac{-7}{8}\right|$, and then arrange in order from smallest to largest.

5. Determine the x- and y-intercepts of $y = -x - 5$.

6. Simplify: $(-1)^2 - \left\{\frac{1}{4}[(-3)(-2) - 2]^2 + \left(\frac{1}{5}\right)\left(\frac{5}{7}\right)\right\}^0$

7. Find the solution set for $-x + 8 \geq -9$.

8. Simplify: $(-2)^5(-1)^{11}$

9. Solve for t: $26 = 0.9 - \frac{t}{5}$

10. Subtract: $(-2.39) - (-0.28)$

11. Complete this table of values for $-y + 2x = 5$, and graph the equation.

x	y
1	
0	
	0

12. Graph $y = 3x + 2$ using the slope and the y-intercept.

13. Evaluate $rt - r^2t + rt^2 + (r - t)^2$ when $r = 0$ and $t = -4$.

14. Simplify: $(2 - 3)^2 + 9(-6)^2$

15. Combine like terms:
    $7v^2x - 5x^2v + 8v^2x - 3xv^2 + 11v^3 + 12x^2$

16. Find the slope of the line that passes through the points $(-3, -4)$ and $(-6, -5)$.

17. Solve for r: $\frac{t}{r} = x - 4$

18. Evaluate $x^3y^2 - 4xy^2 + x$ when $x = -2$ and $y = \frac{1}{2}$.

19. Graph the inequality $-24x + 48 \leq 8y$.

20. Are the graphs of $5y - 4x = 7$ and $4y - 5x = 14$ perpendicular, parallel, or neither?

EXPONENTS AND POLYNOMIALS

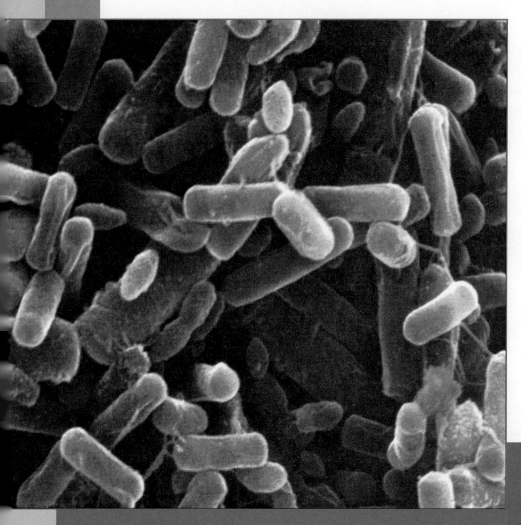

*S*cientists often work with very small or very large numbers. Virologists measure the size of viruses in millimicrons—0.000001 mm. Astronomers measure the vast distances of space using light years—the 9,460,000,000,000 kilometers that light travels in one year. And geologists estimate Earth's age at 4,600,000,000 years. Using exponents, the above numbers can also be written, respectively, as 10^{-6}, 9.46×10^{12}, and 4.6×10^9.

■ *Name quantities that would be easier to write using exponents.*

SKILLS CHECK

Take this short quiz to see how well prepared you are for Chapter 5. The answers follow the quiz.

1. Simplify: $2^0 \cdot 3^1$

2. Add: $2.5 + 3^2$

3. Add: $2n + 3n - 1$

4. Subtract: $1.7 - 0.98$

5. Solve: $x - 7 = -3$

6. Multiply: 0.34×4.2

7. What is the slope of $6y = 55x$?

8. Solve for x: $2 = \frac{t}{x}$

9. Divide: $(-126) \div (-12)$

10. Simplify: $3 + 2 \times 6$

ANSWERS: **1.** 3 [Section 1.4] **2.** 11.5 [Section 1.4] **3.** $5n - 1$ [Section 2.3] **4.** 0.72 [Section 1.2]
5. 4 [Section 3.1] **6.** 1.428 [Section 1.3] **7.** $\frac{55}{6}$ or $9\frac{1}{6}$ [Section 4.3] **8.** $x = \frac{t}{2}$ [Section 3.2]
9. $10\frac{1}{2}$ [Section 1.3] **10.** 15 [Section 1.4]

CHAPTER LEAD-IN

The English Imperial System of Measures used to contain a long sequence of measures in which each measure was double the previous one.

$$1 \text{ tun} = 2 \text{ pipes} = 4 \text{ hogsheads} = 8 \text{ barrels}$$
$$= 16 \text{ kilderkins} = 32 \text{ firkins (or bushels)}$$
$$= 64 \text{ demi bushels} = 128 \text{ pecks} = 256 \text{ gallons}$$
$$= 512 \text{ pottles} = 1024 \text{ quarts} = 2048 \text{ pints}$$
$$= 4096 \text{ chopins} = 8192 \text{ gills*}$$

Note that the numbers from 1 to 8192 are powers of 2 from 2^0 to 2^{13}.

Look at the units digit in each of the numbers and note the repeating pattern

$$1, 2, 4, 8, 6, 2, 4, 8, 6, 2, 4, 8, 6, \ldots$$

If this pattern repeats, what would be the units digit of each of the following

$$2^{17} \quad ?$$
$$2^{23} \quad ?$$
$$2^{40} \quad ?$$
$$2^{100} \quad ?$$

*Source: Keith Devlin, *Guardian*, 20 October 1983.

5.1 **Using Exponents**

SECTION LEAD-IN

Much of the early research in mathematics revolved around exploring patterns formed by the numbers themselves. We are going to do some of that.

There are four digits that have the following interesting characteristic. When they are the units digit in a number that is squared or cubed, they always reappear as the units digit in the result. Two examples appear below.

Zero:

$$10^2 = 100 \qquad 10^3 = 1000$$
$$20^2 = 400 \qquad 20^3 = 8000$$
$$30^2 = 900 \qquad 30^3 = 27{,}000$$

One:

$$11^2 = 121 \qquad 11^3 = 1331$$
$$21^2 = 441 \qquad 21^3 = 9261$$
$$31^2 = 961 \qquad 31^3 = 29{,}791$$

- Find the other two digits that have the same property.

Positive-Integer Exponents

You will recall that we use exponents to indicate repeated multiplication. The **exponent** tells us how many times the base is used as a factor:

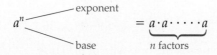

Therefore, 2^3 means $2 \times 2 \times 2$, or 8.

For all real numbers a,
$$a^1 = a$$
For all real numbers a, where a is not equal to zero,
$$a^0 = 1$$

Note: 0^0 is undefined.

■ ■ ■
EXAMPLE 1

Rewrite 4^3 without an exponent.

SOLUTION

We rewrite the expression 4^3 as repeated multiplication and then perform the multiplications. Because the exponent is 3, we know that 4 is used as a factor 3 times.

$$4^3 = 4 \times 4 \times 4 = 16 \times 4 = 64$$

So $4^3 = 64$.

▶ *CHECK* **Warm-Up 1**

In general,
$$(-a)^n = \underbrace{(-a)(-a)(-a)\cdots(-a)}_{n \text{ factors of } -a}$$
$$-a^n = -(a)^n = -\underbrace{[(a)\cdot(a)\cdots\cdot(a)]}_{\substack{\text{the opposite of} \\ n \text{ factors of } a}}$$

■ ■ ■
WRITER'S BLOCK

Write rules that a friend can use when (s)he is working with $(-a)^n$ and $-(a^n)$

■ ■ ■
EXAMPLE 2

Rewrite $(-6)^2$ and -6^2 without exponents.

SOLUTION

The parentheses in $(-6)^2$ tell us that the minus sign is part of the base. Therefore,

$$(-6)^2 = (-6)(-6) = 36$$

In -6^2, the exponent means the opposite of 6^2. Thus

$$-6^2 = (-1)(6)(6) = -36$$

So $(-6)^2 = 36$ and $-6^2 = -36$.

▶ *CHECK* **Warm-Up 2**

■ ■ ■
EXAMPLE 3

Rewrite $(3 + 5)^2$ and $(3 \times 5)^2$ without exponents.

SOLUTION

According to the definition,

$$(3 + 5)^2 = (3 + 5)(3 + 5) = (8)(8) = 64$$
$$(3 \times 5)^2 = (3 \times 5)(3 \times 5) = (15)(15) = 225$$

▶ *CHECK* **Warm-Up 3**

Calculator Corner

You can use your graphing calculator to check your work as you go through each step. You do this by "asking" the calculator if a statement is true or false. Study the calculator steps for Example 3.

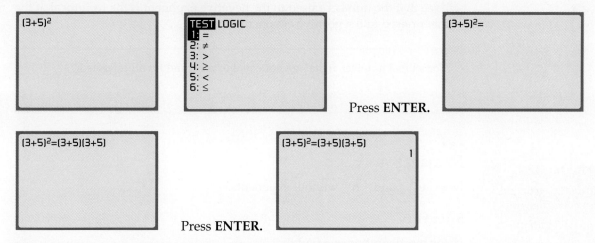

Press **ENTER.**

Press **ENTER.**

The graphing calculator returns an answer of 1 when the statement is true. However, if the statement is false, the calculator returns an answer of 0, as in the following example.

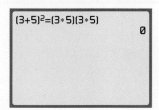

Negative-Integer Exponents

What happens when the exponent is a negative integer? To see, look at the pattern that is formed as we decrease the exponent by 1.

$$3^3 = (3)(3)(3) = 27$$
$$3^2 = (3)(3) = 9 \qquad \text{This is } 27 \div 3$$
$$3^1 = 3 \qquad \text{This is } 9 \div 3$$
$$3^0 = 1 \qquad \text{This is } 3 \div 3$$

Each time we reduce the exponent by 1, we divide the numerical value by the base. If we were to continue the process, we would get

$$3^{-1} = 1 \div 3 = \tfrac{1}{3}$$
$$3^{-2} = \tfrac{1}{3} \div 3 = \tfrac{1}{9}$$
$$3^{-3} = \tfrac{1}{9} \div 3 = \tfrac{1}{27}$$
$$3^{-4} = \tfrac{1}{27} \div 3 = \tfrac{1}{81}$$

If you look back and compare 3^3 and 3^{-3}, 3^2 and 3^{-2}, 3^1 and 3^{-1}, and so on, you will see that the number raised to the negative exponent is the reciprocal of the number raised to the positive exponent. That is,

> For all real numbers a that are not equal to zero, and for all integers n,
> $$a^{-n} = \frac{1}{a^n}$$

ERROR ALERT

Identify the error and give a correct answer.

Incorrect Solution:

$3^{-2} = -9$

∎∎∎

EXAMPLE 4

Rewrite 7^{-2} and -5^{-3} without exponents.

SOLUTION

Using our definition, we find that

$$7^{-2} = \tfrac{1}{7^2} = \tfrac{1}{(7)(7)} = \tfrac{1}{49}$$
$$-5^{-3} = (-1)(5^{-3})$$
$$\qquad = (-1)\tfrac{1}{5^3} \qquad \text{Negative exponent}$$
$$\qquad = (-1)\tfrac{1}{(5)(5)(5)} \qquad \text{Definition of exponent}$$
$$\qquad = (-1)\left(\tfrac{1}{125}\right) \qquad \text{Multiplying}$$
$$\qquad = {}^{-}\tfrac{1}{125} \qquad \text{Simplifying}$$

So $7^{-2} = \tfrac{1}{49}$ and $-5^{-3} = {}^{-}\tfrac{1}{125}$.

▶ CHECK **Warm-Up 4**

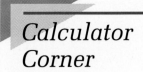

Calculator Corner

The graphing calculator can also be used to check your work with negative exponents and give the answer in fraction form. (*Note:* See the Calculator Corner on page 6 to review how to get an answer in fractional form.)

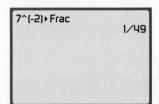

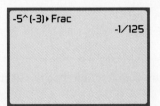

Our definitions apply to variables as well as to real numbers because the variables represent real numbers. So, for example,

$$x^4 = \underbrace{x \cdot x \cdot x \cdot x}_{4 \text{ factors of } x}$$
$$x^1 = x$$
$$x^0 = 1$$

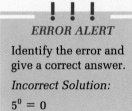

ERROR ALERT

Identify the error and give a correct answer.

Incorrect Solution:

$5^0 = 0$

••

EXAMPLE 5

Simplify: $(3y)^2$

SOLUTION

By the definition,

$$(3y)^2 = 3y \cdot 3y = 3 \cdot 3 \cdot y \cdot y = 9y^2$$

So $(3y)^2 = 9y^2$.

▶ *CHECK* **Warm-Up 5**

••

EXAMPLE 6

Rewrite p^{-3} and $(-y)^{-2}$ with positive exponents.

SOLUTION

By the definition,

$$p^{-3} = \frac{1}{p^3}$$

$$(-y)^{-2} = \frac{1}{(-y)^2}$$

$$= \frac{1}{(-y)(-y)} \quad \text{Definition of exponent}$$

$$= \frac{1}{y^2} \quad \text{Simplifying}$$

So $p^{-3} = \frac{1}{p^3}$ and $(-y)^{-2} = \frac{1}{y^2}$.

▶ CHECK **Warm-Up 6**

ERROR ALERT

Identify the error and give a correct answer.

Incorrect Solution:

$6^0 = 6^1 = 6$

> The reciprocal of a non-zero fraction $\frac{m}{a}$ is the fraction $\frac{a}{m}$. So for fractions,
>
> $$\left(\frac{m}{a}\right)^{-n} = \left(\frac{a}{m}\right)^n$$

■ ■ ■

EXAMPLE 7

Rewrite $\left(\frac{3}{y}\right)^{-2}$ with a positive exponent.

SOLUTION

Because the reciprocal of $\frac{3}{y}$ is $\frac{y}{3}$, we have $\left(\frac{3}{y}\right)^{-2} = \left(\frac{y}{3}\right)^2 = \left(\frac{y}{3}\right)\left(\frac{y}{3}\right) = \frac{y^2}{9}$.

So $\left(\frac{3}{y}\right)^{-2} = \frac{y^2}{9}$.

▶ CHECK **Warm-Up 7**

■ ■ ■

EXAMPLE 8

Rewrite $9\left(-\frac{9}{11}\right)^{-3}$ without an exponent.

SOLUTION

Because the reciprocal of $\left(-\frac{9}{11}\right)$ is $\left(-\frac{11}{9}\right)$,

$$9\left(-\frac{9}{11}\right)^{-3} = 9\left(-\frac{11}{9}\right)^3$$

Using the standard order of operations to simplify, we have

$$9\left(-\frac{11}{9}\right)^3 = 9\left(-\frac{11}{9}\right)\left(-\frac{11}{9}\right)\left(-\frac{11}{9}\right)$$

$$= \frac{9(-11)(-11)(-11)}{(9)(9)(9)}$$

$$= -\frac{1331}{81}$$

So $9\left(-\frac{9}{11}\right)^{-3} = -\frac{1331}{81}$.

▶ CHECK **Warm-Up 8**

Calculator Corner

You can check your work with a graphing calculator.

Raising Terms to Powers

There are two ways to raise a number to a power. One is to multiply out the powers, as in

$$(3)^4 = (3)(3)(3)(3) = 3^4$$

However, there is a simpler way:

> **Power Rule for Variables with Exponents**
> For any real number a and integers m and n,
> $$(a^m)^n = a^{mn}$$

So $(3^2)^4 = 3^{2 \cdot 4} = 3^8$, and $(x^3)^2 = x^{3 \cdot 2} = x^6$.

When we raise a term to a power, we raise each part of the term to that power. For example,

$$(xy)^3 \quad \text{means} \quad (xy)(xy)(xy)$$

Rearranging the factors gives us

$$x \cdot x \cdot x \cdot y \cdot y \cdot y = x^3 y^3$$

As you might suspect, there is again a simpler way:

> **Power Rule for Products with Exponents**
> For any real numbers a and b and any integers m, n, and t,
> $$(a^m b^n)^t = a^{mt} b^{nt}$$

In words, to raise a term that contains exponents to a power, we multiply each exponent of the term by the power.

■■■
EXAMPLE 9

Simplify: $(v^3xy)^{12}$

SOLUTION

The exponent of x and of y is 1, so

$$(v^3xy)^{12} = (v^3x^1y^1)^{12}$$

Then, multiplying each exponent by the power gives us

$$
\begin{aligned}
(v^3x^1y^1)^{12} &= v^{3\cdot12}x^{1\cdot12}y^{1\cdot12} \quad &&\text{Power rule} \\
&= v^{36}x^{12}y^{12} \quad &&\text{Simplifying}
\end{aligned}
$$

So $(v^3xy)^{12} = v^{36}x^{12}y^{12}$.

▶ CHECK **Warm-Up 9**

When a term that is raised to a power contains a coefficient, be sure to raise the coefficient to the given power also.

■■■
EXAMPLE 10

Simplify: $(-3x^4y^3z^9)^2$

SOLUTION

To simplify this expression, we must raise (-3) to the second power *as well as* x^4, y^3, and z^9.

$$
\begin{aligned}
(-3x^4y^3z^9)^2 &= (-3)^{1\cdot2}x^{4\cdot2}y^{3\cdot2}z^{9\cdot2} \quad &&\text{Power rule} \\
&= (-3)^2x^8y^6z^{18} \quad &&\text{Simplifying} \\
&= 9x^8y^6z^{18} \quad &&\text{Squaring}
\end{aligned}
$$

So $(-3x^4y^3z^9)^2 = 9x^8y^6z^{18}$.

▶ CHECK **Warm-Up 10**

Power Rule for Fractions

For any real numbers a and b, where b is not equal to zero, and for any integer m,

$$\left(\frac{a}{b}\right)^m = \frac{a^m}{b^m}$$

■■■
EXAMPLE 11

Simplify: $\left(\dfrac{-2x^3y^2}{5v^4}\right)^3$

SOLUTION

When we simplify this expression, we must be careful to raise both numerator and denominator to the given power, 3. You will remember to do so if you apply the fraction rule as the first step.

$$\left(\frac{-2x^3y^2}{5v^4}\right)^3 = \frac{(-2x^3y^2)^3}{(5v^4)^3} \quad \text{Power rule for fractions}$$

Now, multiplying the exponents by the power 3 in both numerator and denominator yields

$$\frac{(-2x^3y^2)^3}{(5v^4)^3} = \frac{(-2)^{1\cdot3}\,x^{3\cdot3}\,y^{2\cdot3}}{(5)^{1\cdot3}\,v^{4\cdot3}} \quad \text{Power rule}$$

$$= \frac{(-2)^3x^9y^6}{5^3v^{12}} \quad \text{Simplifying}$$

$$= \frac{-8x^9y^6}{125v^{12}} \quad \text{Simplifying}$$

So $\left(\dfrac{-2x^3y^2}{5v^4}\right)^3 = \dfrac{-8x^9y^6}{125v^{12}}$.

▶ *CHECK* **Warm-Up 11**

. . .

EXAMPLE 12

Simplify $(4rt^3)^{-2}$ and write the result using only positive exponents.

SOLUTION

Using the power rule, we have

$$(4rt^3)^{-2} = 4^{1\cdot(-2)}r^{1\cdot(-2)}t^{3\cdot(-2)} \quad \text{Power rule}$$

$$= 4^{-2}r^{-2}t^{-6} \quad \text{Simplifying}$$

$$= \left(\frac{1}{16}\right)r^{-2}t^{-6} \quad \text{Simplifying}$$

Because the problem asked for positive exponents only, we must rewrite $\frac{1}{16}r^{-2}t^{-6}$ as

$$\frac{1}{16}\cdot\frac{1}{r^2}\cdot\frac{1}{t^6} = \frac{1}{16r^2t^6}$$

An alternative method can also be used to solve this problem. Recall that $a^{-n} = \frac{1}{(a^n)}$, where a can be a term as well as a number. That gives us

$$(4rt^3)^{-2} = \frac{1}{(4rt^3)^2}$$

$$= \frac{1}{4^{1\cdot2}r^{1\cdot2}t^{3\cdot2}} \quad \text{Power rule}$$

$$= \frac{1}{4^2r^2t^6} \quad \text{Simplifying}$$

$$= \frac{1}{16r^2t^6} \quad \text{Simplifying}$$

The answers are the same. So $(4rt^3)^{-2} = \dfrac{1}{16r^2t^6}$.

▶ *CHECK* **Warm-Up 12**

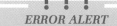

ERROR ALERT

Identify the error and give a correct answer.

Incorrect Solution:

$6x^{-3} = \dfrac{1}{6x^3}$

■■■

EXAMPLE 13

Simplify $\left(\dfrac{2xt^2}{y}\right)^{-4}$ and write the result using only positive exponents.

SOLUTION

We first remove the negative exponent.

$$\left(\frac{2xt^2}{y}\right)^{-4} = \left(\frac{y}{2xt^2}\right)^4 \quad \text{Removing the negative exponent}$$

$$= \frac{y^4}{(2xt^2)^4} \quad \text{Power rule for fractions}$$

$$= \frac{y^4}{2^4 x^4 t^{2\cdot4}} \quad \text{Power rule}$$

$$= \frac{y^4}{16x^4 t^8} \quad \text{Simplifying}$$

So $\left(\dfrac{2xt^2}{y}\right)^{-4} = \dfrac{y^4}{16x^4 t^8}$.

▶ *CHECK* **Warm-Up 13**

Practice what you learned.

SECTION FOLLOW-UP

The digits 5 and 6 work just like 0 and 1. Finish what we started.

Five:

$$15^2 = 225 \qquad 15^3 = 3375$$
$$25^2 = \underline{\hspace{1cm}} \qquad 25^3 = 15{,}625$$
$$35^2 = 1225 \qquad 35^3 = \underline{\hspace{1cm}}$$

Six:

$$16^2 = 256 \qquad 16^3 = 4096$$
$$26^2 = \underline{\hspace{1cm}} \qquad 26^3 = \underline{\hspace{1cm}}$$
$$36^2 = \underline{\hspace{1cm}} \qquad 36^3 = \underline{\hspace{1cm}}$$

If we raised any of these numbers to the 4th, 5th, or any power, would the units digit of the answer be different from the units digit of the original number? Check it out.

5.1 WARM-UPS

Work these problems before you attempt the exercises.

1. Rewrite 5^2 without exponents.

2. Rewrite -5^4 and $(-5)^2$ without exponents.

3. Rewrite $(2 + 9)^2$ without an exponent.

4. Rewrite 8^{-1} and -6^{-4} without exponents.

5. Simplify: $(4z)^3$

6. Rewrite x^{-1} and $(-x)^{-5}$ with positive exponents.

7. Rewrite $\left(-\frac{5}{6}\right)^{-3}$ without an exponent.

8. Simplify: $3\left(-\frac{6}{9}\right)^{-3}$

9. Simplify: $(r^2s)^3$

10. Simplify: $-(9x^5y^8)^2$

11. Simplify: $\left(\frac{5r^3s^2}{9t^5}\right)^2$

12. Simplify $(5r^2t^4)^{-4}$. Write the result using positive exponents.

13. Simplify: $\left[\frac{(3rt^3)}{x}\right]^{-2}$

5.1 EXERCISES

Note: Use your graphing calculator to check your results whenever possible.

In Exercises 1 through 8, simplify.

1. 4^0 2. a^1 3. 3^1 4. m^0

5. x^1 6. t^1 7. $(147x^2y^2z^{15})^0$ 8. $(4,851,252)^0$

In Exercises 9 through 24, rewrite each term without an exponent.

9. -7^2 10. $(-7)^2$ 11. $(-3)^5$ 12. -3^4

13. $\left(\frac{4}{9}\right)^3$ 14. -6^4 15. -0.8^2 16. $\left(\frac{4}{5}\right)^3$

17. $(2 + 5)^2$ 18. $(7 + 3)^3$ 19. $(1 + 8)^3$ 20. $\left(\frac{1}{5} + \frac{3}{5}\right)^2$

21. $(4 \times 0.4)^5$ 22. $(2.4 \times 3)^2$ 23. $(6 - 1.3)^3$ 24. $\left(\frac{2}{7}\right)^3$

In Exercises 25 through 44, rewrite each term without a negative exponent.

25. $(-2)^{-1}$

26. $(-5)^{-3}$

27. $\left(-\frac{1}{3}\right)^{-1}$

28. $\left(-\frac{1}{2}\right)^{-4}$

29. m^{-4}

30. n^{-3}

31. $(-p)^{-2}$

32. $(-r)^{-4}$

33. $(-t)^{-1}$

34. $(-x)^{-2}$

35. $(6y)^{-2}$

36. $6y^{-2}$

37. $-3x^{-3}$

38. $-(3n)^{-1}$

39. $\left(-\frac{1}{x}\right)^{-3}$

40. $\left(-\frac{2}{m}\right)^{-5}$

41. $\left(-\frac{3}{5}\right)^{-2}$

42. $\left(-\frac{4}{7}\right)^{-4}$

43. $-\left(-\frac{4}{9}\right)^{-4}$

44. $-\left(\frac{11}{5}\right)^{-3}$

In Exercises 45 through 48, find the real number represented.

45. $14\left(\frac{7}{4}\right)^{-5}$

46. $8\left(\frac{2}{5}\right)^{-3}$

47. $9\left(\frac{3}{8}\right)^{-2}$

48. $8 + \left(\frac{8}{9}\right)^{-2}$

49. Rewrite the formula $x = \frac{a}{2t^{-2}}$ without a negative exponent.

50. Rewrite the formula $t = dr^{-1}$ without a negative exponent.

51. Rewrite the formula $I(pr)^{-1} = t$ without a negative exponent.

52. Rewrite the formula $C = p^{-2}t$ without a negative exponent.

In Exercises 53 through 100, simplify by raising each term to the indicated power. Write answers with positive exponents only.

53. $(4x)^2$

54. $(7m)^5$

55. $(5n)^3$

56. $(9r)^4$

57. $(-8x)^2$

58. $(-4m)^3$

59. $(-9n)^2$

60. $(-2r)^4$

61. $(w^4xy)^{-3}$

62. $(mnx^6)^2$

63. $(n^8yz)^4$

64. $(r^5tv)^{-5}$

65. $(5x^3y^2)^4$

66. $(2m^2n^4)^2$

67. $(4n^2y^3)^{-3}$

68. $(3r^6t^4)^{-5}$

69. $(-2x^5y^3)^2$

70. $(-4m^2n^4)^3$

71. $(-2n^3y^5)^4$

72. $(-3r^6t)^{-1}$

73. $(2x^{-2}y^3)^{-5}$

74. $(5m^{-4}n^3)^{-2}$

75. $(3ny^{-2})^3$

76. $(2rt^{-4})^2$

77. $(-3x^{-3}y^2)^3$

78. $(-8mn^{-3})^4$

79. $(-6ny^2)^2$

80. $(-5xt^4)^2$

81. $(11x^2y^5)^2$

82. $(-3m^2r^3)^2$

83. $(-4n^{-3}x^3)^0$

84. $(7r^{-3}s^2)^0$

85. $\left(\frac{3}{x}\right)^2$

86. $\left(\frac{4}{m}\right)^3$

87. $\left(\frac{2}{n}\right)^{-4}$

88. $\left(\frac{6}{r}\right)^{-4}$

89. $\left(\frac{2w^3x^4}{3y^4}\right)^5$

90. $\left(\frac{4m^3x^2}{7n^4}\right)^3$

91. $\left(\frac{3n^2z^3}{5y^4}\right)^2$

92. $\left(\frac{2r^3}{3t^6v^2}\right)^4$

93. $\left(\frac{-8x^3}{9y^4}\right)^2$

94. $\left(\frac{-m^3}{2n^2x^5}\right)^3$

95. $\left(\frac{5n^4}{-7y^7z^6}\right)^{-2}$

96. $\left(\frac{9r^3t^2}{13v^5}\right)^{-2}$

97. $\left(\frac{r^3s^{-2}t}{m^{-3}np}\right)^{-2}$

98. $\left(\frac{x^2y^{-1}z^4}{rst}\right)^{-3}$

99. $\left(\frac{m^3n^{-2}p^{-5}}{3x^2yz}\right)^{-7}$

100. $\left(\frac{-4r^2t}{-6mn^{-5}z}\right)^{-2}$

EXCURSIONS

Exploring Patterns

1. Evaluate the following expressions on your graphing calculator's Home Screen. Try to find a pattern in the results that will help you write a rule for working with $(-a)^n$ and $-(a^n)$. The work for part (a) is shown at the right.

   a. $(-3)^2$ **b.** $(-3)^3$ **c.** $(-6)^5$ **d.** $(-6)^6$

   e. $(-11)^3$ **f.** $(-11)^8$ **g.** $(-10)^3$ **h.** $(-10)^8$

   Now evaluate these expressions. Notice the difference in the position of the parentheses and the negation sign in these exercises and the exercises you just completed.

   i. $-(3^3)$ **j.** $-(3^2)$ **k.** $-(6^5)$ **l.** $-(6^6)$ **m.** $-(11^5)$ **n.** $-(11^8)$

   ✏ Can you now write a rule that a friend can use when he or she is working with $(-a)^n$ and $-(a^n)$?

2. Use your graphing calculator to complete the following lists of values. Give your answers for part (b) in fraction form.

   a. $3^1 =$ $3^{11} =$ **b.** $3^{-1} =$ $3^{-11} =$
        $3^2 =$ $3^{12} =$ $3^{-2} =$ $3^{-12} =$
        $3^3 =$ $3^{13} = 1{,}594{,}323$ $3^{-3} =$ $3^{-13} =$
        $3^4 =$ $3^{14} =$ $3^{-4} =$ $3^{-14} =$
        $3^5 = 243$ $3^{15} =$ $3^{-5} =$ $3^{-15} =$
        $3^6 =$ $3^{16} =$ $3^{-6} =$ $3^{-16} =$
        $3^7 =$ $3^{17} =$ $3^{-7} =$ $3^{-17} =$
        $3^8 =$ $3^{18} =$ $3^{-8} =$ $3^{-18} =$
        $3^9 =$ $3^{19} =$ $3^{-9} =$ $3^{-19} =$
        $3^{10} = 59{,}049$ $3^{20} =$ $3^{-10} =$ $3^{-20} =$

   c. ✏ Describe the relationship between positive and negative exponents.

Exploring Numbers

3. Find the number that when raised to the given exponent results in the given answer.

   a. $\underline{\hspace{1cm}}^2 = 169$ **b.** $\underline{\hspace{1cm}}^2 = 225$ **c.** $\underline{\hspace{1cm}}^2 = 324$

   d. $\underline{\hspace{1cm}}^2 = 529$ **e.** $\underline{\hspace{1cm}}^2 = 1024$ **f.** $\underline{\hspace{1cm}}^2 = 2116$

   g. $\underline{\hspace{1cm}}^2 = 2601$ **h.** $\underline{\hspace{1cm}}^2 = 3969$ **i.** $\underline{\hspace{1cm}}^2 = 5329$

   j. $\underline{\hspace{1cm}}^2 = 7225$ **k.** $\underline{\hspace{1cm}}^3 = 12{,}167$ **l.** $\underline{\hspace{1cm}}^3 = 35{,}937$

   m. $\underline{\hspace{1cm}}^4 = 83{,}521$ **n.** $\underline{\hspace{1cm}}^4 = 194{,}481$ **o.** $\underline{\hspace{1cm}}^5 = 161{,}051$

   p. $\underline{\hspace{1cm}}^5 = 7{,}962{,}624$

   q. ✏ Discuss how you found these values. What procedure did you use? How did you modify your "plan" as you worked through these problems?

CONNECTIONS TO *STATISTICS*

Reading Bar Graphs

Bar Graphs

Information is often presented in the form of a **graph,** a diagram that shows numerical data in visual form. Graphs enable us to "see" relationships that are difficult to describe with numbers alone.

The following graph is called **bar graph.** This one shows the amount of electricity used daily, on the average, by a certain customer in each of 13 successive months.

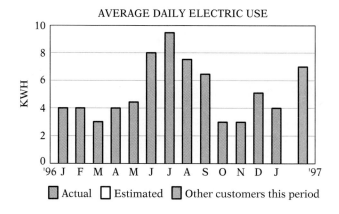

AVERAGE DAILY ELECTRIC USE

■ Actual ☐ Estimated ■ Other customers this period

Information that helps you read the graph is given along the bottom and left side of the graph. Along the bottom are the months covered by the graph, abbreviated, from January 1996 to January 1997. There is one bar for each month. The height of the bar gives the average amount of electricity used per day in kilowatt hours (kwh) for its month. The height is read on the scale at the left of the graph.

■■■

EXAMPLE

a. How many kilowatt hours did this customer use daily during March, to the nearest whole number?

b. During four of the months shown, the customer used the same number of kilowatt hours per day. Which months were these?

SOLUTION

a. The third bar from the left, the bar for March, ends halfway between 2 and 4. Thus, during March this customer used 3 kilowatt hours per day.

b. We must find four bars that are the same height. These occur in January of both years, February, and April. They have a height of 4 kwh.

Double Bar Graphs

Comparative information is often shown on a **double bar graph** such as the following figure. This type of graph usually presents two measurements for each given date or time. The reader then compares the measurements by comparing bars.

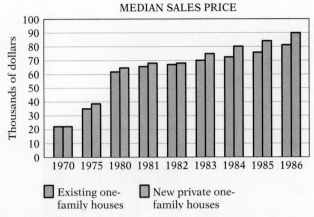

Source: U.S. Bureau of the Census.

The scale on the left side of this graph shows the median sales prices of one-family houses—existing or new—in thousands of dollars. The scale along the bottom shows the years from 1970 through 1986.

▪▪▪

EXAMPLE

In what years was the difference in median sale prices about $10,000?

SOLUTION

The price scale on the left shows that $10,000 is the difference between any two adjacent horizontal lines. The years for which the tops of the two bars are approximately that far apart are 1984, 1985, and 1986.

PRACTICE

In Exercises 1 and 2, use the Average Daily Electric Use graph.

1. How many kilowatt hours were used in October?

2. How many kilowatt hours were used altogether in the months April and June?

In Exercises 3 and 4, use the Median Sales Price graph.

3. In what year was the median sales price the highest for new private one-family houses?

4. In what year was the median sales price of existing houses closest to that of new houses?

Use the Average Daily Electric Use graph to answer Exercises 5 and 6.

5. How many more kilowatt hours were used per day in June than in November?

6. How many kilowatt hours were used altogether in the months of October, November, and December 1996 and January 1997?

Use the following figure to answer Exercises 7 and 8.

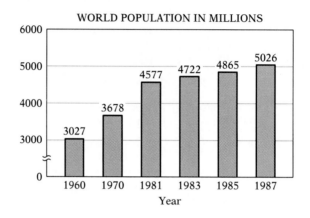

WORLD POPULATION IN MILLIONS

7. Was the increase from 1960 to 1970 more or less than the increase from 1970 to 1981?

8. Which two-year time period had the greatest increase?

In Exercises 9 and 10, use the Median Sales Price graph.

9. What was the difference between the approximate median prices of new and existing one-family houses in 1983?

10. Was the jump in median prices of houses greater from 1970 to 1975 or from 1975 to 1980? How can you tell?

SECTION GOALS

- *To multiply monomials*
- *To divide monomials*

5.2 Multiplying and Dividing Monomials

SECTION LEAD-IN

A particular bacteria will double in number every 10 hours.

- How long would it take (in theory) for 1 bacterium to become 1,000,000?

Multiplication

There is an easy way to multiply terms that contain the same bases.

Product rule for exponents

For any non-zero real number a, and for any integers m and n,

$$a^m \cdot a^n = a^{m+n}$$

To multiply two numbers with the same base, we add their exponents. This rule and the other rules you will learn here also apply to variables (because they also represent numbers).

An expression with only one term is called a **monomial.** Usually, a monomial includes a coefficient and a variable part that contains one or more variables, as in $3x^2y^2$ and $32.7mn^3p^4$.

To multiply two monomials

To multiply two monomials that have the same variables, multiply the coefficients and then multiply the variables according to the product rule.

In the first example, we are asked to simplify. This means to perform all indicated operations.

▪▪▪

EXAMPLE 1

Simplify: $(2x^{12})(3xy^3)$

SOLUTION

Terms are easier to multiply when we can see all coefficients and the exponents. So we rewrite this as

$$2x^{12}(3x^1y^3)$$

We then multiply the coefficients and add the exponents of each like base.

$$(2x^{12})(3x^1y^3) = (2)(3)x^{12+1}y^3 = 6x^{13}y^3$$

So $2x^{12}(3xy^3) = 6x^{13}y^3$.

▶ *CHECK* **Warm-Up 1**

In the next example, one of the terms has no indicated coefficient. But remember, a term without an indicated coefficient has the coefficient 1.

▪ ▪ ▪

EXAMPLE 2

Multiply: $\frac{1}{2}x^4y(x^2y^3)$

SOLUTION

ERROR ALERT

Identify the error and give a correct answer.

Incorrect Solution:
$4x^5 \cdot 8x^3 = 32x^{15}$

Here we rewrite to show all the coefficients and exponents.

$$\frac{1}{2}x^4y(x^2y^3) = \frac{1}{2}x^4y^1(1x^2y^3)$$

Then we multiply the coefficients and add the exponents of each like base. We get

$$\frac{1}{2}(1)(x^4)(x^2)(y^1)(y^3) = \frac{1}{2}x^{4+2}y^{1+3} = \frac{1}{2}x^6y^4$$

So $\frac{1}{2}x^4y(x^2y^3) = \frac{1}{2}x^6y^4$.

▶ CHECK **Warm-Up 2**

The next example contains three monomials. The commutative property enables us to multiply in any order.

▪ ▪ ▪

EXAMPLE 3

Multiply: $(3p^2n^3)(-8pn^4)(-0.5p^2n^5)$

SOLUTION

Rewriting each term to show the exponents, we have

$$(3p^2n^3)(-8p^1n^4)(-0.5p^2n^5)$$

Then, using the commutative property to regroup the coefficients and variable parts, we obtain

$$(3)(-8)(-0.5)(p^2)(p^1)(p^2)(n^3)(n^4)(n^5) \quad \text{Rearranging}$$
$$= 12p^{2+1+2}n^{3+4+5} = 12p^5n^{12} \quad \text{Simplifying}$$

So $(3p^2n^3)(-8pn^4)(-0.5p^2n^5) = 12p^5n^{12}$.

▶ CHECK **Warm-Up 3**

In this last example, two of the variables have negative exponents. We simply add them as we add positive exponents.

▪ ▪ ▪

EXAMPLE 4

Multiply: $(3x^3y^2z^4)(4x^{-2}y^{-3})$

SOLUTION

Regrouping coefficients and variables gives us

$$(3)(4)(x^3)(x^{-2})(y^2)(y^{-3})(z^4)$$

Simplifying, we have

$$12x^{3+(-2)}y^{2+(-3)}z^4 = 12xy^{-1}z^4 = \frac{12xz^4}{y}$$

So $(3x^3y^2z^4)(4x^{-2}y^{-3}) = 12xy^{-1}z^4 = \frac{12xz^4}{y}$.

▶ CHECK **Warm-Up 4**

Division

There is a simple way to divide monomials.

> **Quotient Rule for Exponents**
>
> For any non-zero real number a and any integers m and n,
> $$a^m \div a^n = a^{m-n}$$

In words, to divide numbers or variables with the same bases, subtract their exponents. So, for example,

$$\frac{x^7}{x^5} = x^{7-5} = x^2$$

In the first example, the variable with the larger exponent is in the divisor. We simply apply the quotient rule.

∎∎∎

EXAMPLE 5

Divide: $\frac{45x}{15x^4}$

SOLUTION

By the quotient rule,

$$\frac{45x}{15x^4} = \frac{45}{15}x^{1-4} = 3x^{-3} = \frac{3}{x^3}$$

So $\frac{45x}{15x^4} = 3x^{-3} = \frac{3}{x^3}$.

▶ CHECK **Warm-Up 5**

∎∎∎

EXAMPLE 6

Divide: $\frac{18m^3n^{-6}}{8m^{-5}n}$

SOLUTION

By the quotient rule,

$$\frac{18m^3n^{-6}}{8m^{-5}n} = \frac{18}{8}m^{3-(-5)}n^{-6-1} = \frac{9}{4}m^8n^{-7}$$

!!!
ERROR ALERT

Identify the error and give a correct answer.

Incorrect Solution:

$$\frac{3^8}{3^2} = \frac{3^8}{3^2}$$
$$= \frac{1^8}{1^2}$$
$$= \frac{1}{1}$$
$$= 1$$

Then we eliminate the negative exponent.

$$\frac{9}{4}m^8n^{-7} = \frac{9}{4}m^8 \cdot \frac{1}{n^7} = \frac{9m^8}{4n^7}$$

Thus

$$\frac{18m^3n^{-6}}{8m^{-5}n} = \frac{9m^8}{4n^7}$$

▶ CHECK **Warm-Up 6**

In the next example, there is a variable in the denominator that is not in the numerator.

▪▪▪
EXAMPLE 7

Divide: $\dfrac{14x^3y^3}{7xy^{-4}t^2}$

! ! !
ERROR ALERT

Identify the error and give a correct answer.

Incorrect Solution:

$\dfrac{x^{12}}{x^6} = x^{\frac{12}{6}}$

$\quad = x^2$

SOLUTION

Again, we apply the quotient rule and simplify. But first we need to write the numerator and denominator with all the same variables.

$$\frac{14x^3y^3}{7xy^{-4}t^2} = \frac{14x^3y^3t^0}{7xy^{-4}t^2}$$

$$= \frac{14}{7}x^{3-1}y^{3-(-4)}t^{0-2} \quad \text{Quotient rule}$$

$$= 2x^2y^7t^{-2} \quad\quad\quad \text{Simplifying}$$

$$= 2x^2y^7 \cdot \frac{1}{t^2} = \frac{2x^2y^7}{t^2}$$

Thus

$$\frac{14x^3y^3}{7xy^{-4}t^2} = \frac{2x^2y^7}{t^2}$$

▶ CHECK **Warm-Up 7**

Practice what you learned.

SECTION FOLLOW-UP

To find the number of bacteria, we start by using 2^n, where n changes from 0 to any positive integer.

$$1 = 2^0$$

$$2 \times 1 = 2^1 \cdot 2^0 = 2^1 = 2$$

$$2 \times 2 \times 1 = 2^2 \cdot 2^0 = 2^2 = 4$$

It turns out that

$$2^{19} = 524{,}288$$

and

$$2^{20} = 1{,}048{,}576$$

It also turns out that the time needed to go from 1 bacterium to 2 bacteria is 10 hours. And it takes 10 hours to go from 2 bacteria to 4.

$$2^0 \longrightarrow 2^1 = 10 \text{ hours}$$
$$2^0 \longrightarrow 2^2 = 2 \times 10 = 20 \text{ hours}$$
$$2^0 \longrightarrow 2^3 = 3 \times 10 = 30 \text{ hours}$$

So from

$$2^0 \longrightarrow 2^{19} = 19 \times \underline{} = 190 \text{ hours}$$
$$2^0 \longrightarrow 2^{20} = \underline{} \times 10 = \underline{} \text{ hours}$$

So it would take from 190 to 200 hours. How many days is this? minutes? seconds?

5.2 WARM-UPS

Work these problems before you attempt the exercises.

1. Multiply: $4x^3(2x^2y)$

2. Multiply: $\frac{3}{4}x^5y^2(x^3y^2)$

3. Multiply: $(3x^3y^2)(11x^3y^4)(-4x^3)$

4. Multiply: $(-3.4x^{-2}y^{-3})(11xy)$

5. Divide: $\frac{6r^2}{12r^3}$

6. Divide: $\frac{24x^{-3}}{16x^4}$

7. Divide: $\frac{-2.6m^3n^{-2}}{0.5m^{-4}t^3}$

5.2 EXERCISES

Note: Use your graphing calculator to check your results whenever possible.

In Exercises 1 through 40, multiply the terms in each exercise. Write your answers without negative exponents.

1. $(m^9)(-m)$

2. $(-r)(-r^5)$

3. $(3x)(4x^2)$

4. $(7m)(9m^5)$

5. $(2x)(-5x^{-9})$

6. $(m)(-8m^{-2})$

7. $(-23t)(-2t^{-3})$

8. $(4x^{-2}y)(3x^3)$

9. $(5r^4t)(3rt)$

10. $(2r^2t^5)(5t^3)$

11. $(-4x^{-2}y)(3x^3y^2)$

12. $(-14p^3n^{-3})(9p^{-5}n)$

13. $(8r^2t^3)(28r^{-4}t)$

14. $(-18t^3u^4)(6t^{-7}u)$

15. $(2x^{-3}y)(3x^{-5}yt)$

16. $(7pn)(9p^{-3}n^2)$

17. $(12r^{-1}u)(8r^3)$

18. $(5tv)(4t^4v^{-2})$

19. $(5ru)(-7r^3)(-8r^2u^2)$

20. $(4tv)(-9v^2)(-8tv^2)$

21. $(-2xy)(3x^{-4})(-3r^{-2})$

22. $(-16mn^{-7})(4mn)(-9m^2n^3)$

23. $(7m^4t)(-12mt^{-6})(-12m^{-3}t)$

24. $(-10tu^3)(5tu^{-2})(9tu^3)$

25. $\left(\frac{1}{2}x^5y\right)(3x^2y^3)$

26. $\left(\frac{1}{4}p^2n\right)(3p^4)$

27. $\left(\frac{5}{6}x^2y\right)(-3x^3y^6)$

28. $(-7m^4n^9)\left(\frac{6}{7}n^5m^8\right)$

29. $(-16t^{-2}v^3)\left(\frac{1}{2}tv^2\right)$

30. $\left(\frac{7}{9}t^3v^{-4}\right)(-9v^{-2}t^3)$

31. $\left(\frac{2}{5}x^4y\right)(3x^2y^2)$

32. $\left(\frac{5}{6}n^2p\right)(2p^{-4})$

33. $\left(\frac{1}{2}x^5y\right)\left(\frac{2}{5}x^4y\right)$

34. $\left(\frac{5}{6}n^2p\right)\left(\frac{1}{4}p^2n\right)$

35. $(-4.3x^2y^3)(2.1x^3y^4)$

36. $(-0.7m^5)(1.5mn^{-7})$

37. $(1.5x^2)(0.2x^2y)$

38. $(2.4p^2)(-5pn^{-2})$

39. $(-12mn^{-2})(4mnt^3)$

40. $(22xy^2)(-5tx^{-2}y)$

In Exercises 41 through 80, perform each division. Rewrite all answers with positive exponents.

41. $\dfrac{w^4}{w}$

42. $\dfrac{m^7}{m}$

43. $\dfrac{n}{n^{-7}}$

44. $\dfrac{r}{r^{-8}}$

45. $\dfrac{-n}{n^8}$

46. $\dfrac{-r}{r^{10}}$

47. $\dfrac{-w^{-2}}{w^5}$

48. $\dfrac{-m^{-2}}{m^4}$

49. $\dfrac{27w^7}{w^{-6}}$

50. $\dfrac{36m^5}{m^{-4}}$

51. $\dfrac{48n^3}{n^{-2}}$

52. $\dfrac{32r^8}{r^{-7}}$

53. $\dfrac{w^{-2}x}{wx^3}$

54. $\dfrac{m^{-3}n}{mn^4}$

55. $\dfrac{3w^3}{8w^{-8}}$

56. $\dfrac{5m^4}{12m^{-7}}$

57. $\dfrac{2n^5}{7n^9}$

58. $\dfrac{3r^2}{13r^6}$

59. $\dfrac{-2n}{n^9}$

60. $\dfrac{6r}{-24r^{-5}}$

61. $\dfrac{w^{-2}x}{w^4y}$

62. $\dfrac{m^{-4}n}{m^6r}$

63. $\dfrac{27w^{-3}x}{w^3x^2}$

64. $\dfrac{-48m^2n^{-1}}{26n^2r}$

65. $\dfrac{12n^2ty^4}{18n^3t^2y^5}$

66. $\dfrac{24r^2t^2u}{36r^3t^4u^3}$

67. $\dfrac{27w^2x^{-3}y}{3w^3x^2y^5}$

68. $\dfrac{18m^2n^3r^7}{6m^5n^6r^5}$

69. $\dfrac{63n^4t^2}{-7n^7t^5y}$

70. $\dfrac{-56r^2s^3}{-8r^{-2}s^6t^3}$

71. $\dfrac{4.5w^2x^8y^4}{0.5wx^2y^4}$

72. $\dfrac{-4.9m^3n^8}{-0.7m^8n^2}$

73. $\dfrac{-6.3n^2t^{-6}y^3}{2.1t^3y^7}$

74. $\dfrac{3.8rs^2t}{7.6s^3t^4}$

75. $\dfrac{27.5w^4x^8y^3}{0.5w^2xy^6}$

76. $\dfrac{-11m^{-9}r^{-4}}{-22m^{-3}}$

77. $\dfrac{3.6x^{-2}y^3z^2}{7.2x^5z^{-3}}$

78. $\dfrac{7.4r^{-1}s^2t^{-5}}{3.4r^2s}$

79. $\dfrac{-2.8m^{-8}n^{12}t^{-4}}{-4m^2n^{-1}t}$

80. $\dfrac{-9x^2y^{-5}}{-18x^7y^{-6}t^{-3}}$

81. *Acceleration* A race car travels on a circular track at 120 miles per hour. The radius of the track is 1000 feet. There is a formula that gives the acceleration of the center of gravity of the car. It is

$$a = \frac{v^2}{r}$$

Find a for this race car, where v is the speed and r is the radius in miles. *Hint:* Use $r = \frac{1000}{5280}$.

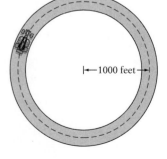

←1000 feet→

82. *Optimal Acoustics* In choosing a room that you wish to use for a concert, one thing you consider is the *reverberation time T*, a measure that describes the acoustical (sound) properties of the room. The optimal value of T for a room that is used for music is 1.5 seconds. The formula is

$$T = \frac{0.049V}{a}$$

where V is the volume of the room in cubic feet and a is the absorption coefficient of the room. Find the value for a that gives $T = 1.5$ seconds for a room that measures 50 feet by 40 feet by 12 feet. (For your information, a is given in units called sabins.)

MIXED PRACTICE

By doing these exercises, you will practice the topics up to this point in the chapter.

83. Rewrite 4^3 without exponents.

84. Simplify this exponential expression: 6^{-2}

85. Multiply: $(9v^2u^{-3})(8v^{-2}u^5)$

86. Simplify: $(-8x^3y^7t^3)^{-2}$

87. Simplify: $(2xy^3)^2$

88. Simplify: $(2x^2y^3)^{-4}$

89. Simplify: $(r^{-2}t^{-4}v^{-5})^5$

90. Simplify: $(-3r^{-2}t^3v)^4$

91. Multiply: $(-8x^6y^4)(6x^{-7}y)$

92. Multiply: $(-2u^2x^9)(5ux^{-3})$

93. Simplify this exponential expression: -2^5

94. Simplify: $(r^3s^{-6}t^2)^{-3}$

95. Multiply: $(9x^{-8}y^3)(4xy^{-8})$

96. Simplify: $\left(\dfrac{-11x^5t^2y^3}{33x^8t^4}\right)^{-5}$

97. Divide: $\dfrac{12m^4r^{-3}}{48m^2r^5}$

98. Multiply: $(-3s^3t^2)(4st^9)$

99. Divide: $\dfrac{-2s^3t^2}{18s^3t^5}$

100. Divide: $\dfrac{-2.8xy^3}{-1.4x^{-2}y^{-3}}$

EXCURSIONS

Class Act

1. a. The following graph shows the percent of businesses reporting that information technology has affected each area (customer service, productivity, and so on). The data was analyzed in 1994. How do you believe this graph might have changed in 1996?

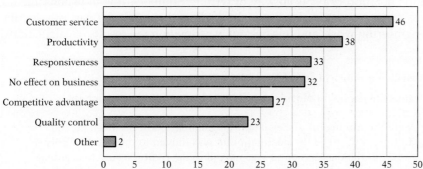

IMPACT OF INFORMATION TECHNOLOGY
ON SMALL AND MID-SIZED BUSINESSES (%)

Source: National Small Business United and Arthur Andersen Enterprises, June 1994, as cited in the *1966 Information Please® Business Almanac* (©1995 Houghton Mifflin Co.), p. 593. All rights reserved. Used with permission by Information Please LLC.

b. Ask and answer four questions using information from this graph.

2. When the Three Tenors come to Giants Stadium (capacity 60,000) in New Jersey on July 20, the average ticket price will be $250, with more than 4000 tickets costing between $1000 and $1500 each.

Some 500 premium tickets that offer a reception with the three stars will be sold for $2500 apiece. Assume 98% attendance.

a. Given this information, determine a set of ticket prices that could result in an average (or mean) ticket price of $250 for the 60,000 tickets.

b. Determine five levels of ticket prices, including the ones given, that result in the average $250.

c. Compare these prices to what the Rolling Stones charged when they played Giants Stadium last year: $55. Specifically, how much would the Stones have made; how much more would be made by the Tenors?

d. *Research* Who are the Three Tenors?

e. *Research* Who are the members of the Rolling Stones?

Exploring Numbers

3. Twenty-five can be written as the sum of two squares:

$$4^2 + 3^2 = 25$$

a. Sixty-five can be written as the sum of two squares in two different ways. Find them.

b. ✏ Describe your "plan" for finding these numbers.

CONNECTIONS TO *MEASUREMENT*

Scientific Notation

Writing Numbers in Scientific Notation

Scientific notation is a special way of writing numbers. It is indispensable for writing the very large and very small numbers that often are part of scientific and engineering work, and it is helpful in estimating the results of calculations.

> **To write a number in scientific notation**
>
> 1. Move the decimal point right or left to obtain a number n such that $1 \leq n < 10$.
> 2. Count the number of places p that the decimal point has been moved.
> 3. Multiply n by 10^p if the decimal point was moved to the left. Multiply n by 10^{-p} if the decimal point was moved to the right. Be sure to eliminate any meaningless zeros.

▪ ▪ ▪

EXAMPLE

Write in scientific notation:

a. 10,300,000   **b.** 0.00089

SOLUTION

a. We need to move the decimal point *to the left* 7 places to get a number *n* such that $1 \leq n < 10$.

$$1.0300000.$$

So we multiply *n* by 10^7. The zeros to the right of the 3 are meaningless, so we eliminate them, getting

$$1.03 \times 10^7$$

b. We need to move the decimal point *to the right* 4 places to obtain a number *n* such that $1 \leq n < 10$. Then we multiply the result by 10^{-4} and eliminate the meaningless zeros on the left.

$$00008.9 \times 10^{-4} = 8.9 \times 10^{-4}$$

Calculator Corner

You can use your graphing calculator's Home Screen to check your answers in scientific notation.

For example, writing the number 10,300,000 in scientific notation would give the answer 1.03×10^7. Now check your answer with your calculator. If your answer is correct, the calculator will return a 1, meaning the statement is true.

If your answer is wrong, the calculator will return a 0, meaning the statement is false. (Note: See the Calculator Corner on page 269 to review how to test a statement.)

```
10300000=1.03*10^7
                  1
```

```
10300000=1.03*10^9
                  0
```

Notice the difference here:

Now try the following on your own.

1. 12,300,000

2. 123,456,000

3. 222,800,000

4. 943,006,000

5. 11,562,300,000

6. 56,123,456,000

Writing Numbers in Standard Notation

We reverse the process when we are given a number in scientific notation and must write it in common, or standard, notation.

To write a number in standard notation

1. Move the decimal point the number of places p in 10^p. Move it to the right if the exponent is positive; move it to the left if the exponent is negative. (Add zeros as necessary.)

2. Eliminate the multiplication sign and power of 10.

■ ■ ■

EXAMPLE

Write in standard notation:

a. 1.206×10^9 **b.** 3.05×10^{-7}

SOLUTION

a. Because the exponent is 9, we move the decimal point 9 places *to the right*.
$$1.206 \times 10^9 = 1.206000000. = 1{,}206{,}000{,}000$$

b. Because the exponent is -7, we must move the decimal point 7 places *to the left*.
$$3.05 \times 10^{-7} = 0.0000003.05 = 0.000000305$$

Calculating with Scientific Notation

■ ■ ■

EXAMPLE

a.   Multiply: $(4.8 \times 10^{15})(6.4 \times 10^{-12})$
b.   Divide the first of these numbers by the second.

SOLUTION

a. To multiply two numbers in scientific notation, multiply decimal numbers and powers of 10.

$$(4.8 \times 10^{15})(6.4 \times 10^{-12})$$
$$= (4.8)(6.4) \times 10^{15+(-12)}$$
$$= 30.72 \times 10^{3}$$

This number is not in scientific notation, however, because $30 > 10$. To write it correctly, we put the decimal part in proper scientific notation and then simplify.

$$30.72 \times 10^{3} = (3.072 \times 10^{1}) \times 10^{3}$$
$$= 3.072 \times 10^{4}$$

b. To divide in scientific notation, we divide the decimal numbers and subtract the powers of ten.

$$\frac{4.8 \times 10^{15}}{6.4 \times 10^{-12}} = \frac{4.8}{6.4} \times 10^{15-(-12)}$$
$$= 0.75 \times 10^{27}$$
$$= (07.5 \times 10^{-1}) \times 10^{27}$$
$$= 7.5 \times 10^{26}$$

◀

PRACTICE

In Exercises 1 and 2, rewrite each number in scientific notation.

1. Number of years that there has been life on Earth: 70,800,000,000,000,000 years.

2. Energy required to use a 10-watt flashlight for 1 minute: 6,000,000,000 ergs

In Exercises 3 and 4, rewrite each number in standard notation.

3. Number of pounds of chocolates eaten per day: 5.8×10^{6} pounds

4. Number of gallons of water used by Americans daily: 4.5×10^{11} gallons

In Exercises 5 through 8, rewrite using scientific notation.

5. 173.6×10^{-5} **6.** 185×10^{-6}

7. 0.003×10^{-2} **8.** 0.13×10^{16}

In Exercises 9 through 12, multiply or divide as indicated.

9. $(1.24 \times 10^{-23})(0.08 \times 10^{2})$

10. $(3.521 \times 10^{5}) \times (2 \times 10^{-3})$

11. $\dfrac{1.08 \times 10^{-12}}{5.4 \times 10^{-11}}$

12. $\dfrac{1.24 \times 10^{-13}}{6.2 \times 10^{20}}$

5.3 Adding and Subtracting Polynomials

SECTION LEAD-IN

What is the U.S. national debt? Let's say, for example, that it is one trillion dollars.

$$\$1,000,000,000,000$$

▪ If we could halve the debt every year, how many years would be necessary to reduce the debt to less than \$1?

Introduction to Polynomials

So far in this chapter, we have dealt only with monomials—that is, single terms such as $5x^2y$. From now on, however, we shall be concerned with sums and differences of monomials such as $5x^2y - 7x^2 + 5y^2$. These expressions are called *polynomials.* A monomial is the simplest form of a polynomial.

> **Polynomial**
>
> A **polynomial** is an algebraic expression that is a monomial or the sum or difference of two or more monomials.

There are also specific names for polynomials that consist of two terms and three terms. A **binomial** is a polynomial that has two terms, and a **trinomial** is a polynomial that has three terms.

$$\left. \begin{array}{ll} 4x^2 - 15x & \text{Binomial} \\ 11x^3 + 4x^2 - 15x & \text{Trinomial} \end{array} \right\} \text{Polynomials}$$

A **polynomial in one variable** is a polynomial that includes only one variable. That variable may appear in several terms of the polynomial, and it may have different exponents in different terms.

> **Degree of a Polynomial**
>
> The **degree of a polynomial in one variable** is the greatest exponent on the variable in any term of the polynomial. The degree of a constant term is zero.

The terms of a polynomial in one variable are written with the exponents in descending order from left to right. Thus, for example, the polynomial whose terms are $3x$, -1, and $2x^2$ is written $2x^2 + 3x - 1$.

▪▪▪
EXAMPLE 1

Determine whether these polynomials are binomials, trinomials, or neither, and give the degree of each.

a. $5x^2 + 4x - 6$ **b.** $2x^{15} - 9x^{31}$

SOLUTION

a. This polynomial has three terms. It is a trinomial. The greatest exponent on the variable is 2, so the degree of the polynomial is 2.

b. This is a binomial because it has two terms. Its degree is 31.

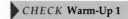 ▶ *CHECK* **Warm-Up 1**

Adding Polynomials

We add polynomials in the same way we combine like terms.

> **To add polynomials**
> 1. Use the commutative and associative properties to group like terms, making sure that you move the signs with the terms.
> 2. Use the distributive property to combine like terms.

! ! !
ERROR ALERT
Identify the error and give a correct answer.
Incorrect Solution:
$4x + 7x = 11x^2$

▪▪▪
EXAMPLE 2

Add: $(4x^2 + 3x - 6) + (-3x^2 - 12x - 8)$

SOLUTION

First we regroup the terms.

$$(4x^2 + 3x - 6) + (-3x^2 - 12x - 8)$$
$$= 4x^2 - 3x^2 + 3x - 12x - 6 - 8$$

Then we combine like terms.

$$4x^2 - 3x^2 + 3x - 12x - 6 - 8$$
$$= (4 - 3)x^2 + (3 - 12)x + (-6 - 8)$$
$$= 1x^2 \qquad -9x \qquad -14$$

So $(4x^2 + 3x - 6) + (-3x^2 - 12x - 8) = x^2 - 9x - 14$.

▶ *CHECK* **Warm-Up 2**

STUDY HINT
You may add and subtract in a vertical format if you wish. Line up the like terms one under the other.

■■■
WRITER'S BLOCK
Explain how the distributive property is used to justify that
$$6x + 2x = 8x$$

A polynomial is said to be *simplified* when all like terms have been combined.

▪ ▪ ▪

EXAMPLE 3

Simplify: $(3x^3 + 2x^2 - x + 4) + (x^2 - 5x) + (7x^3 + 9)$

SOLUTION

First, regroup the like terms.

$$(3x^3 + 2x^2 - x + 4) + (x^2 - 5x) + (7x^3 + 9)$$
$$= \underbrace{3x^3 + 7x^3}_{x^3 \text{ terms}} + \underbrace{2x^2 + x^2}_{x^2 \text{ terms}} - \underbrace{x - 5x}_{x \text{ terms}} + \underbrace{4 + 9}_{\text{constants}}$$

Then combine like terms.

$$= \underset{10x^3}{3x^3 + 7x^3} + \underset{+3x^2}{2x^2 + x^2} - \underset{-6x}{x - 5x} + \underset{+13}{4 + 9}$$

Thus

$$(3x^3 + 2x^2 - x + 4) + (x^2 - 5x) + (7x^3 + 9)$$
$$= 10x^3 + 3x^2 - 6x + 13$$

▶ *CHECK* **Warm-Up 3**

> **STUDY HINT**
>
> *It's a good idea to count the original terms and the regrouped terms to make sure you haven't lost one!*

Subtracting Polynomials

Recall that we subtracted a real number b from a real number a by adding the opposite of b to a; that is,

$$a - b = a + (-b)$$

We do the same for polynomials.

To find the opposite of a polynomial, we change the sign of every term of the polynomial. Thus the opposite of $2x^2 - 5x - 3$ is $-2x^2 + 5x + 3$. Formally,

---

To subtract polynomial b from polynomial a

1. Rewrite the problem, removing the parentheses and changing signs.
2. Use the commutative and associative properties to group like terms.
3. Use the distributive property to combine like terms.

> **WRITER'S BLOCK**
>
> **Explain why**
>
> $-(2x + 4)$
>
> **is equal to**
>
> $-2x - 4$

▪ ▪ ▪

EXAMPLE 4

Subtract: $(6x - 3) - (2x + 4)$

SOLUTION

The opposite of $2x + 4$ is $-2x - 4$. Therefore, by the procedure for subtracting, we have

$$(6x - 3) - (2x + 4) = 6x - 3 - 2x - 4 \quad \text{Rewriting}$$
$$= 6x - 2x - 3 - 4 \quad \text{Regrouping}$$
$$= \quad 4x \quad - \quad 7 \quad \text{Simplifying}$$

So $(6x - 3) - (2x + 4) = 4x - 7$.

▶CHECK **Warm-Up 4**

In the next example, we must be careful when grouping the terms.

▪ ▪ ▪

EXAMPLE 5

Subtract: $(-7n^2 + 5n - 3) - (-3n^2 - 1)$

SOLUTION

We rewrite the problem as an addition after changing the sign of *every term* in the second polynomial.

$$(-7n^2 + 5n - 3) - (-3n^2 - 1)$$
$$= -7n^2 + 5n - 3 + 3n^2 + 1 \quad \text{Rewriting}$$

We then perform the indicated addition.

$$-7n^2 + 5n - 3 + 3n^2 + 1$$
$$= -7n^2 + 3n^2 + 5n - 3 + 1 \quad \text{Regrouping}$$
$$= -4n^2 + 5n - 2 \quad\quad\quad \text{Simplifying}$$

Thus $(-7n^2 + 5n - 3) - (-3n^2 - 1) = -4n^2 + 5n - 2$

▶CHECK **Warm-Up 5**

! ! !
ERROR ALERT
Identify the error and give a correct answer.
Incorrect Solution:

$$(2x^3 - 3xy) - (6x^3 - 2xy)$$
$$= 2x^3 - 3xy - 6x^3 - 2xy$$
$$= -4x^3 - 5xy$$

Practice what you learned.

SECTION FOLLOW-UP

We will be paying off the U.S. national debt for a lot of years:

$\$1,000,000,000,000 \div 2^{10} = 976,562,500$ (amount after 10 years)

$976,562,500 \div 2^{10} = 953,674$ (amount after 20 years)

$953,674 \div 2^{10} = 931$ (amount after 30 years)

$931 \div 2^{10} = \$.90$ (amount after 40 years)

It will take 40 years to reduce the debt if we halve it every year.

■ Suppose we pay $199 a month. How many years will it take us if no additional interest is added to the debt?
■ **Research** Redo the problem using the actual U.S. national debt.

5.3 WARM-UPS

Work these problems before you attempt the exercises.

1. Determine whether these polynomials are binomials, trinomials, or neither.
$$4x^2 + 5, \quad 6x^3 + 4x + 3$$

2. Add: $(8x^2 + 5x + 5) + (7x^2 - 3x + 4)$

3. Add: $(5x^2 + 3x + 9) + (6x^2 - 2) + (12x - 5)$

4. Subtract: $(n - 3) - (n + 4)$

5. Subtract: $(7n^2 + 9) - (-n^2 + 9n + 1)$

5.3 EXERCISES

Note: Use your graphing calculator to check your results whenever possible.

In Exercises 1 through 4, find the degree of the polynomial.

1. $-4y^2 + 3y + 3$ 2. $4x^4 + 2x^2 - 5$ 3. $4y^2 + 5y + 5y^4 - 3y^3$ 4. $17x + 3x^2$

In Exercises 5 through 12, determine whether each of the following is a monomial or a polynomial. State whether each polynomial is a binomial, a trinomial, or neither.

5. $3x + 4x$ 6. $3y + 5 - y^2$ 7. $-7m^2 - 3m + 10$ 8. $5t + 4t^2 - 7$

9. $7x^2 + 3x^2 - 4x^2 - 2$ 10. $5r^2 - 3r^2 + 5r^2$ 11. $6y + 7y - 9y - y^2 + 5$ 12. $6rs + 2rs - 31r^2s$

In Exercises 13 through 40, add the polynomials.

13. $(2x + 8) + (3x - 5)$

14. $(4x - 2) + (6x + 9)$

15. $(18x - 12) + (6 - 9x)$

16. $(23x - 15) + (19 + 7x)$

17. $\left(\frac{1}{3}x - 2\right) + \left(\frac{1}{3}x^2 - x\right)$

18. $\left(\frac{1}{5}x^2 - 2x\right) + \left(\frac{2}{5}x^2 - 5\right)$

19. $\left(\frac{2}{3}x - \frac{1}{2}\right) + \left(\frac{1}{2}x - \frac{2}{3}\right)$

20. $\left(\frac{1}{5} - \frac{2}{3}x\right) + \left(\frac{2}{3} - \frac{1}{5}x\right)$

21. $(2.5x^2 - 3x) + (1.2x^2 + 0.9x)$

22. $(1.2x + 2x^2) + (3.6x^2 - 2x)$

23. $(5x - 3.6x^2) + (2.5x^2 - 2.4x)$

24. $(10.8x^2 - 9.3x) + (15.2x - 16.8x^2)$

25. $5r^2 + 3$
$\underline{\quad r^2 - 6}$

26. $-4t^2 + 7t - 4$
$7t^2 - 11t$
$\underline{-9t^2 + 5t + 6}$

27. $(-3t^2 + 15t) + (-4 + t^2) + (-11t + 5)$

28. $(-1.8m^3 + 2.2m^2 - 0.5m) + (-1.4m^2 + 4.4)$

29. $(1.9n^3 + 2.3n + 0.2) + (-0.7n^3 - 1.8n^2 - 2.9n)$

30. $\left(-\frac{1}{2}m^2 - 5\right) + (5m^2 - 9)$

31. $(-15n - 6) + (14n^2 - 8n + 14)$

32. $(14n^2 - 13n - 8) + (-19n + 7n^2 - 3) + (6n^2 + 5)$

33. $(x^2 + 7x - 4) + (5x^2 - 11x - 2)$

34. $(-2x^2 - 9) + (x^2 - 3x + 2)$

35. $(x^4 - x^3 + 2x^2) + (-x^4 + 2x^3 + 3x)$

36. $(4x^3 + 3x^2 - 2x) + (-2x^3 + 2x^2 + 2)$

37. $(2x^2 - 3x + 7) + (2x - 3x^2 + 4)$

38. $(3x^2 - 2x + 4) + (-3x^2 + 2x - 7)$

39. $(4x^2 + 6x - 2) + (9x^3 - 6x^2 + 3x - 15)$

40. $(15x^3 - 10x^2 - 5x + 20) + (24x^2 - 32x + 16)$

In Exercises 41 through 70, subtract.

41. $(3x + 8) - (2x - 5)$

42. $(4x - 2) - (3x + 8)$

43. $(10x + 3) - (3x - 4)$

44. $(8x - 6) - (16x - 9)$

45. $\left(\frac{1}{3}x - \frac{1}{5}x^2\right) - \left(-\frac{2}{3}x - 1\right)$

46. $\left(\frac{1}{2}x^2 + \frac{2}{3}x\right) - \left(\frac{1}{4}x^2 - \frac{1}{3}x\right)$

47. $\left(1\frac{1}{3}x - 2x^2\right) - \left(\frac{1}{2}x^2 + \frac{2}{3}x\right)$

48. $\left(3\frac{1}{2}x^2 + \frac{1}{4}x\right) - \left(\frac{1}{3}x^2 - 2\frac{1}{2}x\right)$

49. $(5.6x^2 - 2x + 4) - (2.1x^2 + 1.3x)$

50. $(6.8x^2 - 1.2x + 0.6) - (1.2 + 2.3x^2 - 0.3x)$

51. $(2x^3 + 1.1x^2 + x - 0.4) - (3.1x^3 - 2x^2 - 4)$

52. $(8x^3 - 2.6x^2 + 1.2x) - (-7.6x^3 - 1.5x^2 - 2.3x)$

53. $11x^2 - 4x + 3$
$\underline{-(-9x^3 + 6x - 9)}$

54. $23r^3 - 6r + 5$
$\underline{-(-9r^2 + 7r - 8)}$

55. $11z - 9z^2 + 8$
$\underline{-(z^4 + 9z^2)}$

56. $(-17r^2 + 4) - (-35r^2 - 6)$

57. $(-44x^2 + x) - (-17x^2 - 4)$

58. $(6.6r^2 + 4.4r + 0.12) - (1.3r - 6.8)$

59. $(-25x^2 + 26) - (21x + 12x - 9)$

60. $(-6r^2 - 2r - 32) - (9r^2 + 13r - 19)$

61. $(-4t^2 + 36t - 12) - (6t^2 - 4t - 18)$

62. $(3y^2 + 18y + 25) - (12y^2 - y + 3)$

63. $(5x^2 + 3x - 4) - (-2x^2 - 7x + 12)$

64. $(x^3 + 3x^2 - 2x - 9) - (-2x^2 + 5x + 1)$

65. $(-x^2 - 3x - 4) - (-5x^2 - 3x + 4)$

66. $(x^2 - 2xy - y^2) - (2 + 2y - 3y^2)$

67. $(21x^2 - 24x + 12) - (16x^3 - 12x^2 - 20x)$

68. $(4x^2 + 24x - 4) - (3x^2 - 9x + 27)$

69. $(10x^2 - 15xy + 20y^2) - (-6x^2 + 9xy - 9y^2)$ **70.** $(2x^3 - 22x^2 + 2x - 4) - (5x^3 - 40x - 10)$

In Exercises 71 through 80, add or subtract as indicated.

71. $(2r^2 + 4r - 8) - (7r^2 - 8r + 5) + (14r^2 + 16r - 3)$

72. $(-7.2t^2 + 2.5t - 4.6) + (1.06t^2 + 6.4t - 19) - (1.8t^2 + 0.2t - 11)$

73. $(-13m^3 - 8m^2 - 16m - 11) - (12m^2 + 4m + 7) - (-13m^2 - 9m + 17)$

74. $(-12n^2 + 22n - 21) - (15n^2 - 2n - 6) - (20n^2 - 5n - 8)$

75. $(-x^2 - 3x - 4) - (-5x^2 - x - 2) - (2x - 5 + 3x^2 + x^3 + 4x - 9)$

76. $(8 + 48x^5 - 18x - 27x + 60) - (2x^2 - 3xy + 4y^2) + (-6x^2 + 9xy - 9y^2)$

77. $(x^2 - 2xy - y^2) - (5x^2 + xy - y^2) - (x^2 + 6x - 1) - (x^2 - 3x + 9)$

78. $(3x^2 - 2x - 9 + 5x^2 + 2x - 14) - (-2x^2 - 9) + (-4x - 16)$

79. $\left(\frac{1}{5}x + \frac{1}{9} - \frac{2}{3}x + \frac{1}{6}\right) + \left(-\frac{5}{4}x^2 + \frac{1}{3}x - \frac{3}{8}\right) - \left(-\frac{5}{3}x^2 + \frac{1}{3}x - \frac{1}{4}\right)$

80. $\left(\frac{1}{4}x + \frac{1}{6} - \frac{3}{8}x - \frac{2}{9}\right) - \left(\frac{5}{9}x^2 + \frac{1}{4}x - \frac{1}{6}\right) + \left(-\frac{2}{3}x^2 - \frac{1}{12}x + 1\right)$

81. Write two polynomials that add to give

$$3x^3 - 2x^2 + x - 7$$

82. Write two polynomials whose difference is

$$2x^3 + 5x^2 + 6x - 12$$

83. ***Fruit-Producing Trees***   The number of bushels of lemons N_L that are produced by x trees in a small orchard ($x \le 100$) is given by

$$N_L = 30x - 0.1x^2$$

The number of bushels of apples N_A on the same number of trees (x) where $x \le 100$ is given by

$$N_A = 40x - 0.2x^2$$

Find a formula for how many more apples are produced on x trees than lemons by subtracting $N_A - N_L$.

84. ***Total Profit***   The total profit from the production and sale of electronic pocket organizers is the difference between the total revenue from the sale of x units and the cost of producing x units. Say you know that

$$\text{Revenue from sale of } x \text{ units} = 20x^2 - 18$$

and that

$$\text{Cost of producing } x \text{ units} = 2x^2 - 9x - 10$$

Find the total profit (P_T) by finding the difference between these two polynomials.

MIXED PRACTICE

By doing these exercises, you will practice the topics up to this point in the chapter. Write all answers with positive exponents.

85. Divide: $\dfrac{9r^3t^4y^7}{24r^2t^2y^5}$

86. Simplify: $(x^2 + 3x - 11) + (7x^2 - 13x + 9)$

87. Simplify:
$(3t^2 + 4t - 8) + (2t^2 - 6t + 12) - (-5t^2 - 9t + 7)$

88. Multiply: $(xy^2)(-8yx^{-2}r^3)$

89. Divide: $\dfrac{-10x^4y^5}{18x^3y^6}$

90. Multiply: $(xy^{-5}t^{-2})(t^{-2}xy^3)$

91. Simplify: $\left(\dfrac{3x^4y^2}{18x^3y^9}\right)^{-3}$

92. Simplify this exponential expression: $(-3)^3$

93. $(-t)(-t^6)$

94. $(-18r)(-5r^2)$

95. $\left(\dfrac{3}{5}t^3v\right)(5t^{-8}v^5)$

96. $(8p^3t)(4p^{-3}t)$

97. Simplify this exponential expression: $(-9)^{-2}$

98. $\left(\dfrac{2}{3}r^2u^{-3}\right)(6r^3u^4)$

99. $(1.12x^2 - 3.04x - 4.79) + (6.19x^2 + 1.42x - 6.34)$

100. Simplify: $(2xy^3)^{-3}$

101. Multiply: $(-3s^3t^2)(4st^9)^2$

102. Simplify: $\left(\dfrac{-2s^3t^2}{18s^3t^5}\right)^{-2}$

EXCURSIONS

Data Analysis

1. Study the following tables and then answer the questions.

TABLE 1 **Average weekly earnings and relative pay levels of professional and administrative occupations, United States and four regions, September 1994**

Occupation	U.S.	Northeast	South	Midwest	West	Dispersion percent
Accountants III	$ 767	$ 779 (102)	$ 747 (97)	$ 759 (99)	$ 787 (103)	5
Attorneys III	1241	1267 (102)	1172 (94)	1257 (101)	1289 (104)	10
Engineers IV	1094	1092 (100)	1090 (100)	1092 (100)	1104 (101)	1
Nurses II	710	799 (113)	647 (91)	662 (93)	789 (111)	23
Buyers II	642	661 (103)	614 (96)	642 (100)	664 (103)	8
Computer programmers III	743	755 (102)	724 (97)	742 (100)	765 (103)	6
Computer systems analysts II	892	898 (101)	871 (98)	885 (99)	915 (103)	5
Personnel specialists III	768	772 (101)	740 (96)	760 (99)	805 (105)	9

TABLE 2 Average weekly earnings and relative pay levels of technical and protective service occupations, United States and four regions, September 1994

Occupation	U.S.	Northeast	South	Midwest	West	Dispersion percent
Computer operators II	$433	$455 (105)	$ 418 (97)	$ 430 (99)	$446 (103)	9
Drafters III	607	608 (100)	604 (100)	597 (98)	633 (104)	6
Engineering technicians IV	739	705 (95)	734 (99)	766 (104)	749 (101)	9
Licensed practical nurses II	448	522 (117)	411 (92)	440 (98)	504 (113)	27
Nursing assistants II	276	348 (126)	239 (87)	264 (96)	285 (103)	46
Corrections officers	533	651 (122)	422 (79)	508 (95)	671 (126)	59
Firefighters	631	697 (110)	516 (82)	638 (101)	770 (122)	49
Police officers I	660	731 (111)	532 (81)	641 (97)	788 (119)	48

[1]Relative pay levels, shown in parentheses, indicate regional pay as a percent of national pay for the same job. Average pay of accountants III in the Northeast, for example, is approximately 2 percent higher than the nationwide average for the job. [2]The dispersion percent shows the percentage difference between average earnings in the highest paying and the lowest paying region. *Sources: Compensation and Working Conditions* (June 1996).

a. Graph and discuss the differences between salaries for the various categories.

b. Ask and answer four questions abut this data.

Class Act

2. Too many people have problems dealing with credit cards. To see how your purchases can cost more (much more) when you charge, do the following exercise.

Tracy's Department Store advertises an 18% interest rate on purchases with no interest if you pay off your bill within 30 days of purchase.

You spend $100 and decide to pay the minimum of $20. Therefore, during the first month you must pay interest. You have

$$\$100 - 20 = \$80$$

left to pay on your debt, plus interest on the $80.

a. Use d_R to stand for "debt remaining." Write a function (a rule) for calculating each month's remaining debt.

b. A rule that can be used to calculate each month's remaining debt is

$$d_R + 0.18d_R - 20 = \text{remaining debt}$$

Apply the rule repeatedly until the remaining debt is zero.

i.   How many months will it take to pay the debt?

ii. How much interest would you pay?

iii. How much would your $100 charge really cost?

iv. ✏ Would the rule

$$\left(1 + \frac{0.18}{12}\right)d_R - 20 = \text{remaining debt}$$

work also? Explain why.

v. ✏ Explain how you can use the ANS button on your calculator to keep from re-entering the previous month's numbers.

3. Study the following graph and then answer the questions.

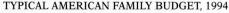

TYPICAL AMERICAN FAMILY BUDGET, 1994

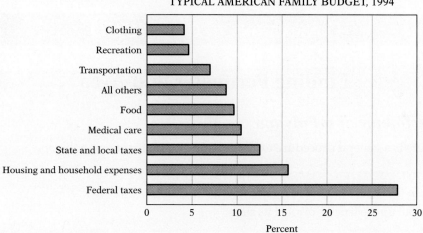

a. Using this graph, estimate the amount of money spent in each category in 1994 by a family earning $32,000 a year.

b. Assume that the typical budget for 1998 is similar to the typical budget for 1994 but that earnings are 17% greater. Using two different methods, estimate the actual amount of money spent in each category in 1998. Describe the two methods.

4. a. Gather data about how students in your class spend their money. Make a bar graph that illustrates this information.

b. What is the mean amount spent by each student on all the various categories?

Exploring Numbers

5. a. ✏ Discuss why every non-negative even number can be represented by

$$2n$$

where n is a non-negative integer and every non-negative odd number can be represented by

$$2n + 1$$

b. Show that the sum of two even numbers is always an even number.

c. Show that the difference of two even numbers is always an even number.

d. Show that the sum of two odd numbers is always an even number.

e. Show that the difference of two odd numbers is always an even number.

f. Show that the sum of an even and an odd number is always an odd number.

CONNECTIONS TO *GEOMETRY*

Finding Perimeter and Area

Perimeter of a Polygon

There is a general procedure for finding the perimeter of any polygon.

> **To find the perimeter of a polygon**
> Add up the lengths of all of its sides.

Note: The measurement of length is given in units: feet, yards, meters, and so on.

■ ■ ■
EXAMPLE

A tomato patch is rectangular in shape. If the patch is 20 feet by 15 feet, how much wire fence will be needed to enclose it?

SOLUTION

To find the length of the fence, we must find the perimeter. By drawing a picture, we see that we need to add the lengths of the sides.

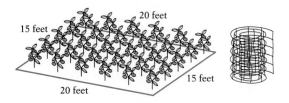

$$20 + 15 + 20 + 15 = 70$$

The perimeter is 70 feet.

Perimeter of a Square

> **To find the perimeter of a square**
> If s is the length of a side, then
> $$\text{Perimeter} = 4s$$

The formula $P = 4s$ can also be written as $P = 4 \times s$ or as $4 \cdot s$. In words, it tells us to multiply the length of the side by 4. The perimeter is given in units—feet, inches, meters, any unit of length.

▪▪▪
EXAMPLE

Find the perimeter of a square with a side that measures 1 foot 7 inches.

SOLUTION

Using the formula for finding the perimeter of a square, we substitute the value of the length of one side. (We will use $1\frac{7}{12}$ feet instead of 1 foot 7 inches):

$$P = 4s$$
$$P = 4\left(1\frac{7}{12}\right) = 4\left(\frac{19}{12}\right) = 6\frac{4}{12}$$

So the perimeter of the square is $6\frac{4}{12}$ feet, or 6 feet 4 inches. ◢

Area of a Square

> **To find the area of a square**
> If s is the length of a side of a square, then
> $$\text{Area} = s^2$$

This formula is what gives the name "squared" to the exponent 2.

▪▪▪
EXAMPLE

Find the area of a square with a side measuring 5 inches.

SOLUTION

Substituting in the formula for area of a square, we have:

$$A = s^2$$
$$= 5^2$$
$$= 25$$

The area is 25 square inches. ◢

WRITER'S BLOCK

Explain how we can use both methods (for a polygon and for a square) to find the perimeter of a square.

STUDY HINT

When possible, always draw a diagram and label it first.

PRACTICE

1. Find the perimeter of an equilateral triangle that has a side of length 4.8 inches.

2. Find the perimeter of the irregular house lot shown in the figure.

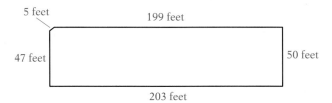

3. Find the perimeter of a regular pentagon that has a side of $12\frac{1}{2}$ feet.

4. A square plot of grass is 1.5 yards on each side. How much fencing would you need to enclose it?

5. A pentagon has a perimeter of 1601.3 meters. Four of its sides are 350 meters, 190.6 meters, 465.2 meters, and 200 meters. What is the length of the fifth side?

6. The perimeter of a quadrilateral is $193\frac{1}{3}$ yards. Three of its sides are $19\frac{1}{4}$ yards, $29\frac{2}{3}$ yards, and 90 yards. What is the length of the fourth side?

7. Find the perimeter of a square surface if its area is 400 square feet (boxing ring). See the accompanying figure.

8. Find the area of a karate exhibition mat that is square and is 26 feet on each side.

9. Find the area of a square judo court 52 feet 6 inches on a side.

10. Find the area of a square baseball "diamond" if its perimeter is 360 feet.

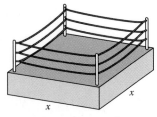

Area = 400 square feet

5.4 Multiplying Polynomials

SECTION GOALS

- To multiply polynomials by monomials and by polynomials
- To use the FOIL method to multiply binomials
- To use the special-products formulas

SECTION LEAD-IN

Many so-called geniuses can wow their friends by applying the rules they will learn in this section.

In your head, multiply

$$(21)(19)$$
$$39 \cdot 39$$

Having trouble? Check in later!

You know how to multiply a monomial by a monomial. That and the distributive property will enable you to multiply any two polynomials together.

Multiplying a Polynomial by a Monomial

The major task here is to make sure that every term of the polynomial gets multiplied by the monomial. For this, we must apply the distributive property carefully.

∎∎∎

EXAMPLE 1

Multiply: $4x(2x - 4)$

SOLUTION

As a first step, we distribute multiplication by $4x$ to each monomial inside the parentheses.

$$4x(2x - 4) = (4x)(2x) + (4x)(-4)$$

Then we multiply monomials as indicated.

$$(4x)(2x) + (4x)(-4)$$
$$= (4)(2)(x^{1+1}) + (4)(-4)x$$
$$= 8x^2 - 16x$$

So $4x(2x - 4) = 8x^2 - 16x$.

▶ *CHECK* **Warm-Up 1**

In this next example, the monomial has a negative coefficient. The minus sign must go everywhere the coefficient goes.

∎∎∎

EXAMPLE 2

Multiply: $-4x^2(2x^2 + 3x - 2)$

SOLUTION

We first distribute multiplication by $-4x^2$ to each term of the polynomial, taking care not to lose a minus sign.

$$-4x^2(2x^2 + 3x - 2)$$
$$= (-4x^2)(2x^2) + (-4x^2)(3x) + (-4x^2)(-2)$$

We then perform the monomial multiplications.

$$(-4x^2)(2x^2) + (-4x^2)(3x) + (-4x^2)(-2)$$
$$= -8x^{2+2} + (-12)x^{2+1} + (8)x^2$$
$$= -8x^4 - 12x^3 + 8x^2$$

So $-4x^2(2x^2 + 3x - 2) = -8x^4 - 12x^3 + 8x^2$.

> **STUDY HINT**
>
> *When you apply the distributive property, be very careful to distribute the sign with each term.*

▶ *CHECK* **Warm-Up 2**

Multiplying a Polynomial by a Polynomial

Here's where the distributive property is really important. When we multiply two polynomials, every term in the first polynomial must multiply every term in the second. We need to apply the distributive property carefully.

▪ ▪ ▪
EXAMPLE 3

Multiply: $(2x + 5)(3x^2 - 3x + 4)$

SOLUTION

We use the distributive property to multiply the second polynomial by each term of the first polynomial.

$$(2x + 5)(3x^2 - 3x + 4) = 2x(3x^2 - 3x + 4) + 5(3x^2 - 3x + 4)$$

Then we use the distributive property again to perform the indicated monomial-polynomial multiplications.

$$2x(3x^2 - 3x + 4) + 5(3x^2 - 3x + 4)$$

$$= 2x(3x^2) + 2x(-3x) + (2x)(4) + 5(3x^2) + 5(-3x) + 5(4)$$ Multiplying

$$= 6x^3 - 6x^2 + 8x + 15x^2 - 15x + 20$$ Regrouping like terms

$$= 6x^3 - 6x^2 + 15x^2 + 8x - 15x + 20$$

$$= 6x^3 + 9x^2 - 7x + 20$$ Simplifying

Thus $(2x + 5)(3x^2 - 3x + 4) = 6x^3 + 9x^2 - 7x + 20$.

▶ CHECK **Warm-Up 3**

Multiplying a Binomial by a Binomial: FOIL

In its most general form, the multiplication of two binomials may be written as

$$(a + b)(c + d)$$

where $a, b, c,$ and d are terms of the binomials and may include variables.

Now, let's label the First and Last terms of the binomials, and the Inner and Outer terms of the multiplication. We obtain

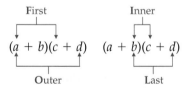

Finally, take a look at the completed multiplication.

$$(a + b)(c + d) = a(c + d) + b(c + d)$$

$$= ac + ad + bc + bd$$

First Outer Inner Last
terms terms terms terms

The product of the two binomials is the sum of the products of the two First terms, the two Outer terms, the two Inner terms, and the two Last terms. The initials FOIL give this method of multiplying binomials its name.

$$(a + b)(c - d)$$

$$\begin{array}{cccc} F & O & I & L \end{array}$$
$$= ac + ad + bc + bd$$

To multiply two binomials by the FOIL method, add the product of the First terms of the binomials, the product of the Outer terms, the product of the Inner terms, and the product of the Last terms.

▪▪▪

EXAMPLE 4

Multiply using FOIL: $(-7x + 3)(5x - 4)$

SOLUTION

The FOIL method gives us

$$(-7x + 3)(5x - 4) = \overset{F}{(-7x)(5x)} + \overset{O}{(-7x)(-4)} + \overset{I}{3(5x)} + \overset{L}{3(-4)}$$

$$= -35x^2 + 28x + 15x - 12 \qquad \text{Multiplying}$$
$$= -35x^2 + 43x - 12 \qquad \text{Simplifying}$$

So $(-7x + 3)(5x - 4) = -35x^2 + 43x - 12$.

▶ *CHECK* **Warm-Up 4**

Calculator Corner

You can check the answer you got using the FOIL method by graphing *each side* of the equation. If your work is correct, then you should get only one graph. If your answer is incorrect, you would have two graphs. For instance, Example 4 could be checked on your calculator as follows.

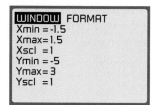

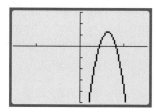

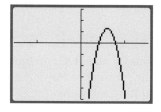

If, however, your use of the FOIL method was incorrect, then you would have two graphs, meaning that the two sides of the equation *are not equal* and, therefore, your equation is not true.

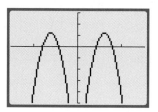

Special Products

Product of the Sum and Difference of Two Terms

Here, we consider two *special products* of binomials. The first is the product of the two expressions—the sum of two terms and the difference of those same two terms.

▪▪▪

EXAMPLE 5

Multiply: $(x + 4)(x - 4)$

SOLUTION

Using the FOIL method, we find that

$$\overset{\text{F}\quad\text{O}\quad\text{I}\quad\text{L}}{(x + 4)(x - 4)} = (x)(x) + (x)(-4) + (4)(x) + (4)(-4)$$
$$= x^2 - 4x + 4x - 16 \qquad \text{Multiplying}$$
$$= x^2 - 16 \qquad \text{Simplifying}$$

So $(x + 4)(x - 4) = x^2 - 16$.

▶ CHECK **Warm-Up 5**

Our results in Example 5 suggest this rule:

The product of the sum and difference of two terms is the difference of the squares of the terms. That is,

$$(a + b)(a - b) = a^2 - b^2$$

For example,

$$(2x + 3)(2x - 3) = (2x)^2 - 3^2$$
$$= 4x^2 - 9$$

Square of a Binomial

Our other special product is the square of a binomial.

▪▪▪

EXAMPLE 6

Simplify: $(x + 5)^2$

SOLUTION

We can rewrite this as the product of identical binomials.

$$(x + 5)^2 = (x + 5)(x + 5)$$

Then, using the FOIL method, we have

$$\overset{\text{F}\quad\text{O}\quad\text{I}\quad\text{L}}{(x + 5)(x + 5) = (x)(x) + (x)(5) + (5)(x) + (5)(5)}$$
$$= x^2 + 5x + 5x + 25$$
$$= x^2 + 10x + 25$$

So $(x + 5)(x + 5) = x^2 + 10x + 25$.

▶ *CHECK* **Warm-Up 6**

Do you see a relationship between the binomial and its square?

The square of a binomial is a trinomial in which the first and last terms are the squares of the terms in the binomial, and the middle term is twice the product of these terms. That is,

$$(a + b)^2 = a^2 + 2ab + b^2$$

So, for example,

First term squared

Twice product of first and second terms

Second term squared

$$(2x - 3)^2 = (2x)^2 + (2)(2x)(-3) + (-3)^2$$
$$= 4x^2 - 12x + 9$$

Simplifying Complex Expressions Using Multiplication

In Examples 7 and 8, we combine multiplication with addition and subtraction.

• • •

EXAMPLE 7

Simplify: $(3x^3 + 2x^2 - x) + x(2x - 3) + 2(x^2 + 5x)$

SOLUTION

First, use the distributive property:

$$3x^3 + 2x^2 - x + 2x^2 - 3x + 2x^2 + 10x$$

Then combine like terms:

$$3x^3 + 2x^2 + 2x^2 + 2x^2 - x - 3x + 10x$$
$$= 3x^3 + 6x^2 + 6x$$

▶ CHECK **Warm-Up 7**

In Example 8, we must use the distributive property before we subtract.

• • •

EXAMPLE 8

Simplify: $(3y^2 - 2) - 2(5y^2 + 6y - 1)$

SOLUTION

$$(3y^2 - 2) - 2(5y^2 + 6y - 1)$$

$= 3y^2 - 2 - 10y^2 - 12y + 2$	Distributing
$= 3y^2 - 10y^2 - 12y - 2 + 2$	Grouping like items
$= -7y^2 - 12y + 0$	Combining like terms
$= -7y^2 - 12y$	

▶ CHECK **Warm-Up 8**

! ! !
ERROR ALERT

Identify the error and give a correct answer.

Incorrect Solution:

$2(3x - 5) - 5(4x - 7)$
$= 6x - 10 - 20x - 35$
$= -14x - 45$

Practice what you learned.

◤ ## SECTION FOLLOW-UP

We can use what we learned in this chapter to do multiplication in our heads.

$$(21)(19)$$

can be rewritten (mentally) as

$$(20 + 1)(20 - 1)$$

Using what we just learned, we can obtain

$$400 - 1 = 399$$

To multiply $39 \cdot 39$, we rewrite the problem (mentally) as

$$(40 - 1)^2$$

We know this is

$$40^2 - 2(40)(1) + 1^2$$
$$= 1600 - 80 + 1$$
$$= 1521$$

■ What other numbers can you multiply this way?

5.4 WARM-UPS

Work these problems before you attempt the exercises.

1. Multiply: $(3x - 5)6x$

2. Multiply: $-2x^2(x^2 - x + 3)$

3. Multiply: $(7x - 4)(8x^2 + 3x - 9)$

4. Multiply using FOIL: $\left(\frac{1}{2}r - 4\right)(6r + 4)$

5. Multiply: $(r - 4)(r + 4)$

6. Simplify: $(z + 3)^2$

7. Add: $(5x^2 + 3x + 9) + 3(6x^2 - 2) + x(12x - 5)$

8. Subtract: $(8x^2 - 7x + 5) - 2(7x^2 + 9)$

5.4 EXERCISES

Note: Use your graphing calculator to check your results whenever possible.

In Exercises 1 through 8, multiply by using the distributive property.

1. $3n(n + 5)$

2. $-6n(-n - 3)$

3. $-12y(-7y^2 - 9y)$

4. $2n(3n^2 + 4n - 3)$

5. $-7n(-3n^2 - 5n + 8)$

6. $-11y(-2y^2 - y + 5)$

7. $(x - 5)(2x^3 + 5x - 3)$ 　　　　　　　　　　**8.** $(x - 8x^2)(x^3 + 7x - 3)$

In Exercises 9 through 12, find the products.

9. 　　$20n^3 + 11n^2 + 7$
　　$\underline{\times \qquad\qquad 4n - 5}$

10. 　　$90y^2 - 9y - 3$
　　$\underline{\times \qquad\qquad 8y + 18}$

11. 　　$-12n^3 + 4n - 12$
　　$\underline{\times \qquad 11n^2 - 8n - 4}$

12. 　　$-13y^3 - 8y^2 - 6y - 5$
　　$\underline{\times \qquad\qquad\qquad y^2 - 1}$

In Exercises 13 through 24, multiply using FOIL.

13. $(7x + 2)(3x + 2)$ 　　**14.** $(6x - 3)(2x + 3)$ 　　**15.** $(5t + 4)(7t - 1)$ 　　**16.** $(2m - 5)(2m + 5)$

17. $(2y - 5)(2y - 11)$ 　　**18.** $(2x - 3y)(2x - 4y)$ 　　**19.** $(3r - 2t)(4r - t)$ 　　**20.** $(6t + 2v)(-3t + 2v)$

21. $(3s - 2m)(7s - 3m)$ 　　**22.** $(rs + t)(rs - 2t)$ 　　**23.** $(2x^2 + 5x)(-3x^2 - 4)$ 　　**24.** $(y^2 + 2y)(y^2 - 3y)$

In Exercises 25 through 32, multiply by applying the special-products formulas. Do *not* use FOIL.

25. $(n + 9)^2$ 　　　　**26.** $(r - 2)^2$ 　　　　**27.** $(3x - 4)(3x + 4)$ 　　**28.** $(3y + 7)^2$

29. $(x + 2)(x - 2)$ 　　**30.** $(y - 5)^2$ 　　　　**31.** $(6x - 2y)^2$ 　　　　**32.** $(-7x - 5)^2$

In Exercises 33 through 56, simplify.

33. $(x^2 + 7x - 4) + 3(5x^2 - 11x - 2)$ 　　　　**34.** $(-2x^2 - 9) + 5(x^2 - 3x + 2)$

35. $(x^4 - x^3 + 2x^2) + 3(-x^4 + 2x^3 + 3x)$ 　　**36.** $(4x^3 + 3x^2 - 2x) + 2(-2x^3 + 2x^2 + 2)$

37. $3x(2x^2 - 3x + 7) + x(2x - 3x^2 + 4)$ 　　　　**38.** $5x(3x^2 - 2x + 4) + 2x(-3x^2 + 2x - 7)$

39. $-\frac{1}{2}x(4x^2 + 6x - 2) + \frac{2}{3}(9x^3 - 6x^2 + 3x - 15)$ 　　**40.** $-\frac{2}{5}(15x^3 - 10x^2 - 5x + 20) + \frac{1}{8}x(24x^2 - 32x + 16)$

41. $2(5x^2 + 3x - 4) - 7(-2x^2 - 7x + 12)$ 　　**42.** $4(x^3 + 3x^2 - 2x - 9) - 5(-2x^2 + 5x + 1)$

43. $8(-x^2 - 3x - 4) - 2x(-5x^2 - 3x + 4)$ 　　**44.** $6x(x^2 - 2xy - y^2) - 4x^2(2 + 2y - 3y^2)$

45. $\frac{1}{3}x(21x^2 - 24x + 12) - \frac{1}{4}(16x^3 - 12x^2 - 20x)$ 　　**46.** $\frac{1}{2}x(4x^2 + 24x - 4) - \frac{1}{3}x(3x^2 - 9x + 27)$

47. $-\frac{1}{5}x(10x^2 - 15xy + 20y^2) - \frac{2}{3}x(-6x^2 + 9xy - 9y^2)$

48. $-\frac{1}{2}x(2x^3 - 22x^2 + 2x - 4) - \frac{2}{5}x(5x^3 - 40x - 10)$

49. $2x(-x^2 - 3x - 4) - (-5x^2 - x - 2) - (2x - 5 + 3x^2 + x^3 + 4x - 9)$

50. $(8 + 48x^5 - 18x^2 + 27x^2 + 60) - x^3(2x^2 - 3x) + (-6x^2 + 9x^5 - 9x^4)$

51. $(x^2 - 2xy - y^2) - (5x^2 + xy - y^2) - 2(x^2 + 6x - 1) - (x^2 - 3x + 9)$

52. $(3x^2 - 2x - 9 + 5x^2 + 2x - 14) - 2(-2x^2 - 9) + (-4x - 16)$

53. $-45\left(\frac{1}{5}x + \frac{1}{9} - \frac{2}{3}x + \frac{1}{45}\right) + 24\left(-\frac{5}{4}x^2 + \frac{1}{3}x - \frac{3}{8}\right) - 12\left(-\frac{5}{3}x^2 + \frac{1}{3}x - \frac{1}{4}\right)$

54. $-72\left(\frac{1}{4}x + \frac{1}{6} - \frac{3}{8}x - \frac{2}{9}\right) - 36\left(\frac{5}{9}x^2 + \frac{1}{4}x - \frac{1}{6}\right) + 12\left(-\frac{2}{3}x^2 - \frac{1}{12}x + 1\right)$

55. $-72\left(\frac{1}{3}x^2 - \frac{3}{8}x - \frac{5}{9} + \frac{5}{6}x^2\right) - \left(\frac{1}{4}x - \frac{1}{3}\right)12 - 12\left[\left(\frac{1}{2}x^2 - \frac{1}{3}x - \frac{1}{4}\right) - \left(\frac{2}{3}x^2 - \frac{3}{4}x - \frac{5}{6}\right)\right]$

56. $240\left[\left(\frac{1}{2}x^2 - \frac{1}{3}x - \frac{3}{4}\right) + \left(\frac{3}{10}x^2 - \frac{5}{6}x - \frac{1}{2}\right)\right] - 120\left[\left(\frac{2}{3}x^2 - \frac{1}{2}x + \frac{3}{5}\right) + \left(\frac{3}{4}x^2 - \frac{1}{8}x + \frac{7}{10}\right)\right]$

In Exercises 57 through 62, translate each expression into algebra and then simplify.

57. The product of the sum of a number and 4, and the difference of the same number and 3.

58. The product of the difference of 5 and the square of a number, and the difference of the same number squared and 5.

59. The square of the sum of a number and twelve.

60. The product of the sum of eleven and the square of a number, and the difference of that number and 3.

61. The dimensions of a certain box are

   length: $n - 3$
   width: $2n - 1$
   height: n

Substitute these values in the formula for finding the volume of this box, which is

$$V = \text{length} \times \text{width} \times \text{height}$$

Multiply and simplify.

62. The teachers at a particular school vote to take a temporary pay cut of p percent. What percent raise r is needed to bring their salaries (s) back up to where they were? To find this out, solve the equation

$$s(1 + r)(1 - p) = s$$

for r. Use the resulting equation to find the raise needed to "undo" a pay cut of 10%. Check your answer by starting with a salary of 25,000.

MIXED PRACTICE

By doing these exercises, you will practice the topics up to this point in the chapter. Write all answers with positive exponents.

63. Divide: $\frac{18r^2t^2v^5}{20r^3t^4}$

64. Simplify: $(xy^2z^3)^2$

65. Multiply: $(2x + 3)(2x - 4)$

66. Simplify: $(t^2 - 7t + 9) + (-8t^2 - 6t - 12)$

67. Simplify: $\left(\dfrac{-24x^3y^2}{32x^5y^4}\right)^{-4}$

68. Simplify: $(7xy^2)^{-5}$

69. Multiply: $-x^2 + 3x - 5$ by $x^2 + 5x$

70. Simplify: $(-5x^{-2}y^3z^3)^3$

71. Multiply: $(-7rt^4)(8r^2t^3)(-2rt^2)(5r^2t^3)$

72. Simplify: $(-7r^2 + 3r - 12) - (8r^2 + 10r + 3)$

73. Divide: $\dfrac{28r^3s^2}{30r^2s^4}$

74. Multiply: $-7r^2(8r^2 - 5r)$

EXCURSIONS

Class Act

1. You start with $10 in a bank account and have the option to let your account (D) double every two years or (T) triple every three years.

   a. Which account will have the most money five years after you start? ten years?

   b. You can remove your money during any year that is divisible by 6 after it has been deposited for 20 years. If you first invest your money in 1992, which account would have the most money at the earliest time you can withdraw?

   c. At some point after 30 years, one account becomes consistently greater than the other. Find the last year that the accounts "switch" status. Which account becomes the larger at that time and stays that way?

Exploring Patterns

2. The following lists were found. On each ticket, the blanks can be filled by one digit and only one digit. Fill in the blanks.

   a. $5\_\_^2 = 313\_\_$

   $52\_\_^2 = 27\_\_,\_\_7\_\_$

   $5\_\_^3 = 175\_\_1\_\_$

   $5\_\_^4 = 9{,}834{,}49\_\_$

   b. $\_\_8\_\_ = 5{,}308,\_\_16$

   $\_\_6\_\_ = \_\_,\_\_77,\_\_56$

   $_____ = 3{,}7\_\_8{,}096$

   $22\_\_ = 23\_\_,256$

   c. $\_\_\_\_^5 = \_\_6\_\_,05\_\_$

   $3\_\_^7 = 27{,}5\_\_2{,}6\_\_4,_____$

   $4\_\_^5 = \_\_\_\_5{,}856{,}20\_\_$

   $_____^3 = \_\_,367{,}63\_\_$

5.5 Dividing Polynomials

SECTION LEAD-IN

When will I ever use this?

I wrote an exam out longhand, but I can't read my own writing. I have a polynomial

$$9x^2 - 3x + \quad\quad \text{I can't read this part.}$$

I know that $x - 1$ is one of the factors that I want in my answer. What does the missing number have to be?

Dividing a Polynomial by a Monomial

You already know how to divide a monomial by a monomial: Write the division as a fraction and use the rules of exponents to simplify. As an example,

$$3x^2 \div 6x = \frac{3x^2}{6x} = \frac{3}{6}x^{2-1} = \frac{x}{2}$$

When the dividend has two or more terms, as in

$$\frac{3x^2 + 12x}{6x}$$

the entire numerator must be divided by the denominator. To see what we must do, we rewrite our division problem as

$$\frac{1}{6x}(3x^2 + 12x)$$

and then use the distributive property to get

$$\frac{1}{6x}(3x^2 + 12x) = \frac{3x^2}{6x} + \frac{12x}{6x}$$

You know how to do such divisions.

$$\frac{3x^2}{6x} + \frac{12x}{6x} = \frac{3}{6}x^{2-1} + \frac{12}{6}x^{1-1} = \frac{1}{2}x + 2$$

> **To divide a polynomial by a monomial**
>
> Divide each term of the polynomial (dividend) by the monomial (divisor) and add the results.

In the following example, we divide by a monomial that has a negative coefficient. We solve the same way, dealing with the signs as soon as possible. Remember that

$$\frac{-a}{b} = \frac{a}{-b} = -\frac{a}{b}$$

▪▪▪

EXAMPLE 1

Divide: $\dfrac{9x^2 - 15x + 3}{-3x}$

SOLUTION

The last term on the right in the numerator, 3, has no variable part. To keep track of the variable, it helps to rewrite it as $3x^0$ because $3x^0 = 3 \cdot 1 = 3$. Then we divide each term in the numerator by the denominator.

$$\frac{9x^2 - 15x + 3}{-3x} = \frac{9x^2}{-3x} + \frac{-15x}{-3x} + \frac{3x^0}{-3x}$$

$$= -\frac{9x^2}{3x} + \frac{15x}{3x} - \frac{3x^0}{3x} \qquad \text{Changing signs to get positive divisors}$$

$$= -3x^{2-1} + 5x^{1-1} - x^{0-1} \quad \text{Dividing}$$

$$= -3x + 5x^0 - x^{-1}$$

$$= -3x + 5 - \frac{1}{x}$$

Thus

$$\frac{9x^2 - 15x + 3}{-3x} = -3x + 5 - \frac{1}{x}$$

▶ CHECK **Warm-Up 1**

In our next example, there are several variables in each term.

▪▪▪

EXAMPLE 2

Divide: $\dfrac{5r^5u^5 - 25u^7 + 5r^3u^2}{5r^4u^5}$

SOLUTION

Rewriting to show the division, we get

$$\frac{5r^5u^5}{5r^4u^5} - \frac{25r^0u^7}{5r^4u^5} + \frac{5r^3u^2}{5r^4u^5}$$

$$= 1r^{5-4}u^{5-5} - 5r^{0-4}u^{7-5} + 1r^{3-4}u^{2-5}$$

$$= r^1 - 5r^{-4}u^2 + r^{-1}u^{-3} = r - \frac{5u^2}{r^4} + \frac{1}{ru^3}$$

Thus

$$\frac{5r^5u^5 - 25u^7 + 5r^3u^2}{5r^4u^5} = r - \frac{5u^2}{r^4} + \frac{1}{ru^3}$$

▶ CHECK **Warm-Up 2**

Dividing a Polynomial by a Polynomial

To divide a polynomial by a polynomial, such as

$$\frac{-x - 20 + x^2}{x + 4}$$

we use a procedure that is similar to long division in arithmetic.

To divide a polynomial by another polynomial

1. Rewrite the division as a long division, with powers of the variable in descending order.
2. If a power of the variable is missing, list that power with the coefficient zero.
3. Perform the long division.

▪▪▪

EXAMPLE 3

Divide: $(-x - 20 + x^2) \div (x + 4)$

SOLUTION

We set the problem up as a long division, with the exponents in descending order in both the dividend and the divisor.

$$x + 4 \overline{)x^2 - x - 20}$$

Then, to determine the first partial quotient, we consider how many times the first term in the divisor, namely x, divides into the first term in the polynomial, namely x^2. It is x times because $\frac{x^2}{x} = x$, so we write x above x^2.

$$
\begin{array}{r}
x \quad\quad\quad\quad\quad \\
x + 4 \overline{)x^2 - x - 20}
\end{array}
$$
First partial quotient

Next we find the product of x and $x + 4$, write each of that product's terms below the like terms of the dividend, and subtract. (Remember that to subtract means to add the additive inverse.)

$$
\begin{array}{r}
x \quad\quad\quad\quad\quad \\
x + 4 \overline{)\ x^2 -\ x - 20} \\
\underline{x^2 + 4x \quad\quad\quad} \\
-5x \quad\quad\quad
\end{array}
$$
Add $-x^2 - 4x$
Difference

Then we "bring down" the next term of the dividend.

$$
\begin{array}{r}
x \quad\quad\quad\quad\quad \\
x + 4 \overline{)\ x^2 -\ x - 20} \\
\underline{x^2 + 4x \quad\quad\quad} \\
-5x - 20 \quad
\end{array}
$$
Bring down -20

We next consider how many times the first term of the divisor will divide into $-5x$. This is -5 times, so -5 is the next term of the quotient. We multiply -5 by the divisor, write the result below the dividend, and subtract.

$$
\begin{array}{r}
x - 5 \quad\quad\quad \\
x + 4 \overline{)\ x^2 -\ x - 20} \\
\underline{-x^2 + 4x \quad\quad\quad} \\
-5x - 20 \quad \\
\underline{-5x - 20} \quad \\
0 \quad
\end{array}
$$
Second partial quotient

Add $5x + 20$
Remainder

Because there are no other terms to bring down, and we have a remainder of zero, we are finished.

Thus $(-x - 20 + x^2) \div (x + 4)$ is $x - 5$.

To check the answer, we multiply.

$$
\begin{array}{cccc}
& \text{F} \quad \text{O} \quad \text{I} \quad \text{L} \\
(x - 5)(x + 4) = & x^2 + 4x - 5x - 20 \\
= & x^2 - x - 20
\end{array}
$$

▶ *CHECK* **Warm-Up 3**

Calculator Corner

You can check your division work with your graphing calculator. Graph each side of the equation. If your result is correct, you should obtain *only one graph* on your graphing calculator screen. If your answer is incorrect, there will be *two graphs* on the screen.

Example 3 is given as: $(-x - 20 + x^2) \div (x + 4)$.

After you performed the division, you should have gotten the answer $(-x - 20 + x^2) \div (x + 4) = x - 5$.

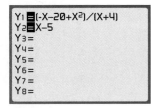
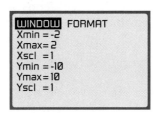
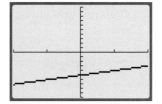

The binomial $(x - 5)$ is the correct answer because you got *only one graph*, meaning the two sides of the equation are equal. If your answer had been incorrect, then you would have obtained two graphs, meaning the two sides of the equation *are not equal!*

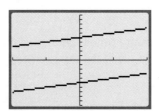

In the next example, there is a non-zero remainder. Follow along to see how to determine when to stop dividing.

▪▪▪

EXAMPLE 4

Divide $r - 4$ into $-19r^2 + 3r^3 + 27r + 8$.

SOLUTION

First we rewrite the problem in long-division form, with powers of the variable in descending order.

$$r - 4 \overline{)\, 3r^3 - 19r^2 + 27r + 8 }$$

Then we determine the first partial quotient by dividing $3r^3$ by r to get $3r^2$. We write this above the first term, multiply $3r^2(r - 4)$, write the result, and subtract.

$$
\begin{array}{r}
3r^2 \qquad\qquad\qquad \\
r - 4 \overline{)\, 3r^3 - 19r^2 + 27r + 8 } \\
\underline{3r^3 - 12r^2 \qquad\qquad} \\
-7r^2 + 27r \qquad
\end{array}
$$

First partial quotient

Add $-3r^3 + 12r$

Bring down $27r$

We bring down $27r$ and find the next partial quotient by dividing r into $-7r^2$, obtaining $-7r$. We again multiply, write the result, subtract, and bring down.

$$
\begin{array}{r}
3r^2 - 7r \qquad\qquad \\
r - 4 \overline{)\, 3r^3 - 19r^2 + 27r + 8 } \\
\underline{3r^3 - 12r^2 \qquad\qquad} \\
-7r^2 + 27r \qquad \\
\underline{-7r^2 + 28r \qquad} \\
-1r + 8
\end{array}
$$

Second partial quotient

Add $7r^2 - 28r$

Bring down 8

We repeat the process again, this time finding the partial quotient -1.

$$
\begin{array}{r}
3r^2 - 7r - 1 \\
r - 4 \overline{)\, 3r^3 - 19r^2 + 27r + 8 } \\
\underline{3r^3 - 12r^2 \qquad\qquad} \\
-7r^2 + 27r \qquad \\
\underline{-7r^2 + 28r \qquad} \\
-1r + 8 \\
\underline{-1r + 4} \\
4
\end{array}
$$

Third partial quotient

Add $1r - 4$

4 is the remainder

There are no additional terms to bring down, but there is still a remainder. *When the degree of the remainder (zero) is less than the degree of the divisor (1), we are finished.*

So $3r^3 - 19r^2 + 27r + 8$ divided by $(r - 4)$ is $3r^2 - 7r - 1$ with remainder 4. We write this quotient as

$$3r^2 - 7r - 1 + \frac{4}{r - 4}$$

▶ *CHECK* **Warm-Up 4**

> **STUDY HINT**
>
> *It may be easier for you to write the "new" sign above the original one and circle the sign you will use to subtract, such as*
>
> $\ominus\, 3r^3 \overset{\oplus}{-} 12r^2.$

In the last example, we will have to insert a term when we arrange the terms in order.

▪ ▪ ▪

EXAMPLE 5

Divide $8t^3 + 14t + 8$ by $2t + 1$.

SOLUTION

We arrange the powers in descending order, inserting $0t^2$ so as to include all powers in the dividend.

$$2t + 1\overline{)8t^3 + 0t^2 + 14t + 8}$$

We then divide step by step until the remainder (if any) has a degree that is less than the degree of the divisor. We get

$$
\begin{array}{r}
4t^2 - 2t\ + 8 \\
2t + 1\overline{)\ 8t^3 + 0t^2 + 14t + 8} \\
\underline{8t^3 + 4t^2} \qquad\qquad \text{Add } -8t^3 - 4t^2 \\
-4t^2 + 14t \\
\underline{-4t^2 -\ \ 2t} \qquad \text{Add } 4t^2 + 2t \\
16t + 8 \\
\underline{16t + 8} \quad \text{Add } -16t - 8 \\
0
\end{array}
$$

So $(8t^3 + 14t + 8) \div (2t + 1) = 4t^2 - 2t + 8$.

▶ *CHECK* **Warm-Up 5**

Practice what you learned.

SECTION FOLLOW-UP

I can use division to find the answer

$$
\begin{array}{r}
9x +\ \ \ 6 \\
x - 1\overline{)\ 9x^2 - 3x + \underline{\quad}} \\
\underline{-(9x^2 - 9x)} \\
6x + \underline{\quad} \\
\underline{-(6x -\ \ 6)} \\
0
\end{array}
$$

The missing digit is 6. The polynomial is

$$9x^2 - 3x - 6$$

5.5 WARM-UPS

Work these problems before you attempt the exercises.

1. Divide: $\dfrac{4x^4 + 12x^2 - 8}{-8x}$

2. Divide: $\dfrac{4x^2y - 18x^2y^2 + 3xy}{6x^3y^2}$

3. Divide: $(x^2 + 5x + 4) \div (x + 1)$

4. Divide: $(3x^3 - 5x^2 + 2x + 4) \div (x - 2)$

5. Divide: $(4x^3 + 4x + 5) \div (x + 2)$

5.5 EXERCISES

Note: Use your graphing calculator to check your results whenever possible.

In Exercises 1 through 38, divide each expression by using the distributive property.

1. $\dfrac{4x^2 + 8x}{2x}$

2. $\dfrac{8m^4 + 16m^2}{2m}$

3. $\dfrac{6x^3 - 3x}{2x}$

4. $\dfrac{18n^9 - 15n^4}{-5n^4}$

5. $\dfrac{22x^2 + 11x^3}{-11x^2}$

6. $\dfrac{30m^4 + 5m^5}{-5m^4}$

7. $\dfrac{36r^5 - 12r^7}{-12r^5}$

8. $\dfrac{32n^3 - 8n^8}{-8n^3}$

9. $\dfrac{16x^8 + 4x^{15}}{4x^{15}}$

10. $\dfrac{24m^{10} + 8m^{12}}{8m^{12}}$

11. $\dfrac{25r^{10} - 5r^{12}}{5r^{12}}$

12. $\dfrac{36n^{11} - 9n^{13}}{9n^{13}}$

13. $\dfrac{15x^3y + 9x^4}{3x^2y}$

14. $\dfrac{12m^5 + 8m^7}{4m^4r}$

15. $\dfrac{49r^7u - 21r^6}{7r^5u}$

16. $\dfrac{24n^6 - 12n^7}{6n^2y}$

17. $\dfrac{11x^3 + 44x^5y}{11x^3y}$

18. $\dfrac{10m^4 + 30m^6r}{10m^3r}$

19. $\dfrac{13r^5 - 39r^7u}{13r^5u}$

20. $\dfrac{20n^4 - 80n^7y}{20n^4y}$

21. $\dfrac{26xy^2 + 22xz}{2x^3y}$

22. $\dfrac{16mr^4 + 8mr}{6m^3r}$

23. $\dfrac{24ru^6 - 18ru}{-3r^7u}$

24. $\dfrac{22ny^9 - 24ny}{-4n^6y}$

25. $\dfrac{18x^2 + 16xy^2 + 14y}{-2y}$

26. $\dfrac{21m^2 + 14r^4 + 28r}{-7r}$

27. $\dfrac{27r^2 - 18ru^2 - 12u}{3ru}$

28. $\dfrac{18n^3 - 36y^9 - 27y}{9ny}$

29. $\dfrac{3x^3y^2 - 9x^3y^4 + 3x^3y^3}{-3x^3}$

30. $\dfrac{3m^5r^3 - 9m^5r^5 + 3m^5r^6}{-3m^5}$

31. $\dfrac{22r^2u^3 - 16r^2u^4 - 12r^2u^5}{2r^2u^4}$

32. $\dfrac{4n^5y^3 - 12n^5y^5 - 24n^5y^6}{4n^5y^2}$

33. $\dfrac{12r^5u^4 - 4ru^4}{-12r^5u^4}$

34. $\dfrac{9n^6y^7 - 2ny^7}{-9n^6y^7}$

35. $\dfrac{14xy^2 - 2x^2y + 21x^2y^2}{-7x^2y^2}$

36. $\dfrac{32m^3r^2 + 6m^2r - 16mr^2}{-8m^2r^2}$

37. $\dfrac{18ru^4 - 3r^3u + 24r^4u^4}{-6r^3u^5}$

38. $\dfrac{40n^4y^6 - 3n^3y + 24ny^6}{-4n^3y^6}$

In Exercises 39 through 64, divide by using long division.

39. $\dfrac{x^2 - 169}{x + 13}$

40. $\dfrac{x^2 - 144}{x - 12}$

41. $\dfrac{4x^2 - 9x - 9}{x + 3}$

42. $\dfrac{10x^2 - 11x - 6}{2x + 3}$

43. $\dfrac{2 + x^2 + 3x}{x + 1}$

44. $\dfrac{x^2 + 15 - 8x}{x - 3}$

45. $\dfrac{x^2 - 5 - 4x}{x - 5}$

46. $\dfrac{-6 + x^2 + x}{x - 2}$

47. $\dfrac{11y + y^2 + 18}{y + 9}$

48. $\dfrac{-13n - 30 + n^2}{n - 2}$

49. $\dfrac{m^2 - 13m - 14}{m - 1}$

50. $\dfrac{y^2 - 5y - 6}{y + 2}$

51. $\dfrac{-30 + 3x^2 - 26x}{x - 13}$

52. $\dfrac{-24x + 2x^2 + 20}{x - 2}$

53. $\dfrac{t^3 - t^2 - 4}{t - 2}$

54. $\dfrac{r^3 - 12r^2 + 8r + 96}{r - 4}$

55. $\dfrac{t^3 - 8}{t - 2}$

56. $\dfrac{-t^3 - 1}{t + 1}$

57. $\dfrac{t^3 + 64}{t + 4}$

58. $\dfrac{27t^3 - 1}{3t - 1}$

59. $\dfrac{x^2 - 64}{x + 8}$

60. $\dfrac{9x^2 - 100}{3x - 10}$

61. $\dfrac{x^4 - 1}{x - 1}$

62. $\dfrac{16x^4 - 1}{2x - 1}$

63. $\dfrac{16x^4 - 81}{2x - 3}$

64. $\dfrac{81x^4 - 1}{3x - 1}$

65. *Heart Rate* The optimal exercise heart rate (H) for an adult of age A is given by the formula

$$\frac{7(220 - A)}{10} = H$$

Solve this equation for A. Then use the formula to find your own optimal heart rate during exercise.

66. *Depreciation* The amount by which a machine decreases in value from the original cost is called depreciation. One way of calculating depreciation is to use the formula

$$\frac{\text{cost} - \text{scrap value}}{\text{expected life}} = \text{yearly depreciation}$$

A certain machine has an expected life of 4 years. Its original cost was $5900, and its yearly depreciation is $1000. What is its value as scrap?

MIXED PRACTICE

By doing these exercises, you will practice the topics up to this point in the chapter. Write all answers with positive exponents.

67. Simplify this exponential expression: -8^{-4}

68. Divide: $\dfrac{16r^2v + 22rv - 8rv^2}{-4r^2v}$

69. Multiply: $-5r(11r^2 + 15rv - 5v)$

70. Divide: $\dfrac{18rt - 27r^2t + 3r^2}{3r}$

71. Divide: $x^3 + 4x + x^2 + 4$ by $x + 1$.

72. Multiply: $(7y - 4)(8y + 9)$

73. Multiply: $(-9rt^2)(-13r^3y^2)$

74. Simplify:
$(3x^2 + 5x - 7) + (8x^2 + 2x - 4) + (7x^2 - 5x + 1)$

75. Multiply: $(11x - 2)(-5x^2 + 4x - 5)$

76. Divide: $\dfrac{26r^3t^2}{13tr^3v^4}$

77. Multiply: $(7x + 2)(3x - 7)$

78. Simplify: $(r^3t^2v^4)^{-2}$

79. Simplify: $(-3n^2 - 8n + 4) - (-8n^2 - 9n - 11)$

EXCURSIONS

Exploring Numbers

1. The numbers 242, 243, 244, and 245 each have 6 divisors. Find them.

2. Show that the product of two even numbers is always even.

3. **a.** A micron is one-thousandth of a millimeter, or one twenty-five thousandth of an inch.

   Given what you know about the relationship between an inch and a centimeter, critique the above statement and give a more precise comparison.

   b. A bacteriophage is one-fortieth of a micron, or *one billionth of an inch.*

   What do you think this statement in italics should really say?

Posing Problems

4. Ask and answer four questions using the data from the table on the following page. Share your questions with a friend.

National Parks

Name, location, and year authorized	Acreage	Name, location, and year authorized	Acreage
Acadia (Maine), 1919	41,951.06	Gates of the Arctic (Alaska), 1980	7,523,888.00
Arches (Utah), 1971	73,373.98	Glacier Bay (Alaska), 1980	3,225,284.00
Badlands (S.D.), 1978	242,755.94	Glacier (Mont.), 1910	1,013,572.42
Big Bend (Tex.), 1935	801,163.02	Grand Canyon (Ariz.), 1919	1,217,159.32
Biscayne (Fla.), 1980	172,924.73	Grand Teton (Wyo.), 1929	309,974.28
Bryce Canyon (Utah), 1924	35,835.08	Great Basin (Nev.), 1986	77,180.00
Canyonlands (Utah), 1964	337,570.43	Great Smoky Mts. (N.C.-Tenn.), 1926	520,269.44
Capitol Reef (Utah), 1971	241,904.26	Guadalupe Mountains (Tex.), 1966	86,415.97
Carlsbad Caverns (N.M.), 1930	46,766.45	Haleakala (Hawaii), 1960	28,099.00
Channel Islands (Calif.), 1980	249,353.77	Hawaii Volcanoes (Hawaii), 1916	229,177.03
Crater Lake (Ore.), 1902	183,223.77	Hot Springs (Ark.), 1921	5,839.24
Denali (Alaska), 1917	4,741,910.00	Isle Royale (Mich.), 1931	571,790.11
Dry Tortugas (Fla.), 1992	64,700.00	Katmai (Alaska), 1980	3,586,000.00
Everglades (Fla.), 1934	1,506,499.40		

Source: 1996 Information Please® Almanac (©1995 Houghton Mifflin Co.), p. 574. All rights reserved. Used with permission by Information Please LLC.

Exploring Problem Solving

5. The moon has a nearly circular orbit about Earth. Use the formula

$$V = \frac{2\pi r}{T} \quad \text{and} \quad a = \frac{V^2}{r}$$

where

V = speed of the moon as it revolves about Earth
r = radius of the orbit so $2\pi r$ is the path (circumference) of the orbit
T = time it takes for the moon to circle Earth
a = acceleration of the moon toward Earth

Find the time T it takes the moon to complete one orbit. Use

a = 0.00272 meters/second^2
V = 1020 meters/second
r = 384,000 km (384,000,000 m)

Your result will be in terms of seconds. Convert your answers to days.

CHAPTER LOOK-BACK

Many patterns exist in nature and in mathematics. You could have used a pattern to determine the units digit of

$$2^{17}$$
$$2^{23}$$
$$2^{40}$$
$$2^{100}$$

Power of 2	Units digit	Power of 2	Units digit
2^1	2**	2^7	8***
2^2	4	2^8	6*
2^3	8***	2^9	2**
2^4	6*	2^{10}	4
2^5	2**	2^{11}	8***
2^6	4	2^{12}	6*

Observations:

*The power of 2 is divisible by 4 and the units digit is 6. So, the units digit of 2^{40} and of 2^{100} is 6.

**The power of 2 is in the form $4n + 1$ and the units digit is 2. So, the units digit of $2^{17} = 2$ because $17 = 4 \cdot 4 + 1$.

***The power of 2 is in the form $4n + 3$ and the units digit is 8. So, the units digit of $2^{23} = 8$ because $23 = 4 \cdot 5 + 3$.

Many biological systems grow at a rate that is exponential.

The area of a square grows exponentially in relation to the growth of its side. Complete the following table.

Length of side n	Area n^2		Difference 1		Difference 2
1	1				
2	4		3		
3	9		5		2
4	16		7		2
5	25		—		2
6	36		—		—
7	49		—		—
8	64		—		—

■ Find some other patterns that involve exponents.

CHAPTER 5
REVIEW PROBLEMS

The following exercises will give you a good review of the material presented in this chapter. Write all answers without negative exponents.

SECTION 5.1

1. Rewrite $(-6)^2$ without exponents.

2. Rewrite -5^3 without exponents.

3. Rewrite 4^{-5} without exponents.

4. Place grouping symbols around -2^4 such that the result is equivalent to -16.

5. Compare $(-2)^3$ and $-(2^3)$ and determine which is greater.

6. Rewrite $\left(-\frac{3}{4}x^2y^{-3}\right)^{-2}$ in simplest terms.

7. Simplify: $(-2x^2y^3)^2$

8. Simplify $(3r^3t^4y)^{-2}$ and rewrite without negative exponents.

9. Simplify $(8r^6s^{-4}t)^{-3}$ and rewrite without negative exponents.

10. Simplify: $(-5m^3n^{-2})^4$

SECTION 5.2

11. Multiply: $(-1.9x^3y^2)(-3.2xy^2)$

12. Find the product of $3\frac{1}{2}x^3y^3$ and $\frac{4}{5}x^2y$.

13. Multiply: $(5m^5n^{-2})(7n^3)(-6m^2)$

14. Find the product: $(26m^2n^3)(2.4n^{-2})$

15. Divide: $\dfrac{-36r^4t^2}{-9rt^3}$

16. Divide: $\dfrac{-5.6m^3n^5}{0.8m^5n^2}$

17. Find the quotient of $24x^2y^3$ and $6xy$.

18. Divide: $\dfrac{-36xt^3}{-24yt^4x^2}$

19. Divide $6.3m^2n^{-3}$ by $0.9m^4n^7r^5$.

20. Multiply: $(-6r^3t^2)(5r^3t^2)$

SECTION 5.3

21. Add: $x^2 + 3x - 5$
$\underline{+5x - x^2 + 4}$

22. Subtract: $x^3 - 5$
$\underline{-(x^2 + 3x - 5 + 4x^3)}$

23. Find the sum:
$(3y^2 - 8y + 2) + (-12y^3 + 4y - 2y^2 + 3)$

24. Find the difference:
$(-t^2 + 7t + 3) - (-12t^3 + 5t + 9t^2 + 4)$

SECTION 5.4

25. Multiply: $(2x - 4)(5x - 3)$

26. Multiply: $(x + 9y)(7x - 4y + 6y^{-2})$

27. Multiply using the FOIL method:
$(-2y - 5)(6y + 3)$

SECTION 5.5

28. Find the quotient: $\dfrac{-27r^3t^2 + 18rt}{-9rt}$

29. Divide by using the distributive property: $\dfrac{-7rt^2 + 8r^3t^4 - 9rt^5}{2rt^5}$

30. Find the quotient:
$(x^3 + 9x^2 + 17x - 12) \div (x + 4)$

31. Divide: $(x^3 + x^2 - 12x) \div (x - 3)$

MIXED REVIEW

32. Combine:
$(-5x^2 + 3x - 7) + (7x^2 - 6x + 12) - (-x^2 + 5)$

33. Multiply: $(27m^3n^{-2})^{-1}(-3m^4n^5)$

34. Divide $x^3 + 3x^2 + 5x + 15$ by $x - 3$.

35. Simplify $(-9x^{-2}y^{-3}z^{-5})^{-3}$ and rewrite it without negative exponents.

36. Simplify: $\dfrac{-12r^5t^3}{-18r^{-2}t^9}$

37. Find the product: $(7t^2 - 4t - 8)(3t^2 + 9)$

CHAPTER 5 TEST

This exam tests your knowledge of the material in Chapter 5.

1. For parts (a) and (b), rewrite the expression without exponents.

   a. -6^4 **b.** 2^{-3}

   c. Simplify the algebraic expression $(-2x^3t^2y^{-5})^3$. Write your answer using only positive exponents.

2. **a.** Divide: $\dfrac{14xy^3}{-7x^2y}$ **b.** Multiply:
$(-5xy^3)(3vx^2y^2)$ **c.** Multiply:
$(-14xy^2)(-6x^{-2}y^2)$

3. Combine as indicated:

   a.
$$\begin{array}{r} 8x^2 - 2x \\ 3x - 5 \\ +\ -6x + 14x^2 + 7 \\ \hline \end{array}$$

   b. $(-11y^2 + 3y) - (-7y^3 - 5y - 4)$

   c. $(5r - 3r^2) + (7r - 7) - (-6r^2 - 12)$

4. **a.** Find the product:
$-4y(5x^2y^2 + 3y^2)$ **b.** Multiply using the FOIL method: $(3x - 4)(9x - 3)$ **c.** Multiply:
$(4x - 5)(2x^2 + 6x - 5)$

5. For parts (a) and (b), find the quotient.

   a. $\dfrac{-36r^2t^3 + 18rt^2}{-2rt^2}$ **b.** $\dfrac{-22x^2y + 5x}{11xy}$

   c. Divide using long division: $(x^2 + 8x + 25)$ by $(x + 5)$

CUMULATIVE REVIEW

CHAPTERS 1–5

The following exercises will help you maintain the skills you have learned in
this and previous chapters.

1. Subtract: $(-6.89) - (-5.25)$

2. Combine like terms:
    $-11ty^2x - 6x^2ty - 7ty^2x + 5xty^2 + 8ty^3 + 12x^2$

3. Solve for x: $-8x + 3 = -6(x - 4)$

4. Rewrite in scientific notation:
    0.00000000000006

5. Find the slope of the line that passes through
    the points $(-6, -2)$ and $(6, 5)$.

6. Multiply: $(-1.73)(-9.198)(1.5)$

7. Solve for x: $2.4 - 3x \geq 6 + x$

8. Solve for t: $0.4t - 5 = -9$

9. Divide: $161.5 \div 9\frac{1}{2}$

10. Divide: $\frac{32t^2r^3 - 16t^3r^4}{-8rt^2}$

11. Add: $\left(-6\frac{2}{3}\right) + \left(-9\frac{1}{2}\right)$

12. Rewrite -5^{-2} without exponents.

13. Simplify: $(0.05 - 0.02)^2 - 3(1.2 - 2.3)$

14. Multiply: $(-12x^3y^2)(-18xy^3t^2)$

15. Multiply: $(3x + 5)(2x - 2)$

16. Simplify: $(-6y^2t^4)^3$

17. Evaluate $xy - xy + xy^2 - x^2y$ when $x = -1$
    and $y = -8$.

18. Multiply: $-6xy(-3xy^2 - 3rxy^2)$

19. Graph $12x - 3y = 9$ using a table of values.

20. Graph the inequality $4x - 3y \geq 8$.

FACTORING POLYNOMIALS

Metalsmiths often need to buy sheet metal cut in specific dimensions. By determining the factors and multiples of the amount to be cut, waste is avoided. For example, a smith might need two sheets of silver, both 54 square inches, but one must be a strip no less than 2 inches wide and the other as square as possible (both measured with whole numbers). Knowing that the prime factorization of 54 is $2 \cdot 3^3$ tells the smith and the supplier that the metal strip will be either 2 inches by 27 inches, or 3 inches by 18 inches wide and the squarer one will be 6 inches by 9 inches.

■ *Under what other circumstances might prime factorization be used to determine the shape of an area?*

SKILLS CHECK

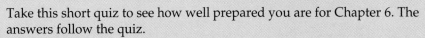

Take this short quiz to see how well prepared you are for Chapter 6. The answers follow the quiz.

1. Simplify: $2(3x + 4)$

2. Add: $-14 + (-23)$

3. Subtract: $-87 - 36$

4. Multiply: $(-12)(2)(4)$

5. Combine: $-4x + (-6x)$

6. Simplify: $x \cdot x^3$

7. Multiply: $(2x + 4)(3x - 5)$

8. Simplify: $x^2 + 6x - 3x + 2x^2 + 2$

ANSWERS: **1.** $6x + 8$ [Section 2.3]   **2.** -37 [Section 1.2]   **3.** -123 [Section 1.2]   **4.** -96 [Section 1.3]   **5.** $-10x$ [Section 2.3]   **6.** x^4 [Section 5.2]   **7.** $6x^2 + 2x - 20$ [Section 5.4]
8. $3x^2 + 3x + 2$ [Section 2.3]

CHAPTER LEAD-IN

Certain genes can be carried in the genetic code of an organism and not always be observable. One such gene determines attached earlobes; the gene for unattached, or hanging, earlobes is **dominant.***

A person with the F gene will have hanging earlobes.

F: hanging earlobes
f: attached earlobes

Each person carries two genes for this particular characteristic.

FF	*Ff*	*fF*	*ff*
Hanging	Hanging	Hanging	Attached

The Hardy-Weinberg equation is used to model the distribution of genes in the population for characteristics that are wholly controlled by two genes. It is

$$p^2 + 2pq + q^2 = 1$$

where

p = proportion of population with Gene I
q = proportion of population with Gene II

■ For our example, if Gene I is F and Gene II is f, what does each term in the Hardy-Weinberg equation mean?

*Source: Gregory Fiore, "An Out of Math Experience: Quadratic Equations and Polynomial Multiplication as used in Genetics." The AMATYC Review, Volume 17, Number 1 (Fall 1995): pp. 20–27.

6.1 Factoring by Removing the Greatest Common Factor

SECTION LEAD-IN

The probability of having children of a certain sex in a family that consists of exactly two children can be found by using the equation

$$p^2 + 2pq + q^2 = 1$$

(What is this equation called?)

The probability of a boy is $\frac{1}{2}$.

The probability of a girl is $\frac{1}{2}$.

- What does each term of the above equation mean in terms of boys and girls?

Finding Integer Factors

Factors are the numbers that yield a product; and that product is a multiple of each of its factors.

> **Factors and Multiples**
>
> Let a, b, and c be integers, and let
> $$a \cdot b = c$$
> Then a and b are **factors**, or **divisors**, of c, and c is a **multiple** of a and of b.

Thus if $5 \times 3 = 15$, then 5 and 3 are factors (or divisors) of 15, and 15 is a multiple of both 5 and 3.

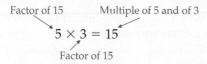

Any number that divides another integer evenly (with no remainder) is a factor of the second integer. Therefore, all the factors of 15 are 1, 3, 5, and 15.

Study the following rules to help you find factors (or divisors).

> **Rule for Divisibility by 2**
>
> Every even integer has 2 as a divisor.
>
> **Rule for Divisibility by 3**
>
> If the sum of the digits of an integer is divisible by 3, the integer is divisible by 3.
>
> **Rule for Divisibility by 5**
>
> Integers that end in 5 or 0 have 5 as a divisor.
>
> **Rule for Divisibility by 10**
>
> Integers that end in 0 have 10 as a divisor.

The integer 24 can be written as the product of two factors in several different ways:

$$24 = 1 \cdot 24 \quad \text{or} \quad 2 \cdot 12 \quad \text{or} \quad 3 \cdot 8 \quad \text{or} \quad 4 \cdot 6$$

Each of these is called a **factorization** of 24. A factorization can have more than two factors, as in

$$24 = 2 \cdot 2 \cdot 6 \quad \text{and} \quad 24 = 2 \cdot 2 \cdot 2 \cdot 3$$

but the product of these factors must always be the original number. The process of finding a factorization of a number is called **factoring.**

Prime Factors

Some integers have many positive-integer factors. Others have only two, 1 and the integer itself. Integers are classified as *prime* or *composite* according to the number of their factors.

> **Prime Number**
>
> If an integer greater than 1 has only the positive-integer factors 1 and itself, that integer is called a **prime number,** or is said to be **prime.**
>
> **Composite Number**
>
> If an integer greater than 1 has positive-integer factors other than itself and 1, that integer is called a **composite number,** or is said to be **composite.**

■■■
WRITER'S BLOCK

Why are 0 and 1 neither prime nor composite?

The prime numbers less than 50 are

2	3	5	7	11	13	17	19
23	29	31	37	41	43	47	

The integers 0 and 1 are neither prime nor composite.

Factorizations that consist only of prime numbers are useful in much of the work of this chapter.

> **Prime Factorization**
>
> The **prime factorization** of an integer is the expression of that integer as the product of only prime numbers, called **prime factors.**

One method for finding the prime factors of an integer makes use of something called a **tree diagram.**

▪▪▪
EXAMPLE 1

Find the prime factorization of 180 using a tree diagram.

SOLUTION

We begin by choosing any two integers whose product is the number to be factored. Here, we'll start with 18 and 10, because we know that $18 \cdot 10 = 180$. We write these integers under the 180, connected to it by "branches."

Then we break the integer on each branch into two factors and write them in the same way, forming a "tree."

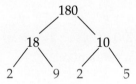

We continue in the same way, factoring each new factor. When an integer cannot be factored further, we circle it and leave it. The process ends when every integer that is on the end of a "branch" is circled.

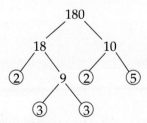

The circled numbers are the prime factors of the original number. We usually write them in increasing order from left to right; so we would write the prime factorization of 180 as

$$2 \cdot 2 \cdot 3 \cdot 3 \cdot 5 \quad \text{or} \quad 2^2 \cdot 3^2 \cdot 5$$

You can check a factorization by multiplying the factors together to get the original number. Try it for the prime factorization of 180.

▶ *CHECK* **Warm-Up 1**

ERROR ALERT

Identify the error and give a correct answer.

Find the factors of 124.

Incorrect Solution:

The factors of 124 are

2   and   31

Calculator Corner

You can use your calculator to check if the solution in Example 1 is correct. (Note: See the Calculator Corner on page 269 to review how to test a statement.) The calculator returns an answer of 1, meaning that your prime factorization of 180 is correct. If your factorization had been incorrect, then the calculator would return a result of 0.

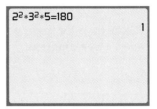

■■■

EXAMPLE 2

Find four factorizations of $6x^2y$.

SOLUTION

Any product that gives us $6x^2y$ is a factorization of that monomial. Some of these products are

$$6 \cdot x^2y \qquad 6 \cdot x^2 \cdot y$$
$$2 \cdot 3 \cdot x^2 \cdot y \qquad 2 \cdot 3 \cdot x \cdot xy$$

You should be able to write several more.

▶ CHECK **Warm-Up 2**

Greatest Common Factors

> **Greatest Common Factor**
>
> The **greatest common factor** of two integers is the greatest positive integer that is a factor of both.

WRITER'S BLOCK

Explain to a friend how the GCF and the greatest common divisor are alike.

Consider the integers -6 and -18. The positive integers 1, 2, 3, and 6 are factors of both -6 and -18. But 6 is the greatest common factor (GCF), because it is the greatest of the "common" factors.

This definition also applies in a general way to variable terms: The greatest common factor of two monomials is the greatest term that is a factor of both monomials. It consists of the GCF of the coefficients, multiplied by each common variable raised to its least power. Thus, for example,

The GCF of $18x^3$ and $6x^5y$ is $6x^3$.
The GCF of x^2y^2 and x^5y is x^2y.

∎∎∎
EXAMPLE 3

Find the GCF of $3x^4z^8$ and $-9x^2y^2z^3$.

SOLUTION

The GCF of the coefficients is 3. The common variables are x and z; raised to their least powers, they are x^2 and z^3.

So the GCF is $3x^2z^3$.

▶ *CHECK* **Warm-Up 3**

Factoring Out a Common Factor

The distributive property states that

$$a(b + c) = ab + ac$$

When we reverse this procedure, we "factor." That is, when we write $ab + ac$ as $a(b + c)$, we have "factored out a." One way to factor an algebraic expression is *by removing the greatest common factor*.

> **To factor an algebraic expression by removing the greatest common factor**
> 1. Find the GCF of the terms of the expression (coefficients and variables).
> 2. Rewrite each term of the expression as the product of the GCF and another factor.
> 3. Use the distributive property to rewrite the result of step 2.

∎∎∎
EXAMPLE 4

Factor: $5xy^2 + 10y^3$

SOLUTION

The GCF of the coefficients 5 and 10 is 5. The GCF of the variable parts xy^2 and y^3 is y^2. Therefore, the GCF of the two terms is $5y^2$.

Now we want to write each term as the product of $5y^2$ and something else.

To do this, we divide each term by $5y^2$,

$$5xy^2: \quad \frac{5xy^2}{5y^2} = x \qquad 10y^3: \quad \frac{10y^3}{5y^2} = 2y$$

Thus the original expression may be written as

$$(5y^2)(x) + (5y^2)(2y) = 5y^2(x + 2y)$$

So in factored form, $5xy^2 + 10y^3$ is $5y^2(x + 2y)$. You can multiply out the last expression to check the result.

▶ *CHECK* **Warm-Up 4**

ERROR ALERT

Identify the error and give a correct answer.

Factor: $10x^2y - 5xy$

Incorrect Solution:

$5xy(2x)$

■ ■ ■
EXAMPLE 5

Factor: $27x^4y^5 - 9x^2y^3 + 18x^2y^5$

SOLUTION

The GCF of the coefficients (27, −9, and 18) is 9. The GCF of the variable parts is x^2y^3 (because both x and y appear in each term, and 2 and 3 are the least powers of each.) Therefore, the GCF of the three terms is $9x^2y^3$. The original expression may be written as

$$(9x^2y^3)(3x^2y^2) + (9x^2y^3)(-1) + (9x^2y^3)(2y^2)$$
$$= 9x^2y^3(3x^2y^2 - 1 + 2y^2)$$

So in factored form, $27x^4y^5 - 9x^2y^3 + 18x^2y^5$ is $9x^2y^3(3x^2y^2 - 1 + 2y^2)$.

▶ CHECK **Warm-Up 5**

■ ■ ■
EXAMPLE 6

Factor: $15xy^3 - 5xy^2 + 10x^2y^6$

SOLUTION

The GCF of the coefficients is 5. The variable parts have x and y^2 in common (because both x and y appear in each term, and 1 and 2 are their least powers, respectively). So the GCF of the three terms is $5xy^2$.

Dividing each term by $5xy^2$ gives us

$$15xy^3 \div 5xy^2 = 3y \quad \text{so} \quad 15xy^3 = (5xy^2)(3y)$$
$$-5xy^2 \div 5xy^2 = -1 \quad \text{so} \quad -5xy^2 = (5xy^2)(-1)$$
$$10x^2y^6 \div 5xy^2 = 2xy^4 \quad \text{so} \quad 10x^2y^6 = (5xy^2)(2xy^4)$$

Thus the original polynomial may be written as

$$(5xy^2)(3y) + (5xy^2)(-1) + (5xy^2)(2xy^4)$$
$$= 5xy^2(3y - 1 + 2xy^4) \qquad \text{Distributive property}$$

▶ CHECK **Warm-Up 6**

ERROR ALERT

Identify the error and give a correct answer.

Factor:
$36x^2y - 6xy - 12xy^2$

Incorrect Solution:
$36x^2y - 6xy - 12xy^2 =$
$6xy(6x - 2y)$

Factoring an Expression by Grouping

Sometimes, when there are several terms, it is not possible to find a common factor for all terms but only for some of the terms. We factor those terms that can be factored and combine factors where possible. This is called *factoring by grouping*.

> **To factor by grouping**
> 1. Remove the GCF if there is one.
> 2. Group terms of the expression that have common factors by using the commutative and associative properties.
> 3. Remove the GCF of each group and rewrite the group as a product.
> 4. Use the distributive property to rewrite the result of step 3.

▪▪▪

EXAMPLE 7

Factor: $2x^2 + 10y - 4xy - 5x$

SOLUTION

There is no GCF. But we can group the terms as follows:

$$(2x^2 - 4xy) + (10y - 5x)$$

Each of these two groups has a GCF. We remove the GCF from each group and rewrite the group as a product.

$$\begin{aligned}(2x^2 - 4xy) + (10y - 5x) &= 2x(x - 2y) + 5(2y - x) \\ &= 2x(x - 2y) - 5(x - 2y) \quad \text{Rewriting } 5(2y - x) \\ &\qquad\qquad\qquad\qquad\qquad\quad \text{as } -5(x - 2y)\end{aligned}$$

Because the two terms now have the common factor $(x - 2y)$, we can use the distributive property to simplify the result to

$$(2x - 5)(x - 2y)$$

Thus

$$2x^2 + 10y - 4xy - 5x = (2x - 5)(x - 2y)$$

You can multiply these binomials with the FOIL method to check this result.

▶ *CHECK* **Warm-Up 7**

WRITER'S BLOCK

Explain to a friend which properties we must use to group and factor.

▪▪▪

EXAMPLE 8

Factor: $3r^2 - 6rt + 2r - 4t$

SOLUTION

There is no GCF. But the first two terms have a common factor, and so do the last two. We group them as follows: $(3r^2 - 6rt) + (2r - 4t)$.

We remove the GCF from each group and rewrite the group as a product: $(3r^2 - 6rt) + (2r - 4t) = 3r(r - 2t) + 2(r - 2t)$.

These two new terms have the common factor $(r - 2t)$, so we simplify the result to $(r - 2t)(3r + 2)$. You can multiply these binomials by the FOIL method to check this result.

LOOK AGAIN

We could have grouped $(3r^2 + 2r) + (-6rt - 4t)$ and factored:

$$r(3r + 2) - 2t(3r + 2) = (3r + 2)(r - 2t)$$

▶ CHECK **Warm-Up 8**

▪ ▪ ▪

EXAMPLE 9

Factor: $20r^2 + 30r + 24r + 36$

SOLUTION

Because there is a common factor, 2, we must first factor it out.

$$20r^2 + 30r + 24r + 36 = 2(10r^2 + 15r + 12r + 18)$$

We could add like terms, but instead we are going to factor. The first two terms have a common factor and so do the last two; we group them together. We then remove the GCF from each group and rewrite the group as a product.

$$(10r^2 + 15r) + (12r + 18) = 5r(2r + 3) + 6(2r + 3)$$

There is now a common factor, $2r + 3$. So we can simplify this result to

$$(2r + 3)(5r + 6)$$

Thus $20r^2 + 30r + 24r + 36$ in factored form is $2(2r + 3)(5r + 6)$.

▶ CHECK **Warm-Up 9**

Practice what you learned.

ERROR ALERT

Identify the error and give a correct answer.

Factor by grouping

$4r^2 + 15t^2 - 20rt^2 - 3r$

Incorrect Solution:

$(4r^2 + 15t^2) - (20rt^2 + 3r)$
$= (4r^2 + 15t^2) - r(20t^2 + 3)$

The expression cannot be factored further.

SECTION FOLLOW-UP

In the equation

$$p^2 + 2pq + q^2 = 1$$

p^2 is the probability of exactly 2 boys.
q^2 is the probability of exactly 2 girls.
$2pq$ is the probability of exactly one boy and one girl.

Those probabilities are

$$p^2 = p \cdot p = \left(\tfrac{1}{2}\right)\left(\tfrac{1}{2}\right) = \tfrac{1}{4} \quad \text{Probability of 2 boys}$$

$$q^2 = q \cdot q = \left(\tfrac{1}{2}\right)\left(\tfrac{1}{2}\right) = \tfrac{1}{4} \quad \text{Probability of 2 girls}$$

$$2pq = 2(p)(q) = 2\left(\tfrac{1}{2}\right)\left(\tfrac{1}{2}\right) = \tfrac{1}{2} \quad \text{Probability of one boy and one girl}$$

▪ In a population of 1000 families that have exactly two children each, what would we expect to find?

6.1 WARM-UPS

Work these problems before you attempt the exercises.

1. Factor 60 into prime factors.

2. Find two factorizations of $3r^2t$.

3. What is the GCF of $14rt^3$ and $21r^3t^2$?

4. Factor: $22x^2y - 10x^3y^5$

5. Factor: $6xy^6 - 20xy^4 + x^6$

6. Factor: $12x^4y^2 - 20xy^3 + 16x^2$

7. Factor by grouping: $x^2 + 10y + 5x + 2xy$

8. Factor by grouping: $r^2 - 3r + 4r - 12$

9. Factor by grouping: $2y^2 - 14y + 6y - 42$

6.1 EXERCISES

Note: Use your graphing calculator to check your results whenever possible.

In Exercises 1 through 4, list the whole-number factors of each number. Then circle the prime factors among them.

1. 60   2. 84   3. 35   4. 49

In Exercises 5 through 8, use a tree diagram to find the prime factorization of each number. Write your answer in exponential notation.

5. 224   6. 136   7. 126   8. 360

In Exercises 9 through 20, find the greatest common factor of the given terms.

9. $24m^2, 6m$ 10. x^3y, xy^2 11. $15m^5t, 25mt$

12. x^5, x^4y^4, x^3y 13. $3x^2y, 9xy, 6y$ 14. $16x^2y^3, 24x^3y, 36xy$

15. $15xy^3, 21x^2y^4, 27x^5y^6$ 16. $9tx^5, 11tx, 19t^7$ 17. $-36mn^6, 60m^8n^2, 56mn$

18. $-8rt^3, 6rt^4, 12r^4t^5$ 19. $-12x^4y^2, -16x^3y, -20xy$ 20. $-3.4u^3v^2, -1.7uv, -5.1u^5$

In Exercises 21 through 24, use your knowledge of divisibility to decide whether the statement is true or false.

21. All numbers divisible by both 2 and 3 are divisible by 6.

22. Any number divisible by 2 and 4 is divisible by 8.

23. A number divisible by 6 is also divisible by 2 and 3.

24. If a number is a multiple of 4, it is divisible by 2.

In Exercises 25 through 42, factor by removing the greatest common factor.

25. $18xy^2 + 24xyz$

26. $2s - s^2$

27. $-3r^2s^3 - 6rs^2$

28. $6x^2y^2 - 9x^3y^2 + xy$

29. $4t^3xy + 8t^2y$

30. $-r + 9r^2$

31. $15rt + 9rt^2$

32. $16rt^2 + 24r^2t$

33. $-18xy^2 + 28x^2y^2$

34. $36x^4y^2 - 18x^3y^3 - 4xy$

35. $2m^4n^2 - m^2n - mnt$

36. $6x(x - 1) - 5(x - 1)$

37. $5x(2 - x) + 4(2 - x)$

38. $3t(t + 5) + 2(5 + t)$

39. $16n(n - 1) + (1 - n)$

40. $r(r - 2) + (2 - r)$

41. $18x(3x - 5) - 7(5 - 3x)$

42. $4x(7x - 8) + 3(8 - 7x)$

In Exercises 43 through 60, factor each polynomial by grouping. If a polynomial cannot be factored using these procedures, state this.

43. $x^2 + 4x + 8 + 2x$

44. $y^2 + 1y + 5 + 5y$

45. $r^2 + 3r + 6 + 2r$

46. $8 + x^2 + x + 8x$

47. $6t^2 - 10t + 20 - 12t$

48. $7x^2 - x + 3 - 21x$

49. $20 - 12x - 6x^2 + 10x$

50. $12t^2 + 21 - 14t - 18t$

51. $10x^2 - 15x + 12x - 18$

52. $9x^2 + 12x + 6x + 8$

53. $24x - 4 - 30x^2 + 5x$

54. $9x + 6x - 24x^2 - 16x^2$

55. $20 + 12v + 6v^2 + 10v$

56. $24 + 3x^2 + 3x + 24x$

57. $y^2 - 3y + 6 - 2y$

58. $60x^2 + 105 - 60x - 105x$

59. $7y^2 - 7y + 35 - 35y$

60. $14x^2 - 2x + 6 - 42x$

EXCURSIONS

Class Act

1. **a.** Replace *A*, *B*, *C*, and *D* with different digits chosen from 0 to 9 to give a correct result. You will know when you have a correct answer.

$$
\begin{array}{r}
A\,B\,C\,D \\
\times\ 9 \\
\hline
D\,C\,B\,A
\end{array}
$$

   b. Replace *A*, *B*, *C*, *D*, and *E* with different digits chosen from 0 to 9 to give a correct result. You will know when you have a correct answer.

$$
\begin{array}{r}
A\,B\,C\,D\,E \\
\times\ 4 \\
\hline
E\,D\,C\,B\,A
\end{array}
$$

   c. ✏ Write a set of directions for a friend who is trying to solve the two previous problems.

Exploring Numbers

2. 325 is the smallest number that can be written as the sum of two squares in three different ways. Find them.

Posing Problems

3. Use the following graph to ask and answer four questions.

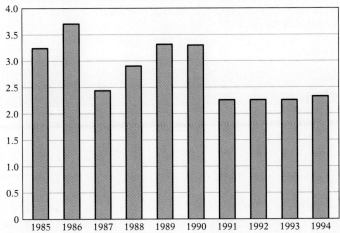

NUMBER OF CARS LEASED FOR BUSINESS USE (MILLIONS)

SECTION GOALS

■ To factor trinomials of the form $x^2 + bx + c$

■ To factor such trinomials by first removing the greatest common factor

■ To factor trinomials of the form $ax^2 + bx + c$

6.2 Factoring Trinomials of the Form $ax^2 + bx + c$

SECTION LEAD-IN*

The Punnett square has been used to calculate and display genetic information.

There are three genes for blood types A, B, and O. The ways that these three genes can combine can be found by using the Punnett square:

	A	B	O
A	AA	AB	AO
B	BA	BB	BO
O	OA	OB	OO

Possible genotypes

Genes A and B are dominant. When they are paired with O, the dominant genes determine the blood type.

Blood types are A, B, O, or AB in phenotype. (The phenotype is the physical appearance, or, in this case, the chemical appearance.)

What are the blood types in each of the above cells?

*Source: Gregory Fiore, "An Out of Math Experience: Quadratic Equations and Polynomial Multiplication as used in Genetics." *The AMATYC Review,* Volume 17, Number 1 (Fall 1995): pp. 20–27.

Factoring when $a = 1$

When we multiply two binomials $x + m$ and $x + n$, we get a trinomial.

$$\overset{\text{F}\quad\text{O}\quad\;\text{I}\quad\;\text{L}}{(x + m)(x + n) = x^2 + nx + mx + mn}$$
$$= x^2 + (m + n)x + mn$$

Note that the coefficient of x is the sum of m and n and that the constant is the product of m and n. The procedure for squaring a binomial (described in Section 5.4) is a special case of FOIL.

Therefore, if we have a trinomial $x^2 + bx + c$, we can write it as the product of two binomial factors.

$$x^2 + bx + c = (x + m)(x + n)$$
$$\underset{(m + n)}{\qquad} \underset{(mn)}{\qquad}$$

We can do this by finding m and n such that

$$m + n = b \quad \text{and} \quad m \cdot n = c$$

To factor a trinomial in the form x² + bx + c

1. List all pairs of integers whose product is c.
2. Of these, find the pair m and n whose sum is b.
3. Write the factorization of the trinomial as $(x + m)(x + n)$.

Note: This procedure works only when the coefficient of the x^2 term is 1. We will discuss factoring of other trinomials later in this section.

▪ ▪ ▪

EXAMPLE 1

Factor: $x^2 + 3x + 2$

SOLUTION

The coefficient of x^2 is 1, so we can use our process. We first want to find m and n such that $m \cdot n = 2$. Only two pairs of integers have the product 2. They are

$$2 \text{ and } 1 \qquad \text{because} \quad 2 \cdot 1 = 2$$
$$-2 \text{ and } -1 \quad \text{because} \quad -2 \cdot -1 = 2$$

We want $m + n$ to be equal to 3. To determine which of our pairs of numbers has the sum 3, we add:

$$2 + 1 = 3 \quad \text{Good}$$

Thus m and n are 2 and 1 (we stopped as soon as we found them). The factors of the trinomial are $(x + 2)(x + 1)$.

We check by multiplying the factors using FOIL.

$$\begin{array}{cccc} \text{F} & \text{O} & \text{I} & \text{L} \end{array}$$
$$(x + 2)(x + 1) = x^2 + 1x + 2x + 2$$
$$= x^2 + 3x + 2$$

So $x^2 + 3x + 2$ is $(x + 2)(x + 1)$ in factored form.

▶ *CHECK* **Warm-Up 1**

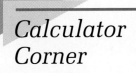

Calculator Corner

You can use graphing to check your answers when factoring trinomials. For instance, check Example 1 using your graphing calculator.

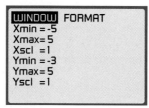

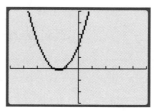

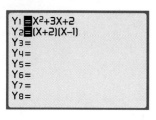

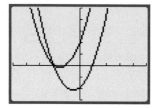

Two different graphs mean that the two sides of the equation *are not equal.*

STUDY HINT

For $x^2 + bx + c = (x + m)(x + n)$, when b and c are positive, then m and n are both positive.

STUDY HINT

For $x^2 + bx + c = (x + m)(x + n)$, when c is positive and b is negative, then m and n are both negative.

! ! !

ERROR ALERT

Identify the error and give a correct answer.

Factor: $x^2 + 4x + 3$

Incorrect Solution:

$x^2 + 4x + 3 = (x + 3)(x + 3)$

In factoring a trinomial, we do not always have to check each possible pair of integers m and n. We can eliminate some pairs by considering the signs of the constants b and c. For instance, in Example 1 we did not have to check the negative pair -2 and -1.

Thus when the constant c in the trinomial is positive, both m and n have the same sign as b.

We use this rule in the next example.

▪ ▪ ▪

EXAMPLE 2

Factor: $y^2 - 8y + 12$

SOLUTION

The coefficient of y is 1, so we can use the procedure. Because 12 is positive and -8 is negative, we consider only negative values of m and n. We want m and n such that $m \cdot n$ is 12 and $m + n = -8$.

Pairs, m and n where $m \cdot n = 12$	Sums, $m + n$ where $m + n = -8$	
-12 and -1	$-12 + (-1) = -13$	Not good
-6 and -2	$-6 + (-2) = -8$	Good
-4 and -3	not needed	

So the factors are $(x - 6)(x - 2)$.

Check them with a FOIL multiplication.

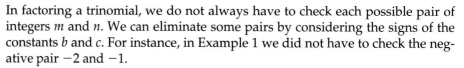

▶ *CHECK* **Warm-Up 2**

Removing a Common Factor

Most trinomials do not have 1 as the coefficient of the x^2 term. For some, however, we can remove a common factor to obtain x^2 as the first term.

▪▪▪

EXAMPLE 3

Factor: $2r^2 - 4r - 16$

SOLUTION

Upon inspection, we can see that the factor 2 is common to all three terms. So we first factor out the 2, obtaining

$$2r^2 - 4r - 16 = 2(r^2 - 2r - 8)$$

The trinomial in parentheses can be factored further. We want to find m and n such that $m \cdot n = -8$ and $m + n = -2$.

Thus m and n are -4 and 2, and the factors of $r^2 - 2r - 8$ are $(r - 4)$ and $(r + 2)$. But we must remember that we factored out a 2 earlier.

So $2r^2 - 4r - 16$ is $2(r - 4)(r + 2)$ in factored form. You should check this by multiplying.

▶ CHECK **Warm-Up 3**

STUDY HINT

When factoring a trinomial, always check for common factors first.

Trinomials in Two Variables

We can use our procedure to solve certain trinomials in *two* variables. They must have the form $x^2 + bxy + cy^2$. For these trinomials, the factors are in the form $(x + my)(x + ny)$.

▪▪▪

EXAMPLE 4

Factor: $x^2 + 2xy - 8y^2$

SOLUTION

We want m and n such that $m \cdot n = -8$ and $m + n = 2$. The choices are

Pairs, m and n	Sums, $m + n$
-1 and 8	$-1 + 8 = 7$
-2 and 4	$-2 + 4 = 2$ Good

So the correct pair is -2 and 4, and we find that the trinomial can be factored as $(x - 2y)(x + 4y)$. You should check it to verify this.

▶ CHECK **Warm-Up 4**

Prime Trinomials

Certain trinomials in the form $x^2 + bx + c$ are called prime trinomials. These trinomials cannot be factored in the form $(x + m)(x + n)$ such that b is the sum of m and n, c is the product of m and n, and m and n are integers. One prime trinomial is $x^2 + x + 6$. Many more exist.

Factoring Trinomials of the Form $ax^2 + bx + c$ when $a \neq 1$

Even after we remove a common factor from a trinomial, the x^2 term may have a coefficient other than 1. In that case, we factor the trinomial using a method that is partly trial and error and partly the method described previously in this section.

To factor a trinomial of the form $ax^2 + bx + c$ when a, b, and c have no common factors

1. If the term ax^2 is negative, remove the common factor -1 to make it positive.
2. List the possible pairs of positive factors of a.
3. List the possible pairs of factors of c (both positive and negative).
4. Combine pairs of factors from steps 2 and 3, in turn, to form trial binomial factors. Determine the middle terms that these trial factors would produce when multiplied together.
5. The trial factors that produce the proper middle term bx are the factors of the original trinomial.

In the first example, both the first and last terms have more than one possible pair of factors. We may have to go through quite a bit of trial and error to find the correct pairs.

■■■

EXAMPLE 5

Factor: $9x + 9 - 4x^2$

SOLUTION

We write the trinomial in standard form and remove the common factor -1, obtaining

$$9x + 9 - 4x^2 = -4x^2 + 9x + 9$$
$$= (-1)(4x^2 - 9x - 9)$$

There are no other common factors.

Now, the coefficient of x^2 is 4, which has the possible positive factors 4 and 1 or 2 and 2. The constant term is negative, so we must give it one positive factor and one negative factor. The possibilities are -9 and 1, 9 and -1, and 3 and -3. So we have

Factors of 4	Factors of -9
4, 1	-9, 1
2, 2	9, -1
	-3, 3
	3, -3
	1, -9
	-1, 9

$\left. \begin{array}{l} \\ \\ \end{array} \right\}$ Reverses of first three pairs of factors

STUDY HINT

If the absolute value of a trial middle term is not the same as the absolute value of the middle term of the trinomial being factored, you may also eliminate the additive inverses of these trial factors.

Note that we have listed the reverses of the first three pairs of factors of -9 to be sure we don't forget them.

Now we must combine each pair of factors of 4 with each pair of factors of -9 to obtain trial binomial factors. We start at the top and work our way down the list, checking the value of the middle term for each trial. We want a middle term of $-9x$.

Possible Factors	Middle Term
$(4x - 9)(x + 1)$	$4x - 9x = -5x$
$(4x + 9)(x - 1)$	$-4x + 9x = 5x$
$(4x + 3)(x - 3)$	$-12x + 3x = -9x$ Good

STUDY HINT

If the terms of a trinomial do not have a common factor, then the terms of its trial factors cannot have a common factor. Accordingly, you should first remove the common factors from the binomial. Then eliminate every trial binomial factor that has a common factor.

The factors $(4x + 3)(x - 3)$ give us the middle term $-9x$ of the trinomial. We need not continue, because only one pair of factors will give us a match. We must, however, multiply the binomial factors by the -1 that we factored out originally.

So $9x + 9 - 4x^2$ in factored form is $(-1)(4x + 3)(x - 3)$. You should check this by multiplying.

▶ *CHECK* **Warm-Up 5**

Note that in Example 5, the first and second trials led to results with the same absolute value, $-5x$ and $5x$. Once we saw that one did not work, we did not have to try the second. Apply the Study Hints in this section to reduce the number of trials.

Practice what you learned.

SECTION FOLLOW-UP

	A	B	O
A	A	AB	A
B	AB	B	B
O	A	B	O

Possible phenotypes   ◀——— (What you see is not necessarily what you got!)

a. A mother and father are both type O. What blood type is their son?

b. A mother is type A; the father is type B. Their daughter is type O. What are the genotypes of the parents?

c. Ask and answer two other questions about these tables.

d. Use the table to multiply $(a^2 + bc - 3)^2$.

6.2 WARM-UPS

Work these problems before you attempt the exercises.

1. Factor: $n^2 + 6n + 5$

2. Factor: $n^2 - 12n + 20$

3. Factor: $4x^2 - 12x - 72$

4. Factor: $x^2 - 15y^2 + 2xy$
   (*Hint*: Rewrite the terms in form $ax^2 + bxy + cy^2$.)

5. Factor: $6x^2 + 43x + 20$

6.2 EXERCISES

Note: Use your graphing calculator to check your results whenever possible.

In Exercises 1 through 12, factor the trinomial.

1. $y^2 + 11y + 18$
2. $x^2 + 13x + 40$
3. $x^2 + x - 12$

4. $n^2 + 2n - 8$
5. $r^2 - 13r + 40$
6. $n^2 - 6n + 8$

7. $r^2 - r - 2$
8. $t^2 - 2t - 15$
9. $m^2 + 3m - 40$

10. $x^2 + 5x - 24$
11. $n^2 - 3n + 2$
12. $t^2 - 10t + 9$

In Exercises 13 through 24, factor each trinomial by first removing the greatest common factor.

13. $3r^2 - 51r + 156$
14. $2t^2 + 16t - 18$
15. $2x^2 - 24x + 40$

16. $4r^2 + 12r - 72$
17. $3t^2 + 30t + 27$
18. $5y^2 - 30y - 80$

19. $4x^2 - 4x - 24$
20. $3n^2 - 21n + 30$
21. $5x^2 - 30x - 200$

22. $2xy^2 + 22xy - 84x$ **23.** $10t^2u - 110tu + 300u$ **24.** $r^2t - rt - 30t$

In Exercises 25 through 36, factor the trinomial, if possible. If factoring is not possible, label the trinomial prime.

25. $3t^2 + 3t - 168$ **26.** $5r^2 + 85r + 260$ **27.** $y^2 - 8y - 33$

28. $x^2 - 6x - 72$ **29.** $-42 + t^2 + t$ **30.** $3 + x^2 - 3x$

31. $-56 + y^2 + y$ **32.** $-24 + r^2 - 2r$ **33.** $5x^2 - 50x - 140$

34. $4r^2 + 24r - 364$ **35.** $3x^2 + 6xy - 9y^2$ **36.** $6r^2 - 6rt - 36t^2$

In Exercises 37 through 78, factor each trinomial.

37. $18r^2 - 20r + 2$ **38.** $12t^2 - 2t - 2$ **39.** $15y^2 + 10y - 5$

40. $18w^2 + 27w + 9$ **41.** $16n^2 - 24n + 8$ **42.** $9tx^2 - 15tx - 6t$

43. $16v^2 - 4v - 20$ **44.** $10n^2 - 9n - 9$ **45.** $3r^2t + 14rt + 8t$

46. $8rt^2 - 6rt + r$ **47.** $7y^2 - 5y - 2$ **48.** $5n^2 + n - 22$

49. $5nx^2 - 25nx - 30n$ **50.** $16mx^2 - 16mx + 4m$ **51.** $3x^2 - 12x + 9$

52. $9y^2 - 22y + 8$ **53.** $56v^2 - 59v - 30$ **54.** $2mn^2 - 19mn - 21m$

55. $12tx^2 - 45tx - 12t$ **56.** $12r^2 + 4r - 8$ **57.** $64t^2 - 48t - 16$

58. $2y^2 + 14y - 88$ **59.** $15w^2 + 48w + 33$ **60.** $4x^2 + 8x - 12$

61. $6x^2 + 12x - 18$ **62.** $7rx^2 - 28rx + 21r$ **63.** $9nx^2 - 45nx + 54n$

64. $15n^2 - 10n - 25$ **65.** $4m^2 + 14m - 30$ **66.** $9m^2 - 30m + 21$

67. $6n^2 + 30n - 36$ **68.** $12x^2y + 30xy + 18y$ **69.** $4mv^2 + 6mv - 54m$

70. $6x^2 + 23x + 7$ **71.** $2y^2 - 24y + 64$ **72.** $16w^2 - 20w - 6$

73. $35n^2 - 21n - 56$ **74.** $9n^2 + 27n - 90$ **75.** $28x^2 + 56x + 28$

76. $9x^2y + 54xy + 45y$ **77.** $3tx^2 + 6tx - 45t$ **78.** $40 + 4x^2 + 44x$

MIXED PRACTICE

By doing these exercises, you will practice the topics up to this point in the chapter.

79. Factor: $30x - 300 + 6x^2$ **80.** Factor: $-63n + 54 + 27n^2$

81. Factor: $-16r + 40 - 24r^2$ **82.** Factor: $5y^2 - 15y - 200$

83. Factor: $xt^2 - 17tx + 66x$

84. Factor: $y(a + 4x) + (a + 4x)$

85. Factor: $7x^2 + 10x - 8$

86. Factor: $12t - 12t^2 + 9$

87. Factor: $3y^2 - 7y + 6yt - 14t$

88. Factor: $r^2 - 4r - 12$

89. Factor: $9x^2y - 27x^2y^3$

90. Factor: $8tx^2 + 10t^2x^3 - 16t^5x^3$

91. Find the prime factorization of 265.

92. Factor: $2y^2 - 24y + 64$

EXCURSIONS

Data Analysis

1. Study the following table about speed limits, then answer the questions.

STATES THAT RAISED SPEED LIMITS DURING 1996

State	Rural interstates	Urban interstates
Alabama	70 mph	70 mph
Arizona	75 mph	55 mph
California	70 mph[1]	65 mph
Colorado	75 mph	65 mph
Delaware	65 mph	55 mph
Florida	70 mph	65 mph
Georgia	70 mph	65 mph
Idaho	75 mph	65 mph
Kansas	70 mph	70 mph
Michigan	70 mph[1]	65 mph
Mississippi	70 mph	70 mph
Missouri	70 mph	60 mph
Montana	Unlimited[2]	Unlimited
Nebraska	75 mph	65 mph
Nevada	75 mph	65 mph
New Mexico	75 mph	55 mph
New York	65 mph	65 mph
North Carolina	70 mph	65 mph
North Dakota	70 mph	55 mph
Oklahoma	75 mph	70 mph
Rhode Island	65 mph	55 mph
South Dakota	75 mph	65 mph
Texas	70 mph[3]	70 mph
Utah	75 mph	65 mph
Washington	70 mph[3]	60 mph
Wisconsin	65 mph	65 mph
Wyoming	75 mph	60 mph

[1]Trucks 55 mph   [2]Trucks 65 mph   [3]Trucks 60 mph

Source: Advocates for Highway and Auto Safety, Insurance Institute for Highway Safety, Association of International Automobile Manufacturers.

a. Graph and compare the states' speed limits on rural interstates and on urban interstates.

b. Ask and answer four questions about this data.

Exploring Numbers

2. **a.** Assign the digits 1, 2, 3 to the letters a, b, c so that the following statement is true. *Note:* Each letter stands for a single digit; that is, abc is a three-digit number.

$$abc = ab + ac + ba + bc + ca + cb$$

b. In words, tell what this statement means.

3. 204^2 is the sum of three consecutive cubes. Find them.

4. The number 1201 can be expressed in the form

$$x^2 + ny^2$$

for all values of n from 1 to 10. Find the values that x and y take on for each n.

CONNECTIONS TO *GEOMETRY*

Perimeter and Area

Perimeter of a Rectangle or Parallelogram

A rectangle has two pairs of equal sides—its length and its width. The same is true for a parallelogram.

To find the perimeter of a rectangle or parallelogram

If ℓ is the length and w is the width, then

$$P = 2\ell + 2w \quad \text{or} \quad P = 2(\ell + w)$$

The length ℓ and width w are shown in the accompanying figure. (Actually, the labels ℓ and w could be reversed and the formula would still hold.)

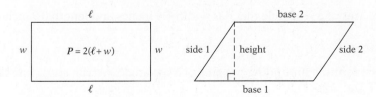

Note that in order to use either formula, we need to be familiar with the order of operations. Both formulas give the same result when applied properly.

▪▪▪

EXAMPLE

Find the perimeter of a rectangle with length 6 feet and width 7 feet.

SOLUTION

We use the second formula, substituting the actual length and width.

$$P = 2(\ell + w) = 2(6 + 7)$$

Then, using the standard order of operations, we have

$$P = 2(13) = 26 \text{ feet}$$

As a check, we use the first formula, obtaining

$$P = 2\ell + 2w$$
$$= 2(6) + 2(7)$$
$$= 12 + 14 = 26$$

Therefore, the perimeter is 26 feet.

Area of a Rectangle

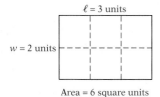

$\ell = 3$ units

$w = 2$ units

Area = 6 square units

A rectangle with a length of 3 units and a width of 2 units has an area of 6 square units. The accompanying figure shows that this is true.

> **To find the area of a rectangle**
> If ℓ is the length and w is the width, then
> $$\text{Area} = \ell w$$

Remember that the measurement of area is always expressed in *square units*.

> **Square Unit**
> A **square unit** (square inch, square foot, square mile, and so on) is a square each of whose sides has a length of 1 such unit (inch, foot, mile, and so on).

▪▪▪

EXAMPLE

Find the area of a rectangle that has width 1.6 inches and length 0.9 inches.

SOLUTION

Using the formula for finding the area of a rectangle, we can substitute the values for ℓ and w:

$$A = \ell w = 0.9(1.6) = 1.44$$

So, the area is 1.44 square inches.

Please note that the length can be shorter, longer, or equal to the width.

Area of a Parallelogram

We use multiplication to find the area of a parallelogram.

To find the area of a parallelogram

If b is the length of the base and h is the height, then

$$\text{Area} = bh$$

The base and height are shown in the accompanying figure.

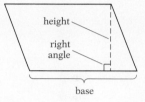

▪▪▪

EXAMPLE

Find the area of a parallelogram with height 1.9 feet and base 12 feet. Give your answer in square meters. Use the exact relationship—2.54 centimeters = 1 inch—to convert.

SOLUTION

First, we will change feet to meters.

$$1.9 \text{ ft} \cdot \frac{12 \text{ in.}}{1 \text{ ft}} \cdot \frac{2.54 \text{ cm}}{1 \text{ in.}} \cdot \frac{1 \text{ m}}{100 \text{ cm}}$$

We cancel the units:

$$1.9 \,\cancel{\text{ft}} \cdot \frac{12 \,\cancel{\text{in.}}}{1 \,\cancel{\text{ft}}} \cdot \frac{2.54 \,\cancel{\text{cm}}}{1 \,\cancel{\text{in.}}} \cdot \frac{1 \text{ m}}{100 \,\cancel{\text{cm}}}$$

We multiply and divide:

$$\frac{(1.9)(12)(2.54)(1)\text{m}}{(1)(1)(100)} = 0.57912 \text{ m}$$

We change 12 feet to meters similarly:

$$12 \,\cancel{\text{ft}} \cdot \frac{12 \,\cancel{\text{in.}}}{1 \,\cancel{\text{ft}}} \cdot \frac{2.54 \,\cancel{\text{cm}}}{1 \,\cancel{\text{in.}}} \cdot \frac{1 \text{ m}}{100 \,\cancel{\text{cm}}} = 3.6576 \text{ m}$$

Substituting in the formula, we get:

$$A = b \times h = (3.6576)(0.57912) \approx 2.12$$

So the area of the parallelogram is approximately 2.12 square meters.

◀

> ■▪■
> **WRITER'S BLOCK**
> Explain why
> $$\frac{2.54 \ cm}{1 \ in.}$$
> and
> $$\frac{1 \ in.}{2.54 \ cm}$$
> are equivalent.

PRACTICE

The playing areas described in Exercises 1 through 6 are rectangular. Find their perimeters and areas. Express your answers in meters using the exact equivalency—2.54 centimeters = 1 inch—in your conversions. (*Hint:* Convert lengths before calculating area.) Write your answers to the nearest whole number.

1. U.S. football field: 120 yards long, 53 yards 1 foot wide

2. Fencing piste: length 46 feet, width 6 feet 6 inches

3. Basketball court: 28 yards by 15 yards 9 inches

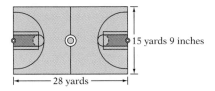
15 yards 9 inches
28 yards

4. Netball court: 33 yards 1 foot by 16 yards 2 feet

5. Ice hockey rink: length 66 yards 2 feet and width 33 yards 1 foot

6. Judo court: 52 feet 6 inches on a side

7. Find the perimeter and area of the double-parallelogram shown below. Give the results in meters or square meters as necessary. Treat the figure as one large surface.

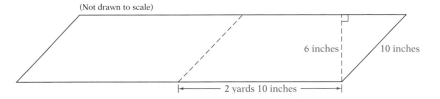
(Not drawn to scale)
6 inches   10 inches
2 yards 10 inches

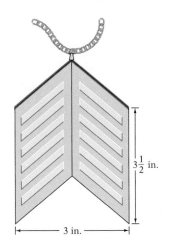
$3\frac{1}{2}$ in.
3 in.

8. An arrow-shaped pendant was designed by an artist. See the accompanying figure. The two parallelograms that form the arrow are each $3\frac{1}{2}$ inches long. The distance across the pendant is 3 inches and the outside long edges are parallel to each other. The point of the arrow (emphasized in color in the figure) has a total "length" of 4 inches. Find the perimeter and area in centimeters and square centimeters. (Disregard its "depth.") Round your answers to the nearest whole number.

9. A rectangle with a height of 2 feet and an area of 8.2 square feet was sat on until its height was only 18.5 inches. What is its new perimeter (in meters) and area (in square meters)?

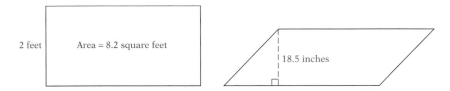
2 feet   Area = 8.2 square feet
18.5 inches

10. Four parallelograms form a trivet (see the accompanying figure). All four sides are equal. The perimeter of the trivet is 12 inches. The area is 6 square inches. Find the perimeter (in centimeters) and the area (in square centimeters) of each parallelogram in the trivet. Round your answers to the nearest whole number. (*Hint:* There are three different sizes of parallelograms. How many of each size are there?)

6.3 Factoring Differences of Two Squares and Perfect-Square Trinomials

SECTION LEAD-IN

We can use the Punnett square to find $(p + q)^6$. We simply have to use it several times, adapting it to accomplish the last step of $(p + q)^6$.

$(p + q)^2$:

	p	q
p	p^2	pq
q	pq	q^2

$= p^2 + 2pq + q^2$

$(p + q)^4$:

	p^2	$2pq$	q^2
p^2			
$2pq$			
q^2			

$(p + q)^6$:

	p^4			
$2pq$				

▪ Complete these Punnett squares. In a family of exactly 6 children, what is the probability of having 3 boys and 3 girls?

In this section, we factor two special kinds of polynomials. Both can be factored with the methods of preceding sections. However, their special forms enable us to use factoring short cuts.

Difference of Two Squares

Suppose we wanted to factor $x^2 - 64$. This expression is a difference of two squares: The term on the left is the square of x, and the term on the right is the square of 8. If you're not sure how we find this, see Section 5.4, where we found a product in this form. To factor it the long way, we would write it as

$$x^2 + 0x - 64$$

Then we would have to find two numbers whose product is -64 and whose sum is 0. Those numbers are $+8$ and -8, because

$$(8)(-8) = -64 \quad \text{and} \quad 8 + (-8) = 0$$

Thus

$$x^2 - 64 = (x + 8)(x - 8)$$

If we write the left side as a difference of squares,

$$x^2 - 8^2 = (x + 8)(x - 8)$$

you may recognize the short-cut method.

To factor the difference of two squares, $x^2 - n^2$

Write the factors as the sum and difference of the values that were squared.

$$x^2 - n^2 = (x + n)(x - n)$$

▪▪▪

EXAMPLE 1

Factor: $x^2 - 169$

SOLUTION

Inspecting the binomial, we see that both x^2 and 169 are squares (because $169 = 13 \times 13$, or 13^2). This binomial is a difference of squares, and we factor it as follows:

$$(x + 13)(x - 13)$$

We check by multiplying.

$$(x + 13)(x - 13) = x^2 - 13x + 13x - 169$$
$$= x^2 - 169$$

So $x^2 - 169$ in factored form is $(x + 13)(x - 13)$.

▶ *CHECK* **Warm-Up 1**

In this next example, there are two variables and a common factor.

▪▪▪

EXAMPLE 2

Factor: $32y^4 - 200n^2$

SOLUTION

The coefficients have a common factor of 8, which we factor out first.

$$32y^4 - 200n^2 = 8(4y^4 - 25n^2)$$

The expression in parentheses is a difference of two squares, because

$$4y^4 = (2y^2)^2 \quad \text{and} \quad 25n^2 = (5n)^2$$

We factor it as

$$(2y^2 + 5n)(2y^2 - 5n)$$

Don't forget the common factor 8. So $32y^4 - 200n^2$ in factored form is $8(2y^2 + 5n)(2y^2 - 5n)$. You should check this.

▶ CHECK **Warm-Up 2**

Perfect-Square Trinomials

Suppose we square the expression $x + n$.

$$(x + n)(x + n) = x^2 + nx + nx + n^2$$
$$= x^2 + 2nx + n^2$$

The square of a binomial is called a **perfect-square trinomial.** Its first and last terms are the squares of the terms of the binomial, and its middle term is twice the product of the terms of the binomial.

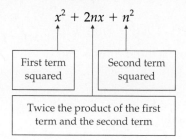

Is $4x^2 + 16x + 4$ a perfect-square trinomial? Well, both $4x^2$ and 4 are perfect squares, because $4x^2$ is $(2x)^2$ and 4 is $(2)^2$. But twice the product of $2x$ and 2 is $(2)(2x)(2) = 8x$, not $16x$. So $4x^2 + 16x + 4$ is *not* a perfect-square trinomial.

To factor a perfect-square trinomial, we simply write it as the original binomial squared. For example, we write $x^2 + 6x + 9$ as

$$(x + 3)^2 \quad \text{or} \quad (x + 3)(x + 3)$$

To factor a perfect-square trinomial ($x^2 + 2nx + n^2$)

1. Find the x and n of the original trinomial.
2. Give n the sign of the middle term of the trinomial.
3. Write the factors as the square of $x + n$.

$$x^2 + 2nx + n^2 = (x + n)^2$$

STUDY HINT

A perfect-square trinomial always has a positive lead coefficient and a positive constant term.

•••

EXAMPLE 3

Factor: $n^2 + 10n + 25$

SOLUTION

We first make sure that this is a perfect-square trinomial. The first and last terms are squares, n^2 and 5^2. Twice the product of n and 5 is $(2)(n)(5) = 10n$, the middle term. Therefore, this is a perfect-square trinomial.

The middle term of the trinomial is positive, so our factors are $(n + 5)(n + 5)$.

Thus $n^2 + 10n + 25$ in factored form is $(n + 5)^2$. Check this using FOIL.

▶ *CHECK* **Warm-Up 3**

▪▪▪

EXAMPLE 4

Factor: $2nt^2 + 28n^2t + 98n^3$

SOLUTION

First we must realize that there is a common factor, $2n$. We factor it out:

$$2nt^2 + 28n^2t + 98n^3 = 2n(t^2 + 14nt + 49n^2)$$

The first and last terms are perfect squares, t^2 and $(7n)^2$. Twice $t \cdot 7n$ is equal to $14nt$, the middle term. Therefore, this is a perfect-square trinomial. Its factors are

$$(t + 7n)(t + 7n)$$

So $2nt^2 + 28n^2t + 98n^3$ in factored form is $2n(t + 7n)^2$.

▶ *CHECK* **Warm-Up 4**

Practice what you learned.

SECTION FOLLOW-UP

The probability of having exactly 3 boys and 3 girls in a family of 6 children can be found using

$$20p^3q^3$$

where p is the probability of having a girl and q is the probability of having a boy. For any single birth, each of these probabilities is $\frac{1}{2}$. So, the probability of having 3 boys and 3 girls in a family of 6 children is

$$20\left(\frac{1}{2}\right)^3\left(\frac{1}{2}\right)^3 = 20\left(\frac{1}{8}\right)\left(\frac{1}{8}\right) = \frac{20}{64} = \frac{5}{16}$$

a. What is the probability of having 3 boys and 2 girls in a family of 5 children? (*Hint:* Find the results of $(p + q)^5$ and use the p^3q^2 term, with $p = \frac{1}{2}$ and $q = \frac{1}{2}$.)

b. Ask and answer two more questions using this technique.

6.3 WARM-UPS

Work these problems before you attempt the exercises.

1. Factor: $x^2 - 49$

2. Factor: $128x^2 - 18$

3. Factor: $x^2 + 8x + 16$

4. Factor: $4m^2y - 64my + 256y$

6.3 EXERCISES

Note: Use your graphing calculator to check your results whenever possible.

In Exercises 1 through 12, factor each difference of two squares.

1. $t^2 - 9$
2. $y^2 - 100$
3. $4y^2 - 25$

4. $100x^2 - 81$
5. $t^4 - 16$
6. $n^{12} - 625$

7. $5x^2 - 125$
8. $64y^8 - 36$
9. $81t^{16} - 49t^8$

10. $(x + 2)^2 - y^2$
11. $(x - 3)^2 - y^2$
12. $x^2 - (2y - 1)^2$

In Exercises 13 through 24, factor each perfect-square trinomial.

13. $t^2 - 16t + 64$
14. $y^2 - 26y + 169$
15. $y^2 + 4y + 4$

16. $r^2 + 8r + 16$
17. $t^2 - 22t + 121$
18. $y^2 - 18y + 81$

19. $y^2 + 10y + 25$
20. $9t^2 - 24t + 16$
21. $4n^2 - 44n + 121$

22. $49t^2 - 28t + 4$
23. $25y^2 - 120y + 144$
24. $9y^2 + 48y + 64$

In Exercises 25 through 36, factor each expression. Don't forget to factor out any common factors first.

25. $xr^2 - 900x$
26. $mn^2 - 49m$
27. $3y^2z - 108z$

28. $t^2 + 12tx + 36x^2$
29. $4x^2 + 12xy + 9y^2$
30. $r^2 - 4rt + 4t^2$

31. $9y^2 - 18y + 9$
32. $4y^2 + 8y + 4$
33. $50n^2 + 60n + 18$

34. $9y^2 - 36y + 36$
35. $36n^2 - 48n + 16$
36. $49x^2 - 14x + 1$

In Exercises 37 through 48, the trinomials shown are perfect-square trinomials. In printing them, one term was lost. Supply the missing term and give both the correct trinomial and its factors.

37. $x^2 - \underline{\hspace{1cm}} + 16$

38. $x^2 + \underline{\hspace{1cm}} + 36$

39. $\underline{\hspace{1cm}} - 26x + x^2$

40. $\underline{\hspace{1cm}} + 18x + x^2$

41. $25x^2 - \underline{\hspace{1cm}} + 9$

42. $100x^2 - \underline{\hspace{1cm}} + 1$

43. $49x^2 - 42x + \underline{\hspace{1cm}}$

44. $64x^2 + 64x + \underline{\hspace{1cm}}$

45. $4 - 72x + \underline{\hspace{1cm}}x^2$

46. $36 - 84x + \underline{\hspace{1cm}}x^2$

47. $\underline{\hspace{1cm}}x^2 + 48x + 36$

48. $\underline{\hspace{1cm}}x^2 - 66x + 9$

MIXED PRACTICE

By doing these exercises, you will practice the topics up to this point in the chapter.

49. Factor: $t^2 - 16t + 64$

50. Factor: $rx^2 - 18rx + 77r$

51. Factor: $-3t^2 + 48$

52. Factor: $2x^2 - 17x + 30$

53. Factor: $27x^2y^2 - 9x^5y^3 - 81x^2y^3$

54. Factor: $y^2 - 26y + 169$

55. Factor: $3t^2 - 39t + 126$

56. Factor: $6y^2 + 13y + 6$

EXCURSIONS

Class Act

1. For parts (a) through (d), assume that the cost of the sign is determined by two major factors—the cost per square inch of the metal and the cost per letter. If the most expensive sign costs about $52 for the metal and about $5 for the lettering, determine the approximate cost for each of the four signs. Justify all the numbers you use for your calculations.

a.

24 in. × 30 in.

b.

24 in. × 48 in.

c.

CENTER LANE

BUSES AND CAR POOLS ONLY

6 AM-9 AM MON-FRI

30 in. × 42 in.

d.

◊ BUSES AND 4 RIDER CAR POOLS ONLY

6 AM-9 AM   MON-FRI

72 in. × 60 in.

2. a. For the figure at the left below, calculate the area of the sign that we can see. (That is, don't include the part of the sign between rail and road.)

b. For the figure at the right below, calculate the area of the "3 tracks" sign.

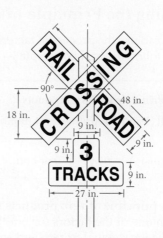

Exploring Numbers

3. a. Assign the first three non-negative odd integers to the letters a, b, and c in some order so that the following statement is true. Treat abc as a three-digit number.

$$abc = a^a + b^2 + c^b$$

b. In words, tell what this statement says.

4. Replace a, b, c, d, and n with the digits 1, 2, 3, 6, and 9 to make these two statements true. Treat abc and cba as three-digit numbers; treat ad and da as two-digit numbers.

$$abc = ad^n$$
$$cba = da^n$$

6.4 Using Factoring to Solve Equations and Applications

SECTION GOALS

- *To solve equations with the principle of zero products*
- *To solve equations by factoring*
- *To solve word problems that involve trinomials in equations*

SECTION LEAD-IN

Mathematics can be used to explore many natural phenomena. Entomologists, for example, use mathematics to study the rhythms of cicadas.

Cicadas are insects. Many cicadas have life cycles that are prime. There is a 13-year cicada and a 17-year cicada as well as others. (You could research this.) Neither of these insects has natural predators for a good reason. A natural predator needs a regular food supply. The 13-year and 17-year cicadas only appear every 13 or 17 years. During the rest of the time, what are they doing?

Using the Principle of Zero Products

We can use the process of factoring to solve certain equations. But we also need the following basic principle.

Principle of Zero Products

If A and B represent real numbers, and

$$A \cdot B = 0$$

then $A = 0$ or $B = 0$ or both.

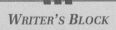

WRITER'S BLOCK

What do we mean by zero products?

The **principle of zero products** tells us that if the product of two factors is zero, then one of the factors (or both) must equal zero. To use this principle, we set *both* factors equal to zero and solve the resulting two equations.

■ ■ ■

EXAMPLE 1

Solve the equation $(x)(x + 2) = 0$.

SOLUTION

Because the product of x and $x + 2$ is zero, by the principle of zero products, both these factors may be equal to zero. Therefore, we set each factor equal to zero and solve both equations.

$x = 0$	$x + 2 = 0$
(already in	$x = 0 - 2$
solution form)	$x = -2$

We should check the solutions by substituting each into the original equation.

For $x = 0$: $(0)(0 + 2) = (0)(2) = 0$
For $x = -2$: $(-2)(-2 + 2) = (-2)(0) = 0$

Both solutions check.

▶ *CHECK* **Warm-Up 1**

Calculator Corner

Use your graphing calculator to solve the equation $(x)(x + 2) = 0$.

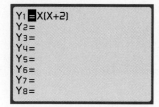 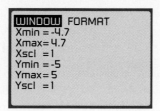 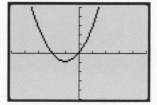

Now use **TRACE** to find out where the equation crosses the x-axis (because that is where $y = 0$). The values of x where $y = 0$ are called the **roots,** or solutions, of the equation.

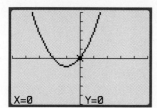 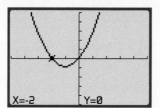

Some graphing calculators also have the ability to find the root, or zero, of an equation. On the *TI-82* it would look like the following screens.

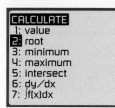

Move the cursor to
the left of the root.
Press **ENTER.**

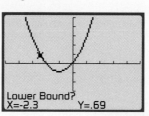

Move the cursor to
the right of the
root. Press **ENTER.**

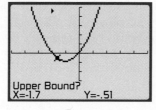

Press **ENTER.**

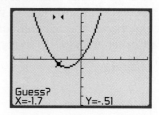

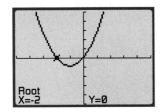

Press **ENTER.**

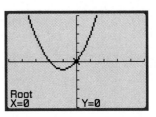

Repeat to find the second root.

▪▪▪

EXAMPLE 2

Solve: $20x^2 + 22x = 12$

SOLUTION

First, we have to set one side of the equation equal to zero. Then we factor and apply the principle of zero products.

$$20x^2 + 22x - 12 = 0 \quad \text{Adding } -12$$
$$2(5x - 2)(2x + 3) = 0 \quad \text{Factoring}$$
$$(5x - 2)(2x + 3) = 0 \quad \text{Multiplying by } \tfrac{1}{2}$$

Now, setting each factor equal to zero and solving, we get

$5x - 2 = 0$	$2x + 3 = 0$
$5x = 2$	$2x = -3$
$x = \frac{2}{5}$	$x = -\frac{3}{2}$

You will find that $x = \frac{2}{5}$ or $x = -\frac{3}{2}$ are both solutions of $20x^2 + 22x = 12$.

▶ CHECK **Warm-Up 2**

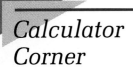

!!! ERROR ALERT

Identify the error and give a correct answer.

Solve:
$x^2 + 10x + 25 = 0$

Incorrect Solution:

$x^2 + 10x + 25 = 0$
$(x + 5)(x + 5) = 0$
$x = 5$

Calculator Corner

You can use your graphing calculator to solve $20x^2 + 22x = 12$ by graphing each side of the equation and finding where the two graphs intersect. Some graphing calculators can find the point of intersection of two graphs. (Note: See the Calculator Corner on page 125 to review how to find the points of intersection of two graphs.)

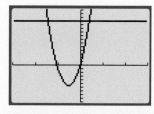

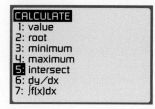

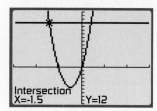

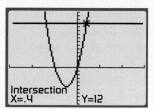

In Example 2, we multiplied by the inverse of a constant factor in order to remove it. We cannot do that with a variable factor but instead must retain it in the equation.

▪▪▪
EXAMPLE 3

Solve: $10y^3 - 49y = y^3$

SOLUTION

We add $-y^3$ to both sides of this equation to obtain zero on the right side. The resulting equation has a common factor, which we remove. Now we see that the expression in parentheses is the difference of two squares. We factor it and get

$$y(3y + 7)(3y - 7) = 0$$

Although there are three factors here, we can still apply the principle of zero products to find the solutions.

$$
\begin{array}{c|c|c}
y = 0 & 3y + 7 = 0 & 3y - 7 = 0 \\
 & 3y = -7 & 3y = 7 \\
 & y = \dfrac{-7}{3} & y = \dfrac{7}{3}
\end{array}
$$

So, $y = 0$, $y = -\frac{7}{3}$, or $y = \frac{7}{3}$ are (possible) solutions of $10y^3 - 49y = y^3$. We leave it to you to check these solutions.

▶ *CHECK* **Warm-Up 3**

Calculator Corner

We can find where the graphs of the two sides of the equation have the same values for x and y by finding where they intersect. Those will be the values that solve the equation. (Note: See the Calculator Corner on page 125 to review how to find the points of intersection of two graphs.)

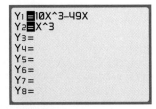 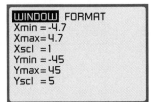 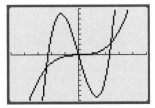

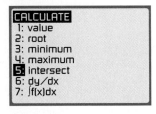

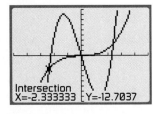

TRACE close to another intersection point and repeat the process twice more.

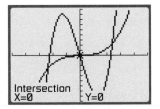 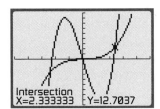

The intersections of the graphs of the two sides of the equation intersect at values of x when y is equal to zero. These are the zeros, or solutions, of the original equation. Here the solutions are $x = 0$, $x = -2\frac{1}{3}$, and $x = 2\frac{1}{3}$. See the intersection points on the graph. Check in the original equation.

Applying Factoring to Word Problems

Now you will use what you know about factoring to solve some word problems. The first problem is from geometry.

▪▪▪

EXAMPLE 4

If a polygon has n sides, then it has d diagonals, where d and n are related by the formula $2d = n^2 - 3n$. How many sides does a figure have if it has 14 diagonals?

SOLUTION

We insert any given information in the equation and simplify. Here we can substitute 14 for d. We get

$$2d = n^2 - 3n \qquad \text{Original equation}$$
$$2(14) = n^2 - 3n \qquad \text{Substituting 14 for } d$$
$$28 = n^2 - 3n \qquad \text{Multiplying}$$
$$0 = n^2 - 3n - 28 \quad \text{Adding } -28$$

This equation can be solved by factoring. We need two numbers whose product is -28 and whose sum is -3. They are -7 and 4, so we have

$$(n - 7)(n + 4) = 0$$

We set each factor equal to zero, getting

$$
\begin{array}{c|c}
n - 7 = 0 & n + 4 = 0 \\
n = 7 & n = -4
\end{array}
$$

Because we are looking for a number of sides, the answer must be a positive number; only $n = 7$ can be a solution to this problem.

A polygon with 14 diagonals has 7 sides.

▶ *CHECK* **Warm-Up 4**

The next example is an application from physics.

▪▪▪
EXAMPLE 5

A projectile is fired straight up into the air. Its height h in meters at t seconds after it is fired upward is given by the formula $h = 96t - 16t^2$. At what two times is the projectile on the ground?

SOLUTION

When the projectile is on the ground, its height is zero, so we substitute $h = 0$ into the formula and solve.

$$
\begin{array}{ll}
h = 96t - 16t^2 & \text{The formula} \\
0 = 96t - 16t^2 & \text{Substituting 0 for h} \\
0 = 16t(6 - t) & \text{Factoring}
\end{array}
$$

$$
\begin{array}{c|c}
16t = 0 & 6 - t = 0 \\
t = 0 & -t = -6 \\
 & t = 6
\end{array}
$$

Thus the projectile is on the ground at time 0 (just when it is fired) and at time 6 seconds (when it has fallen back to the ground).

▶ *CHECK* **Warm-Up 5**

Practice what you learned.

SECTION FOLLOW-UP

You can research the answer to the question posed in the Section 6.4 Lead-In. Here are two more questions.

1. Two (mythical) cicadas, a 13-year and a 17-year cicada, decide to meet again to compare notes. If they first met in 1998, when will they meet again?

2. A third insect, an 11-year cicada, joins up with the first two. When will they all meet if they first met in 1943?

6.4 WARM-UPS

Work these problems before you attempt the exercises.

1. Solve: $(x)(x - 3) = 0$

2. Solve: $4r^2 + 12r = 72$

3. Solve: $(n)(n - 3)(2n + 7) = 0$

4. How many sides does a polygon have if it has 35 diagonals? (Use the formula in Example 4.)

5. When will the projectile in Example 5 have a height of 128 feet?

6.4 EXERCISES

Note: Use your graphing calculator to check your results whenever possible.

In Exercises 1 through 12, solve the equations by applying the principle of zero products.

1. $(x)(x + 1) = 0$

2. $(r)(25 - r) = 0$

3. $(6n + 3)(n - 65) = 0$

4. $(m - 729)(4m) = 0$

5. $(2x + 3)(x - 3) = 0$

6. $(r + 125)(-r) = 0$

7. $(2r - 8)(r - 5) = 0$

8. $(p + 3.7)(3p) = 0$

9. $(5r - 15)(r + 28) = 0$

10. $(4t)(2.5 - t) = 0$

11. $\left(y - \frac{1}{2}\right)\left(\frac{3}{4} + y\right) = 0$

12. $(x - 6.3)(2 + x) = 0$

In Exercises 13 through 42, solve each equation by factoring and using the principle of zero products.

13. $t^2 - 3t = 0$

14. $5r^2 + r = 0$

15. $x^2 + 13x = 0$

16. $18u^2 - 72u = 0$

17. $t^2 - 4t = 0$

18. $11t^2 + 4t = 0$

19. $30n - 6n^2 = 0$

20. $20t^2 + 27t = 0$

21. $t^2 - 100 = 0$

22. $3t^2 - 39t + 120 = 0$

23. $81t^2 + 144t + 64 = 0$

24. $t^2 - 10t + 9 = 0$

25. $y^2 + 24y + 144 = 0$

26. $x^2 + 2x + 1 = 0$

27. $9t^2 - 100 = 0$

28. $16n^2 = 24n - 8$

29. $r^2 + 8r + 16 = 0$

30. $12r^2 + 4r = 8$

31. $9y^2 - 18y + 9 = 0$

32. $r^2 + 20r + 100 = 0$

33. $4r^2 + 16r = 180$

34. $64y^2 - 36 = 0$

35. $t^2 - 52t + 100 = 0$

36. $t^2 + 3t - 10 = 0$

37. $63t^2 = 8t + 16$

38. $m^2 - 13m = 14$

39. $6n^2 + n = 15$

40. $10n^2 - 9n = 9$

41. $16x^2 - 16x = -4$

42. $32x^2 = 8x + 40$

In Exercises 43 through 58, solve each of the following word problems. Be sure to read each problem carefully before starting.

43. Two classrooms sit next to each other. Their room numbers are two consecutive positive odd numbers. The product of those two numbers is 255. What are the numbers?

44. In a certain hotel, the doors are numbered consecutively. The product of the numbers of two rooms located next to each other is 210. What are the room numbers?

45. The sum of the square of a number and the opposite of that number is 110. What is the number? (There can be two answers.)

46. The length of a box is one inch more than the width. The product of the length and one less than the width is eight. What are the length and width of this box?

47. In a certain chess tournament, each person plays every other person exactly two times. The number M of matches played in a chess tournament with p players is $M = p^2 - p$. How many players are there in a tournament of 156 matches?

48. The number of diagonals d in an n-sided polygon is given by $2d = n^2 - 3n$. What kind of figure has no diagonals?

49. A picture measures 15 inches by 20 inches and is framed with a mat. The mat has a uniform width w (see the accompanying figure), and the mat and picture together have an area of 456 square inches. What is the width of the mat?

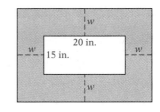

50. Use the figure from Exercise 49. Suppose a new mat for the picture has an area 6 square inches less than that of the picture. What is the area of the mat? What are its outside dimensions?

51. The length of each side of a square is decreased by 5 inches to produce a smaller square. Nine times the area of the smaller square is 4 times that of the original square. What are the areas of the two squares?

52. Two teams are playing ball. One more than the square of the score of the first team is five times the score of the second. The second team scored five points more than the first. What are the two scores?

53. DeNean and Michelle have matching hockey jerseys with consecutive numbers. The product of their jersey numbers is 506. What are their numbers?

54. The number of possible handshakes N in a group of n people is given by the formula $2N = n(n - 1)$. How many people are needed for 300 handshakes?

55. The length of the base of a triangle is 5 less than the height. If the base and height are both reduced by 2 inches, the area is 33 square inches. What is the original area of this triangle?

56. If the sides of a square are lengthened by 6 feet, the area is 81 square feet (see figure). What was the area of the original square?

57. The square of a number minus 5 times the number is 150. What is the number? (There are two possible answers.)

58. The area of a square table is 25 square feet. What will the new area be if each side is increased by n feet? Write your answer as an algebraic expression.

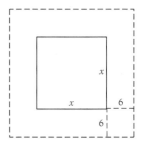

MIXED PRACTICE

By doing these exercises, you will practice all the topics up to this point in the chapter.

59. Factor: $4a(2y + 1) - 3(2y + 1)$

60. Solve: $5x^2 - 125 = 0$

61. Factor: $6x^2 + 66x - 30$

62. Factor: $p^2r^2 + 14p^2r + 49p^2$

63. Solve: $y^2 + 14y = 120$

64. Factor: $27n^2 + 63n - 54$

65. The product of two consecutive integers is 110. What are the numbers?

66. Factor: $120y^2 - 16y - 192$

67. Solve: $4r^2 - 18r + 8 = 0$

68. Solve: $12n^2 - 7n = 10$

69. Factor: $y^2 - 484$

70. Factor: $3m^2n^2 - 6m^2n - 9m^2$

EXCURSIONS

Exploring Numbers

1. Show that the product of two numbers—one odd and one even—is always even.

2. Show that the product of two odd numbers is always odd.

Data Analysis

3. Make a bar graph of the U.S. Presidents' birth dates by month from the following table. Ask and answer four questions about this data.

Presidents

Name and (party)	Term	State of birth	Born	Name and (party)	Term	State of birth	Born
1. Washington (F)	1789–1797	Va.	2/22/1732	22. Cleveland (D)	1885–1889	N.J.	3/18/1837
2. J. Adams (F)	1797–1801	Mass.	10/30/1735	23. B. Harrison (R)	1889–1893	Ohio	8/20/1833
3. Jefferson (DR)	1801–1809	Va.	4/13/1743	24. Cleveland (D)	1893–1897	—	—
4. Madison (DR)	1809–1817	Va.	3/16/1751	25. McKinley (R)	1897–1901	Ohio	1/29/1843
5. Monroe (DR)	1817–1825	Va.	4/28/1758	26. T. Roosevelt (R)	1901–1909	N.Y.	10/27/1858
6. J. Q. Adams (DR)	1825–1829	Mass.	7/11/1767	27. Taft (R)	1909–1913	Ohio	9/15/1857
7. Jackson (D)	1829–1837	S.C.	3/15/1767	28. Wilson (D)	1913–1921	Va.	12/28/1856
8. Van Buren (D)	1837–1841	N.Y.	12/5/1782	29. Harding (R)	1921–1923	Ohio	11/2/1865
9. W. H. Harrison (W)	1841	Va.	2/9/1773	30. Coolidge (R)	1923–1929	Vt.	7/4/1872
10. Tyler (W)	1841–1845	Va.	3/29/1790	31. Hoover (R)	1929–1933	Iowa	8/10/1874
11. Polk (D)	1845–1849	N.C.	11/2/1795	32. F. D. Roosevelt (D)	1933–1945	N.Y.	1/30/1882
12. Taylor (W)	1849–1850	Va.	11/24/1784	33. Truman (D)	1945–1953	Mo.	5/8/1884
13. Fillmore (W)	1850–1853	N.Y.	1/7/1800	34. Eisenhower (R)	1953–1961	Tex.	10/14/1890
14. Pierce (D)	1853–1857	N.H.	11/23/1804	35. Kennedy (D)	1961–1963	Mass.	5/29/1917
15. Buchanan (D)	1857–1861	Pa.	4/23/1791	36. L. B. Johnson (D)	1963–1969	Tex.	8/27/1908
16. Lincoln (R)	1861–1865	Ky.	2/12/1809	37. Nixon (R)	1969–1974	Calif.	1/9/1913
17. A. Johnson (U)	1865–1869	N.C.	12/29/1808	38. Ford (R)	1974–1977	Neb.	7/14/1913
18. Grant (R)	1869–1877	Ohio	4/27/1822	39. Carter (D)	1977–1981	Ga.	10/1/1924
19. Hayes (R)	1877–1881	Ohio	10/4/1822	40. Reagan (R)	1981–1989	Ill.	2/6/1911
20. Garfield (R)	1881	Ohio	11/19/1831	41. Bush (R)	1989–1993	Mass.	6/12/1924
21. Arthur (R)	1881–1885	Vt.	10/5/1830	42. Clinton (D)	1993–	Ark.	8/19/1946

1. The following party abbreviations are used: F = Federalist, DR = Democratic-Republican, D = Democratic, R = Republican, W = Whig, and U = Union. *Source: 1996 Information Please® Almanac* (©1995 Houghton Mifflin Co.), p. 633. All rights reserved. Used with permission by Information Please LLC.

> ◣ **CHAPTER LOOK-BACK**
>
> In the Hardy-Weinberg equation, we have seen that the terms mean something specific.
>
> $p^2 = $ the proportion of the population with 2 Gene Is
>
> $q^2 = $ the proportion of the population with 2 Gene IIs
>
> $2pq = $ the proportion of the population with both Gene I and Gene II
>
> ▪ **Research**   Find other ways that mathematics can be used in biology or in nature. Write some questions that can be answered by mathematics.

CHAPTER 6
REVIEW PROBLEMS

The following exercises will give you a good review of the material presented in this chapter.

SECTION 6.1

1. Factor: $20r^2t^2 - 10r^2t + 5rt^2$

2. Factor: $18xy^2 - 6xy + 9x^2y^2$

3. Factor: $36x^2y^2 + 18x^2y - 9xy^2$

4. Group and factor: $4xt + 8t + 12mx + 24m$

5. Factor: $6a(x + 5y) + 7(x + 5y)$

6. True or false: All numbers divisible by both 2 and 3 are divisible by 6.

SECTION 6.2

7. Factor: $x^2 + 4x - 5$

8. Factor: $-12 + y^2 + y$

9. Factor: $-21t + 3t^2 + 36$

10. Factor: $4r^2 + 8r - 96$

11. Factor: $xr^2 - 3rx - 40x$

12. Factor: $5t^2 + 3 - 8t$

13. Factor: $44 + 11t^2 + 4t$

14. Factor: $8y^2 - 44y - 84$

15. Factor: $27x^2 - 6xy - 5y^2$

16. Factor: $-26x + 24x^2 - 28$

SECTION 6.3

17. Factor: $-25 + 625x^2$

18. Factor: $48x^2 - 3$

19. Factor: $24y + 24 + 6y^2$

20. Factor: $-9x^2 + 24x - 16$

21. Factor: $4x^2 + 4x + 1$

22. Factor: $3tr^2 - 48rt + 192t$

SECTION 6.4

23. Solve: $r^2 - 10r + 21 = 0$

24. Solve: $2000 - 5n^2 = 0$

25. Solve: $3x^2 - 6x - 105 = 0$

26. Solve: $48 - 22x + 2x^2 = 0$

27. Solve: $5r^2 - 4r - 12 = 0$

28. Solve: $20m^2 + 68m + 56 = 0$

29. The product of two consecutive even numbers is 440. What are the numbers?

30. A negative number and its square have a sum of 110. What is the number?

31. The length of a rectangle is 3 times the width. The area is 108 square inches. What are the dimensions?

32. The lengths of two rectangles are the same. The width of one is 3 more than the length, and the width of the other is 3 less than the length. The difference in their areas is 60 square inches. What are the dimensions of the rectangles?

33. Factor: $-15 + 8x^2 + 2x$

34. Factor: $4r^2 + 64r - 369$

35. Factor: $-13x - 3 + 56x^2$

36. Factor: $-6 + 20w - 16w^2$

MIXED REVIEW

37. One number is 12 more than another. The product of the two numbers is -35. What are the numbers? (There are two possible number pairs.)

38. Factor: $24x^2 - 41x + 12$

39. Factor: $48x^2 + 10x - 3$

40. Factor: $3x^2y + 9xy^2 - x^2y^2$

41. Factor: $x^2 + 2x - 15$

42. Solve: $r^2 + 22r + 117 = 0$

43. Solve: $y^2 - 3y - 18 = 0$

44. Solve: $9x^2 - 16 = 0$

45. Solve: $y^2 - 4y - 96 = 0$

46. Factor: $8y + 4y^2 - xy - 2x$

CHAPTER 6 TEST

This exam tests your knowledge of the material in Chapter 6.

1. Factor:

    a. $r^2t^2 - rt$

    b. $t(m - 12) + (12 - m)$

    c. $9w^6x^3 - 12w^4x^2 - 18w^2x^3$

2. Factor:

    a. $x^2 + 11x + 28$

    b. $-14x - 15 + 8x^2$

    c. Factor: $2x^2 + 8 - 10x$

3. Solve each of the following problems.

    a. The product of two consecutive positive even integers is 360. What are the integers?

    b. $24 + t^2 - 14t = 0$

    c. The length of each side of a square is decreased by 6 inches. Four times the area of the smaller square is equal to that of the original square. What are the areas of the two squares?

4. Factor:

    a. $x^2 - 9y^2$ **b.** $49y^2 - 144x^2$ **c.** $-30y + 9 + 25y^2$

CUMULATIVE REVIEW

CHAPTERS 1–6

The following exercises will help you maintain the skills you have learned in this and previous chapters.

1. Factor: $-3x^2 + 3$

2. Multiply: $(-5rt^2)(-15r^2t^3v^4)$

3. Factor: $2x^2 - 12x - 14$

4. Write in slope-intercept form the equation of the line that passes through the point $(-6, 4)$ and has slope $\frac{1}{3}$.

5. Find the slope of the line that passes through the points $(-8, 2)$ and $(2, 3)$.

6. Simplify: $(-3x^2y)^4$

7. Multiply: $\left(-2\frac{2}{3}\right)\left(6\frac{1}{4}\right)\left(-5\frac{2}{5}\right)$

8. Write in slope-intercept form the equation of the line that passes through the points $(2, -3)$ and $(5, 1)$.

9. Evaluate $xy + 2xy - 3xy^2 + 2x^2y$ when $x = -3$ and $y = -6$.

10. Solve for t: $2.4 - 8t \le 3 + t$

11. Solve: $2x^2 = 8$

12. Solve for r: $17.5r - 5 = -8.5$

13. Factor: $x^2 - 144$

14. Multiply: $(6x - 3)(x - 2)$

15. Divide: $\dfrac{-24xy^2 + 36x^2y}{-3xy}$

16. Solve for b: $mx + \frac{b}{c} = d$

17. Solve: $-9.8x + 8.3x = 15$

18. Solve: $4x^2 - 20x + 24 = 0$

19. Graph $6y - 5x = -36$ using a table of values.

20. Graph the inequality $-3x + 4y > 9$.

RATIONAL EXPRESSIONS

*L*ow fat means less than three calories of fat to every ten calories. Sometimes food labels can be confusing. To determine the fat calories in a particular food item, you can use the following procedure. Take the number of grams of fat per serving listed on the package. Multiply by 9 (each gram of fat burns 9 calories) and divide by the total calories. This is the ratio of fat to total calories.

■ *What other situations use ratios?*

SKILLS CHECK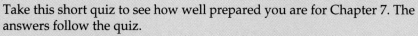

Take this short quiz to see how well prepared you are for Chapter 7. The answers follow the quiz.

1. Factor: $x^2 + 14x + 49$

2. Add: $\frac{2}{3} + \frac{6}{7}$

3. Divide: $\frac{12}{25} \div \frac{4}{5}$

4. Simplify:
$$-2(x - 5) - 3(x^2 - 2) + x(x - 4)$$

5. Solve: $3x - 5 = 13$

6. Solve for x: $\frac{x}{a} = \frac{b}{c}$

7. Find n: $\frac{3}{19} = \frac{9}{n}$

8. Find the GCF of 12 and 90.

ANSWERS: **1.** $(x + 7)^2$ [Section 6.3]   **2.** $\frac{32}{21} = 1\frac{11}{21}$ [Section 1.2]   **3.** $\frac{3}{5}$ [Section 1.3]
4. $-2x^2 - 6x + 16$ [Section 2.3]   **5.** $x = 6$ [Section 3.1]   **6.** $x = \frac{ab}{c}$ [Section 3.2]
7. $n = 57$ [Section 3.1]   **8.** GCF = 6 [Section 6.1]

CHAPTER LEAD-IN

Some fitness experts say that no more than 10% of one's diet should be from fat.

Keep a record of what you eat during a typical day. Using the food labels on foods you prepare or purchase and eat, estimate your fat intake for the day. What percent of your calories come from fat? What is your fat ratio?

7.1   Defining Rational Expressions

SECTION LEAD-IN

Rational expressions occur in many formulas that can be calculated by graphing calculators. You never see the formula—you simply push a button and get an answer. One such formula follows.

This is a financial formula for computing the present value of money in a certain transaction.

$$PV = \left(\frac{PMT \times G}{i} - FV \right) \times \frac{1}{(1 + i)^N} - \frac{PMT \times G}{i}$$

when $i \neq 0$.

- What does PV equal when $i = 1$?

Another formula is given for this same calculation when $i = 0$.

$$PV = -(FV + PMT \times N)$$

- Why do we need another formula for this special case?

Rational expressions are fractions that are made up of polynomials.

Rational Expressions

A fraction of the form $\frac{p}{q}$, where p and q are polynomials and q does not equal zero, is called a **rational expression.**

Some examples are

$$\frac{1}{x + 2} \qquad \frac{-2m}{n} \qquad \frac{17x^5}{x^2 + 5x + 2} \qquad \frac{x^2 + 5x + 2}{17x^5}$$

All the properties that hold for rational numbers, such as the commutative and associative properties, also hold for rational expressions.

Meaningful Expressions

Recall from arithmetic that a fraction bar indicates a division and that division by zero is undefined. This is why our definition of a rational expression includes the condition that q *is not equal to zero*. When q, the denominator of a rational expression, is equal to zero, the rational expression is said to be **undefined,** or **not meaningful.** Values of the variable that make the denominator zero must be excluded from consideration.

···

EXAMPLE 1

For which real numbers are the following expressions not meaningful?

a. $\dfrac{y^2 + 2y}{5x}$ **b.** $\dfrac{3n - 4}{x - 6}$ **c.** $\dfrac{x - 5}{x^2 - 9}$

SOLUTION

A rational expression is not meaningful for all real numbers that make the denominator 0. We find those numbers by setting the denominator equal to zero and solving for the variable.

a. The denominator here is $5x$.

$$\text{Let} \qquad 5x = 0$$

$$\text{Then} \qquad x = 0 \quad \text{Multiplying both sides by } \tfrac{1}{5}$$

So the expression $\dfrac{y^2 + 2y}{5x}$ is not meaningful when $x = 0$. It is **defined** (or **meaningful**) for all other real-number values of x and y.

b. The denominator is $x - 6$.

$$\text{Let} \quad x - 6 = 0$$

$$\text{Then} \qquad x = 6 \quad \text{Adding 6 to both sides}$$

So this expression is not meaningful when x is 6.

c. The denominator is $x^2 - 9$.

$$\text{Let} \qquad x^2 - 9 = 0$$

We can solve this quadratic equation by factoring the expression on the left and setting each factor equal to zero.

$$(x + 3)(x - 3) = 0 \qquad \text{Factoring}$$

$$x + 3 = 0 \quad \bigg| \quad x - 3 = 0 \quad \text{Setting factors equal to 0}$$

$$x = -3 \quad \bigg| \quad x = 3 \quad \text{Solving}$$

So when x is either -3 or 3, the expression is not meaningful.

▶ *CHECK* **Warm-Up 1**

Evaluating Rational Expressions

We evaluate a rational expression just as we evaluate an algebraic expression— by substituting the given values and simplifying.

To evaluate a rational expression

1. Substitute the given values for the variable or variables.
2. Simplify the numerator and denominator using the standard order of operations.
3. Simplify the resulting fraction.

▪▪▪

EXAMPLE 2

Evaluate the following rational expressions when $x = -2$ and $y = 4$.

a. $\dfrac{xy^3 - x}{xy}$ **b.** $\dfrac{(x - y)^2}{x + 2}$

SOLUTION

a. We first substitute the numerical values for the variables.

$$\frac{xy^3 - x}{xy} = \frac{(-2)(4)^3 - (-2)}{(-2)(4)}$$

Then, following the standard order of operations, we simplify the numerator and denominator.

$$\frac{(-2)(4)^3 - (-2)}{(-2)(4)} = \frac{(-2)(64) - (-2)}{(-2)(4)} = \frac{-128 + 2}{-8} = \frac{-126}{-8} = \frac{126}{8}$$

Finally, we simplify the fraction.

$$\frac{126}{8} = 15\frac{6}{8} = 15\frac{3}{4}$$

So $\frac{xy^3 - x}{xy} = 15\frac{3}{4}$ when $x = -2$ and $y = 4$.

b. We first substitute the value for each variable.

$$\frac{(x - y)^2}{x + 2} = \frac{[(-2) - 4]^2}{(-2) + 2}$$

Looking at the denominator, we can see that $-2 + 2$ will result in 0 in the denominator. There is no sense in continuing, because the result will not be meaningful.

▶ *CHECK* **Warm-Up 2**

> **■ ■ ■**
> **W**RITER'S **B**LOCK
>
> **Explain to a friend why zero is unacceptable as a denominator.**

Calculator Corner

You can use your graphing calculator's Home Screen to evaluate rational expressions by **STO**ring in each variable its given numerical value. In Example 2, for instance, **STO**re the value -2 for x and the value 4 for y, then evaluate each expression and give the results in fraction form. (Note: See the Calculator Corner on page 85 to review how to store a value for a variable.)

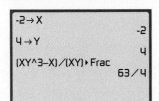
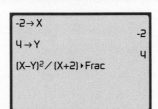

Press **ENTER.**

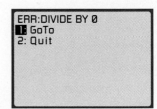

What is the calculator telling you? You cannot divide by 0, therefore, the calculator gives you an **ERR**or message.

Simplifying Rational Expressions

To simplify a fraction, we cancel (or divide out) any factor that appears in both the numerator and the denominator. So, for example,

$$\frac{3 \cdot \cancel{2}}{4 \cdot \cancel{2}} = \frac{3}{4}$$

We can also simplify rational expressions.

WRITER'S BLOCK

Restate the *fundamental property of fractions* in your own words.

> **Fundamental Property of Fractions**
>
> If p, q, and r are integers, and q and r are not equal to zero, then
>
> $$\frac{pr}{qr} = \frac{p}{q}$$

This property applies to rational expressions as well as to fractions. If we divide the numerator and denominator of a rational expression by the same non-zero factor, then the value of the rational expression is not changed. This procedure is called **reducing a rational expression to lowest** (or simplest) **terms.**

> **To reduce a rational expression to lowest terms**
>
> 1. Factor the numerator and denominator.
> 2. Divide numerator and denominator by the same factors (cancel common factors).

▪▪▪

EXAMPLE 3

Reduce $\dfrac{3x^2y + xy}{3x + 1}$ to lowest terms.

SOLUTION

Because xy appears in both terms of the numerator, we check to see if we can factor them. We can.

$$\frac{3x^2y + xy}{3x + 1} = \frac{xy(3x + 1)}{3x + 1} \quad \text{Factoring out } xy$$

WRITER'S BLOCK

How can you tell when a rational expression is reduced to lowest terms?

We then cancel the common factors.

$$\frac{xy\cancel{(3x + 1)}^{1}}{\cancel{3x + 1}_{1}} = \frac{xy}{1} = xy$$

Thus

$$\frac{3x^2y + xy}{3x + 1} = xy$$

▶ CHECK **Warm-Up 3**

■ ■ ■

EXAMPLE 4

Simplify: $\dfrac{n^2 - 49}{7 - n}$

SOLUTION

First we factor the numerator.

$$\frac{n^2 - 49}{7 - n} = \frac{(n - 7)(n + 7)}{7 - n}$$

Notice that $n - 7$ appears in the numerator and $7 - n$ is in the denominator. These two expressions have n and 7, but the signs of each in the numerator are opposite from the denominator.

In fact

$$n - 7 = -1(7 - n)$$

and

$$7 - n = -1(n - 7)$$

We rewrite the numerator and simplify.

$$\frac{(n - 7)(n + 7)}{7 - n} = \frac{-1(7 - n)(n + 7)}{7 - n}$$

$$= -1(n + 7)$$

$$= -n - 7$$

So

$$\frac{n^2 - 49}{7 - n} = -n - 7$$

▶ CHECK **Warm-Up 4**

Practice what you learned.

SECTION FOLLOW-UP

We start with

$$PV = \left(\frac{PMT \times G}{i} - FV\right) \times \frac{1}{(1 + i)^N} - \frac{PMT \times G}{i}$$

When $i = 1$, this formula simplifies to

$$PV = \frac{1}{2^N}(PMT \times G - FV) - PMT \times G$$

Verify this. Now can you answer the question, "Why do we need another formula for the special case $i = 0$?"

7.1 WARM-UPS

Work these problems before you attempt the exercises.

1. Determine the values of x for which the expression $\dfrac{3}{3x - 2}$ has meaning.

2. Evaluate $\dfrac{xy^3}{-9y}$ for $x = -3$ and $y = 4$.

3. Reduce $\dfrac{x^2 - x}{x - 1}$ to lowest terms.

4. Simplify: $\dfrac{9n^2 - 25}{5 - 3n}$

7.1 EXERCISES

Note: Use your graphing calculator to check your results whenever possible.

In Exercises 1 through 8, determine the values of the variables for which each algebraic expression has meaning.

1. $\dfrac{y}{y}$

2. $\dfrac{5}{r - 2}$

3. $\dfrac{x - 4}{4 + x}$

4. $\dfrac{t + 8}{5t}$

5. $\dfrac{y - 4}{y^2 - 9}$

6. $\dfrac{4r}{r^2 + 2r}$

7. $\dfrac{8 - x}{x^2 - 3x + 2}$

8. $\dfrac{2t}{t^2 + 4t + 4}$

In Exercises 9 through 16, evaluate each expression when $x = -3$ and $y = -2$.

9. $\dfrac{y - 8}{y^2 x - 1}$

10. $\dfrac{2y + 6}{x^2 - 9}$

11. $\dfrac{(y - 3)^2}{y^2 - 6x + 9}$

12. $\dfrac{y^2 - 2x}{xy^2 + 2x}$

13. $\dfrac{(3x - 5)^2}{y + 2}$

14. $\dfrac{x^2 + 5}{3y - 6}$

15. $\dfrac{x - y^2}{3y - 2x}$

16. $\dfrac{x^2 + 3x + 2}{2x + 2}$

In Exercises 17 through 52, reduce each rational expression to lowest terms.

17. $\dfrac{x}{xy}$

18. $\dfrac{r}{r^3 t}$

19. $\dfrac{y^3}{y^2 r}$

20. $\dfrac{t^5}{t^3}$

21. $\dfrac{x^2 y^2}{x^3 y}$

22. $\dfrac{r^4 t^4}{r^2 t^6}$

23. $\dfrac{y^5 t^5}{y^3 t^7}$

24. $\dfrac{r^5 y^2}{r^2 y^4}$

25. $\dfrac{x^2}{x^3 + x}$

26. $\dfrac{y^5}{y^6 + 5y}$

27. $\dfrac{r^6}{r^3 - 3r}$

28. $\dfrac{t^4}{t^2 + 8t}$

29. $\dfrac{x + 5}{5x + 25}$

30. $\dfrac{3y - 18}{y - 6}$

31. $\dfrac{2x - 4}{x - 2}$

32. $\dfrac{4x - 7}{20x - 35}$

33. $\dfrac{x + 4}{x^2 + 3x - 4}$

34. $\dfrac{x - 3}{x^2 - 9}$

35. $\dfrac{x + 7}{x^2 + 2x - 35}$

36. $\dfrac{x + 1}{x^2 + 5x + 4}$

37. $\dfrac{x^2 + 5x + 6}{x^2 + 7x + 10}$

38. $\dfrac{x^2 + 5x - 6}{x^2 + 4x - 5}$

39. $\dfrac{x^2 - 9}{x^2 - x - 6}$

40. $\dfrac{x^2 - 6x - 7}{x^2 + 3x + 2}$

41. $\dfrac{y^2 + 15y + 26}{y + 13}$

42. $\dfrac{y^2 - 13y + 42}{y - 6}$

43. $\dfrac{y^2 + 4y + 4}{y^2 - 4}$

44. $\dfrac{2y^2 - 4y}{y^2 - 17y + 30}$

45. $\dfrac{243 - 3t^2}{81 - t^2}$

46. $\dfrac{15r^2 - 15}{r^2 + 14r + 13}$

47. $\dfrac{36y^2 - 93y + 60}{12y^2 - 31y + 20}$

48. $\dfrac{6x^2 - 32x - 24}{3x + 2}$

49. $\dfrac{70x^3 + 32x^2 - 6x}{25x + 15}$

50. $\dfrac{8x^2 + 20x - 12}{10x^3 + 25x^2 - 15x}$

51. $\dfrac{2x^2 + 3xy + y^2}{3x + 3y}$

52. $\dfrac{2n^2 - 7nt + 3t^2}{2n - 6t}$

53. *Proportions*   For any proportions where

$$\frac{a}{A} = \frac{b}{B} = \frac{c}{C} = K$$

where K is not zero, and for *any* non-zero numbers p, q, and r, it is true that

$$\frac{ap + bq + cr}{Ap + Bq + Cr} = K$$

Consider the proportions

$$\frac{1}{5} = \frac{2}{10} = \frac{14}{70} = 0.2$$

and let $p = 8$, $q = 10$, and $r = 12$. Substitute and show that

$$\frac{1p + 2q + 14r}{5p + 10q + 70r} = 0.2$$

is a true statement. We will learn about ratio and proportion in Section 7.2.

54. *Compound Interest*   The compound interest formula is

$$A = p\left(1 + \frac{r}{n}\right)^{nt}$$

It is used to compute the balance (A) on some savings accounts that give compound interest.

p is the principal or amount in the bank.
r is the interest rate (written as a decimal).
n is the number of times interest is paid annually.
t is the length of time the money remains in the bank.

Find A when

$p = \$1000$
$r = 5\%$
$n = 4$ times a year
$t = 2$ years

55. *Efficiency of a Jack*   The efficiency E of a jack is determined from the pitch of the jack's thread with the formula

$$E = \frac{\frac{p}{2}}{p + \frac{1}{2}}$$

What is the efficiency of a jack with $p = 0.85$ millimeter?

56. *Recording on Tape* For a 7-inch reel of magnetic recording tape, the remaining playing time t (in minutes) can be determined by measuring the "depth" d (in inches) of tape remaining on the reel. The formula is

$$\frac{1}{d^2} = \frac{9.38 + \frac{2.25}{d}}{t}$$

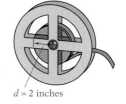

$d = 2$ inches

Solve this formula for t when $d = 2$ inches.

57. *Windmill Energy* A formula to determine the power P generated by a windmill is

$$\frac{P}{K} = v^3$$

where P is in watts and v is in miles per hour. Solve this equation for K, and then find the value of K when $P = 9.41$ and $v = 8.6$.

58. *Photographic Lens*

$$\frac{1}{f} = k\left(\frac{1}{r_1} - \frac{1}{r_2}\right)$$

is known as the lensmaker's equation, where f is the focal length of a thin lens, k is a constant, and r_1 and r_2 are the radii of the lens' spherical surfaces. Solve the equation for k when $r_1 = -r_2 = 2$ and $f = 3.5$.

For Exercises 59 through 80, evaluate each of the expressions when $x = 0$. In mathematics we can also say, "Evaluate each expression as x approaches zero." Substitute $x = 0$ and simplify each rational expression.

59. $\dfrac{2x - 4}{x^2 - 4}$ **60.** $\dfrac{5x - 15}{3x - 9}$ **61.** $\dfrac{6x + 18}{9x + 27}$ **62.** $\dfrac{27 - 3x^2}{27 - x^3}$

63. $\dfrac{8x^2 - 32}{x^3 - 8}$ **64.** $\dfrac{8x + 5x^2}{101x + 501}$ **65.** $\dfrac{x + 1}{x^2 + 5x + 4}$ **66.** $\dfrac{x + 7}{x^2 + 2x - 35}$

67. $\dfrac{x^2 - 13x + 42}{x - 6}$ **68.** $\dfrac{x^2 + 15x + 26}{x + 13}$ **69.** $\dfrac{2x^2 - 4x}{x^2 - 17x + 30}$ **70.** $\dfrac{80x - 14x + 3}{200x - 35x}$

71. $\dfrac{x^2 - 6x - 7}{x^2 + 3x + 2}$ **72.** $\dfrac{x^2 + 5x - 6}{x^2 + 4x - 5}$ **73.** $\dfrac{15x^2 - 15}{x^2 + 14x + 13}$ **74.** $\dfrac{6x^2 + 24x}{x^2 + 3x - 4}$

75. $\dfrac{x^2 - 9}{x^2 - x - 6}$ **76.** $\dfrac{x^2 + 4x + 4}{x^2 - 4}$ **77.** $\dfrac{4x - 5}{25 - 40x + 16x^2}$ **78.** $\dfrac{x^2 - 100}{10 - x}$

79. $\dfrac{(3x - 12)^2}{12 - 3x}$ **80.** $\dfrac{-2 + 7x}{4 - 28x + 49x^2}$

EXCURSIONS

Posing Problems

1. "China's population is about 1.2 billion people; that is 22% of the world's population on only 9% of the world's *arable* land. [Do you know what that word means? Research it.] China's fertile land covers 130 million hectares—only 0.11 hectare for each person in China, compared with 0.73 hectare for each American."*

*Source: Roy L. Prosterman, Tim Hanstad, and Li Ping, "Can China Feed Itself?" *Scientific American*, November 1996, 90.

a. ***Research*** Find the size of a hectare in a unit you are familiar with.

b. Ask and answer two questions using the information from the quote.

Exploring Numbers

2. According to *USA Today*, the number of animators that worked on the movie *Space Jam* is large.

> It has 4 digits.
> It is the product of two squares.
> The sum of its digits is 18.
> No two digits are alike.

How many animators worked on *Space Jam*?

7.2 Multiplying and Dividing Rational Expressions

SECTION LEAD-IN

For a 7-inch reel of magnetic recording tape, the remaining playing time t (in minutes) can be determined by measuring the "depth" d (in inches) of tape remaining on the reel. A formula for finding t, when you are given d is

$$\frac{1}{d^2} = \frac{9.38 + \frac{2.25}{d}}{t}$$

a. When $d = 1$, what is the value of t?

b. As t gets larger, what happens to the "depth" of the tape?

c. As d gets smaller, what happens to the amount of time left?

d. What does the answer to part (a) tell you about the tape? Write a sentence that explains what we found out.

SECTION GOALS

- To multiply rational expressions
- To divide rational expressions
- To simplify complex fractions using division
- To use cross products to solve rational equations
- To find the least common multiple of two or more polynomials

Multiplying Rational Expressions

In this section and the next, we apply the four arithmetic operations to rational expressions. We begin here with multiplication.

Multiplying Rational Expressions

If p, q, r, and s are polynomials, and q and s are not equal to zero, then

$$\frac{p}{q} \cdot \frac{r}{s} = \frac{pr}{qs}$$

Therefore, to multiply rational expressions, we multiply the numerators, multiply the denominators, and then simplify.

▪ ▪ ▪
EXAMPLE 1

Multiply: $\dfrac{3}{2} \cdot \dfrac{2n + 1}{3n + 3}$

SOLUTION

Multiplying the numerators and the denominators gives us

$$\frac{3}{2} \cdot \frac{2n + 1}{3n + 3} = \frac{3(2n + 1)}{2(3n + 3)}$$

Then we use the distributive property.

$$\frac{3(2n + 1)}{2(3n + 3)} = \frac{6n + 3}{6n + 6}$$

Finally, we divide out a common factor of 3.

$$\frac{6n + 3}{6n + 6} = \frac{\cancel{3}(2n + 1)}{\cancel{3}(2n + 2)} = \frac{2n + 1}{2n + 2}$$

▶ CHECK **Warm-Up 1**

STUDY HINT

Factor first, cancel, and then multiply or divide.

It is a good idea to factor the numerators and denominators of rational expressions, to the extent possible, before multiplying them. That can reduce the amount of work involved.

▪ ▪ ▪
EXAMPLE 2

Multiply: $\dfrac{x^2 + 2x + 1}{(x + 1)^3} \cdot \dfrac{2x^2 - 5x + 3}{x - 1}$

SOLUTION

The numerator in each fraction can be factored.

$$\frac{x^2 + 2x + 1}{(x + 1)^3} \cdot \frac{2x^2 - 5x + 3}{x - 1} = \frac{(x + 1)(x + 1)}{(x + 1)^3} \cdot \frac{(x - 1)(2x - 3)}{(x - 1)}$$

Now we can simplify each fraction by cancelling.

$$\frac{\cancel{(x + 1)}\cancel{(x + 1)}}{\cancel{(x + 1)^3}} \cdot \frac{\cancel{(x - 1)}(2x - 3)}{\cancel{(x - 1)}}$$
$$\frac{}{(x + 1)^2} \qquad \frac{}{1}$$
$$(x + 1)$$

This leaves us with the multiplication.

$$\frac{1}{x + 1} \cdot \frac{(1)(2x - 3)}{1} = \frac{2x - 3}{x + 1}$$

So

$$\frac{x^2 + 2x + 1}{(x + 1)^3} \cdot \frac{2x^2 - 5x + 3}{x - 1} = \frac{2x - 3}{x + 1}$$

▶ CHECK **Warm-Up 2**

Dividing Rational Expressions

The second operation on rational expressions that we will discuss is division.

Dividing Rational Expressions

If p, q, r, and s are polynomials, and q, r, and s are not equal to zero, then

$$\frac{p}{q} \div \frac{r}{s} = \frac{p}{q} \cdot \frac{s}{r} = \frac{ps}{qr}$$

Thus to divide rational expressions, we multiply the first fraction by the reciprocal of the second fraction.

■■■

EXAMPLE 3

Divide: $\dfrac{2x + 2}{x + 5} \div \dfrac{x + 1}{4x + 20}$

SOLUTION

First we invert the divisor and write the problem as a multiplication.

$$\frac{2x + 2}{x + 5} \div \frac{x + 1}{4x + 20} = \frac{2x + 2}{x + 5} \cdot \frac{4x + 20}{x + 1}$$

Then we factor.

$$\frac{2x + 2}{x + 5} \cdot \frac{4x + 20}{x + 1} = \frac{2(x + 1)}{x + 5} \cdot \frac{4(x + 5)}{x + 1}$$

Now we can multiply, cancelling any common factors.

$$\frac{2\overset{1}{\cancel{(x + 1)}} \cdot 4\overset{1}{\cancel{(x + 5)}}}{\underset{1}{\cancel{(x + 5)}} \cdot \underset{1}{\cancel{(x + 1)}}} = 2 \cdot 4 = 8$$

Thus

$$\frac{2x + 2}{x + 5} \div \frac{x + 1}{4x + 20} = 8$$

▶ *CHECK* **Warm-Up 3**

Using Division to Simplify Complex Fractions

When a rational expression has a fraction or rational expression in the numerator or denominator or both, it is said to be a **complex fraction.** Complex fractions appear in some formulas and often must be simplified before they can be used. That is, the fractions or rational expressions must be removed from both the numerator and the denominator.

Rewriting as Division

In this example, the numerator and denominator both contain one fraction. Therefore, we rewrite the complex fraction as a division problem and then solve.

▪▪▪
EXAMPLE 4

Simplify: $\dfrac{\dfrac{x+3}{x}}{\dfrac{2x+5}{x}}$

SOLUTION

We first rewrite this as a division problem.

$$\frac{\dfrac{x+3}{x}}{\dfrac{2x+5}{x}} = \frac{x+3}{x} \div \frac{2x+5}{x}$$

We then do the division. That is, we multiply by the reciprocal of the divisor.

$$\frac{x+3}{x} \div \frac{2x+5}{x} = \frac{x+3}{x} \cdot \frac{x}{2x+5} = \frac{(x+3)\overset{1}{\cancel{(x)}}}{\cancel{(x)}(2x+5)}$$

So,

$$\frac{\dfrac{x+3}{x}}{\dfrac{2x+5}{x}} = \frac{x+3}{2x+5}$$

▶ *CHECK* **Warm-Up 4**

If p, q, and r are polynomials, and q and r are not zero, then

$$\frac{\dfrac{p}{q}}{\dfrac{r}{q}} = \frac{p}{r}$$

We will see complex fractions again in Section 7.3.

Solving Equations Containing Rational Expressions

Using Ratio and Proportion

A **ratio** is a comparison of two or more measurements with the same units. Examples of ratios are

3 doctors/4 doctors   4 students/5 students

The first ratio can be written as 3:4, 3 to 4, or as the fraction $\frac{3}{4}$.

A **rate** is a special type of ratio. It is a ratio in which the measurements have different units. Examples of rates are

55 miles/hour   27 women/14 men

A **proportion** is an equation that shows that two ratios or rates are equal. An example of a proportion is

$$\frac{4 \text{ boys}}{3 \text{ girls}} = \frac{16 \text{ boys}}{12 \text{ girls}}$$

This means that there is the same relationship between 4 boys and 3 girls as there is between 16 boys and 12 girls. A proportion is a true statement.

Cross Products

In a true proportion $\frac{a}{b} = \frac{c}{d}$, the **cross products** are equal. That is,

$$ad = bc$$

This is sometimes referred to as the **cross-products test.** The process of finding the cross products is called **cross-multiplication.** It can be used to find a missing value in a proportion.

We can further extend the concept of cross products to rational expressions. In this section and the next, we will be solving equations that involve rational expressions. When there is exactly one rational expression on each side of the equation, we use cross-multiplication to solve for the variable. More formally,

---

Cross Products of Rational Expressions

If p, q, r, and s are polynomials with q and s not equal to zero, and if $\frac{p}{q} = \frac{r}{s}$, then $ps = qr$.

■ ■ ■

EXAMPLE 5

Solve for x: $\dfrac{3}{5} = \dfrac{x + 10}{27}$

SOLUTION

Using the cross-products test, we obtain

$$3(27) = 5(x + 10)$$

We solve for x.

$$81 = 5x + 50 \quad \text{Multiplying}$$

$$31 = 5x \quad\quad \text{Adding } -50 \text{ to both sides.}$$

$$6\tfrac{1}{5} = x \quad\quad \text{Multiplying both sides by } \tfrac{1}{5}$$

So $x = 6\tfrac{1}{5}$.

We check by substituting $x = 6\tfrac{1}{5}$ and by cross-multiplying.

$$\frac{3}{5} = \frac{\left(6\tfrac{1}{5}\right) + 10}{27}$$

$$\frac{3}{5} = \frac{16\tfrac{1}{5}}{27} \quad\quad \text{Simplifying numerator}$$

$$(3)(27) = \left(16\tfrac{1}{5}\right) \cdot 5 \quad\quad \text{Cross-multiplying}$$

$$81 = \left(\tfrac{81}{5}\right) \cdot 5$$

$$81 = 81 \quad\quad \text{Correct} \qquad\qquad \blacktriangleright CHECK \textbf{ Warm-Up 5}$$

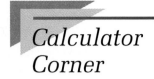

Calculator Corner

Equations containing rational expressions can be solved on your graphing calculator by graphing each side of the equation. The solution to the equation will be the point of intersection of the two graphs. Try Example 5 on your graphing calculator as follows. The **WINDOW** chosen here is a "friendly window" for the *TI-82* graphing calculator. Check your calculator manual for the dimensions of a "friendly screen" for your particular graphing calculator. (A "friendly window" is one that will give *nice numbers* when you **TRACE**.)

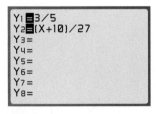
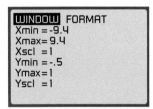
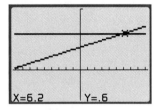

Notice that when you **TRACE,** the two lines intersect at the point $x = 6.2$ and $y = 0.6$. Some graphing calculators have the ability to find the point of intersection for you. (Note: See the Calculator Corner on page 125 to review how to find the points of intersection of two graphs.)

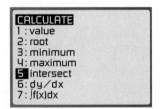
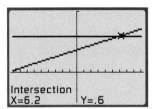

Least Common Multiple of Polynomials

To find the least common multiple (LCM) of two or more positive integers, we first factor each integer and write the factorization in exponential form. Then we list all the bases and attach to each the greatest exponent it has in any of the factored integers. The product of these bases is the LCM of the integers. For example to find the LCM of 24 and 36, use the factors of 24 and 36.

$$24 = 2^3 \cdot 3 \qquad 36 = 2^2 3^2$$
$$\text{LCM} = 2^3 3^2 = 72$$

When we work with polynomials, we use a similar method.

> **To find the least common multiple (LCM) of two or more polynomials**
>
> 1. Factor the polynomials completely and write each in exponential form.
> 2. Multiply together all the bases found in any of the exponential forms.
> 3. Give each of those bases the greatest exponent it has in any of the exponential forms.

▪ ▪ ▪

EXAMPLE 6

Find the LCM of x^2, $x^2 - 9$, $2x + 6$, and $3x^2 - 18x + 27$.

SOLUTION

We begin by factoring these polynomials completely and writing the products in exponential form.

$$x^2 = x^2$$
$$x^2 - 9 = (x - 3)(x + 3)$$
$$2x + 6 = 2(x + 3)$$
$$3x^2 - 18x + 27 = 3(x - 3)(x - 3) = 3(x - 3)^2$$

The least common multiple is the product of all the bases $x(x - 3)$, $(x + 3)$, 2, and 3, each with its greatest exponent.

The least common multiple is

$$x^2(x - 3)^2(x + 3) \cdot 2 \cdot 3 = 6x^2(x - 3)^2(x + 3)$$

▶ *CHECK* **Warm-Up 6**

Practice what you learned.

SECTION FOLLOW-UP

The questions asked in the Section Lead-In explore the meaning of the formula

$$\frac{1}{d^2} = \frac{9.38 + \frac{2.25}{d}}{t}$$

a.   When $d = 1$, $t = 11.63$ minutes.

b.   As t gets larger, d gets larger.

c.   As d gets smaller, t gets smaller.

d.   When the "depth" of the tape is 1 inch, there are 11.63 minutes of playing time left.

You should choose some very large and very small positive values to check the results in parts (b) and (c). (Why are we only interested in positive values here?)

7.2 WARM-UPS

Work these problems before you attempt the exercises.

1. Multiply: $\dfrac{t}{3} \cdot \dfrac{3t - 3x}{t + x}$

2. Multiply: $\dfrac{r^2 + 6r + 9}{r - 1} \cdot \dfrac{r^2 - r}{r + 3}$

3. Divide: $\dfrac{3x + 9}{x - 7} \div \dfrac{x + 3}{3x - 21}$

4. Simplify: $\dfrac{\frac{x + 5}{x}}{\frac{3x + 15}{y}}$

5. Solve for x: $\dfrac{2}{3} = \dfrac{x + 4}{18}$

6. Find the LCM of $(x + 2)^2$, $x^2 - 4$, and $2x$.

7.2 EXERCISES

Note: Use your graphing calculator to check your results whenever possible.

In Exercises 1 through 24, multiply or divide as indicated.

1. $\dfrac{x^2}{x^8} \cdot \dfrac{x^3}{n^6}$

2. $\dfrac{x^5}{n^6} \cdot \dfrac{8n^9}{4x^2}$

3. $\dfrac{7r^{11}}{x^3} \div \dfrac{x^6}{14r^3}$

4. $\dfrac{3x + 9}{5} \cdot \dfrac{2x + 8}{3}$

5. $\dfrac{5x + 15}{2} \cdot \dfrac{11x - 33}{x + 3}$

6. $\dfrac{3x - 12}{x - 4} \div \dfrac{x - 4}{2}$

7. $\dfrac{4x - 28}{4} \div \dfrac{x - 7}{7}$

8. $\dfrac{2n^2 - 8n + 6}{2y^2 + 22y - 84} \div \dfrac{2n - 2}{3y - 9}$

9. $\dfrac{2y^2 + 16y - 18}{y + 9} \div \dfrac{y^2 - 81}{y - 9}$

10. $\dfrac{x}{6x} \cdot \dfrac{2x + 12}{2x + 6}$

11. $\dfrac{n}{n - 4} \cdot \dfrac{n - 4}{4n}$

12. $\dfrac{4t}{t} \div \dfrac{t - 5}{2t + 10}$

13. $\dfrac{6t}{t} \div \dfrac{12t - 84}{6t - 42}$

14. $\dfrac{x^2 + x - 12}{x^2 + 12x + 35} \cdot \dfrac{x^2 + 13x + 40}{2x^2 + 2x - 24}$

15. $\dfrac{y^2 + 11y + 18}{y^2 + 5y - 6} \cdot \dfrac{y^2 - 6y - 72}{y^2 - 10y - 24}$

16. $\dfrac{r^2 - 169}{r + 13} \cdot \dfrac{r^2 + 26r + 169}{2r - 26}$

17. $\dfrac{r^2 - 4r + 4}{49r^2 - 14r + 1} \div \dfrac{r - 2}{7r - 1}$

18. $\dfrac{w^2 - 144}{w - 12} \div \dfrac{w + 12}{w^2 - 24w + 144}$

19. $\dfrac{y^2 - 18y + 81}{y - 9} \div \dfrac{y + 9}{y^2 - 81}$

20. $\dfrac{n^2 + 16n + 64}{n^2 - 2n + 1} \div \dfrac{n^2 + 9n + 8}{n^2 - 1}$

21. $\dfrac{t^2 + 4t + 4}{t^2 + 18t + 81} \div \dfrac{t^2 - 7t - 18}{t^2 - 81}$

22. $\dfrac{x^2 + 3x - 10}{x^2 - 3x + 2} \cdot \dfrac{x^2 + 6x - 7}{x^2 + 12x + 35}$

23. $\dfrac{10x^2 + 13x - 3}{2x^2 + x - 3} \div \dfrac{5x^2 + 4x - 1}{x^2 - 1}$

24. $\dfrac{n^2 - 13n + 36}{n^2 - 4} \div \dfrac{n^2 - 5n - 36}{n^2 + 2n - 8}$

For Exercises 25 through 34, simplify.

25. $\dfrac{\frac{xy}{x}}{\frac{x^2}{y}}$

26. $\dfrac{\frac{r^3t}{r^3}}{\frac{rt^3}{t^2}}$

27. $\dfrac{\frac{x+4}{4}}{\frac{5x+20}{8}}$

28. $\dfrac{\frac{x-5}{6}}{\frac{4x-20}{12}}$

29. $\dfrac{\frac{m+n}{6}}{\frac{m^2-n^2}{m}}$

30. $\dfrac{\frac{xy}{x^2y^2}}{\frac{x}{x-y}}$

31. $\dfrac{\frac{r-6}{3}}{\frac{3r-18}{9}}$

32. $\dfrac{\frac{x-8}{7}}{\frac{2x-16}{21}}$

33. $\dfrac{\frac{7r}{r+t}}{7r+12t}$

34. $\dfrac{\frac{4y}{x-y}}{6x-y}$

For Exercises 35 through 38, solve for the variable.

35. $\dfrac{-2}{x+1}=\dfrac{x-4}{3}$

36. $\dfrac{-4}{r}=\dfrac{r-4}{-3}$

37. $\dfrac{2x}{x+2}=\dfrac{1+2x}{x}$

38. $\dfrac{6}{r-2}=\dfrac{r^2+r}{r-2}$

For Exercises 39 and 40, solve.

39. *Hem Tape* The ratio of the lengths of two pieces of hem tape is 5:8. Their total length is 65 inches. If each piece of hem tape is decreased by n inches, what will the new ratio be?

40. *Scale Drawings* In a scale drawing depicting the distances between the sun, the moon, and Earth, 1 centimeter is equivalent to 2000 miles. If the sun is 93,000,000 miles away from Earth, and the moon is 240,000 miles away from Earth, find how far away from Earth they each have to be in the scale drawing.

In Exercises 41 through 58, find the LCM of the given terms.

41. $76xy$ and $24x^2$

42. $77y^2$ and $49x^2$

43. $50xy$ and $20x^2$

44. $88xy$ and $16y^2$

45. $14xy$ and $21x^2$

46. $42xy$ and $22xy$

47. $24x^2$ and $28xy$

48. $98y^2$ and $49x^2$

49. $32y^2$ and $40xy$

50. $25xy$ and $35x$

51. $52xy$ and $13y$

52. $30xy$ and $45y^2$

53. 82 and $41xy$

54. $96xy$ and 63

55. $58w^2xy$, $29x^2y$, $4y^2$

56. $38x^2y$, $57xy^4$, $19x^5$

57. $86x^2y^5$, 43, $4y^2$

58. $40w^2xy^2$, $64y^2$, $36xy$

MIXED PRACTICE

By doing these exercises, you will practice the topics up to this point in the chapter.

59. Divide: $\dfrac{x^2+10x+9}{x^2+7x-18} \div \dfrac{x^2-6x-7}{x^2+5x-14}$

60. Determine when $\dfrac{x-y}{y+4}$ is not meaningful.

61. Reduce $\dfrac{x-4}{x^2-8x+16}$ to lowest terms.

62. Divide: $\dfrac{y^2+y-2}{xy^2+8xy} \div \dfrac{y-1}{y^2+8y}$

63. Reduce $\dfrac{x^2 - 16}{x^2 - 4x}$ to lowest terms.

64. Determine when $\dfrac{x^2 - 5x + 6}{x^2 - 6x + 8}$ is not meaningful.

65. Multiply: $\dfrac{r^2 + 15r + 50}{y^2 - 64} \cdot \dfrac{y^2 - 18y + 80}{r^2 + 13r + 40}$

66. Reduce $\dfrac{8x^2y^3}{14xy^5}$ to lowest terms.

EXCURSIONS

Exploring Problem Solving

1. The smallest meaningful segment of language is the *phoneme*. Combinations of 44 phonemes produce every word in the English language. To determine the number of *possible* three-phoneme words that can be formed out of the 44 phonemes, we can use the formula

$$\binom{n}{3} = \frac{n!}{3!(n - 3)!}$$

where $n! = n(n - 1)(n - 2) \cdots (n - n + 1)$. (For example, $5! = 5 \cdot 4 \cdot 3 \cdot 2 \cdot 1$.)

Using $n = 44$, find the number $\binom{n}{3}$. (*Hint:* $\dfrac{n!}{n!} = 1$.)

2. We can use the equations

$$\frac{1}{d_o} + \frac{1}{d_i} = \frac{1}{f} \qquad f = \frac{r}{2} \qquad m = \frac{-d_i}{d_o}$$

where

d_o = the distance from a mirror to an object
d_i = the distance the image appears from the mirror
f = the focal length of the mirror
r = the radius of curvature of the mirror
m = the magnification of the image

to answer the following question.

A convex rearview car mirror has a radius of curvature of 50 centimeters. Determine the actual location of the image and its magnification for an object 1000 centimeters from the mirror.

Here $r = 50$ centimeters and $d_o = 1000$ centimeters. Find d_i and m.

Class Act

3. In an article in a scientific journal, we are told that

> "A 12-inch test span [of a particular type of material] can support 650 pounds."

a. Criticize this statement. What do we need to know? What might the article mean?

b. Suppose the article had said, "A piece of this material (of a standard thickness) 3 inches by 2 inches can support 2100 pounds." How many pounds would be supported by 1 square inch? one square centimeter?

7.3 Adding and Subtracting Rational Expressions

SECTION LEAD-IN

We do not need to understand the meaning or use of a formula to be able to investigate how that formula reacts when the value of a variable changes. The following formula is used by a certain calculator to find the degrees of freedom (df) for a two-sample t-test in statistics when the variances are "pooled." What does the formula simplify to if

$$Sx_1^2 = Sx_2^2 \quad \text{The variances are equal.}$$

and

$$n_1 = n_2 \quad \text{The sample sizes are equal.}$$

$$df = \frac{\left(\dfrac{Sx_1^2}{n_1} + \dfrac{Sx_2^2}{n_2}\right)^2}{\dfrac{1}{n_1 - 1}\left(\dfrac{Sx_1^2}{n_1}\right)^2 + \dfrac{1}{n_2 - 1}\left(\dfrac{Sx_2^2}{n_2}\right)^2}$$

SECTION GOALS

- To add rational expressions
- To subtract rational expressions
- To rewrite rational expressions as equivalent expressions with their least common denominator
- To simplify complex fractions using the LCD
- To solve rational equations involving addition or subtraction

Introduction

We add and subtract rational expressions very much as we add fractions.

Adding Rational Expressions

If p, q, and r are polynomials, and q is not equal to zero, then

$$\frac{p}{q} + \frac{r}{q} = \frac{p + r}{q}$$

To add two rational expressions with the same denominator, we add the numerators and place the sum over the denominator. Thus, for example,

$$\frac{1}{1 + x} + \frac{x^2}{1 + x} = \frac{1 + x^2}{1 + x}$$

Subtracting Rational Expressions

If p, q, and r are polynomials, and q is not equal to zero, then

$$\frac{p}{q} - \frac{r}{q} = \frac{p - r}{q}$$

To subtract two rational expressions with the same denominator, we subtract the numerators and place the difference over the denominator. For example,

$$\frac{x^2}{x + 1} - \frac{2}{x + 1} = \frac{x^2 - 2}{x + 1}$$

Equivalent Rational Expressions

If the denominators of two rational expressions are different, before we can add or subtract the expressions we must rewrite them as equivalent expressions using their *least common demoninator* (LCD) as the denominator.

Two rational expressions are equivalent if they have the same value. According to the fundamental property of fractions

$$\frac{p}{q} = \frac{pr}{qr}$$

where q and r are polynomials not equal to zero.

In other words, we can multiply (or divide) the numerator and denominator of a rational expression by the same factor without changing its value. We do this to rewrite rational expressions as equivalent expressions with their least common denominator.

WRITER'S BLOCK

How are the LCD and LCM alike?

> The **least common denominator** (LCD) of two (or more) rational expressions is the least common multiple of the two (or more) denominators.

▪▪▪

EXAMPLE 1

Rewrite the following rational expressions as equivalent rational expressions with their LCD. Leave your result in factored form.

$$\frac{a}{3(a^2 + 2ab + b^2)} \quad \text{and} \quad \frac{b}{a^2 - b^2}$$

SOLUTION

We need to find the LCD, so we factor the denominators.

$$3(a^2 + 2ab + b^2) = 3(a + b)^2 \quad \text{Square of a binomial}$$
$$a^2 - b^2 = (a + b)(a - b) \quad \text{Difference of two squares}$$

Writing the factors with their highest exponents gives us $3(a + b)^2(a - b)$ as the LCD.

Now we must rewrite each of the rational expressions with the LCD as the denominator. We work with factored forms so that we can more easily decide what factor of the LCD is missing from each original denominator.

For $\frac{a}{3(a + b)^2}$, the missing factor is $(a - b)$. So we multiply numerator and denominator by $(a - b)$.

STUDY HINT

If a polynomial factors into identical factors, such as $(x - 3)(x - 3)$, you must use the factor twice in the denominator.

$$\frac{a}{3(a + b)^2} = \frac{a(a - b)}{3(a + b)^2(a - b)}$$

For $\frac{b}{(a + b)(a - b)}$, the missing factor is $3(a + b)$.

$$\frac{b}{(a + b)(a - b)} = \frac{3b(a + b)}{3(a + b)(a - b)(a + b)} = \frac{3b(a + b)}{3(a + b)^2(a - b)}$$

▶ *CHECK* **Warm-Up 1**

Adding and Subtracting Expressions with Different Denominators

Now we are ready to add or subtract any two rational expressions.

▪▪▪

EXAMPLE 2

Add: $\dfrac{3}{(n-4)^2} + \dfrac{5n}{2n-8}$

SOLUTION

To add, we need common denominators. The first denominator is already in factored form; factoring the second denominator gives us $2(n-4)$. So the LCD is $2(n-4)^2$.

Writing equivalent expressions with the LCD as the denominator, and then adding the numerators, we obtain

$$\dfrac{3(2)}{(n-4)^2(2)} + \dfrac{5n(n-4)}{2(n-4)(n-4)} \quad \text{Writing equivalent expressions}$$

$$= \dfrac{6 + 5n(n-4)}{2(n-4)^2} \quad \text{Adding the numerators}$$

$$= \dfrac{6 + 5n^2 - 20n}{2(n-4)^2} \quad \text{Distributing}$$

$$= \dfrac{5n^2 - 20n + 6}{2(n-4)^2} \quad \text{Rewriting the numerator}$$

This cannot be factored further, so

$$\dfrac{3}{(n-4)^2} + \dfrac{5n}{2n-8} = \dfrac{5n^2 - 20n + 6}{2(n-4)^2}$$

▶ CHECK **Warm-Up 2**

▪▪▪

EXAMPLE 3

Simplify: $\dfrac{x}{x-2} + \dfrac{1}{x-5} - \dfrac{3}{x^2-7x+10}$

SOLUTION

Factoring the denominators so that we can see if there are any common factors, we obtain

$$\dfrac{x}{x-2} + \dfrac{1}{x-5} - \dfrac{3}{(x-5)(x-2)}$$

The LCD is then $(x-2)(x-5)$.

Making equivalent expressions with this LCD gives us

$$\dfrac{x(x-5)}{(x-2)(x-5)} + \dfrac{1(x-2)}{(x-5)(x-2)} - \dfrac{3}{(x-5)(x-2)}$$

Combining the numerators yields

$$\dfrac{x(x-5) + 1(x-2) - 3}{(x-5)(x-2)} = \dfrac{x^2 - 5x + x - 2 - 3}{(x-5)(x-2)} = \dfrac{x^2 - 4x - 5}{(x-5)(x-2)}$$

STUDY HINT

Using the least common denominator insures that the resulting computation will be the least complicated.

❗❗❗

ERROR ALERT

Identify the error and give a correct answer.

Incorrect Solution:

$\dfrac{4x-5}{x+2} - \dfrac{x+1}{x+2}$

$= \dfrac{4x - 5 - x + 1}{x+2}$

$= \dfrac{3x - 4}{x+2}$

However, we are not finished. The numerator of this result can be factored.

$$\frac{x^2 - 4x - 5}{(x - 5)(x - 2)} = \frac{(x + 1)\overset{1}{\cancel{(x - 5)}}}{\underset{1}{\cancel{(x - 5)}}(x - 2)}$$

So our final result is $\frac{x + 1}{x - 2}$.

▶ CHECK Warm-Up 3

Using the LCD to Simplify Complex Fractions

In Section 7.2 you learned one way to simplify a complex fraction. Now you will learn another method, in which you multiply each term in the numerator and denominator by the LCD of the fractions in both numerator and denominator. This can simplify the fraction immensely. It is equivalent to multiplying by $\frac{\text{LCD}}{\text{LCD}}$, or 1.

■ ■ ■

EXAMPLE 4

Simplify: $\dfrac{\dfrac{6}{x + 3} + \dfrac{1}{6x}}{x + \dfrac{x}{2}}$

SOLUTION

The denominators are $x + 3$, $6x = 2 \cdot 3 \cdot x$, and 2. Their LCD is $6x(x + 3)$.

Multiplying each term of the numerator and denominator by the LCD gives us

> **STUDY HINT**
>
> *Multiply every term in the numerator and denominator by the LCM.*

$$\frac{(6x)(x + 3)\dfrac{6}{x + 3} + \dfrac{1}{6x}(6x)(x + 3)}{(6x)(x + 3)\dfrac{x}{1} + \dfrac{x}{2}(6x)(x + 3)} \qquad \text{Multiplying each term by } 6x(x + 3)$$

$$= \frac{(6x)\cancel{(x + 3)}\dfrac{6}{\cancel{x + 3}} + \dfrac{1}{\cancel{6x}}\cancel{(6x)}(x + 3)}{(6x)(x + 3)\dfrac{x}{1} + \dfrac{x}{\cancel{2}}\overset{3x}{\cancel{(6x)}}(x + 3)} \qquad \text{Cancelling}$$

$$= \frac{6x(6) + (x + 3)}{6x(x + 3)(x) + x(3x)(x + 3)} \qquad \text{Simplifying}$$

$$= \frac{36x + x + 3}{6x^3 + 18x^2 + 3x^3 + 9x^2} \qquad \text{Multiplying}$$

$$= \frac{37x + 3}{9x^3 + 27x^2} \qquad \text{Combining like terms}$$

▶ CHECK Warm-Up 4

Solving Rational Equations

In Section 7.2, we used cross-multiplication to solve simple equations involving rational expressions.

A procedure exists that can be used to solve *all* equations in one variable containing rational expressions.

> To solve an equation that includes rational expressions.
>
> **1.** Multiply each term of the equation by the least common multiple of all the denominators (this is the LCD).
> **2.** Simplify each side and solve for the variable as usual.
> **3.** Check to be sure that the solution is meaningful.

▪▪▪

EXAMPLE 5

Solve for x: $\dfrac{4}{x-2} + 8 = \dfrac{42}{5}$

SOLUTION

The denominators here, from left to right, are $x - 2$, 1, and 5. Their LCM is $5(x - 2)$.

Multiplying by the LCM will clear the equation of fractions.

$$5(x-2)\left(\frac{4}{x-2}\right) + 5(x-2)8 = 5(x-2)\frac{42}{5} \quad \text{Multiplying by } 5(x-2)$$

$$\frac{\overset{1}{\cancel{5(x-2)}}(4)}{\underset{1}{\cancel{x-2}}} + 5(x-2)8 = \frac{\overset{1}{\cancel{(5)}}(x-2)(42)}{\underset{1}{\cancel{5}}} \quad \text{Cancelling}$$

$$5(4) + (5)(8)(x-2) = 42(x-2) \quad \text{Rewriting}$$

Simplifying the equation and solving, we get

$$20 + 40x - 80 = 42x - 84 \qquad \text{Multiplying}$$
$$40x - 60 = 42x - 84 \qquad \text{Combining like terms}$$
$$40x + 24 = 42x \qquad \text{Adding 84 to both sides}$$
$$24 = 2x \qquad \text{Adding } -40x \text{ to both sides}$$
$$12 = x \qquad \text{Multiplying by } \tfrac{1}{2}$$

The only denominator with a variable is $x - 2$. Because substituting $x = 12$ in the denominator does not make it zero, $x = 12$ is a meaningful solution.

So $x = 12$ in $\dfrac{4}{x-2} + 8 = \dfrac{42}{5}$.

▶ CHECK **Warm-Up 5**

ERROR ALERT

Identify the error and give a correct answer.

Solve: $\frac{1}{x} + \frac{3x}{2} = -3.5$

Incorrect Solution:

$$2\left(\frac{1}{x} + \frac{3x}{2}\right) = -3.5(2)$$
$$2 + 3x = -7$$
$$3x = -9$$
$$x = -3$$

ERROR ALERT

Identify the error and give a correct answer.

Solve:
$$\frac{x+3}{2} - 4 = \frac{-3}{x}$$

Incorrect Solution:

$$2x\left(\frac{x+3}{2}\right) - 4 = \frac{-3}{x}(2x)$$
$$x(x+3) - 4 = -6$$
$$x^2 + 3x - 4 = -6$$
$$x^2 + 3x + 2 = 0$$
$$(x+1)(x+2) = 0$$
$$x = -1 \text{ or } x = -2$$

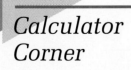

Calculator Corner

Note that when we graph Example 5, the equation for Y1 *does not* result in a line. Rather, what you see is a graph that appears to be in *two pieces!* At what value of x does the graph for Y1 seem to *break apart?* (Another way of saying this would be: For what value of x is there no value of y?) What is the denominator in the first term of Y1? What value of x in the denominator would result in dividing by zero?

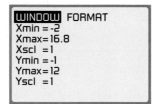
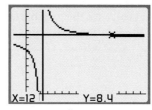

If your graphing calculator has a **CALC**ulate **INTERSECTION** utility, you could also use it to find the point of intersection. (Note: See the Calculator Corner on page 125 to review how to find the points of intersection of two graphs.)

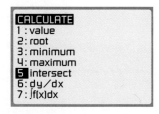
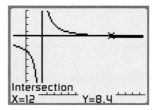

It might be interesting to get a *closer view* of the point of intersection for these two equations. Change the **WINDOW** dimensions in order to get a better view.

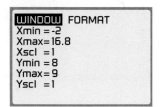
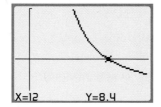

Practice what you learned.

SECTION FOLLOW-UP

The formula

$$df = \frac{\left(\dfrac{Sx_1^2}{n_1} + \dfrac{Sx_2^2}{n_2}\right)^2}{\dfrac{1}{n_1 - 1}\left(\dfrac{Sx_1^2}{n_1}\right)^2 + \dfrac{1}{n_2 - 1}\left(\dfrac{Sx_2^2}{n_2}\right)^2}$$

simplifies to

$$df = 2n - 2$$

when $Sx_1^2 = Sx_2^2$ and $n_1 = n_2$. Verify this.

7.3 WARM-UPS

Work these problems before you attempt the exercises.

1. Rewrite $\dfrac{7}{9x - 27y}$ and $\dfrac{1}{6x - 18y}$ as equivalent rational expressions with their least common denominator.

2. Add: $\dfrac{y}{y - 3} + \dfrac{y - 8}{y}$

3. Add: $\dfrac{x - 7}{x^2 - x - 2} + \dfrac{x + 1}{x - 2}$

4. Simplify: $\dfrac{\dfrac{1}{y - 5} - \dfrac{y}{6}}{\dfrac{y}{3} + \dfrac{2}{y - 5}}$

5. Solve for t: $\dfrac{6}{t + 3} - 9 = \dfrac{27}{8}$

7.3 EXERCISES

Note: Use your graphing calculator to check your results whenever possible.

In Exercises 1 through 8, rewrite the given rational expressions as equivalent rational expressions with their least common denominator.

1. $\dfrac{1}{xy}$ and $\dfrac{3}{5y^2}$

2. $\dfrac{11}{12x + 12y}$ and $\dfrac{5}{7x + 7y}$

3. $\dfrac{x + 2}{x^2 + 3x + 2}$ and $\dfrac{x - 1}{x + 1}$

4. $\dfrac{7x}{12y}$ and $\dfrac{5y}{6x}$

5. $\dfrac{x}{8y}$ and $\dfrac{3y}{4x^2 y^2}$

6. $\dfrac{4x - 8y}{15xy^2}$ and $\dfrac{7}{10y^3}$

7. $\dfrac{4}{75x}$, $\dfrac{7}{10y}$, and $\dfrac{17}{20w}$

8. $\dfrac{7}{80x^2 y}$, $\dfrac{9}{16xy^2}$, and $\dfrac{4}{5wxy}$

For Exercises 9 through 28, add or subtract as indicated.

9. $\dfrac{y}{15} + \dfrac{y}{9}$

10. $\dfrac{r}{11} - \dfrac{r}{9}$

11. $\dfrac{y}{9} + \dfrac{y}{5}$

12. $\dfrac{r}{7} - \dfrac{r}{6}$

13. $\dfrac{5}{3y} + \dfrac{2y}{4y}$

14. $\dfrac{8}{4r} + \dfrac{6r}{8r}$

15. $\dfrac{3x}{9x} - \dfrac{3}{3x}$

16. $\dfrac{11t}{6t} - \dfrac{7}{9t}$

17. $\dfrac{3y - 4}{9y} - \dfrac{6}{4y}$

18. $\dfrac{9 - 2r}{7r} - \dfrac{5}{2r}$

19. $\dfrac{6}{5x} + \dfrac{5 - 4x}{3x}$

20. $\dfrac{2 - 5t}{4t} + \dfrac{2t - 5}{3}$

21. $\dfrac{x + 3}{12x + 9} + \dfrac{x - 4}{8x + 6}$

22. $\dfrac{3t}{6t + 12} + \dfrac{2t - 5}{t + 2}$

23. $\dfrac{x}{x - 3} + \dfrac{1}{2 - x} - \dfrac{x^2}{x^3 - 5x^2 + 6x}$

24. $\dfrac{12}{x^2 - 9} + \dfrac{2}{x + 3} + \dfrac{4}{x - 3}$

25. $\dfrac{-6x}{x^2 - 4} + \dfrac{3}{x - 2} + \dfrac{9}{x + 2}$

26. $\dfrac{y - 7}{y^2 - 14y + 49} + \dfrac{4}{y^2 - 18y + 77}$

27. $\dfrac{x}{x - 2} + \dfrac{6}{x^2 - 7x + 10} - \dfrac{x - 1}{x - 5}$

28. $\dfrac{4x - 4}{x - 2} - \dfrac{1}{x + 1} - \dfrac{3x + 6}{(x - 2)(x + 1)}$

For Exercises 29 through 42, simplify.

29. $\dfrac{\frac{y}{6} + 3}{\frac{y}{9} - 4}$

30. $\dfrac{\frac{x}{4} - 5}{\frac{x}{8} + 6}$

31. $\dfrac{\frac{r}{3} + 9}{\frac{r}{7} + 8}$

32. $\dfrac{\frac{t}{7} - 7}{\frac{t}{6} + 5}$

33. $\dfrac{\frac{m}{9} + m}{\frac{m}{8} + m}$

34. $\dfrac{\frac{n}{7} + n}{\frac{n}{6} - n}$

35. $\dfrac{\frac{x}{3} + x}{\frac{3x}{2} - 5}$

36. $\dfrac{\frac{r}{6} - r}{\frac{2r}{7} + 3}$

37. $\dfrac{\frac{12 - x}{2x}}{\frac{4x}{x - 8}}$

38. $\dfrac{\frac{x - y}{x} + 5}{x - y}$

39. $\dfrac{\frac{4}{n} - \frac{3}{2n}}{\frac{5}{3n} + \frac{9}{4n}}$

40. $\dfrac{\frac{6}{5n} + \frac{4}{3n}}{\frac{7}{7n} - \frac{6}{2n}}$

41. $\dfrac{\frac{3x}{2} - \frac{1}{2x}}{3x^2 - 1}$

42. $\dfrac{\frac{r}{4(r - 2)} + \frac{3}{(r - 2)(r - 8)}}{\frac{r(r - 10) + 24}{2(r - 4)(r - 8)}}$

For Exercises 43 through 64, solve for the variable.

43. $\dfrac{r - 2}{5} + 6 = \dfrac{r + 4}{20}$

44. $\dfrac{t - 3}{5} = \dfrac{t}{7} - 15$

45. $\dfrac{x}{3} = \dfrac{x + 6}{5} + x$

46. $\dfrac{3}{4m} - \dfrac{3}{6m} = \dfrac{1}{2}$

47. $\dfrac{y - 5}{2} = \dfrac{3\frac{1}{2}}{y + 1}$

48. $\dfrac{23}{3x} + \dfrac{3}{x} = \dfrac{2}{3}$

49. $\dfrac{1}{y} + \dfrac{3}{4} = \dfrac{10}{4y}$

50. $\dfrac{x + 3}{4} + \dfrac{x + 4}{6} = \dfrac{7}{2}$

51. $\dfrac{1}{x^2 - 4} + \dfrac{1}{x + 2} = \dfrac{2}{x - 2}$

52. $\dfrac{10}{2t + 6} + \dfrac{2}{t + 3} = \dfrac{1}{2}$

53. $\dfrac{35}{t^2 + 2t + 1} - \dfrac{3}{t + 1} = \dfrac{4}{t + 1}$

54. $\dfrac{3}{2(t - 4)} - \dfrac{2}{t + 4} = \dfrac{35}{t^2 - 16}$

55. $\dfrac{3}{n - 5} = \dfrac{2}{n - 2} + \dfrac{7}{n^2 - 7n + 10}$

56. $1 - \dfrac{5}{t + 10} = \dfrac{3}{t - 6}$

57. $\dfrac{3}{x + 7} + 3 = \dfrac{4}{x + 5}$

58. $\dfrac{11}{v^2 - 9} - \dfrac{7}{2v + 6} = \dfrac{2}{v + 3}$

59. $\dfrac{1}{3r + 15} + \dfrac{5}{4r + 20} = \dfrac{19}{24}$

60. $\dfrac{4}{r^2 - 1} = \dfrac{1}{r - 1} + \dfrac{1}{r + 1}$

61. $\dfrac{16 - 3x}{x + 6} + \dfrac{x + 3}{5} = \dfrac{3x - 2}{15}$

62. $\dfrac{3}{n - 4} - \dfrac{4}{n^2 - 3n - 4} = \dfrac{1}{n + 1}$

63. $\dfrac{5}{x + 3} - \dfrac{7}{3(x + 3)} = 2$

64. $\dfrac{2}{x + 4} - \dfrac{3x}{2x + 8} = 6$

MIXED PRACTICE

By doing these exercises, you will practice the topics up to this point in the chapter.

65. Add: $\dfrac{3x - 4}{x - 2} + \dfrac{x - 7}{5x - 10} + \dfrac{9 + x}{x^2 - 4}$

66. Multiply: $\dfrac{x + 4}{x^2 - 5x - 36} \cdot \dfrac{x^2 + 9x - 52}{x + 13}$

67. Subtract: $\dfrac{3}{x + 1} - \dfrac{x - 4}{x^2 - x - 2}$

68. Divide: $\dfrac{x^2 + 17x + 66}{x + 2} \div \dfrac{x^2 + 13x + 22}{x^2 - 36}$

69. Reduce $\dfrac{x^2yt^3}{y^2x^3t^2}$ to lowest terms.

70. Simplify: $\dfrac{3 + \dfrac{4}{x-5}}{\dfrac{2}{x} + \dfrac{x}{7}}$

71. Determine when $\dfrac{3x^2 - 25x + 6}{4x^2 - 16x}$ has meaning.

72. Divide: $\dfrac{x^2 - 8x - 9}{x^2 - 7x - 18} \div \dfrac{x^2 - 8x + 7}{x^2 - 5x - 14}$

73. Divide: $\dfrac{r}{r^2 - 4r + 4} \div \dfrac{2}{r^2 - 2r}$

74. Reduce $\dfrac{26x^2y^4}{13x^4y^5}$ to lowest terms.

75. Add: $\dfrac{x-3}{2x} + \dfrac{x+2}{5x}$

76. Determine when $\dfrac{y-6}{y+2}$ is meaningful.

77. What is the least common multiple of $r + 4$ and $r^2 + 2r - 8$?

78. Subtract: $\dfrac{4}{x+2} - \dfrac{3-x}{x^2-4}$

79. Reduce $\dfrac{x+4}{x^2 + 8x + 16}$ to lowest terms.

80. Determine when $\dfrac{8-y}{y-9}$ has meaning.

EXCURSIONS

Exploring Problem Solving

1. a. In 1996 archaeologists from the University of Cape Town in South Africa discovered cave paintings that were 500 (plus or minus 150) years old. What is the *percent of error* here? That is, what percent of the number 500 is 150?

   b. If the same percent of error persisted, by how many years might paintings 19,000 years ago be misdated?

   c. A researcher claims that a cave painting dated 19,000 years old might really have been painted 26,000 years ago. Is this possible? By what percent of error might this be true?

2. We can use the lens equation

$$\frac{1}{d_o} + \frac{1}{d_i} = \frac{1}{f}$$

where

   d_o = distance from the object
   d_i = distance that the image appears
   f = focal length of the lens

to answer the following question.

Where must a small cricket be placed if a 25-centimeter focal length diverging lens is to form a virtual image 20 centimeters in front of the lens?

Use $f = -25$ centimeters and $d_i = -20$ centimeters. Find d_o. (Note: We know that lengths and distances cannot be negative. In optics, the signs indicate where the image and object are in relation to each other and to the lens.)

3. We know there are 180° in a triangle. You can convince yourself of this using the following procedure:

- Cut out a triangle.
- Number the angles.
- Tear the angles off and arrange them so the vertices touch.

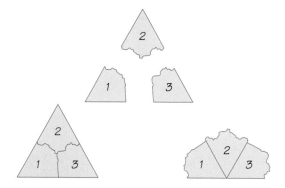

Because the three angles form a straight line (we know a straight angle is 180°), the triangle's angles total 180°.

Use this fact (the angles of a triangle total 180°) to find the number of degrees in each of the following figures.

a. square   b. rectangle   c. trapezoid   d. pentagon   e. hexagon

Data Analysis

4. Study the following table, then answer the questions. (*Note:* A calculator may be useful.)

WORLD WEATHER IN JULY			
City	Average High/Low (°F)	City	Average High/Low (°F)
Beijing	88/70	Madrid	87/63
Boston	80/63	Miami	88/76
Budapest	82/62	Moscow	73/55
Chicago	81/66	New York	82/66
Delhi	96/81	Paris	76/58
Dublin	67/52	Phoenix	104/77
Frankfurt	77/58	Rio de Janeiro	75/63
Geneva	77/58	Rome	87/67
Hong Kong	87/78	San Francisco	65/53
Houston	92/74	Stockholm	71/57
Jerusalem	87/63	Sydney	60/46
Johannesburg	63/39	Tokyo	83/70
London	71/56	Toronto	79/59
Los Angeles	81/60	Washington	87/68

Source: The Times Books World Weather Guide.

a. Which city has the greatest high to low temperature ratio?

b. Which city has the greatest temperature range?

c. What is the median low temperature?

d. Find the mean temperature (for high and for low temperatures) for the nine U.S. cities listed in the table.

Finding Area

Area of a Triangle

To find the area of a triangle, we must first identify the *base* and the *height* of the triangle. Then:

> **To find the area of a triangle**
>
> If b is the length of the base and h is the height, then
> $$\text{Area} = \tfrac{1}{2} \times b \times h, \quad \text{or } A = \tfrac{1}{2}bh$$

Any side can be chosen as the base of a triangle. The height is then the shortest distance from the angle opposite that base. If the triangle is a right triangle, then the height can be the length of any side except the hypotenuse. Usually, though, the height is not the same as the length of a side. But the height always makes a right (90°) angle with the base.

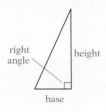

Right triangle

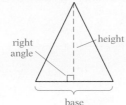

Acute triangle

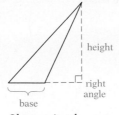

Obtuse triangle

Let's look at an example.

▪ ▪ ▪

EXAMPLE

Find the area of the triangle shown in the accompanying figure. The unit of length is feet.

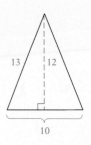

SOLUTION

We choose the 10-foot side as the base; then the height is equal to 12 feet, not 13 feet. Using the formula, we substitute the known values for b and h.

$$A = \tfrac{1}{2}bh$$
$$= \tfrac{1}{2} \times 10 \times 12$$
$$= \frac{120}{2}$$
$$= 60$$

> **WRITER'S BLOCK**
>
> Explain why 13 feet cannot be the base.

So the area of the triangle is 60 square feet. (Remember, area is expressed in *square* units.)

Calculator Corner

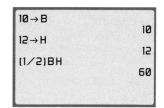

Use your graphing calculator's Home Screen to help you do the work for the geometry example. (Note: See the Calculator Corner on page 85 to review how to **STO**re values for variables.)

Area of a Trapezoid

In this figure, trapezoid *ABCD* has been divided into two triangles.

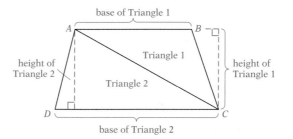

To find the area of the trapezoid, we can add the areas of the triangles. That is,

Area of trapezoid = area ① + area ②

$$= \tfrac{1}{2} \text{ base } ① \times \text{ height } ① + \tfrac{1}{2} \text{ base } ② \times \text{ height } ②$$

The heights of the triangles are equal, so we can write this as

Area of trapezoid = $\tfrac{1}{2}$ height (base ① + base ②).

In symbolic form:

> **To find the area of a trapezoid**
> If h is the height, and b_1 and b_2 are the lengths of the bases, then
> $$A = \tfrac{1}{2}h(b_1 + b_2)$$

▪ ▪ ▪
EXAMPLE

Find the area of a trapezoid with bases of 12 and 36 units and height 10 units.

SOLUTION

We substitute the values for b_1, b_2, and h into the formula

$$A = \frac{1}{2}h(b_1 + b_2)$$
$$= \frac{1}{2}(10)(12 + 36)$$
$$= \frac{1}{2}(10)(48)$$
$$= 5(48)$$
$$= 240$$

So the area is 240 square units.

◢

Calculator Corner

Use your graphing calculator's Home Screen to help you find the area of the trapezoid in the example. (Note: See the Calculator Corner on page 85 to review how to **STO**re values for variables.)

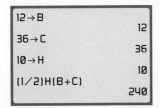

```
12→B
            12
36→C
            36
10→H
            10
(1/2)H(B+C)
           240
```

▪ ▪ ▪
EXAMPLE

The perimeter of a triangle is 19 centimeters. The length of the second side of the triangle is twice that of the first, and the length of the third side is three centimeters greater than that of the first. Find the lengths of all three sides of the triangle.

SOLUTION

We let a be the length of the first side. Then the second side has length $2a$, and the third side has length $3 + a$. We show this in the accompanying figure, where the sides are called a, b, and c, respectively.

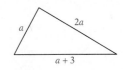

We substitute the known values in the formula for the perimeter of a triangle and solve.

$$P = a + b + c \qquad \text{Formula}$$
$$19 = a + (2a) + (3 + a) \quad \text{Substituting values}$$
$$19 = a + 2a + 3 + a \qquad \text{Removing parentheses}$$
$$19 = 4a + 3 \qquad \text{Combining like terms}$$
$$19 + (-3) = 4a + 3 + (-3) \qquad \text{Addition principle}$$
$$16 = 4a \qquad \text{Simplifying}$$
$$\left(\tfrac{1}{4}\right)16 = \left(\tfrac{1}{4}\right)(4a) \qquad \text{Multiplication principle}$$
$$4 = a \qquad \text{Simplifying}$$

Going back to the original problem and substituting $a = 4$ gives us the three sides:

$$a = 4$$
$$b = 2a = 8$$
$$c = 3 + a = 7$$

So the sides of the triangle are 4, 7, and 8 centimeters. The check is left for you to do.

◢

PRACTICE

1. ✏ Two triangles have equal bases and equal heights, but one is a right triangle and one is a scalene triangle. Which has the larger area? Justify your answer.

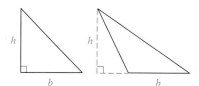

2. **a.** Which has the greater area, a triangular sail whose base is $3\frac{1}{5}$ yards and whose height is $7\frac{3}{5}$ yards, or a $5\frac{1}{2}$-yard square sail?

    b. ✏ Explain the difference between a four-foot square sail and a 4 square-foot sail.

3. Find the area of the accompanying triangle.

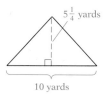

5¼ yards

10 yards

4. The length of a rectangular brass door knocker is one-half of the width. If the perimeter is 14.4 inches, find the dimensions of the door knocker.

5. The width of a rectangular nameplate is 5 centimeters more than 4 times the length. If the perimeter is 40 centimeters, find the dimensions of the nameplate.

6. The perimeter of a small triangular brace is $13\frac{4}{5}$ inches. The second side is 1 inch longer than the first, and the third side is 1 inch longer than the second. Find the lengths of the sides.

7. The area of a sign shaped like a parallelogram is 34.6 square inches. If the length of the base is 5 inches, find the height.

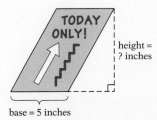

TODAY ONLY!

height = ? inches

base = 5 inches

Area = 34.6 square inches

8. A base for a sculpture is a trapezoid and has a perimeter of 100 inches. One base is 34 inches long. The other base and the two sides are equal in length. What is this length?

9. Find the perimeter of a basketball court that has a length of 28 yards and an area of 3843 square feet. Give your answer in yards.

10. Find the area of a trapezoid with bases of $12\frac{1}{2}$ and $15\frac{1}{2}$ inches, sides of 8 inches and height of 10 inches.

7.4 Applying Rational Expressions to Word Problems

SECTION LEAD-IN

The efficiency E of a jack is determined from the pitch of the jack's thread with the formula

$$E = \frac{\frac{p}{2}}{p + \frac{1}{2}}$$

The pitch is the distance between the threads.

1. When $p = 1$, what is E?

2. When the efficiency is 1, what is the (approximate) value of p?

3. Keeping p positive, as p gets smaller, what happens to the value of E?

In this section, we solve word problems that lead to equations involving rational expressions. You have already seen how to solve such equations.

▪▪▪
EXAMPLE 1

A plane travels 2500 miles at a certain speed in still air (no wind). On a second flight, the plane is slowed by a head wind of 30 miles per hour. In the same travel time, it now covers only 2000 miles. What is its speed in still air?

SOLUTION

INTERPRET This is a distance problem, and we can use the formula

$$d = rt \quad \text{distance} = (\text{rate})(\text{time})$$

PLAN Because the time for the two flights is identical, we want to use the formula in the form

$$t = \frac{d}{r}$$

APPLY

	First Flight	**Second Flight**
Let	x = plane's speed in still air	$x - 30$ = plane's speed in 30 mph head wind
	$d = 2500$	$d = 2000$
	$t = \frac{2500}{x}$	$t = \frac{2000}{x - 30}$

The time t is the same for both trips, so we can set the results equal to each other. We obtain the equation

$$\frac{2500}{x} = \frac{2000}{x - 30}$$

REASON This is the equation we must solve. We note that x cannot be 30 or 0, because either value would make the equation meaningless.

APPLY Cross-multiplying, we get

$$2500(x - 30) = 2000(x) \quad \text{Cross-multiplying}$$
$$2500x - 75000 = 2000x \quad \text{Multiplying}$$

Solving for x gives us

$$2500x - 75000 = 2000x$$
$$-75000 = -500x \quad \text{Adding } -2500x \text{ to both sides}$$
$$150 = x \quad \text{Multiplying by } -1/500$$

REASON So the speed of the plane in still air is 150 miles per hour.

▶ CHECK **Warm-Up 1**

In the next example, we introduce a new type of word problem. It is called a work problem, even though it may not concern work specifically. Such problems are solved by applying the **work principle.**

Work Principle

If it takes t minutes to complete 1 task, then it takes 1 minute to complete $\frac{1}{t}$ of the task.

■ ■ ■
EXAMPLE 2

The hot water faucet fills a tub in 20 minutes. The cold water faucet fills the tub in 15 minutes. How long will it take to fill the tub if both faucets are open?

SOLUTION

INTERPRET AND PLAN   We apply the work principle to each faucet.

If the hot water faucet fills the tub in 20 minutes ($t_h = 20$), then in 1 minute it fills $\frac{1}{20}$ of the tub $\left(\text{because } \frac{1}{t_h} \text{ is } \frac{1}{20}\right)$.

If the cold water faucet fills the tub in 15 minutes ($t_c = 15$), then in 1 minute, it fills $\frac{1}{15}$ of the tub $\left(\text{because } \frac{1}{t_c} \text{ is } \frac{1}{15}\right)$.

APPLY   Suppose that together the faucets can fill the tub in x minutes; then in 1 minute they can fill $\frac{1}{x}$ of the tub. Thus

$$\frac{1}{x} = \frac{1}{20} + \frac{1}{15}$$

This is the equation we have to solve.

The LCD of the fractions is $2^2 \cdot 3 \cdot 5 \cdot x = 60x$, so we have

$$(60x)\frac{1}{x} = (60x)\frac{1}{20} + (60x)\frac{1}{15}$$

We solve for x.

$$(60\cancel{x})\frac{1}{\cancel{x}} = (\cancel{60}x)\frac{1}{\cancel{20}} + (\cancel{60}x)\frac{1}{\cancel{15}}$$

$$60 = 3x + 4x \qquad \text{Simplifying}$$

$$60 = 7x$$

$$8\tfrac{4}{7} = x \qquad \text{Multiplying by } \tfrac{1}{7}$$

REASON   So the tub can be filled in $8\frac{4}{7}$ minutes. You should check this solution.

▶ *CHECK* **Warm-Up 2**

In the next example, our problem gives an equation that is a proportion.

• • •

EXAMPLE 3

A baker makes 30 raisin cookies for every 50 chocolate chip cookies. He wants to increase his production of these two types of cookies in equal amounts, so that the new ratio of raisin cookies to chocolate chip cookies is 3 to 4. How many of each cookie type must he make?

SOLUTION

INTERPRET   He makes 30 raisin cookies for every 50 chocolate chip cookies, so we can write his current production as a fraction.

$$\frac{\text{Raisin}}{\text{Chocolate chip}} = \frac{30}{50}$$

He wants to increase his production of the two types in equal amounts; call those equal amounts x. Then the new production fraction will be

$$\frac{\text{Raisin}}{\text{Chocolate chip}} = \frac{30 + x}{50 + x} \quad \text{New fraction}$$

PLAN The result of this action will be to give a ratio of raisin cookies to chocolate chip cookies of $\frac{3}{4}$. So we have

$$\frac{30 + x}{50 + x} = \frac{3}{4}$$

APPLY Using cross-multiplication and then solving, we obtain

$$4(30 + x) = 3(50 + x) \quad \text{Cross-multiplying}$$
$$120 + 4x = 150 + 3x \quad \text{Multiplying out}$$
$$4x = 30 + 3x \quad \text{Adding } -120 \text{ to both sides}$$
$$x = 30 \quad \text{Adding } -3x \text{ to both sides}$$

Check:

$$\frac{30 + 30}{50 + 30} = \frac{60}{80} = \frac{3}{4}$$

REASON So he must bake an additional 30 raisin cookies and an additional 30 chocolate chip cookies to change the ratio of raisin cookies to chocolate chip cookies to $\frac{3}{4}$.

▶ *CHECK* **Warm-Up 3**

Practice what you learned.

SECTION FOLLOW-UP

The questions asked in the Section Lead-In investigate the formula

$$E = \frac{\frac{p}{2}}{p + \frac{1}{2}}$$

1. When $p = 1$, $E = \frac{1}{3}$.

2. When the efficiency is 1, $p \approx -1$.

3. For $p > 0$, as p gets smaller, E gets smaller.

7.4 WARM-UPS

Work these problems before you attempt the exercises.

1. *Travel* A plane travels 1800 miles at a certain speed with no wind. On a second trip, the plane travels with a head wind of 40 miles per hour. In the same time, the plane now travels 1400 miles. What is the plane's speed with no wind?

2. *Construction* A construction worker does a job in 5 hours. A second worker can do the same job in 3 hours. How long will it take them to do the job if they work together?

3. *Art*   Zoren had 30 drawings of which 18 were in pen and ink. After he sold some of the pen-and-ink drawings, the resulting ratio of pen-and-ink drawings to total drawings was $\frac{1}{4}$. How many pen-and-ink drawings did he sell?

7.4 EXERCISES

Note: Use your graphing calculator to check your results whenever possible.

For Exercises 1 through 28, solve the word problem.

1. Yoshi washes his car in 45 minutes. Tad takes only 30 minutes. How fast could they wash the car if they did it together?

2. I am thinking of a number. If you add 5 to it and divide that sum by twice my number, you will get 8. Find my number.

3. A vanilla soda dispenser fills a mug in 20 seconds. The chocolate soda dispenser fills the mug in 30 seconds. How long will it take to fill the mug if both dispensers are turned on simultaneously?

4. *Construction*   Jameka and Amalia each do construction work. Jameka can put in a window in 3 hours, and Amalia can do the same job in 2 hours. How long will it take them to put in a window if they work together?

5. *Administrative Work*   If I type 45 words per minute and Kevin types 65 words per minute, how long will it take both of us working together (on two different keyboards) to type 1000 words?

6. One-third of a number added to $\frac{1}{6}$ of the same number is 18. Find the number.

7. One hose fills a bucket three times as fast as another. If together they fill the bucket in 24 minutes, how long does each hose take to fill the bucket?

8. *Travel*   An airplane that is flying 650 miles per hour in calm air can cover 2600 miles with the wind in the same time that it can cover 2000 miles against the wind. Find the approximate speed of the wind.

With wind: $650 + x$
Against wind: $650 - x$

9. The sum of a woman's age and that of her older sister is 49. The quotient of their ages is $\frac{3}{4}$. Find how old each woman is.

10. The reciprocal of the difference of a number and eight is two times the reciprocal of that number. Find the number.

11. **Construction** Pine weighs about $23\frac{1}{2}$ pounds per cubic foot, and mahogany weighs about 34 pounds per cubic foot. Which weighs more, a mahogany board that measures $4\frac{1}{2}$ inches by 1 inch by $7\frac{1}{3}$ feet or a pine board that measures 7 feet by $1\frac{1}{2}$ inches by $4\frac{2}{3}$ inches?

12. The denominator of a fraction is four more than the numerator. The denominator is increased by six and the numerator is decreased by eight. This results in a fraction equivalent to $\frac{1}{7}$. Find the original fraction.

13. A swimming pool can be emptied in 6 hours using hose A. If a second hose (B) is added, the pool can be emptied in 4 hours. How many hours will it take hose B, working alone, to empty the pool?

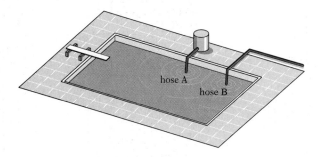

14. The numerator of a fraction is twice the denominator of that fraction. When the numerator is increased by five and the denominator is tripled, the resulting fraction is $\frac{13}{12}$. Find the original fraction.

15. **Geometry** The perimeter of a rectangle is equivalent to twice the sum of its length and width. If the length is three times the width and the perimeter is $4\frac{2}{5}$ feet, find the length and the width of the rectangle.

16. **Travel** The distance between New York and San Francisco is approximately 2600 miles.

    a. If the scale on the map is 0.2 centimeters per 50 miles, how many centimeters is it from New York to San Francisco on the map?

    b. If the map scale has an error of 10%, what are the least and greatest distances that could be represented by 20 centimeters?

17. **Roll Call** In Patrick's class of 80 students, 40 were girls. Some of the girls were transferred, and the new ratio of girls to total students in the class was $\frac{1}{5}$. How many girls were transferred?

18. In a miniature candy box, there were 18 caramels and 12 cremes. Equal amounts of each were eaten. How many were eaten if the new ratio is $\frac{3}{1}$?

19. *Restaurant Work* In an ice cream parlor, there are 42 fruit flavors out of 60 possible flavors. If fruit flavors are added until the ratio of fruit flavors to total flavors is $\frac{9}{10}$, how many additional flavors must be added?

20. *Rate of Current* Justin swims at a rate of 2 miles per hour in still water. After he swam downstream for $\frac{1}{2}$ mile, returning took him 4 times as long. Find the rate of the current.

21. *Road Race* In a 3200-meter race, Ashley takes 6 minutes longer than her sister Shanelle, who runs 6 kilometers per hour. How long did it take Ashley to finish the race? Find Ashley's speed in meters per hour.

22. *Mathematics Education* In an elementary math materials box, there were 40 attribute blocks. Half of the blocks were triangular. After some were given away, the ratio of triangular blocks to total blocks was $\frac{1}{5}$. How many were given away?

23. *Travel* Pokie's boat goes 60 miles downstream, with a current of 4 miles per hour, in the same time it takes to go 45 miles upstream. How fast does the boat travel in still water?

24. The denominator of a fraction is twice the numerator, plus 2. When the numerator is decreased by 9 and the denominator is increased by 8, the result is $\frac{6}{40}$. Find the original fraction.

25. *Shoveling Snow* Mr. Primosch shovels snow twice as fast as his daughter. If they both shovel, they can clear their property in $2\frac{1}{2}$ hours. How long would it take his daughter to do the shoveling alone?

26. One number is four more than the second. The reciprocal of the first number is two-thirds the reciprocal of the second. What are the two numbers?

27. A post has a 30-inch shadow at the same time a man $1\frac{1}{2}$ feet taller than the post has a shadow $3\frac{1}{3}$ feet tall. How tall are the post and the man?

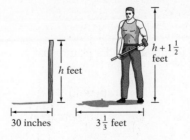

h feet $h + 1\frac{1}{2}$ feet

30 inches $3\frac{1}{3}$ feet

28. *Plumbing* A plumber takes 3 hours to do a job by herself. When a second person helps, the job is done in $1\frac{1}{2}$ hours. How many hours would the second person have taken if she had worked by herself?

MIXED PRACTICE

By doing these exercises, you will practice the topics up to this point in the chapter.

29. *Cooking* In a certain recipe, the ratio of the number of cups of pumpkin to the number of cups of sugar is 4 to $1\frac{1}{2}$. The same amount of pumpkin and of sugar is added until the ratio becomes 5 to 2. How much of each is added?

30. Divide: $\dfrac{2x}{x+2} \div \dfrac{x}{2x^2-8}$

31. Solve: $\dfrac{9}{t^2+t-2} = \dfrac{1}{t-1} + \dfrac{2}{t+2}$

32. Add: $\dfrac{5x}{x^2-64} + \dfrac{3}{x-8} + \dfrac{9}{x+8}$

33. Solve: $\dfrac{1}{n} + \dfrac{4}{n+6} = \dfrac{2}{n}$

34. Solve: $\dfrac{24-2x}{4x} = \dfrac{x-12}{8}$

35. Subtract: $\dfrac{5}{6} - \dfrac{6y+4}{15y-10}$

36. Subtract: $\dfrac{8}{x-5} - \dfrac{6}{x+5}$

37. The numerator of a fraction is 5 more than the denominator. When 10 is added to each, the result is equivalent to $1\frac{1}{5}$. Find the original fraction.

38. Simplify: $\dfrac{\dfrac{2t}{7} - 14}{\dfrac{2t}{6} + 10}$

39. Multiply: $\dfrac{2}{4x+2} \cdot \dfrac{8x^2+8x+2}{2x}$

EXCURSIONS

Exploring Geometry

1. Figures that tessellate form a pattern with no holes about (around) a point. One example of a tessellating shape is a semicircle (half circle).

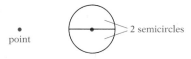

Tell if each of the following figures could tessellate around a point. Justify your answer geometrically.

a. right triangle   **b.** square   **c.** trapezoid

d. regular hexagon   **e.** equilateral triangle   **f.** regular pentagon

2. How might one estimate the volume of a stream? What geometric figure might model a cross section of a stream? Justify your choice and explain your reasoning.

3. We can use our knowledge of triangles and a method called *triangulation* to find the area of the following irregular figure.

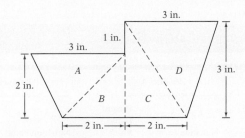

$\triangle A = \frac{1}{2}bh = \frac{1}{2}(3)(2) = 3$

$\triangle B = \frac{1}{2}bh = \frac{1}{2}(2)(2) = 2$

$\triangle C = \frac{1}{2}bh = \frac{1}{2}(2)(3) = 3$

$\triangle D = \frac{1}{2}bh = \frac{1}{2}(3)(3) = 4\frac{1}{2}$

$12\frac{1}{2}$ square inches

a. Justify the measurements used for $\triangle C$ and $\triangle D$.

b. Draw some irregular figures using a ruler. Use your method to find the area.

4. Regular figures are figures with the lengths of all sides equal and all angles equal. Find the number of degrees in each angle of the following figures. To find the answers, you may have to first determine the total number of degrees in the figure.

a. regular triangle (What is this called?)

b. regular hexagon

c. regular pentagon

d. regular dodecahedron (Research to find the number of sides in this shape.)

Exploring Numbers

5. Barbells are used to support weight discs. Discs are marked with their weights. They are loaded on the barbell with the largest disc inside. They are locked onto the bar with a collar. Using two of each size weight (one on each end of the bar so that the sides are equivalent in weight), how many different weights can be formed? Assume that the bar weighs 10 pounds.

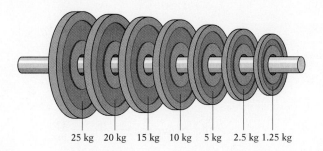

25 kg   20 kg   15 kg   10 kg   5 kg   2.5 kg 1.25 kg

CHAPTER LOOK-BACK

	Product A		Product B	
	Nutrition Facts	Percent Daily Value	Nutrition Facts	Percent Daily Value
Serving size	1/4 cup (28g)		1/2 cup (56g)	
Servings per container	about 4		about 6	
Amount per serving				
Calories	80		95	
Calories from fat	45		54	
Total fat	5g	8%	6g	9.6%
Saturated fat	3.5g	18%	3.5g	18%
Cholesterol	15mg	5%	24mg	8%

1. Which food item (A or B) has the greatest fat ratio per package? per cup?

2. Give an example of a fat ratio less than 10%.

CHAPTER 7
REVIEW PROBLEMS

The following exercises will give you a good review of the material presented in this chapter.

SECTION 7.1

1. Reduce to lowest terms: $\dfrac{(3x^2)(2x)}{4x^5}$

2. Determine when $\dfrac{3x - 4}{2x + 3}$ is not meaningful.

3. Reduce $\dfrac{x^2 + x}{x + 1}$ to lowest terms.

4. Evaluate $\dfrac{(x - y)^3}{x - y}$ when $x = 4$ and $y = -6$.

5. Reduce: $\dfrac{y^2 - 11y + 28}{2y - 14}$

6. Determine when $\dfrac{2x - 3}{x^2}$ is not meaningful.

SECTION 7.2

7. Divide: $\dfrac{3x - 1}{x^2 - 9} \div \dfrac{6x^2 + 7x - 3}{2x + 6}$

8. Multiply: $\dfrac{y^2 - 6y}{4y + 12} \cdot \dfrac{y^2 - 9}{2y - 12}$

9. Divide: $\dfrac{y - 5}{3y^2 - 13y - 10} \div \dfrac{y^2 - 10y + 25}{3y + 2}$

10. Multiply: $\dfrac{x + 2}{2x + 8} \cdot \dfrac{4x + 16}{x^2 + 3x + 2}$

11. Divide: $\dfrac{x^2 + 4x + 3}{3x - 9} \div \dfrac{x + 1}{x^2 - 9}$

12. Multiply: $\dfrac{x + 3}{2x + 6} \cdot \dfrac{x^2 - 9}{x - 3} \cdot \dfrac{4x - 12}{x^2 - 2x - 15}$

13. Solve for x: $\dfrac{14}{5} = \dfrac{4}{x - 6}$

14. Solve for x: $\dfrac{18}{1 - 3x} = \dfrac{12}{5x}$

15. Solve for t: $\dfrac{6}{t} = \dfrac{9}{t-4}$

16. Solve for x: $\dfrac{x-2}{x+1} = \dfrac{2x-8}{x-1}$

SECTION 7.3

17. Rewrite these fractions with their least common denominator.

$$\dfrac{11}{15} \text{ and } \dfrac{2}{3x^2y} \text{ and } \dfrac{9}{100xy^3}$$

18. What is the least common multiple of $x^2 - 25$ and $x + 5$?

19. Rewrite these fractions with their least common denominator.

$$\dfrac{4}{(x+1)(x-4)} \text{ and } \dfrac{16}{(x-4)(x-3)}$$

20. What is the least common multiple of $16x^2y$, $24wx^3$, and $48xy$?

21. What is the least common multiple of $10w^2y$, $20x^2y^2$, and $400wxy^3$?

22. What is the least common multiple of $10wxy$, $12x^2y^2$, and $36w^2y^2$?

23. Add: $\dfrac{4}{x-5} + \dfrac{x+5}{x^2-25} + \dfrac{4x}{x+5}$

24. Subtract: $\dfrac{4}{5x} - \dfrac{3-x}{10x^2}$

25. Add: $\dfrac{2}{4y} + \dfrac{3y}{6y^2}$

26. Subtract: $\dfrac{6}{3x-4} - \dfrac{-12}{2x-3}$

27. Add: $\dfrac{3}{3+x} + \dfrac{3x-4}{x^2-9}$

28. Subtract: $\dfrac{3x-2}{6x^2-29x+28} - \dfrac{4}{2x-7}$

29. Simplify: $\dfrac{3x + \frac{2}{x}}{4 + \frac{1}{x}}$

30. Simplify: $\dfrac{y + \frac{5}{y}}{\frac{2}{y} + y}$

31. Simplify: $\dfrac{y^2 + \frac{y}{2}}{x + \frac{x}{2}}$

32. Simplify: $\dfrac{\frac{3-x}{12x}}{\frac{1+y}{12x^2-36x}}$

33. Solve for x: $\dfrac{3x}{2x} + \dfrac{4+x}{6x} = \dfrac{8x}{12x}$

34. Solve for x: $\dfrac{15}{2x+5} - \dfrac{1}{x} = \dfrac{21}{2x^2+5x}$

35. Solve for x: $\dfrac{1}{x+3} - \dfrac{2}{x-3} = \dfrac{x}{x^2-9}$

36. Solve for x: $\dfrac{9}{x^2-2x-24} - \dfrac{1}{4(x-6)} = \dfrac{3}{x+4}$

SECTION 7.4

37. A rectangular cafe has a perimeter of 96 feet and one side of 13 feet. If each side is reduced to 0.75 of its original length, how much smaller is the cafe's perimeter?

38. **Repair Job** A mechanic can do a repair job in 8 hours. Together, he and his apprentice can do the job in 5 hours. How long will it take to do the job alone?

200 miles
$r - 15$ mph

39. **Speed** Ms. Allen is traveling 15 miles per hour slower than Ms. Appel. Ms. Allen can travel 200 miles in the same time as Ms. Appel can travel 500 miles. How fast is each traveling?

500 miles
r mph

40. One fraction has a numerator that is 2 less than 4 times the denominator. When the numerator is multiplied by 3 and then 5 is added, the result is $\frac{95}{8}$. Find the original fraction.

MIXED REVIEW

41. Solve for y: $\dfrac{2}{y-6} + \dfrac{3}{y-2} = \dfrac{6}{y^2 - 8y + 12}$

42. Solve for x: $\dfrac{x-2}{x+1} = \dfrac{2x-4}{x+8}$

43. A number has a numerator that is 3 more than the denominator. If 3 is added to the numerator of the number and 5 is subtracted from the denominator of the number, the result is 3.5. Find the original fraction.

44. Evaluate $\dfrac{2x-y}{(x+y)^2}$ when $x = -2$ and $y = -8$.

45. Multiply: $\dfrac{x^2 - 4x}{x^2 + 3x + 2} \cdot \dfrac{x+2}{x-4}$

46. Evaluate $\dfrac{y^2 + \frac{y}{2}}{x + \frac{x}{2}}$ when $x = -1$ and $y = -2$.

47. Multiply: $\dfrac{4r + 4r}{2r^6} \cdot \dfrac{12r}{r}$

48. Simplify: $\dfrac{\frac{x+y}{x} + 6}{\frac{7x+y}{2x}}$

CHAPTER 7 TEST

This exam tests your knowledge of the material in Chapter 7.

1. **a.** When is $\dfrac{3x-2}{6x}$ undefined? **b.** Reduce $\dfrac{x^3 y}{xy^2}$ to lowest terms.

   c. Evaluate $\dfrac{r^1 - rt^3}{ut^2}$ when $r = -2$, $t = 4$, and $u = 6$.

2. Simplify the following.

   a. $\dfrac{\frac{9}{x} - 1}{\frac{1}{x-9}}$

   b. $\dfrac{x^2 + 8x + 12}{x + 6} \div \dfrac{x+2}{x^2 + 12x + 36}$

   c. $\dfrac{x^2 - x - 42}{x + 6} \cdot \dfrac{x^2 + 7x}{x^2 - 49}$

3. **a.** Solve for x: $\dfrac{x-3}{2x} - 6 = \dfrac{11}{x}$

   b. Subtract: $\dfrac{7}{y+2} - \dfrac{y+3-4y}{(y+2)(y+1)} - \dfrac{2}{y+1}$

   c. Rewrite the fractions $\dfrac{x+5}{9}$ and $\dfrac{x}{5}$ with their least common denominator.

4. Solve the following word problems.

   a. The ratio of the lengths of two pieces of ribbon is 5 to 8. Their total length is 65 inches. If each piece of ribbon is decreased by n inches, the new ratio is 1 to 4. What is n?

b. Working together, a girl and her younger brother can complete a complicated jigsaw puzzle in 5 hours. By herself, it takes the girl 8 hours. How long will her brother take to do the puzzle if he works alone?

c. The larger of two numbers is two less than twice the smaller number. The quotient of the larger and smaller number is equal to $\frac{13}{7}$. Find the two numbers.

CUMULATIVE REVIEW

CHAPTERS 1–7

The following exercises will help you maintain the skills you have learned in this and previous chapters.

1. Solve: $\dfrac{4}{x + 1} + \dfrac{2}{x^2 - 4x - 5} = \dfrac{3}{x - 5}$

2. Find the slope and the y-intercept of $x - 3y = 6$.

3. Simplify: $-2\frac{2}{9} - 4\frac{3}{7}$

4. Multiply: $\dfrac{2x + 6}{x^2 - 25} \cdot \dfrac{x - 5}{x + 3}$

5. Simplify: $-18(27) - 27 \div (-3) + 9(6 - 4)$

6. Find the equation of the line that passes through the points $(-6, 1)$ and $(-2, 3)$.

7. Combine: $-3(y - 7) + y(y + 4) - 4y + 7 - 8y$

8. Factor: $2x^2 - 4x - 6$

9. Solve: $-3(7 + 3y) - (y + 5) = 11$

10. Solve: $\frac{2}{3} - \frac{x}{4} = 9$

11. Simplify: $\dfrac{\frac{3}{rt} - 1}{\frac{4}{r^2 t^2}}$

12. Solve: $3x - 7 < -9(x + 5)$

13. Add: $\dfrac{3x^2 y^3}{x^2 - 4} + \dfrac{5xy^3}{x - 2}$

14. Solve: $\dfrac{x + 3}{x - 7} = 6$

15. Subtract: $\dfrac{7}{y - 7} - \dfrac{3}{y + 7}$

16. Divide $x^3 + 2x^2 - 6x - 4$ by $x - 2$.

17. Multiply: $\dfrac{x^2 - 4x - 5}{4x - 20} \cdot \dfrac{2x - 4}{x^2 - x - 2}$

18. Simplify: $(2x + 5)^2$

19. Graph: $-y = 2x + 5$

20. Graph the inequality: $7x - 2y \leq 8$

SYSTEMS OF LINEAR EQUATIONS AND INEQUALITIES

*T*he United States Constitution calls for a national census every ten years in order to reassess representation. Sociologists and statisticians look forward to working with the collected population data. Often the data are organized into tables or graphed; some appear as a straight line and may have been predicted by a linear equation. When predicting how populations change, statisticians may also work with systems of linear equations and inequalities. Predicted solutions are usually obtained through graphing, which illustrates trends.

■ *In what other fields, or in what type of situation, might a linear equation indicate the direction of a trend?*

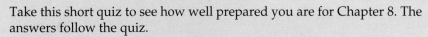

SKILLS CHECK

Take this short quiz to see how well prepared you are for Chapter 8. The answers follow the quiz.

1. Combine: $4x + 3x - 2x - 5x$

2. Solve: $-4x - 8 = 24$

3. $\frac{2}{3} + \frac{x}{5} = 4$

4. Solve for x: $-5x - 7 > 8$

5. Determine the value of y in the equation $4x + y = 9$ when x is 1, when x is 0, and when x is -1.

6. Graph: $3x + 5 = y$

7. Find the slope of $2x + 4 - y = 0$.

8. Are the graphs of $2y - x + 2 = 0$ and $y = \frac{1}{2}x$ parallel, perpendicular, or neither?

9. What is the opposite of the reciprocal of $1\frac{1}{2}$?

10. Is the origin a solution of $-y - 5 < 2x - 1$?

ANSWERS: **1.** 0 [Section 2.3] **2.** $x = -8$ [Section 3.1] **3.** $x = 16\frac{2}{3}$ [Section 3.1] **4.** $x < -3$ [Section 3.4] **5.** 5, 9, and 13 [Section 3.5] **6.** [Section 4.1] **7.** 2 [Section 4.2] **8.** parallel [Section 4.3] **9.** $-\frac{2}{3}$ [Section 1.3] **10.** yes [Section 3.4]

CHAPTER LEAD-IN

Almost every decision we make in our personal or professional lives is made under some sort of limiting conditions: We have only so much money to spend, only so much time available, or only so much raw material and machinery to use in producing goods. We can use a mathematical technique called **linear programming** in such decision-making situations.

Using linear programming, we can solve problems such as the following:

A baker makes both pan and shortbread cookies. Each batch of pan cookies requires 4 cups of sugar and 2 cups of butter. Each batch of shortbread cookies requires 3 cups of sugar and 3 cups of butter. The baker makes a $2.50 profit on each batch of pan cookies and a $2.75 profit on each batch of shortbread cookies. He has 76 cups of sugar, 56 cups of butter, and as much of the other ingredients as is needed. How many batches of each type of cookie should he make to maximize his profits?

8.1 Solving Systems of Linear Equations and Inequalities by Graphing

SECTION LEAD-IN

An experimental car gets 35 miles per gallon on the highway and 30 miles per gallon in town. The gas tank holds 20 gallons. The equation that represents the number of miles (y) that the car can travel when x of the 20 gallons are used on the highway and $(20 - x)$ gallons are used in town is

$$y = 35x + 30(20 - x)$$

- How many miles can the car travel on one tank of gas?
- What are the limiting factors in solving this equation?
- Write all inequalities that this problem implies.

SECTION GOALS

- *To determine whether a given ordered pair is a solution of a system of two linear equations*
- *To solve a system of linear equations by graphing*
- *To find the solution of systems of linear inequalities by graphing.*

Recall from Chapter 4 that a linear equation in two variables has at most one variable in each term and that each variable appears only to the first power.

> **System of Linear Equations**
>
> A **system of linear equations** is two or more linear equations that contain the same variables.

The **solution of a system of two linear equations** in the variables x and y is an ordered pair (x, y) that makes both equations true.

▪▪▪
EXAMPLE 1

Is the ordered pair $(-2, 3)$ a solution of the system $2x + 5y = 11$ and $y + x = 1$?

SOLUTION

$(-2, 3)$ is a solution of the system if it makes both equations true. To see if it does, we substitute -2 for x and 3 for y in each equation and simplify.

$$
\begin{array}{c|c}
2x + 5y = 11 & y + x = 1 \\
2(-2) + 5(3) = 11 & 3 + (-2) = 1 \\
-4 + 15 = 11 & 1 = 1 \quad \text{True} \\
11 = 11 \quad \text{True} &
\end{array}
$$

Because the substitutions yield two true statements, $(-2, 3)$ is a solution.

▶ *CHECK* **Warm-Up 1**

ERROR ALERT

Identify the error and give a correct answer.

Is $(-2, 4)$ a solution of
$y = -2x$ and
$y = x + 5$?

Incorrect Solution:

$y = -2x$
$4 = -2(-2)$
$4 = 4$

Yes, it is a solution.

STUDY HINT

When checking a solution of a system of equations, always check it in both *equations. An ordered pair must make both equations true to be a solution.*

Calculator Corner

Use your graphing calculator's Home Screen to help you decide if the ordered pair $(-2, 3)$ is a solution of the system $2x + 5y = 11$ and $y + x = 1$. (Note: See the Calculator Corner on page 85 to review how to store values for variables.)

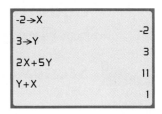

Graphs of Systems of Linear Equations

We can graph a system of two equations by drawing the graphs of both equations on the same coordinate axes. The point(s) at which the graphs intersect (touch or cross) provide information about the solution(s) of the system.

■■■

WRITER'S BLOCK

Describe generally the *solution* of a system of linear equations in two variables. Comment on how many solutions are possible and on how you can tell the number of solutions without graphing.

The lines in the accompanying figure have different slopes. They intersect at one point, so the system they represent has *one solution.* That solution is given by the point of intersection.

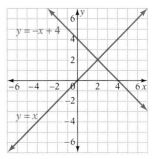

The lines in this figure have the same slope but different y-intercepts. They do not intersect but instead are parallel. Thus the system they represent has *no solution.*

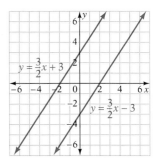

The lines in our final figure have the same slope and y-intercept. They are really the same line. They intersect at an infinite number of points, so the system they represent has an *infinite number of solutions.* Any ordered pair that satisfies one equation also satisfies the other equation.

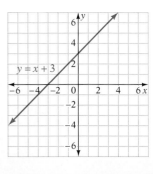

Solving Linear Systems by Graphing

One way to solve a system of linear equations is to graph the equations. If the graphs intersect, the coordinates of the point of intersection provide the solution. This graphical solution then should be checked by substituting the ordered pair into the equations of the system.

■■■
WRITER'S BLOCK

What is meant by a *system* of equations or inequalities?

> **To solve a system of linear equations in two variables by graphing**
>
> 1. Graph and label both equations.
> 2. Estimate the coordinates of the point of intersection of the two lines, if it exists. This is the solution.
> 3. Check by substituting into the original equations.

■ ■ ■
EXAMPLE 2

Solve this system of equations by graphing.

$$x - 2y = -6$$
$$y - 3x = 3$$

SOLUTION

First we solve both equations for y.

$x - 2y = -6$	$y - 3x = 3$
$-2y = -6 - x$	$y = 3 + 3x$
$y = 3 + \frac{x}{2}$	$y = 3x + 3$
$y = \frac{x}{2} + 3$	

■■■
WRITER'S BLOCK

In Example 2, we state that "the graphs of these two equations are not parallel and are not identical." How do we know?

The graphs of these two equations are not parallel and are not identical, so we will graph them to find the solution.

We let $x = 2, 4,$ and 6 for the first equation and obtain a table of values at the left below. We let $x = 0, 1,$ and -1 for the second equation and obtain a table of values at the right below.

x	y	x	y
2	4	0	3
4	5	1	6
6	6	−1	0

We now graph the two equations, using the coordinates in the tables (see the accompanying figure). The lines appear to intersect at $(0, 3)$. We check by substituting these coordinates into the two original equations.

$x - 2y = -6$	$y - 3x = 3$
$(0) - 2(3) = -6$	$(3) - 3(0) = 3$
$-6 = -6$ True	$3 = 3$ True

So $(0, 3)$ is the single solution of the system.

▶ CHECK **Warm-Up 2**

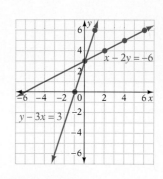

Calculator Corner

In order to graph the two equations in Example 2 you must first get them into the "$y =$ " form. Then graph the two equations and **TRACE** to find the point of intersection. You could also find the intersection point by using the **CALC**ulation utility if your graphing calculator has that feature.

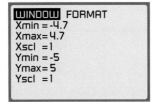
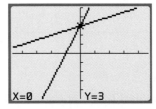

You can check your results by using your graphing calculator's Home Screen to evaluate each equation at the value $x = 0$. If the point $(0, 3)$ is indeed the point of intersection, then y should be equal to 3 in each equation.

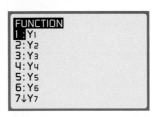
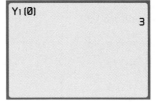

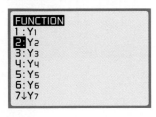
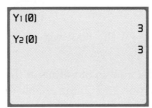

Because both equations give a result of 3 when $x = 0$, the point of intersection is $(0, 3)$.

Solving Systems of Linear Inequalities

A **system of linear inequalities** consists of two or more linear inequalities that contain the same variables. The most straightforward way to solve these systems is by graphing, just as we solved simple inequalities. In fact, we actually graph each inequality individually. The solution of the system is the portion of the coordinate plane that is common to all the individual inequalities in the system.

Let's look at some examples.

EXAMPLE 3

By graphing, find the solution set of the following system of inequalities.

$$x + y < 5$$
$$3x - 2y > 4$$

SOLUTION

We shall graph both inequalities and then find the portion of the plane where the two graphs overlap. That portion consists of all points whose coordinates satisfy both inequalities.

To graph the inequalities, we first must graph the lines they suggest.

Those lines have the equations

$$x + y = 5 \quad \text{and} \quad 3x - 2y = 4$$

Reasonable tables of values are

x	y		x	y
1	4		1	$-\frac{1}{2}$
0	5		0	-2
-1	6		-1	$-3\frac{1}{2}$

We begin by graphing $x + y = 5$ with a broken line to show that the line is not part of the graph. (That's because the inequality is of the "less than" type.)

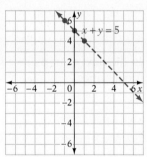

We now must determine which region of the plane is part of the inequality. To do so, we choose a test point and substitute its coordinates into the inequality. Choosing the "easy" point $(0, 0)$, the origin, we obtain

$$x + y < 5$$
$$0 + 0 < 5$$

Because the substitution leads to a true statement, the origin is part of the graph of the inequality. So we shade the region below the broken line—the region that contains the origin.

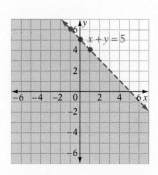

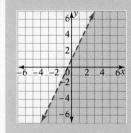

If we check this by substituting the coordinates of a point above the line, say (5, 6), we get a false statement. This shows that we have shaded the figure correctly.

We then graph the inequality $3x - 2y > 4$ in exactly the same way, on the same set of axes.

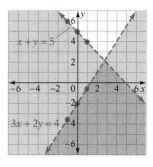

Recall that the graph of an inequality is also the graph of its solution. So every shaded point is part of the solution of one of the inequalities. Points that are doubly shaded are part of the solutions of both inequalities and therefore are the solution of the system.

So the solution of the system is the doubly shaded region in the figure above. It is shown more clearly in the figure at the right.

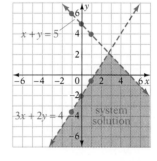

▶ *CHECK* **Warm-Up 3**

In the next example, we will use the slope-intercept form of a line to graph the inequalities.

■■■

EXAMPLE 4

Graph the solution of the system of inequalities $3y - 2x \geq 6$ and $5x - y > 2$.

SOLUTION

We write the associated equations in slope-intercept form, obtaining

$$
\begin{array}{l|l}
3y - 2x = 6 & 5x - y = 2 \\
\quad 3y = 2x + 6 & \quad -y = -5x + 2 \\
\quad\ \ y = \frac{2}{3}x + 2 & \qquad y = 5x - 2
\end{array}
$$

The line $3y - 2x = 6$ has slope $\frac{2}{3}$ and y-intercept (0, 2). In the following figure at the left, it is graphed as a solid line because its inequality ("greater than or equal to") includes the equality. We graph $5x - y = 2$ dashed ($>$ does *not* include the equality), making use of its slope of 5 and its y-intercept of (0, −2). Then, using test points, we shade the proper region of the plane for each inequality. The solution of the system is the doubly shaded region shown in the following figure at the right, including part of the solid line.

! ! !
ERROR ALERT

Identify the error and give a correct answer.

Solve this system by graphing:

$2y > 3x + 2$

$4y < 2x - 6$

Incorrect Solution:

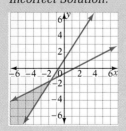

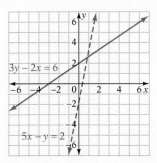 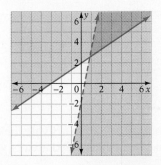

▶ CHECK **Warm-Up 4**

Calculator Corner

Look again at Example 4. If we solve both inequalities for y, the results would be

$$y \geq \frac{2}{3}x + 2 \quad \text{and} \quad y < 5x - 2$$

If we graphed the y's as linear functions, the graphs would look as follows.

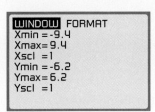 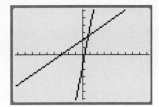

Now use your graphing calculator to **Shade** that area of the graph that we are investigating. The lower bound of the system is Y_1 (we want to shade the area *greater* than this) and the upper bound of the system is Y_2 (we want to shade the area *less* than this). The command for shading is found under the **DRAW** feature of the *TI-82*. The command on the Home Screen would be written as

Shade(lower bound, upper bound, resolution)   or

Shade$\left(\frac{2}{3}X + 2, 5X + 2, 2\right)$ or the equivalent statement

Shade(Y₁, Y₂, 2)

(**Resolution** simply describes the density of the shading. The shading capabilities of graphing calculators vary widely. You may want to consult your calculator manual for the **Shad**ing instructions for your particular model.)

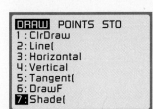 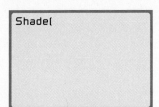

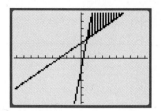

Press **ENTER.**

Now press any of the arrow keys on your calculator to activate the **screen cursor.** Move the cursor to the point (3, 5), which is in the shaded area of your graph.

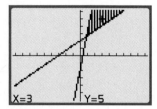

Points within the shaded area satisfy the conditions of both linear inequalities *at the same time.* To verify this, use the **STO**re function to substitute these values for *x* and *y* into the *original inequalities* as shown on the screen below. Nine is definitely greater than or equal to 6, so the point (3, 5) satisfies the first inequality. Ten is greater than 2, so the point (3, 5) also satisfies the second inequality.

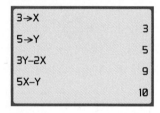

Now move the **screen cursor** to a point that *is not* in the shaded area.

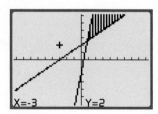

Evaluate each of the *original inequalities* for the point not in the shaded area, (−3, 2) as shown on the screen below. Twelve is definitely greater than or equal to 6, so the point (−3, 2) satisfies the first inequality. However, −17 *is not* greater than 2, so the point (−3, 2) *does not* satisfy the second inequality. Therefore, the point (−3, 2) does not belong in the solution set for this system of inequalities.

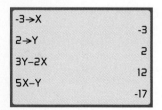

Now let's look at a system of inequalities with the same inequality sign:

$$y > 0.5x + 2 \quad \text{and} \quad y > -3x - 4$$

Notice that both inequalities are *greater than* situations. How would you shade this on your graphing calculator?

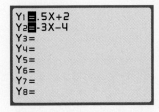 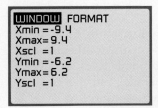 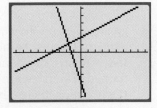

Remember that the **Shade** command is **Shade(lower bound, upper bound, resolution)**. Put this onto your Home Screen and use an arbitrary large number for the upper bound, say 15.

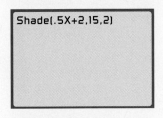 Press **ENTER**.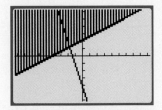

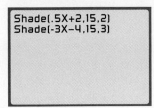 Press **ENTER**.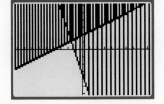

Notice that the area of the graph that satisfies *both inequalities* is the area that has been shaded twice—once with a resolution of 2 and then again with a resolution of 3.

■■■

EXAMPLE 5

Find the solution of the following system of inequalities.

$$5y - 4x > 5$$
$$y < \frac{4}{5}x - 3$$

SOLUTION

The equations associated with this system are

$$5y - 4x = 5$$
$$y = \frac{4}{5}x - 3$$

The second equation is already in slope-intercept form, so we rewrite the first in that form, obtaining

$$5y - 4x = 5 \qquad\qquad y = \tfrac{4}{5}x - 3$$
$$5y = 4x + 5$$
$$y = \tfrac{4}{5}x + 1$$

WRITER'S BLOCK

Do *parallel lines* have a *point of intersection?*

Explain why or why not? What about perpendicular lines?

Next we graph the equations, using their slopes and y-intercepts. Because neither of the inequalities contains an equality, both lines are drawn dashed. The lines are parallel, but the inequalities may nonetheless have a solution.

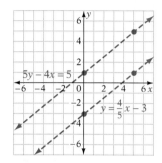

Testing the point (0, 0) for each, we get

$$5y - 4x > 5 \qquad\qquad y < \left(\tfrac{4}{5}\right)x - 3$$
$$5(0) - 4(0) > 5 \qquad\qquad 0 < \left(\tfrac{4}{5}\right)(0) - 3$$
$$0 - 0 > 5 \qquad\qquad 0 < 0 - 3$$
$$0 > 5 \quad \text{False} \qquad\qquad 0 < -3 \qquad\qquad \text{False}$$

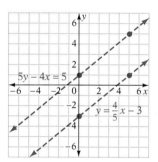

No area of the plane is doubly shaded. The system has no solution.

▶ *CHECK* **Warm-Up 5**

Practice what you learned.

SECTION FOLLOW-UP

The inequalities implied by the problem are

Highway miles possible on one tank of gas ≤ 700 miles $\quad (35)(20)$ miles

Town miles possible on one tank of gas ≤ 600 miles $\quad (30)(20)$ miles

Gas left in tank ≤ 20 gallons

$600 \leq$ total miles ≤ 700

8.1 WARM-UPS

Work these problems before you attempt the exercises.

1. Is the ordered pair (2, 3) a solution to the system of equations $x + y = 5$ and $2x + y = 7$?

2. Solve system by graphing: $\begin{aligned} 6x - 3y &= 12 \\ 2x + y &= -4 \end{aligned}$

3. Solve by graphing: $\begin{aligned} x - y &> 3 \\ 2x - 5y &> 8 \end{aligned}$

4. Graph $y \le 3x + 4$ and $x \le 5$.

5. Find the solution of $x + y < 5$ and $3x > 6 + y$ by graphing.

8.1 EXERCISES

Note: Use your graphing calculator to check your results whenever possible.

In Exercises 1 through 8, determine whether the ordered pair shown is a solution to the given system of linear equations.

1. (3, 5): $\begin{aligned} x + 2 &= y \\ 2x + y &= 11 \end{aligned}$

2. (2, 6): $\begin{aligned} x + y &= 8 \\ 3x - y &= 0 \end{aligned}$

3. (1, 7): $\begin{aligned} y - 6 &= x \\ 5x + 2 &= y \end{aligned}$

4. (−2, −4): $\begin{aligned} 6y &= 4x - 16 \\ x + y &= 2 \end{aligned}$

5. (−1, 0): $\begin{aligned} 16y + 2x &= 3 \\ 5x + 4y &= 9 \end{aligned}$

6. (6, −2): $\begin{aligned} 3x - y &= 16 \\ 4y - x &= 2 \end{aligned}$

7. (−5, 12): $\begin{aligned} 2x - y &= 2 \\ 6y + 5x &= 10 \end{aligned}$

8. (−7, −5): $\begin{aligned} x - y &= -2 \\ -y + 2x &= -9 \end{aligned}$

In Exercises 9 through 20, solve each system by graphing. For systems with only one solution, check your result in the original equations.

9. $\begin{aligned} y &= x - 9 \\ -2y &= 2x - 6 \end{aligned}$

10. $\begin{aligned} 5y &= -10x - 10 \\ y - x &= 1 \end{aligned}$

11. $\begin{aligned} y &= 2x + 5 \\ 3y &= 6x - 18 \end{aligned}$

12. $\begin{aligned} \tfrac{1}{2}y &= 2x + 4 \\ 2y &= 4x + 8 \end{aligned}$

13. $\begin{aligned} 3x &= 2y + 5 \\ 6x &= 4y + 10 \end{aligned}$

14. $\begin{aligned} 4x + 2y &= 6 \\ x + y &= 5 \end{aligned}$

15. $\begin{aligned} 4y &= 4x + 8 \\ 2y &= -2x + 8 \end{aligned}$

16. $\begin{aligned} 5y &= x - 10 \\ 5y &= x + 10 \end{aligned}$

17. $\begin{aligned} 2y &= -x - 10 \\ \tfrac{1}{2}x - 3 &= y \end{aligned}$

18. $y = 3x + 4$
$y = -x + 4$

19. $x = 2$
$y = 3$

20. $x = 4$
$y = -4$

In Exercises 21 through 36, solve each system of inequalities by graphing.

21. $x + 2y < 4$
$2x - y < 0$

22. $2x - y < 6$
$3x + y > -1$

23. $y \leq 2x - 3$
$2y - 3 \leq x$

24. $y \geq x + 3$
$3x + y \geq 11$

25. $x \leq 7 - y$
$y - 3 \leq x$

26. $4x - 5 \geq y$
$3y + 7x \leq 0$

27. $2x - 5 > y$
$4y - x < 1$

28. $3x - y \leq 6$
$y \geq -2x$

29. $8x - 4y \geq 12$
$\frac{y}{3} - x \geq -5$

30. $5x - y < 13$
$4x + y > -17$

31. $y - x < 4$
$x + y \leq 2$

32. $x + 3y \geq 8$
$y \geq 1$

33. $2x + 4 \leq y$
$y < 4$

34. $x - y > 8$
$2x + 4y \geq -12$

35. $3x + 5y \leq 25$
$4x + 2 > y$

36. $2x + 2y < 6$
$3y \geq -9$

In Exercises 37 through 40, match each system of inequalities with the appropriate graph. The lines have been graphed correctly.

(A)

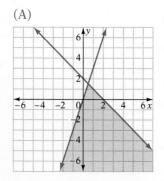

(B)

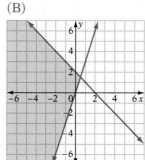

(C)

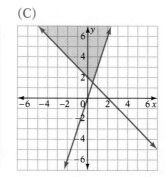

(D)

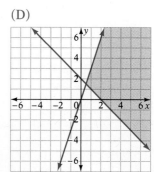

37. $3x \leq y$
    $x + y \leq 2$

38. $3x \leq y$
    $x + y \geq 2$

39. $3x \geq y$
    $x + y \geq 2$

40. $3x \geq y$
    $x + y \leq 2$

In Exercises 41 through 44, match each system of inequalities with the appropriate graph. The lines have been graphed correctly.

(E)
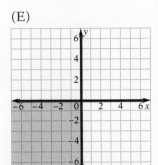

(F)
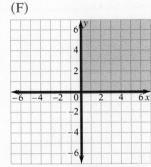

(G)
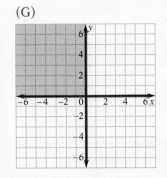

(H)
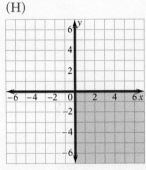

41. $x \leq 0$
    $y \geq 0$

42. $x \leq 0$
    $y \leq 0$

43. $x \geq 0$
    $y \leq 0$

44. $x \geq 0$
    $y \geq 0$

In Exercises 45 through 48, match each system of inequalities with the appropriate graph. The lines have been graphed correctly.

(I)
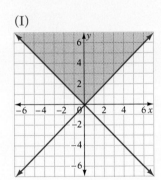

(J)
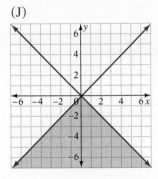

(K)
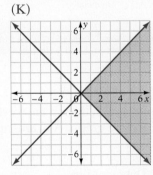

(L)
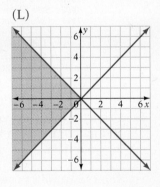

45. $y \geq x$
    $-x \geq y$

46. $y \geq x$
    $-x \leq y$

47. $y \leq x$
    $-x \leq y$

48. $y \leq x$
    $-x \geq y$

EXCURSIONS

Data Analysis

1. Use the table on the following page to graph the Olympic discus distance records by year in a double bar graph. Next to each discus record, graph that year's javelin record. Compare the bars. Ask and answer four questions about these graphs.

Summer Olympics Records

Women's Discus Throw

1928	Helena Konopacka	Poland		129 ft	11 7/8 in.
1932	Lillian Copeland	United States		133 ft	2 in.
1936	Gisela Mauermayer	Germany		156 ft	3 3/16 in.
1948	Micheline Ostermeyer	France		137 ft	6 1/2 in.
1956	Olga Fikotova	Czechoslovakia		176 ft	1 1/2 in.
1960	Nina Ponomareva	U.S.S.R.		180 ft	8 1/4 in.
1964	Tamara Press	U.S.S.R.		187 ft	10 3/4 in.
1968	Lia Manoliu	Romania		191 ft	2 1/2 in.
1972	Faina Melnik	U.S.S.R.		218 ft	7 in.
1976	Evelin Schlaak	East Germany	(69.0 m)	226 ft	4 in.
1980	Evelin Jahl	East Germany		229 ft	6 1/2 in.
1984	Ria Stalman	Netherlands		214 ft	5 in.
1988	Martina Hellmann	East Germany		237 ft	2 1/4 in.
1992	Maritza Marten	Cuba		229 ft	10 1/4 in.
1996	Ilke Wyludda	Germany		228 ft	6 1/2 in.

Women's Javelin Throw

1932	Mildred Didrikson	United States		143 ft	4 in.
1936	Tilly Fleischer	Germany		148 ft	2 3/4 in.
1948	Herma Bauma	Austria		149 ft	6 in.
1952	Dana Zatopek	Czechoslovakia		165 ft	7 in.
1956	Inessa Janzeme	U.S.S.R.		176 ft	8 in.
1960	Elvira Ozolina	U.S.S.R.		183 ft	8 in.
1964	Mihaela Penes	Romania		198 ft	7 1/2 in.
1968	Angela Nemeth	Hungary		198 ft	0 in.
1972	Ruth Fuchs	East Germany		209 ft	7 in.
1976	Ruth Fuchs	East Germany	(65.94 m)	216 ft	4 in.
1980	Maria Colon	Cuba		224 ft	5 in.
1984	Tessa Sanderson	Britain		228 ft	2 in.
1988	Petra Felke	East Germany		245 ft	0 in.
1992	Silke Renke	Germany		224 ft	2 1/2 in.
1996	Heli Rantanen	Finland		222 ft	11 in.

Source: 1997 Information Please® Almanac (©1995 Houghton Mifflin Co.), p. 900. All rights reserved. Used with permission from Information Please LLC.

2. Using the following tables, graph the top ten world's fastest outdoor miles and indoor miles on the same graph. Ask and answer four questions using your data.

Top Ten World's Fastest Outdoor Miles

Time	Athlete	Country	Date	Location
3:44.39	Noureddine Morceli	Algeria	Sept. 5, 1993	Rieti, Italy
3:46.31	Steve Cram	England	July 27,1985	Oslo, Norway
3:47.33	Sebastian Coe	England	Aug. 28, 1981	Brussels, Belgium
3:47.69	Steve Scott	United States	July 7, 1982	Oslo, Norway
3:47.79	Jose Gonzalez	Spain	July 27, 1985	Oslo, Norway
3:48.40	Steve Ovett	England	Aug. 26, 1981	Koblenz, W. Germany
3:48.53	Sebastian Coe	England	Aug. 19, 1981	Zurich, Switzerland
3:48.53	Steve Scott	United States	June 26, 1982	Oslo, Norway
3:48.8	Steve Ovett	England	July 1, 1980	Oslo, Norway
3:48.83	Sydney Maree	United States	Sept. 9, 1981	Rieti, Italy

Note: Professional marks not included. *Source: USA Track & Field,* as cited in the *1996 Information Please® Almanac* (©1995 Houghton Mifflin Co.), p. 958. All rights reserved. Used with permission from Information Please LLC.

Top Ten World's Fastest Indoor Miles

Time	Athlete	Country	Date	Location
3:49.78	Eamonn Coghlan	Ireland	Feb. 27, 1983	East Rutherford, N.J.
3:50.6	Eamonn Coghlan	Ireland	Feb. 20, 1981	San Diego, California
3:50.7	Noureddine Morceli	Algeria	Feb. 20, 1993	Birmingham, England
3:50.94	Marcus O'Sullivan	Ireland	Feb. 13, 1988	East Rutherford, N.J.
3:51.2	Ray Flynn[1]	Ireland	Feb. 27, 1983	East Rutherford, N.J.
3:51.66	Marcus O'Sullivan	Ireland	Feb. 10, 1989	East Rutherford, N.J.
3:51.8	Steve Scott[1]	United States	Feb. 20, 1981	San Diego, California
3:52.28	Steve Scott[2]	United States	Feb. 27, 1983	East Rutherford, N.J.
3:52.30	Frank O'Mara	Ireland	Feb. 1986	New York, New York
3:52.37	Eamonn Coghlan	Ireland	Feb. 9, 1985	East Rutherford, N.J.

1. Finished second. 2. Finished third. *Source: USA Track & Field,* as cited in the *1996 Information Please® Almanac* (©1995 Houghton Mifflin Co.), p. 958. All rights reserved. Used with permission from Information Please LLC.

Posing Problems

3. Dr. Edwards always gave his grades as ordered pairs (x, y), where x was the number of questions on the test and y was the number correct. Passing grades were solutions of

$$10y \geq 7x$$
$$x \geq y$$

   a. Graph the equation $10y = 7x$. Give three solutions that are not on the line and that "make sense."

   b. What other school-type situations can be modeled by a system of linear equations or inequalities? Pose a problem and solve it.

4. To qualify for an academic prize, a student must have a combination of two grades x and y, shown as an ordered pair (x, y), that falls within the solution set of

$$y \geq -x + 150$$
$$y \geq 50$$
$$x \geq 50$$

   a. Sketch a graph of this system of inequalities, and choose three solutions that are not on the line $y = -x + 150$ and that "make sense." (These solutions must be positive.)

   b. What other school-type situations can be described using a system of linear inequalities? Write a problem for each situation and solve it.

CONNECTIONS TO *STATISTICS*

Reading a Histogram

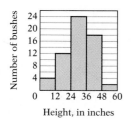

Height, in inches

Some kinds of graphs use the widths and lengths of vertical bars to represent information. The graph shown here is such a **histogram.** The width of each bar indicates a height *interval* of about 12 inches. Specifically, the intervals are 0–11.9, 12.0–23.9, and so on. The length of each bar indicates how many bushes have heights that fall in the interval. The scale at the left indicates the number of bushes in the height interval.

• • •

EXAMPLE

In the histogram shown, what height interval contains the most bushes? How many bushes are in that height interval?

SOLUTION

To find the interval with the most bushes, we look for the tallest bar. That bar is the one representing the 24-to-35.9 inch interval. Next we look across the top of this tallest bar to the left scale, and we find that its height represents 24 bushes on that scale. Therefore, 24 bushes are 24.0 to 35.9 inches tall.

PRACTICE

In the following histogram, the number of tenants who live in a small apartment building are graphed by age intervals, 0–14, 15–29, and so on. (A person who is exactly 15, 30, and so on is included in the older group.) Use this information to answer Exercises 1 through 5.

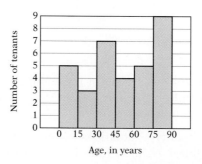

Age, in years

1. What is the width of each age interval in this graph? How many tenants are there altogether?

2. What percent of the tenants in this building are less than 60 years of age? Round to the nearest whole percent.

3. What percent of the tenants are at least 30 but less than 60 years of age? Round to the nearest whole percent.

4. What percent of the tenants in this building are less than 90 years of age but at least 60 years old. Round to the nearest whole percent.

5. What is the ratio of the number of tenants under 30 to the number of tenants under 75 years of age?

Use the histogram in the Example to answer Exercises 6 and 7.

6. How many bushes are shorter than 24 inches or taller than or equal to 48 inches?

7. What percent of the bushes are 24 to 47.9 inches high? Round to the nearest ten percent.

8. ✏ What are the differences between bar graphs and histograms?

8.2 Solving Systems of Linear Equations

SECTION LEAD-IN

Living within a budget involves a number of limiting factors. Make a budget and write a series of equations and inequalities that describe the limits of your spending in categories such as

- food
- transportation
- savings

Solving by Addition

We can add the *same term* to both sides of a linear equation without changing the solution of the equation. We can also add two equal *expressions* to the two sides of an equation without changing the solutions. That is,

---

Addition Principle of Equality

If $a = b$ and $c = d$, then $a + c = b + d$.

We can use this principle to eliminate one of the variables, forming an equation we can solve for the other variable.

▪▪▪

EXAMPLE 1

Solve this system of equations.

$$2x + 3y = 10$$
$$4x - 3y = 8$$

SOLUTION

If we add the left sides and the right sides of these equations, the y terms will drop out. We will be left with an equation in x only, which we can solve easily.

$$\begin{array}{ll} 2x + 3y = 10 & \\ 4x - 3y = 8 & \\ \hline 6x + 0 = 18 & \text{Adding left sides and right sides} \\ x = 3 & \text{Multiplying both sides by } \frac{1}{6} \end{array}$$

We now know that $x = 3$ is part of the solution of the system. We substitute 3 for x in either equation and solve for y.

$$\begin{array}{ll} 2x + 3y = 10 & \text{The first equation} \\ 2(3) + 3y = 10 & \text{Substituting 3 for } x \\ 6 + 3y = 10 & \text{Simplifying} \\ 3y = 4 & \text{Adding } -6 \text{ to both sides} \\ y = \frac{4}{3} & \text{Multiplying by } \frac{1}{3} \end{array}$$

So $\left(3, \frac{4}{3}\right)$ is the solution of this system.

▶ *CHECK* **Warm-Up 1**

In the next example, we cannot just add left and right sides to eliminate one variable. It doesn't work. However, we can first multiply both sides of one equation by an appropriately chosen number (without changing its solutions). And then we can add the result to the first equation to eliminate y. Follow along.

▪▪▪

EXAMPLE 2

Solve: $4y - 3x = 7$
$2y - 5x = 9$

SOLUTION

To eliminate the y-term, we need to multiply the second equation by -2. That gives us the following equivalent system:

$$4y - 3x = 7$$
$$-4y + 10x = -18$$

Now we can add left and right sides to eliminate y and solve for x.

$$4y - 3x = 7$$
$$-4y + 10x = -18$$
$$\overline{ 0 + 7x = -11} \quad \text{Adding left and right sides}$$
$$x = -\frac{11}{7} \quad \text{Multiplying by } \frac{1}{7}$$

Thus $x = -\frac{11}{7}$ is part of the solution. Substituting $-\frac{11}{7}$ for x in the first equation gives us $y = \frac{4}{7}$. Check this answer in both equations.

▶ *CHECK* **Warm-Up 2**

■■■
WRITER'S BLOCK
How is "addition" a part of the *addition* method of solving a system of linear equations?

To solve the next system, we must multiply *both* equations by constants. Watch how we choose the constants.

■■■
EXAMPLE 3

Solve: $2x + 9y = 49$
$5y = 31 - 3x$

SOLUTION

First we rewrite the equations with the variables in the same order on the same side. That makes everything easier.

$$2x + 9y = 49$$
$$3x + 5y = 31$$

In order to be able to eliminate one variable, we want the coefficients of x or those of y to be additive inverses. The coefficients of x will be inverses if we multiply the first equation by 3 and the second equation by -2. Then we can add left and right sides, eliminating x, and solve for y. We get

$$6x + 27y = 147 \quad \text{Multiplying by 3}$$
$$-6x - 10y = -62 \quad \text{Multiplying by } -2$$
$$\overline{ 17y = 85} \quad \text{Adding left and right sides}$$
$$y = 5 \quad \text{Multiplying by } \frac{1}{17} \text{ to solve}$$

So $y = 5$ is part of the solution. We substitute 5 for y in the first equation and solve for x.

$$2x + 9(5) = 49 \quad \text{Substituting for } y$$
$$2x + 45 = 49 \quad \text{Simplifying}$$
$$2x = 4 \quad \text{Adding } -45 \text{ to both sides}$$
$$x = 2 \quad \text{Multiplying by } \frac{1}{2}$$

The complete solution is $(2, 5)$. You should check it by substituting into the original equations.

▶ *CHECK* **Warm-Up 3**

STUDY HINT
In Example 3, to eliminate y *instead of* x, *we could have multiplied the first equation by 5 and the second equation by* -9. *First decide which variable to remove. The choice is yours. Then select the proper multipliers.*

▪▪▪
EXAMPLE 4

Solve: $2x + 3y = 8$
$\quad\quad 4x + 6y = 16$

SOLUTION

We need only multiply the first equation by -2 and add.

$$\begin{array}{rl} -4x - 6y = -16 & \text{Multiplying by } -2 \\ \underline{4x + 6y = 16} & \\ 0 + 0 = 0 & \text{Adding left and right sides} \end{array}$$

Both variables are eliminated when we add left and right sides, and we are left with $0 = 0$, a *true* statement. This means that *every* solution of one equation is a solution of the other equation. In other words, the two original equations represent the same line.

So we find that this system has an infinite number of solutions.

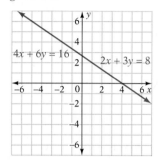

▶ CHECK **Warm-Up 4**

▪▪▪
EXAMPLE 5

Solve: $5x - 2y = 10$
$\quad\quad 15x - 6y = 26$

SOLUTION

Multiplying the first equation by -3 and adding give us

$$\begin{array}{rl} -15x + 6y = -30 & \text{Multiplying by } -3 \\ \underline{15x - 6y = 26} & \\ 0 + 0 = -4 & \text{Adding left and right sides} \end{array}$$

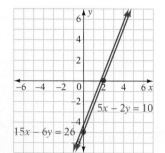

As in Example 4, both variables are eliminated here, too, when we add left and right sides. But now we are left with a *false* statement. This means that there is no solution to this system; the graphs of the two equations are parallel lines (see the accompanying figure).

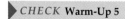

▶ CHECK **Warm-Up 5**

Solving by Substitution

In the addition method for solving a system of two linear equations, we add the equations to eliminate one variable. In the substitution method, we substitute an expression for one variable to eliminate that variable.

Let's look at an example.

▪▪▪
EXAMPLE 6

Solve: $y = 2x + 4$
$\quad\quad 6y + 3x = 54$

SOLUTION

The first equation tells us that y is equal to $2x + 4$, so we can substitute $2x + 4$ for y in the second equation.

$$6y + 3x = 54 \quad \text{The second equation}$$
$$6(2x + 4) + 3x = 54 \quad \text{Substituting for } y$$
$$12x + 24 + 3x = 54 \quad \text{Multiplying out}$$
$$15x + 24 = 54 \quad \text{Combining like terms}$$

This is an equation with one variable, x. We solve it easily.

$$15x = 30 \quad \text{Adding } -24 \text{ to both sides}$$
$$x = 2 \quad \text{Multiplying by } \tfrac{1}{15}$$

We now know that x is equal to 2. Thus we can substitute 2 for x in one of the original equations and solve it for y. The first equation seems easiest.

$$y = 2x + 4$$
$$y = 2(2) + 4 \quad \text{Substituting for } x$$
$$y = 8 \quad \text{Simplifying}$$

So $(2, 8)$ is the solution of the system of equations. Check these results.

▶ *CHECK* **Warm-Up 6**

<aside>
! ! !
ERROR ALERT

Identify the error and give a correct answer.

Solve this system by substitution:

$x = -y + 4$
$y = 2x + 3$

Incorrect Solution:

$x = -y + 4$
$x = 2x + 3 + 4$
$-x = 7$
$x = -7$

$x = -y + 4$
$-7 = -y + 4$
$-11 = -y$
$11 = y$
</aside>

▪▪▪
EXAMPLE 7

Solve: $x + y = 9$
$\quad\quad 2y - 7x = 27$

SOLUTION

We want the ordered pair that is the solution of the given system. We can easily solve the first equation for x, obtaining

$$x = 9 - y$$

(We could just as easily solve for y. The choice is yours.)

Substituting this expression for x in the second equation gives us

$$2y - 7x = 27 \quad \text{The second equation}$$
$$2y - 7(9 - y) = 27 \quad \text{Substituting for } x$$
$$2y - 63 + 7y = 27 \quad \text{Multiplying out}$$
$$9y - 63 = 27 \quad \text{Combining like terms}$$
$$9y = 90 \quad \text{Adding 63 to both sides}$$
$$y = 10 \quad \text{Multiplying by } \tfrac{1}{9}$$

<aside>
▪▪▪
WRITER'S BLOCK

How is "substitution" a part of the *substitution* method of solving a system of linear equations?
</aside>

Part of the system solution, then, is $y = 10$. We substitute this value for y in the original first equation and solve for x.

$$x + y = 9 \qquad \text{The first equation}$$
$$x + 10 = 9 \qquad \text{Substituting for } y$$
$$x = -1 \quad \text{Adding } -10 \text{ to both sides}$$

This gives us the ordered pair $(-1, 10)$. A check would show that $(-1, 10)$ is the solution of this linear system.

▶ *CHECK* **Warm-Up 7**

Practice what you learned.

SECTION FOLLOW-UP

Here is a sample budget based on an income of $1500 per month.

Rent (R) $R = \$500$
Utilities (U) $0 < U \le \$50$
Food (F) $\$150 < F \le \$400$
Transportation (T) $0 \le T \le \$90$
Savings (S) $0 \le S \le \$300$
Miscellaneous (M) $0 \le M \le \$200$
Entertainment (X) $0 \le X \le \$200$

$$R + U + F + T + S + M + X \le \$1500$$

8.2 WARM-UPS

Work these problems before you attempt the exercises.

1. Solve: $4x + 9y = 10$
             $14x - 9y = 8$

2. Solve: $-7x - 2y = 7$
            $14x + 3y = -7$

3. Solve: $0.4y - x = 8$
            $2.4y + 2x = 16$

4. Solve: $7y - 3x = 6$
            $-6x + 14y = 12$

5. Solve: $4x - 2y = 8$
            $8x - 4y = 10$

6. Where do $x = \frac{1}{2}y$ and $5y - 4x = 6$ intersect?

7. Find the point of intersection of $x + y = 6$ and $3x + 4y = 22$.

8.2 EXERCISES

Note: Use your graphing calculator to check your results whenever possible.

In Exercises 1 through 24, solve each system of equations using addition.

1. $3x + 2y = 7$
$4x - 2y = 14$

2. $8x - 2y = 6$
$-2x - 6y = -8$

3. $7y - 2x = 12$
$6y + 2x = 14$

4. $y = 2x - 1$
$y - x = 5$

5. $x + y = 9$
$6x - 5y = 10$

6. $x - y = 15$
$3x + 5y = 21$

7. $4x - 9y = 8$
$11x - 9y = 29$

8. $12x - 12y = 22$
$18x - 3y = 13$

9. $7x - 3 = y$
$y + 5 = 9x$

10. $11y = -8x + 47$
$3y = 8x - 5$

11. $5x = 9y + 7$
$-2x = -9y - 4$

12. $6x = 3y - 9$
$-5x = -3y + 8$

13. $-7x - 6y = 5$
$11x + 8y = -10$

14. $x - 2y = 6$
$-3x + 2y = 6$

15. $15x = y + 19$
$3x = -y + 17$

16. $\frac{9}{20}x = \frac{7}{20}y$
$\frac{2}{3}x = 9y$

17. $5x - y = 9$
$25x - 5y = 45$

18. $-3x + 5y = 7$
$-18x + 30y = 42$

19. $\frac{1}{8}y + \frac{1}{4}x = 3$
$2x + y = 9$

20. $\frac{3}{5}y - \frac{2}{5}x = \frac{7}{5}$
$\frac{3}{7}y - \frac{2}{7}x = \frac{5}{7}$

21. $2x + 5y = 13$
$3x + 6y = 12$

22. $4x + 3y = 9$
$6x + 4y = 18$

23. $7x - 9y = -24$
$5x - 2y = 5$

24. $2x - 5y = -13$
$5x + 6y = 97$

In Exercises 25 through 45, find the solution of the system of equations by substitution.

25. $2x + 5y = 12$
$y = 3x - 1$

26. $6x - 8y = 34$
$y = 3x - 2$

27. $4x - 3y = 16$
$y = 7x + 6$

28. $7x + 9y = -34$
$y = 8x + 5$

29. $3x + 2y = 20$
$x + y = 8$

30. $x + 3y = 8$
$6x + 5y = 9$

31. $4x + 6y = 26$
$6x - 2y = 28$

32. $y = x + 1$
$y = -2x + 1$

33. $y = -x + 7$
$y = x + 1$

34. $y = \frac{1}{4}x + \frac{5}{4}$
$y = x - 1$

35. $4y - 3x = 17$
$y = 5$

36. $y = -3x - 4$
$y = x - 1$

37. $x + y = 18$
$y = x - 4$

38. $y = 2x - 4$
$y = x - 1$

39. $3x + y = 7$
$3x - y = 5$

40. $y - 3x = 9$
     $y = 3$

41. $4x - 8y = 8$
     $2x - 5y = 7$

42. $8x + y = -12$
     $3x + 4y = 10$

43. $y = 2x + 1$
     $y = -2x + 1$

44. $2x = 3y - 1$
     $y = 5$

45. $7x + 3y = 4$
     $y = 3x - 4$

In Exercises 46 through 66, use any method to solve each system of equations.

46. $y = 2x + 6$
     $2x + 4y = 14$

47. $y = 2x + 4$
     $2x - 2 = 4y$

48. $11x + 3y = 2$
     $y = 2x - 5$

49. $x = 9$
     $2x + y = 8$

50. $x = 4$
     $x + y = 5$

51. $x = 2y + 4$
     $x = y - 1$

52. $2x - 3y = 9$
     $y = 2x - 4$

53. $2x - 4y = -6$
     $3y + x = 7$

54. $x = 2y - 1$
     $x = 5$

55. $x = -3$
     $3x + y = 9$

56. $2x - 3 = y$
     $5y - 7 = 6x$

57. $y + 2x = 4$
     $-3x - 2y = -7$

58. $x = -4$
     $5x - y = 10$

59. $3x - 7 = y$
     $y = 3x + 5$

60. $x + y = -9$
     $2x + 2y = 18$

61. $0.25y = 0.75x + 5$
     $0.5x - 7 = 0.25y$

62. $\frac{1}{4}x + \frac{1}{4}y = 2$
     $x + y = 8$

63. $y = 3x + 2$
     $x + y = 4$

64. $5x = 3y - 9$
     $-1.5y = -3.0x - 4.5$

65. $3x + \frac{1}{3}y = 21$
     $2x - 3y = 43$

66. $4x - y = 7$
     $5y + x = 7$

MIXED PRACTICE

By doing these exercises, you will practice the topics up to this point in the chapter.

67. Graph $3y = 2x + 3$ and $x + y = -4$, and identify their point of intersection.

68. Solve using addition: $y + x = 1$
                        $5x - y = 17$

69. Solve by graphing: $x + y = 7$
                        $x - 2y = 4$

70. Solve using substitution: $0.3x + 0.3y = 3.6$
                               $0.2x = 3.4$

71. Solve using substitution: $x + y = 4$
                               $2x + y = 6$

72. Graph $y = 6x - 9$ and $y = -6x + 15$, and identify their point of intersection.

73. Find the point of intersection of the graphs of these equations by addition.

$$y = 7x - 9$$
$$3x + y = 1$$

74. Solve the following system to find where the graphs of the equations intersect. Do not graph.

$$x + y = 6$$
$$3x + 2y = 15$$

75. Solve using addition: $-4x + 2y = 6$
$$6x - 3y = 2$$

76. Solve by graphing: $8y + 8x = 8$
$$y = 2x + 4$$

EXCURSIONS

Class Act

1. Television technologists use the following formula to calculate cathode resistance.

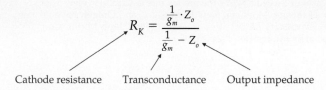

$$R_K = \frac{\frac{1}{g_m} \cdot Z_o}{\frac{1}{g_m} - Z_o}$$

Cathode resistance   Transconductance   Output impedance

a. As Z_o decreases, what happens to R_K?

b. Solve this equation for g_m.

c. *Research* Find an equation used in a field you are interested in.

Exploring Numbers

2. When numbers are raised to the first power or the fifth power, the last digit of the base is always the same as the last digit of the result. For example,

$$9^1 = 9 \qquad 2^5 = 32$$
$$8^1 = 8 \qquad 3^5 = 243$$

List all other powers that give this same result.

3. Using only the digits 1, 2, and 4, replace a, b, and c to make these two statements true:

$$ab^b = acc$$
$$ba^b = cca$$

Posing Problems

4. Ask and answer four questions using the following data. Share your questions with a classmate.

State General Sales and Use Taxes[1]

State	Percent Rate	State	Percent Rate	State	Percent Rate
Alabama	4.0	Louisiana	4.0	Oklahoma	4.5
Arizona	5.0	Maine	6.0	Pennsylvania	6.0
Arkansas	4.5	Maryland	5.0	Rhode Island	7.0
California	6.0	Massachusetts	5.0	South Carolina	5.0
Colorado	3.0	Michigan	6.0	South Dakota	4.0
Connecticut	6.0	Minnesota	6.5	Tennessee	6.0
D.C.	5.75	Mississippi	7.0	Texas	6.25
Florida	6.0	Missouri[2]	4.225	Utah	4.875
Georgia	4.0	Nebraska	5.0	Vermont	5.0
Hawaii	4.0	Nevada	6.5	Virginia[4]	3.5
Idaho	5.0	New Jersey	6.0	Washington	6.5
Illinois	6.25	New Mexico	5.0	West Virginia	6.0
Indiana	5.0	New York[3]	4.0	Wisconsin	5.0
Iowa	5.0	North Carolina	4.0	Wyoming	4.0
Kansas	4.9	North Dakota	5.0		
Kentucky	6.0	Ohio	5.0		

1. Local and county taxes, if any, are additional. 2. State use tax 5.725% in areas where local tax and state meet or exceed 5.725%. In areas that do not meet or exceed 5.725%, the use tax rate is 4.225%. 3. New York City, 8.25%. 4. Local rate 1%. Note: Alaska, Delaware, Montana, New Hampshire, and Oregon have no state-wide sales and use taxes. *Source: 1996 Information Please® Almanac* (©1995 Houghton Mifflin Co.), p. 47. All rights reserved. Used with permission by Information Please LLC.

CONNECTIONS TO *GEOMETRY*

Circles

> **Circle**
>
> A **circle** is the set of points that are equidistant from a fixed point called the **center.**

This means that any point on a circle is the same distance from the center as any other point. That distance is the measure of the radius.

Three radii are shown in the accompanying figure.

Radius

A **radius** (plural: radii) is a line segment drawn from the center of a circle to any point on the circle.

Diameter

A **diameter** is a line segment drawn from one point on a circle through the center to another point on the circle.

In the next figure, we can see that a diameter has measure equivalent to that of two radii of the same circle. The radius of a circle is equal in measure to one-half that of the diameter. In symbols,

If a circle has radius of measure r and diameter of measure d, then
$$r = \frac{d}{2} \quad \text{and} \quad d = 2r$$

Diameter

Circumferences of Circles

Circumference

The **circumference** of a circle is the distance around the circle—its perimeter.

WRITER'S BLOCK

The circumference of any circle divided by its diameter always gives the same constant value. Why is this an important idea?

To calculate the circumference C, we use the fact that $C/d = \pi$ (the Greek letter pi) for any circle. This fact gives us the following formulas.

To find the circumference C of a circle

If d is the diameter or r is the radius of the circle, then
$$C = \pi d \quad \text{or} \quad C = 2\pi r$$

The symbol π represents a constant—a fixed number like 7 or 342—but its value cannot be determined exactly. To ten decimal places, it is 3.1415926536. This is usually approximated as 3.14 or $\frac{22}{7}$.

Circumferences have units of length—that is, feet, inches, miles, and such.

Areas of Circles

The formula for the area of a circle also includes the constant π.

Area A

To find the area A of a circle

If r is the radius of the circle, then
$$A = \pi r^2$$

Areas are measured in square units.

▪▪▪

EXAMPLE

A unit circle is inscribed within a square. Find the area of the square that lies outside of the circle.

SOLUTION

A unit circle is a circle with a radius of one. By *inscribed* we mean that the circle is entirely within the square and touches it on four sides. We make a sketch and note that the diameter of the circle is the same as the length of one side of the square.

The diameter is twice the length of the radius. (All the formulas we need for solving this problem are on the inside covers of this text.)

$$d = 2r$$
$$d = 2(1)$$
$$d = 2 \text{ units}$$

Next we calculate the area of the square. (Area of the square $= s^2$.)

$$A = 2^2$$
$$A = 4 \text{ square units}$$

Now, we calculate the area of the circle.

$$A = \pi r^2$$
$$A = (3.14)(1^2)$$
$$A = 3.14 \text{ square units}$$

To complete this problem, we find the difference in the area of the square and the area of the circle:

$$4 - 3.14 = 0.86 \text{ square inches}$$

◢

PRACTICE

Use the formulas you learned in this Connection to work the following problems. If you need an estimate for pi, use either 3.14 or $\frac{22}{7}$. Write all answers as decimals and round to the nearest hundredth.

1. When baseball was first played, a regulation baseball had a diameter of $2\frac{3}{84}$ inches. Find the radius of a circle with the same diameter.

2. Find the radius of a quarter if its diameter is $\frac{11}{12}$ inches.

3. The diameter of the moon is approximately 2160 miles. If a space vehicle were to move in a straight path all the way around the moon, how far would it travel?

4. The widest point on a baseball bat can be no larger than 2.75 inches in diameter. What is the largest that the circumference can be?

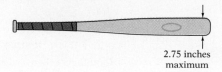

2.75 inches
maximum

5. The largest sunspot recorded was 124,274 miles across. What was the radius of this sunspot?

6. Find the area of a Diamond Jubilee New York City subway token: diameter 0.90 inch

7. One of the first bicycles made in 1876 had a front wheel that measured 5.19 feet in diameter. What was the radius of the wheel?

8. Earth's equator is a circle with a radius of 3963.49 miles. What is its diameter?

$d = 5.19$ ft

9. The actual wrestling area in a wrestling ring has a diameter of 29.5 feet. What does this area measure?

10. A regulation basketball hoop measures 18 inches in diameter. What is the measure of the distance around the rim of the hoop? Use $\frac{22}{7}$ as an estimate for pi in your calculations.

CONNECTIONS TO *STATISTICS*

Reading a Circle Chart

A common type of data display in business reports, newspapers, and magazines is a **circle chart,** or **pie chart.** In a circle chart, the total is represented as a circle, and the parts that make up the total are shown as "slices." The numbers are usually given in the form of percents; the total is 100%.

▪▪▪

EXAMPLE

The following chart shows how a family spends its yearly income of $31,000. How much money does this family spend on transportation?

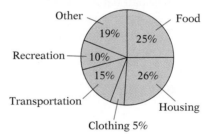

SOLUTION

The chart indicates that 15% of the income is spent on transportation. We must answer the question: 15% of $31,000 is what?

Writing as an equation and solving, we get

$$n = 0.15 \times 31,000 = 4650$$

So the family spends $4650 on transportation yearly.

◢

PRACTICE

Pie chart A represents the way a third-world government intends to spend money received on the sale of oil to the United States. (The total amount of money received is $2 billion.) Use this information to answer Exercises 1 and 2.

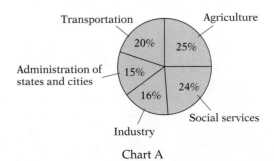

Chart A

1. How much will the country spend on social services and agriculture together?

2. How much more will be spent on industry than on the administration of states and cities?

Circle chart B shows how individuals spent their money for recreation in 1986. Use this information to answer Exercises 3 and 4 for a total annual recreation budget of $2000.

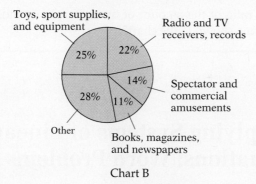

Chart B

3. How much money is left each month after the money for spectator and commercial amusements is spent? Assume the same amount is spent each month.

4. How much is spent per week on radio and TV receivers and records if the same amount is spent each week?

Pie chart C shows how a farmer in a certain third-world country spends his time. Assume that the entire circle represents a 16-hour day (he has to sleep), and use this information to answer Exercises 5 and 6.

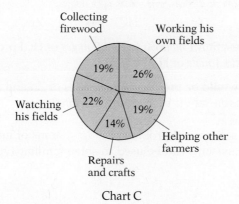

Chart C

5. How many hours does he spend working his own fields and helping other farmers?

6. How many minutes does he spend collecting firewood?

7. See pie chart A. What is the ratio of the amount of money spent on social services to that spent on transportation?

8. See the circle chart given in the Connections example. What is the ratio of the amount of money spent on transportation to that spent on food?

For Exercises 9 and 10, see pie chart C.

9. What is the ratio of the amount of time spent on repairs and crafts to the time spent working the farmer's own fields and helping other farmers?

10. What is the ratio of the amount of time a farmer spends sleeping to the amount of time shown in the chart?

8.3 Applying Systems of Linear Equations: Word Problems

SECTION LEAD-IN

In chemistry we may have limiting factors when we mix chemical elements and compounds to create another compound.

In such a case, only one of the reactants may be completely consumed. This reactant (or reagent as it is sometimes called) is the limiting reactant, or limiting reagent.

When zinc metal reacts with hydrochloric acid, zinc chloride and hydrogen are produced.

$$Zn + 2HCl \longrightarrow ZnCl_2 + H_2$$
$$\underline{1}\ Zn + \underline{2}\ HCl \longrightarrow \underline{1}\ ZnCl_2 + \underline{1}\ H_2$$

In this reaction 1 *mole* of Zn and 2 *moles* of HCl produce 1 *mole* of $ZnCl_2$ and 1 *mole* of H_2.

▪ What would be produced by 8 moles of Zn and 5 moles of HCl?
▪ Which reactant is the limiting reactant?

In this section, we solve word problems using systems of linear equations. As you will see, these systems can be used to solve familiar types of word problems more easily.

▪▪▪
EXAMPLE 1

The Traffic Violations Bureau issues tickets for $45 and $78; the larger ticket is for towed vehicles. On one day, a team of parking officials issued 198 tickets worth a total of $11,088. How many of each type of ticket was issued?

SOLUTION

We let x be the number of \$45 tickets. We let y be the number of \$78 tickets.

	at \$45	at \$78	Total
Number of tickets	x	y	198
Cost of tickets	$45x$	$78y$	11,088

Our equations are

$$x + y = 198$$
$$45x + 78y = 11,088$$

To solve, we multiply the first equation by -45 and solve the resulting system:

$$
\begin{aligned}
-45x - 45y &= -8910 & &\text{Multiplying by } -45 \\
\underline{45x + 78y} &= \underline{11,088} & & \\
33y &= 2178 & &\text{Adding} \\
y &= 66 & &\text{Multiplying by } \tfrac{1}{33} \\
x + (66) &= 198 & &\text{Substituting} \\
x &= 132 & &\text{Solving for } x
\end{aligned}
$$

So there were 66 \$78-tickets and 132 \$45-tickets.

▶ *CHECK* **Warm-Up 1**

> **STUDY HINT**
>
> *Always check the answer to a word problem in the original problem statement, rather than in the equation you wrote.*

•••

EXAMPLE 2

Two trains leave a station at the same time, traveling in opposite directions. One travels 10 miles per hour slower than the other. In 6 hours, they are 1200 miles apart. How fast is each train traveling?

SOLUTION

We want to find the speeds of the trains, so we assign the variables to the speeds.

$$\text{Let } x = \text{speed of first train}$$
$$\text{Let } y = \text{speed of second train}$$

> **STUDY HINT**
>
> *Write out what each variable represents.*

The best way to solve this problem, with these variables, is to make a table showing the distance formula (*rate* times *time* equals *distance*) for the two trains. After they travel 6 hours, we have

	Rate	$\times$	Time	$=$	Distance
First train	x	$\cdot$	6	$=$	$6x$
Second train	y	$\cdot$	6	$=$	$6y$

Now we need to write two equations. The problem states that the speed of the first train is 10 miles per hour less than that of the second, so

$$x = y - 10$$

The problem further states that the distance between the two trains is 1200 miles, so the sum of the distance traveled by the two trains must be 1200 miles. (Remember, the trains are traveling in opposite directions.)

Our second equation is

$$6x + 6y = 1200$$

The system of equations is

$$x = y - 10$$
$$6x + 6y = 1200$$

We solve this system using substitution.

$$6(y - 10) + 6y = 1200$$

$6y - 60 + 6y = 1200$	Distributive property
$12y - 60 = 1200$	Combining like terms
$12y = 1260$	Adding 60 to both sides
$y = 105$	Multiplying by $\frac{1}{12}$

In the first equation, we substitute for y and solve for x.

$$x = 105 - 10$$
$$x = 95$$

To check, we note that after 6 hours, the trains must be 1200 miles apart.

The first train travels

$$95 \text{ miles per hour} \times 6 \text{ hours} = 570 \text{ miles}$$

The second travels

$$105 \text{ miles per hour} \times 6 \text{ hours} = 630 \text{ miles}$$

They are then $570 + 630 = 1200$ miles apart.

So the first train travels at 95 miles per hour, and the second train travels at 105 miles per hour.

▶ *CHECK* **Warm-Up 2**

In the next example, we solve a new type of *mixture* problem. This type of problem involves mixing two different solutions of a certain ingredient to get a desired concentration of the ingredient. (We did some mixture problems involving coins in an earlier chapter.)

■ ■ ■
EXAMPLE 3

Fatima's chemistry lab stocks an 8% acid solution and a 20% acid solution. How many ounces of each must she combine to produce 60 ounces of a mixture that is 10% acid?

SOLUTION

We want to find how much of each solution must be in the mixture. We assign variables to these two unknowns:

Let x = amount of 8% solution needed

Let y = amount of 20% solution needed

Now we can make a table showing what we know about the two solutions and the mixture.

	Amount of Solution	Amount of Acid
8% solution	x	$0.08x$
20% solution	y	$0.20y$
10% mixture	60	$(0.10)(60)$

The right column tells us how much acid is in the ingredients and the mixture. We obtain the right column by remembering that "an 8% acid solution" means that 8% of the solution is acid. Thus 8% of x ounces of solution is $0.08x$ ounces of acid. Similarly, 20% of y ounces is $0.20y$ ounces of acid; and 10% of 60 is $(0.10)(60)$ ounces of acid.

Now we need to write two equations. We do so by noting that the amounts put into any mixture must add up to the total amount that is in the mixture. Therefore, for the solutions (middle column),

$$x + y = 60$$

and for the acid only (right column),

$$0.08x + 0.20y = (0.10)(60)$$

We have a system of two equations in two unknowns that we can solve by substitution. We rewrite the first equation and substitute in the second.

$y = 60 - x$	First equation
$0.08x + 0.20(60 - x) = 6$	Substituting for y
$0.08x + 12 - 0.20x = 6$	Multiplying out
$-0.12x + 12 = 6$	Combining like terms
$-0.12x = -6$	Adding -12 to both sides
$x = 50$	Multiplying by $-\frac{1}{0.12}$

Now, substituting 50 for x in the first equation, we have

$$50 + y = 60$$
$$y = 10$$

We then need to check the values for x and y to see that they total 60 ounces.

$$50 + 10 = 60$$

And we need to check that the concentration of the mixture is correct.

$$\text{Acid concentration} = \frac{\text{amount of acid}}{\text{amount of solution}} \times 100\%$$

$$= \frac{0.08x + 0.20y}{60} \times 100\%$$

$$= \frac{0.08(50) + 0.20(10)}{60} \times 100\%$$

$$= \frac{4 + 2}{60} \times 100\%$$

$$= \frac{6}{60} \times 100\%$$

$$= \frac{1}{10} \times 100\% = 10\% \qquad \text{Correct}$$

So the 10% mixture contains 50 ounces of 8% acid solution and 10 ounces of 20% acid solution.

▶ *CHECK* **Warm-Up 3**

SECTION FOLLOW-UP

The reaction we are considering is

$$Zn + 2HCl \longrightarrow ZnCl_2 + H_2$$

This chemical equation works in a fashion similar to other equations

$$1Zn + 2HCl \longrightarrow 1ZnCl_2 + 1H_2$$
$$2Zn + 4HCl \longrightarrow 2ZnCl_2 + 2H_2$$
$$3Zn + 6HCl \longrightarrow 3ZnCl_2 + 3H_2$$

The amount of $ZnCl_2$ and H_2 produced by 8 moles of Zn and 5 moles of HCl is limited by the HCl.

According to our chemical equations, the amount of H_2 and $ZnCl_2$ that can be produced by HCl is in the ratio of 1 to 2. That is, if we start with x moles of HCl, we can obtain only $\frac{x}{2}$ moles of H_2 and $\frac{x}{2}$ moles of $ZnCl_2$.

$$2 \text{ moles HCl} \longrightarrow \frac{1}{2} \cdot 2 = 1 \text{ mole } H_2 \text{ (and } ZnCl_2)$$

$$4 \text{ moles HCl} \longrightarrow \frac{1}{2} \cdot 4 = 2 \text{ moles } H_2 \text{ (and } ZnCl_2)$$

and

$$5 \text{ moles HCl} \longrightarrow \frac{1}{2} \cdot 5 = \frac{5}{2} \text{ moles } H_2 \text{ (and } ZnCl_2)$$

HCl is the limiting reactant in this equation. Only $\frac{5}{2}$ moles of Zn would be used for this reaction. To use all 8 moles of Zn, we would need 16 moles of HCl.

8.3 WARM-UPS

Work these problems before you attempt the exercises.

1. *Selling Tickets* The ticket prices were raised a total of $1. With the new prices, the same number of tickets in Example 1 would bring in an additional $89.10. What are the new ticket prices? To answer the question, solve the equations

$$n + t = 1$$
$$132n + 66t = 89.10$$

where n is the amount the $45 ticket is raised and t is the amount the $78 ticket is raised.

2. *Travel Problem* A train and a car leave the same location and travel in opposite directions. The train travels 24 miles per hour faster than the car. In 8 hours they are 1000 miles apart. How fast is each traveling?

3. *Mixture Problem* How many milligrams of a 4% aspirin mixture must be combined with how many milligrams of an 8% aspirin mixture to yield 20 milligrams of a 5% aspirin mixture?

8.3 EXERCISES

Note: Use your graphing calculator to check your results whenever possible.

In Exercises 1 through 30, solve each word problem.

1. *Coin Problem* The coins in a bank have a value of $5.94 and consist of pennies and nickels. There are 358 coins. How many of each coin are in the bank?

2. *Taxi Fares* In one city, the cost for taxi service is $1.25 plus 25 cents per mile; in a second city, the cost is $3.75 plus 15 cents per mile. For what number of miles would the price be the same?

3. *Mixture Problem* How many ounces of a certain chemical that is 40% alcohol must be mixed with a second chemical that is 60% alcohol to get 40 ounces of a mixture that is 55% alcohol?

4. *Interest on Deposits* Walter made two deposits in his account—one check deposited at a 6% annual interest rate and a second check deposited at an 11% annual rate. The total amount of money deposited in one year was $2500. The interest accumulated from the two deposits was $200. Find how much was invested at each rate.

5. *Interest on Deposits*   A certain amount of money was deposited at an 8% annual interest rate, and a second amount was deposited at a 5% annual rate. The total amount of money deposited for one year was $1500 and the interest accumulated from the two investments was $99. Find how much was deposited at each rate.

6. *Walking*   Buzzie starts walking to a town 20 miles away at a rate of 4 miles per hour. After a short time he is met by Diane, and they drive to town at a speed of 28 miles per hour. They arrive in town exactly 2 hours after Buzzie started walking. How far did they drive?

7. *Dirt Bikes*   Celeste rides a dirt bike to a lake in the mountains, 240 miles from her home, at a rate of 35 miles per hour. At the same time, her friend Paul starts for the same lake from a point 60 miles nearer to the lake along the same road, traveling 25 miles per hour. How many hours after she starts will Celeste overtake Paul, and how far is this from the lake?

8. *School Supplies*   Jolene bought 100 erasers for $22. Some were 29 cents each, and the rest were 15 cents each. How many of each did she buy?

9. *Mailing Costs*   At a mall the cost of mailing something is $1.60 for the first item and 80 cents for each additional item. At a post office, the same items can be mailed for one dollar each. When does it become more economical to mail from the mall?

10. *Hiking*   A Girl Scout troop hiked to town at the rate of 4 miles per hour. They came back in a car that averaged 32 miles per hour. If the total travel time was $3\frac{1}{2}$ hours, find the time they traveled each way and the distance to town.

11. *Coupon Clipping*   My desk drawer has 20-cent coupons and 25-cent coupons. The total number of coupons in my drawer is 30. How many do I have of each if the total value of the coupons is $6.40?

12. *Number Problem*   The sum of two numbers is 840. The larger number minus 8 times the smaller is 84. Find the numbers.

13. *Supplementary Angles*   Supplementary angles are two angles whose sum is 180°. If the measure of one angle is 20 less than 4 times the other, find the measures of the angle and its supplement.

14. *Chopped Meat*   How many ounces of chopped meat that is 40% lean must a butcher mix with chopped meat that is 60% lean to get 65 ounces of chopped meat that is 52% lean?

15. *Number Problem*   One number is five times the second number. When I add 16 to each number, their sum is 122. Find the numbers.

16. *Jewelry Making*   An artist sells some niobium earrings for $35 a pair and some for $50 a pair. If the artist sold 15 pairs and collected $660, how many of each did he sell?

17. *Mixture Problem* How many liters of a 25% solution of a drug must be mixed with a 55% solution of the drug to produce 50 liters of a 46% solution of this drug?

18. *Coin Problem* A pocketful of coins includes dimes and quarters. The total value of the 36 coins is $5.10. Find the number of each coin.

19. *Bike Speed* Stanley rides his bicycle at 15 miles per hour for 50 miles. Laurel rides her bike for the same amount of time and travels 10 miles further. At what speed did she travel?

20. *Travel* A car leaves a train depot at 1:00 P.M. and travels at 60 miles per hour. A train leaves the same station $1\frac{1}{2}$ hours later and travels at 80 miles per hour in the same direction as the car. At what time will the train overtake the car?

21. *Geometry* You have a rectangular piece of cloth. The length is three times the width. The perimeter is 448 inches. What is the area?

22. *Number Problem* David and Martin are brothers who play hockey. In 1992, the sum of three times David's jersey number and 5 times Martin's number was their father's age. The difference between David's number and Martin's number was −7. If their father was 59 that year, what were their hockey numbers?

23. *Coffee Exports* Coffee exports in 1992 were bountiful, running into the millions of bags. The sum of the number of millions of bags exported that year and the weight per bag is 204. The difference between half the weight per bag and the number of millions of bags exported is −6. How many millions of bags were exported? How much did each bag weigh?

24. *Number Problem* Professor Churchill teaches MATH 150 and Professor Williams teaches STAT 311 in classrooms on different floors in the college. Professor Churchill's room number is 205 greater than Professor Williams'. Three times the sum of their room numbers is 4281. In what rooms do these professors teach?

25. *Number Problem* One number is five more than another. Their sum is 47. Find the numbers.

26. *Age Problem* Five times Mario's age, plus six, is equal to Charles' age. The sum of their ages is 54. Find Mario's and Charles' ages.

27. *Travel Problem* Two cars travel in the same direction, the first at 40 miles per hour and the second at 60 miles per hour. The first car leaves three hours before the second. In how many hours does the second car overtake the first? How far did they travel?

28. *Interest on Deposits* Troy deposited a total of $9375 in two accounts, one yielding 6% interest annually and the other yielding 8% annually. How much did he deposit in each if he received $600 interest in one year?

29. **Coin Problem** A certain number of quarters must be added to a collection of nickels so that the total number of coins is 19 and the total value of the coins is $2.75. Find the number of each coin.

30. **Ticket Sales** Tickets to a certain show are on sale for $45 and $30. How many of each did the theater sell if it sold 150 tickets and received $5400 in sales?

31. **Ballet Tickets** Admission to the *Nutcracker* was $12.50 for adults and $9 for children. There were 625 tickets sold, and $7130 was collected. How many tickets were sold to adults and how many to children? Solve the system using substitution. Let x = adults' tickets, and let y = children's tickets.

$$12.50x + 9y = 7130$$
$$x + y = 625$$

32. **Fishing** You burn calories no matter whether you fish or cut bait. If you fish, you burn y calories per hour. If you cut bait, you burn x calories per hour. One day I fished for 45 minutes and cut bait for 15 minutes, burning 100 calories. The next day I fished for 15 minutes and cut bait for 45 minutes, burning 140 calories. Solve the following system to find the calories I burned per hour during each activity. Use the addition method.

$$\frac{45}{60}x + \frac{15}{60}y = 100$$
$$\frac{15}{60}x + \frac{45}{60}y = 140$$

MIXED PRACTICE

By doing these exercises, you will practice the topics up to this point in the chapter.

33. The sum of two numbers is 38, and the difference of twice the first number and the second is 13. Find the numbers.

34. Solve by graphing: $5x - 2y = 8$
$$6y = 3x$$

35. Solve using substitution: $y = 6x$
$$x + y = -6$$

36. **Mixture Problem** 40 ounces of a 25% salt solution is in a tank. A solution that is 75% salt is added until the final solution is 55% salt. How many ounces of the 75% solution are added?

37. Is $(6, -4)$ the point of intersection of the graphs of the equations $2x = -3y$ and $x = 6$?

38. Solve by graphing: $y = 3$
$$x + y = 5$$

39. Find the point of intersection of $x - y = 0$ and $2x - 2y = 0$.

40. *Stationary Bicycle*   A stationary bicycle has two fixed speeds that you can pedal, x and y miles per hour. One morning you "travel" 4 miles by pedaling 15 minutes at speed x and 8 minutes at speed y. The next morning you pedal 24 minutes at speed x and 10 minutes at speed y and "cover" 5.7 miles. Solve the following system of equations, using the addition method to find x and y.

$$\frac{15x}{60} + \frac{8y}{60} = 4$$

$$\frac{24x}{60} + \frac{10y}{60} = 5.7$$

41. Solve by graphing: $2x - 8y = 4$
$$5x + 2y = 10$$

42. *Motorcycles and Bicycles*   On a small country road, a motorcycle traveling at 25 miles per hour traveled the same amount of time as a bicycle traveling at 15 miles per hour. If they started at the same place and traveled in the same direction, how long did they travel before they were 80 miles apart? How long would it take if they traveled in opposite directions?

43. Solve by graphing: $x - y = -5$
$$12x - 6y = -30$$

44. Solve using addition: $y = x + 1$
$$y = 3 - x$$

EXCURSIONS

Data Analysis

1. Use this picture and caption to estimate the answers to parts (a) and (b).

Nearly two million people attended the Paris Exposition of 1889 to see and perhaps ride the Ferris Wheel. Invented by George Ferris, the steel circle towered 265 feet in the air. Thirty-six cars revolved slowly, allowing passengers to see newly built skyscrapers in the northern part of the city.

a. What is the circumference of the wheel?

b. If one complete trip (without a stop) took 10 minutes, at what speed would you travel?

c. **Research** How many people could be carried at one time?

2. Using these two pie charts, discuss how the Japanese and Americans are alike and different in their savings and investments strategies.

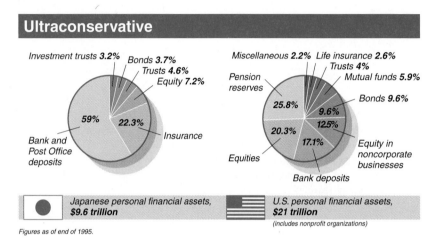

Ultraconservative

Investment trusts **3.2%** Bonds **3.7%**
Trusts **4.6%**
Equity **7.2%**
59% 22.3%
Insurance
Bank and Post Office deposits

Miscellaneous **2.2%** Life insurance **2.6%**
Trusts **4%**
Pension reserves Mutual funds **5.9%**
Bonds **9.6%**
25.8% 9.6%
12.5%
20.3%
17.1%
Equity in noncorporate businesses
Equities
Bank deposits

Japanese personal financial assets, **$9.6 trillion**

U.S. personal financial assets, **$21 trillion**
(includes nonprofit organizations)

Figures as of end of 1995.

Source: Reprinted by permission of *FORBES Magazine* © Forbes Inc., 1996.

Exploring Geometry

3. a. How does the circumference of a circle change when the radius is doubled?

   b. How does the square of the radius of a circle (that is, r^2) change when the area is doubled? (Use the fact that $r^2 = \frac{A}{\pi}$. Consider circles with areas of 31.4 and 62.8 square units.)

   c. How does the circumference of a circle change when the diameter is doubled?

   d. How does the area of a circle change when the diameter is doubled?

4. Use the data in the three pie charts that follow and construct tables. We started the first for you. Interpret the results you find in the tables and in the pie charts.

Number of Programs

Category	Number of People Reporting out of 7000	Equation
1–2	420	6% of 7000 = 0.06 × 7000 = 420
3–4		
5–6		
7–8		
9 or more		
Total	7000	

> **WRITER'S BLOCK**
>
> The Number of Programs pie chart does not sum to 100%. How is this possible?

For frequent flyer tickets and ticket type, choose your own base number of respondents.

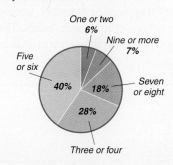

Number of Programs

"In how many frequent flyer programs are you enrolled?"

One or two 6%
Nine or more 7%
Five or six
Seven or eight
40%
18%
28%
Three or four

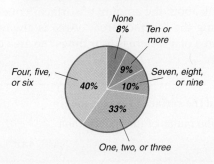

Frequent Flyer Tickets

"How many free tickets did you actually claim in the last twelve months?"

None 8%
Ten or more
Four, five, or six
9%
Seven, eight, or nine
40%
10%
33%
One, two, or three

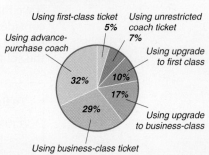

Ticket Type

"How do you fly most frequently when traveling internationally?"

Using first-class ticket 5%
Using unrestricted coach ticket 7%
Using advance-purchase coach
Using upgrade to first class
32%
10%
17%
29%
Using upgrade to business-class
Using business-class ticket

Source: Frequent Flyer, December 1994, as cited in the *1996 Business Information Please® Almanac* (©1995 Houghton Mifflin Co.), pp. 539–540. Reprinted by permission of *Frequent Flyer.*

CONNECTIONS TO *GEOMETRY*

Spheres

Volume of a Sphere

A sphere is shown in the accompanying figure; it is a three-dimensional circle. Every point on the sphere is at a distance r (the radius) from the center.

The volume of a sphere is found with a formula:

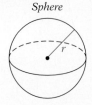

Sphere

> **To find the volume of a sphere**
> If r is the radius of the sphere, then
> $$V = \frac{4}{3}\pi r^3$$

▪▪▪

EXAMPLE

Find the volume of a sphere with a radius of 6 inches.

SOLUTION

To find the volume, we substitute and simplify.

$$V = \frac{4}{3}\pi r^3$$
$$= \frac{4}{3}(3.14)(6)^3$$
$$= \frac{4}{3}(3.14)(216)$$
$$= 904.32$$

So the volume of the sphere is about 904 cubic inches. Note that volume is always measured in cubic units.

◤

Surface Area of a Sphere

If we think about the amount of surface "covering" a sphere or cylinder, we are considering the "surface area." Area is always expressed in square units. The surface area of a sphere is four times the area of a circle with the same radius as the sphere.

To find the surface area of a sphere

If r is the radius of the sphere, then

$$SA = 4\pi r^2$$

▪ ▪ ▪

EXAMPLE

Find the surface area of a sphere with a radius of 2.1 units.

SOLUTION

To find the surface area, we substitute in the formula and simplify.

$$SA = 4\pi r^2$$
$$= 4(3.14)(2.1)^2 = 12.56(4.41)$$
$$= 55.3896$$

The surface area of this sphere is about 55 square units.

◤

PRACTICE

Use the following information to answer Exercises 1 and 2. Neither the sun nor the planets are perfect spheres; they are flattened somewhat at their poles. Use their given diameters to determine their approximate surface areas. Give your answers to the nearest 10 million square miles. Use π on your calculator.

1. **a.** Sun: diameter 865,500 miles   **b.** Pluto: diameter 3700 miles

    c. Venus: diameter 7521 miles   **d.** Neptune: diameter 30,800 miles

    e. Saturn: diameter 74,600 miles

2. **a.** Mercury: diameter 3032 miles   **b.** Mars: diameter 4217 miles

    c. Earth: diameter 7926 miles   **d.** Uranus: diameter 32,200 miles

    e. Jupiter: diameter 88,700 miles

3. Find the approximate volume of the sun and of each planet in Exercise 1. Give your answers in scientific notation.

4. Find the approximate volume of each planet in Exercise 2. Give your answers in scientific notation.

In Exercises 5 and 6, find the surface area and the volume of each ball. Round answers to the nearest ten. Use your calculator if you wish.

5. **a.** A basketball has a   **b.** A cricket ball has a
      circumference of 78 centimeters.   circumference of 9 inches.

6. **a.** A soccer ball has a   **b.** A volleyball has a
      circumference of 28 inches.   circumference of 67 centimeters.

7. The surface of a sphere with a radius of $\frac{1}{5}$ unit is to be painted. How much paint will be needed if 1 gallon of the paint covers exactly 12.5 square units?

$r = 12$ in.

8. You double the radius of a sphere. How does its surface area change?

9. A beach ball with a radius of 12 inches is filled with air. Another beach ball with a radius of 15 inches is also filled with air. What is the difference in volume between the two balls?

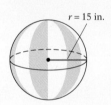

$r = 15$ in.

10. Find the surface area of (a) a sphere with a radius of 0.001 unit and (b) a sphere whose radius is 4 times greater. How do these surface areas compare?

CHAPTER LOOK-BACK

We first assign variables to the amounts of cookies baked:

    Let x = number of batches of shortbread cookies baked

    Let y = number of batches of pan cookies baked

We then translate the constraints into inequalities, reasoning that because we cannot have negative numbers of batches of cookies,

$$x \geq 0 \quad \text{and} \quad y \geq 0$$

To bake x batches of shortbread cookies, the baker needs $3x$ cups of sugar; to bake y batches of pan cookies requires $4y$ cups of sugar. The baker has 76 cups of sugar, so the sugar constraint is

$$3x + 4y \leq 76$$

Similarly, x batches of shortbread cookies require $3x$ cups of butter; y batches of pan cookies require $2y$ cups of butter. The baker has 56 cups of butter, so the butter constraint is

$$3x + 2y \leq 56$$

These are the only constraints.

For x batches of shortbread and y batches of pan cookies, the baker makes a profit of P.

$$P = 2.75x + 2.5y$$

We are asked to maximize this profit, so P is the objective function. The linear programming problem is then to maximize the objective function

$$P = 2.75x + 2.5y$$

subject to the constraints

$$3x + 4y \leq 76$$
$$3x + 2y \leq 56$$
$$x \geq 0$$
$$y \geq 0$$

We first graph the system of inequalities.

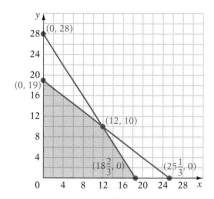

Every point within the shaded area (including the lines) is a solution of the system. However, only points at the intersections of the boundary lines can give a maximum or minimum of the objective function. (You can check this statement by substituting other values in the objective function.) We substitute the coordinates of each of these corner points in the objective function to see which produces a maximum.

$$P = 2.75x + 2.5y$$

For $(0, 0)$: $\quad P = 0$

For $(0, 19)$: $\quad P = 2.75(0) + 2.5(19) = \$47.50$

For $(12, 10)$: $\quad P = 2.75(12) + 2.5(10) = \$58$

For $\left(18\frac{2}{3}, 0\right)$: $\quad P = 2.75\left(18\frac{2}{3}\right) + 2.5(0) = \$51.34$

> Because 58 is the largest value of P, the point $(12, 10)$ maximizes
> the objective function. The baker should make 12 batches of
> shortbread and 10 batches of pan cookies. His profit will be $58.

CHAPTER 8
REVIEW PROBLEMS

The following exercises will give you a good review of the material presented
in this chapter.

SECTION 8.1

1. Is $(-3, 0)$ the point of intersection of $2x + 3y = 6$ and $x + y = -3$?

2. Determine whether there is one solution, no solution, or an infinite number of solutions to the following system.

$$7y + 12 = x$$
$$\frac{1}{2}x - \frac{7}{2}y = 6$$

3. Solve $x = 3$ and $x + y = 3$ by graphing.

4. Solve $x + y = 3$ and $2y + 2x = 10$ by graphing.

5. Graph $y = -4$ and $6x = 3y + 12$, and find their point of intersection.

6. Find the point of intersection of $x - y = 7$ and $y + x = 5$ by graphing.

7. Graph to find the solution of this system of inequalities.

$$x < 5$$
$$x + y \geq 6$$

8. Find the solution of $y \geq -3$ and $y + 2 \leq x$ by graphing.

9. Graph $y \leq 7$ and $3y + x \leq 12$, and find the solution.

10. Find the area of intersection of $y \geq 4$ and $3x - 6 > 9$.

11. Is $(1, 5)$ a point in the solution of $x \geq 8$ and $4x - 3y < 5$? Do not graph.

12. Graph $x + y \geq 2$ and $2x + 2y \geq -4$, and check to see if the point $(-1, 3)$ lies in the solution set.

SECTION 8.2

13. Solve for x and y by addition:
$$y - x = 9$$
$$2y + 3x = 6$$

14. Solve $x - y = -3$ and $2x + y = -3$ by addition.

15. Use addition to solve $7x - 6y = -6$ and $24 + 12y = -14x$

16. Use the addition method to find the point of intersection of $2x - y = 0$ and $5 - 3y = x$.

17. Use substitution to find the solution to $7x + y = 3$ and $x = y + 2$.

18. Use substitution to find x and y in $2x + y = 6$ and $x - y = 4$.

19. Use substitution to find the point of intersection of $3y + x = 7$ and $3y - 12 = -x$.

20. Find x and y in the system $4x + 10y = -30$ and $x + y = -6$ by substitution.

SECTION 8.3

21. The sum of two numbers is 19.4. The difference of the two numbers is 0.2. Find the numbers.

22. The difference of two numbers is 5 and their sum is $11\frac{1}{2}$. Find the numbers.

23. How many pounds of candy at 30 cents per pound must be mixed with 9 pounds of candy at 40 cents per pound to produce a mixture at 36 cents per pound?

24. A train leaves the station and travels at 85 miles per hour. A second train leaves and travels in the opposite direction at 100 miles per hour. How long will it take before they are 1110 miles apart?

MIXED REVIEW

25. Find the solution to the equations $2y - 5x = 8$ and $y - x = 3$.

26. Solve by addition:
$$0.4y - 3.6x = 1.4$$
$$2.4y + 1.8x = 4.5$$

27. Find the solution of $x + y = -2$ and $5x + 4 = y$.

28. Solve by substitution:
$$\tfrac{2}{5}x - 6 = y$$
$$\tfrac{1}{2}y + \tfrac{1}{5}x = 17$$

29. Find the point of intersection of $5y - x = 8$ and $x = y - 5$ by substitution.

30. Solve by graphing:
$$y > 2x - 6$$
$$-4x < y + 5$$

CHAPTER 8 TEST

This exam tests your knowledge of the material in Chapter 8.

1. Determine whether the system of linear equations in part (a) has one solution, no solution, or an infinite number of solutions. For parts (b) and (c), graph and find the solution.

   a. $x - y = 2$
   $y - x = 3$

   b. $3x + 4y = 1$
   $x - y = 5$

   c. $4x + 5y \geq 25$
   $2x > 5$

2. Solve part (a) by addition. For parts (b) and (c), solve by substitution.

   a. $2x + 3y = -4$
   $-2x + 4y = 18$

   b. $12x - 5y = 16$
   $y = x + 1$

   c. $x = y + 2$
   $8y - 7x = -9$

3. Solve these word problems.

   a. The sum of two numbers is 8 and their difference is 2. Find the numbers.

   b. Grade A and Grade B potatoes are sold at two different prices: A costs 59 cents per pound, and B costs 69 cents per pound. How many pounds of Grade A and Grade B potatoes should I buy in order to have 30 pounds of potatoes that cost $20.20 total?

   c. Tickets to a show are $5 for children and $9 for adults. How many tickets of each kind were sold if the total revenue for the show was $4350 and 550 tickets were sold altogether?

CUMULATIVE REVIEW

CHAPTERS 1–8

The following exercises will help you maintain the skills you have learned in this and previous chapters.

1. Factor: $8x^2 - 24x + 16$

2. Add: $\dfrac{3}{x + 4} + \dfrac{1}{3x - 5} + \dfrac{9}{3x^2 + 7x - 20}$

3. Factor: $3x^2 - 243$

4. Divide $27x^2y^{-3}$ by $-3x^5y^{-2}$.

5. Graph: $4x - 9y = 12$

6. Solve for t: $3rt - 4t = 5r$

7. Solve by substitution:
   $12x - 5y = 16$
   $2y = 2x + 2$

8. Add:
   $(x^2 - 3x + 5) + (5x^2 - 3) + (-9x^2 - 7x + 4)$

9. Simplify: $3(-2x^2y)^4$

10. Find the equation of the line with slope -5 that passes through the point $(3, 4)$.

11. Solve by addition: $\begin{aligned} 5x - 4y &= 17 \\ x - y &= -5 \end{aligned}$

12. Simplify: $\dfrac{\frac{x-3}{x-6}}{\frac{x}{x-6}+1}$

13. Subtract: $\dfrac{x}{x^2 - 9} - \dfrac{1}{x - 3}$

14. Multiply: $(-7.4r^2t^3)(-9r^{-3}t^2)$

15. Subtract: $(11x^2 - 9x + 5) - (-18x^2 + 3x - 2)$

16. Solve: $-3x^2 - 42x = 147$

17. Solve for r: $22.4r - 6 \le -7$

18. Solve:
$-6(t - 1) + 7t - 2 + 4(5t - 7) = -(t + 2)$

19. Simplify: $(2x + 5)^2 + (x + 2)^2$

20. Graph the solution:
$y - x \ge -5$
$y + 5 < 2x$

ROOTS AND RADICALS

*I*n the past decade, legislation has led to more equal access for the disabled. Ramps permit equal access for those who use wheelchairs or who have difficulty climbing stairs. The ramps must fit the available space, yet must be usable—a ramp too long or too steep could be dangerous to use. Planning a safe ramp may require the use of the Pythagorean theorem—the square of the ramp's length is equal to the sum of the squares of the height and the base of the ramp— and radicals. The radical, or square root, of the squared length gives the ramp's actual length.

■ *Examine your surroundings for right triangles.*

477

SKILLS CHECK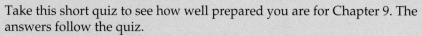

Take this short quiz to see how well prepared you are for Chapter 9. The answers follow the quiz.

1. Evaluate: $2^2 \cdot 3^2 \cdot 5^2$

2. Simplify: $(3xy^2)^2$

3. Find the prime factorization of 240.

4. Simplify by collecting like terms: $3xy - 5x + 4yx - x + xy$

5. Factor: $36x^2y + 6x^2y^2 + 12xy$

6. Reduce to lowest terms: $\dfrac{6ab^2c^4}{12a^2b^3c}$

7. How many whole-number factors does 120 have?

8. Find the square of 103.

ANSWERS: **1.** 900 [Section 1.4] **2.** $9x^2y^4$ [Section 5.1] **3.** $2^4 \cdot 3 \cdot 5$ [Section 6.1] **4.** $8xy - 6x$ [Section 2.3] **5.** $6xy(6x + xy + 2)$ [Section 6.2] **6.** $\dfrac{c^3}{2ab}$ [Section 5.2] **7.** 16 [Section 6.1] **8.** 10,609 [Section 1.4]

CHAPTER LEAD-IN

Architecture makes use of many familiar geometric shapes.

- The triangle is a "rigid" shape that is often used as a brace for other shapes. What characteristics of a triangle make it useful for this purpose?
- What shapes other than squares or rectangles can you find in building structures?

9.1 Finding Roots

SECTION LEAD-IN

A 30-foot water tower sits in back of a farmhouse. It is in bad repair. What length of metal is needed for the four legs of this tower? In this section, we will learn how to calculate such measures.

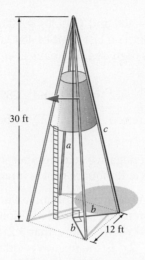

SECTION GOALS

- *To find the square roots of perfect squares*
- *To estimate square roots*
- *To apply the Pythagorean theorem*
- *To apply the distance formula*

Introduction

Multiplying a number by itself is called **squaring** the number. For example, we square 6 by writing

$$6^2 = (6)(6) = 36 \qquad \text{Square of 6}$$

Thus 36 is the square of 6. A number (like 36) that is obtained by squaring an integer is called a **perfect square.**

The operation that is opposite to squaring a number is called **finding a square root of a number.** A **square root** is a number that can be multiplied by itself to produce the square. Thus 6 and −6 are both square roots of 36. Every positive real number has a positive real square root and a negative real square root.

Square Roots

If a and b are positive real numbers and

$$a \cdot a = b$$

then a and $-a$ are square roots of b.

Negative numbers (such as −36) do not have real square roots. The square root of zero is zero.

▪ ▪ ▪

EXAMPLE 1

Find the positive and negative real square roots of each of the following squares.

a. 484   **b.** -484 **c.** $1.44x^4$ **d.** $(-8)^2$

SOLUTION

We must find a number (or term) that, when it is squared, produces each given term.

a. Because $(20)(20) = 400$ and $(30)(30) = 900$, the square root of 484 must be between 20 and 30. The last digit of 484 is 4, so we look for a digit whose square has a units digit of 4. We have $2^2 = 4$ and $8^2 = 64$. So we try 2<u>2</u> and 2<u>8</u>.

$$22^2 = 484 \quad \text{and} \quad 28^2 = 784$$

The positive square root of $484 = 22$, so both 22 and -22 are square roots of 484.

b. Because -484 is a negative real number, it has no real square root, either positive or negative. Its square root is **imaginary.**

c. Here we have a number multiplying a variable. However, the entire term represents a real number, so our definition of square root applies: Because $(1.2x^2)(1.2x^2) = 1.44x^4$, we know that $1.2x^2$ is the positive square root of $1.44x^4$ and that $-1.2x^2$ is the negative square root. We can tell that $1.2x^2$ is positive because the square of any real number x is positive.

d. Because $(-8)^2 = (-8)(-8)$, our definition tells us that -8 is a square root of $(-8)^2$. The other square root is $-(-8) = 8$.

▶ *CHECK* **Warm-Up 1**

Principal Square Root

The *positive* real square root of a number b is called the **principal square root** of b and is written

$$\sqrt{b}$$

The negative real square root of b is then $-\sqrt{b}$. For example,

$$\sqrt{9} = 3 \quad \text{and} \quad -\sqrt{9} = -3$$

If we square either square root of a number, we get the number, so

$$(\sqrt{b})^2 = b \quad \text{and} \quad (-\sqrt{b})^2 = b$$

The sign $\sqrt{}$ is called the **radical sign,** or **square root sign.** A term or expression under the sign is called a **radicand.** Together, they are called a **radical.**

Radical sign $\sqrt{521}$ ◀——— Radicand

Radical

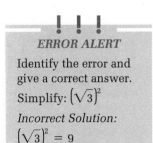

ERROR ALERT

Identify the error and give a correct answer.

Simplify: $(\sqrt{3})^2$

Incorrect Solution:

$(\sqrt{3})^2 = 9$

We know that the square roots of a number x^2 are x and $-x$, because

$$(x)(x) = x^2 \quad \text{and} \quad (-x)(-x) = x^2$$

But which is the principal square root? If we don't know whether x is positive or negative, we must write the principal square root as

$$\sqrt{x^2} = |x|$$

to be sure that the principal square root is positive.

▪▪▪
EXAMPLE 2

Find the principal square root of each of the following.

a. $16a^2$ **b.** $0.49x^2y^2$ **c.** $x + 1$ **d.** $(-5)^2$

SOLUTION

a. The principal square root may be $4a$ or $-4a$, depending on whether the variable a represents a positive or a negative number. We must use absolute value signs to ensure that $\sqrt{16a^2}$ is positive, so $\sqrt{16a^2} = 4|a|$.

b. The principal square root may be $0.7xy$ or $-0.7xy$. Here again we must use absolute value signs, but it is not necessary to put 0.7, a positive real number, within them. We write $\sqrt{49x^2y^2} = 0.7|xy|$.

c. We do not know the value of x, so we cannot evaluate this square root. We simply indicate the principal, or positive, square root as $\sqrt{x + 1}$.

d. Because $(-5)^2 = (-5)(-5) = 25$ and the principal square root of 25 is 5, the principal square root of $(-5)^2$ is also 5.

▶ *CHECK* **Warm-Up 2**

Irrational Square Roots

Positive integers that are not perfect squares do not have integers or rational numbers as square roots. Their square roots are irrational numbers instead. Recall from Section 1.1 that an irrational number is a decimal number whose decimal part does not terminate or repeat. An example is $\sqrt{3} = 1.7320508.\ldots$

▪▪▪
EXAMPLE 3

Determine whether the following square roots are real or not real and whether they are rational or irrational.

a. $\sqrt{36}$ **b.** $-\sqrt{2}$ **c.** $\sqrt{x}$

SOLUTION

a. Because 36 is a perfect square, $\sqrt{36}$ is real and rational.

b. Because 2 is positive, $\sqrt{2}$ is real and so then is $-\sqrt{2}$. Because 2 is not a perfect square, 2 is also irrational.

c. We do not know whether x is positive or negative, so we cannot say whether $\sqrt{x}$ is real or not. Nor do we know whether it is rational or irrational.

▶ CHECK **Warm-Up 3**

Unlike 9 and 16, such numbers as 6 and 12 do not have integer square roots. A rough approximation of the square roots of such whole numbers can be determined by simply finding which two integers the square root is between.

Calculator Corner

You can use your graphing calculator's Home Screen to find the square root of a number.

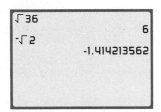

The calculator will also tell you when you try to do the impossible.

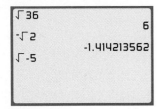   

Press **ENTER.**

Notice that the calculator is telling you there is a *domain error.* You *cannot* take the square root of a negative number.

▪▪▪

EXAMPLE 4

Approximate each of the following numbers.

a. $\sqrt{23}$ **b.** $-\sqrt{103}$ **c.** $3\sqrt{5}$

Solution

a. 23 lies between the perfect squares 16 and 25. Thus $\sqrt{23}$ must lie between $\sqrt{16}$ and $\sqrt{25}$—that is, between 4 and 5.

b. Because 103 is between perfect squares 100 and 121, $\sqrt{103}$ is between 10 and 11. Thus $-\sqrt{103}$ is between -10 and -11.

c. Because 5 is between perfect squares 4 and 9, $\sqrt{5}$ is between 2 and 3. Thus $3\sqrt{5}$ is between $3 \cdot 2$ and $3 \cdot 3$, or between 6 and 9.

> ■■■
> **Writer's Block**
> Use $\sqrt{a^b}$ to give examples of a *radical sign, radical, radicand, base,* and *exponent.*

▶ *CHECK* **Warm-Up 4**

Practice what you learned.

SECTION FOLLOW-UP

The calculations needed for the water tower require simplifying radicals. We will use the Pythagorean theorem:

> In a right triangle with legs of lengths a and b and hypotenuse of length c, $a^2 + b^2 = c^2$.

(You can learn more about the Pythagorean theorem on page 486.) We want to find c in the figure. To find that, we must first find b. We use $c^2 = a^2 + b^2$. We know that $a = 30$ feet. So $c^2 = 30^2 + b^2$. But also, $12^2 = b^2 + b^2$. (Explain.) So

$$2b^2 = 12^2$$
$$2b^2 = 144$$
$$b^2 = 72$$

Substituting in $c^2 = 30^2 + b^2$, we get $c^2 = 30^2 + 72$. So

$$c^2 = 972$$
$$c \approx 31.2$$

We need about 125 feet of this metal. (Explain.)

9.1 WARM-UPS

Work these problems before you attempt the exercises.

1. Find the positive and negative square roots of 625.

2. Find the principal square roots of $169y^2$ and $(-13)^2$.

3. Classify $-\sqrt{225}$ as real or not real and (if it is real) as rational or irrational.

4. Between what two integers does the square root of 983 lie?

9.1 EXERCISES

Note: Use your graphing calculator to check your results whenever possible.

In Exercises 1 through 8, all the numbers are perfect squares. Use the method of Example 1, part (a) to find the principal square root of each. Check by multiplying.

1. 324

2. 484

3. 841

4. 2116

5. 3969

6. 4356

7. 11,025

8. 42,025

In Exercises 9 through 20, determine between what two integers each radical lies.

9. $\sqrt{150}$

10. $\sqrt{173}$

11. $\sqrt{315}$

12. $\sqrt{700}$

13. $\sqrt{1100}$

14. $\sqrt{1700}$

15. $\sqrt{1900}$

16. $\sqrt{2550}$

17. $\sqrt{1436}$

18. $\sqrt{2222}$

19. $\sqrt{10,102}$

20. $\sqrt{62,431}$

In Exercises 21 through 28, use the method of Example 4, part (c) to approximate each number.

21. $5\sqrt{2}$

22. $3\sqrt{12}$

23. $-6\sqrt{6}$

24. $-8\sqrt{8}$

25. $-2\sqrt{27}$

26. $-5\sqrt{310}$

27. $-3\sqrt{566}$

28. $-9\sqrt{838}$

In Exercises 29 through 36, determine whether the number is real or not real and (if it is real) whether it is rational or irrational.

29. $\sqrt{-36}$

30. $\sqrt{17}$

31. $\sqrt{12}$

32. $\sqrt{13}$

33. the square of -2

34. the square root of -27

35. the square of -3

36. the square of -5

In Exercises 37 through 48, find the principal square root. Assume that all variables represent positive real numbers.

37. the square root of $4m^8n^8$

38. the square root of $100x^2y^2$

39. $\sqrt{2^2x^2y^2}$

40. $\sqrt{0.16m^2y^2}$

41. $\sqrt{5^2 \cdot 6^2x^2y^2}$

42. $\sqrt{2^2 \cdot 3^2n^2y^2}$

43. the square root of $900m^2n^8$

44. the square root of $10,000x^{14}y^6$

45. $\sqrt{0.2025x^{44}y^{56}}$

46. $\sqrt{0.0256m^2y^{44}}$

47. $\sqrt{1.0609s^{34}t^{52}}$

48. $\sqrt{420.25r^{28}}$

EXCURSIONS

Exploring Numbers

1. Which is larger? Guess, then evaluate.

   a. 12^3 or 7^4 **b.** 625^2 or 15^4 **c.** 7^3 or 3^7 **d.** 4^2 or 2^4

2. **a.** Try these:

   i. $\square^3 = 157{,}464$ **ii.** $\square^2 = 4225$

   iii. $\square^5 = 7776$ **iv.** $\square^4 = 31{,}640{,}625$

   v. $\square^7 = \underline{\quad\quad\quad\quad}2$

   b. Make a guess and check with a calculator.

   i. $\square^4 = 81$ **ii.** $\square^7 = 823{,}543$

   iii. $\square^5 = 759{,}375$ **iv.** $\square^{10} = 1024$

Data Analysis

3. Study the figure of college expenses, then answer the questions.

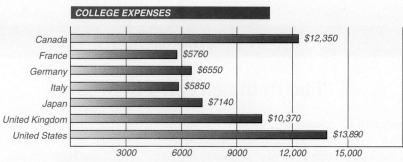

College/university expenditure per student, 1992

Source: OECD, U.S. dollars converted using purchasing power equivalents.
Figures reflect public and private spending.

   a. Represent this data in a table.

   b. ✏ Discuss this information with a classmate. Write a paragraph to be included in a brochure on international education.

   c. Use a pie chart to present this data.

Posing Problems

4. Ask and answer four questions about the figure on the following page.

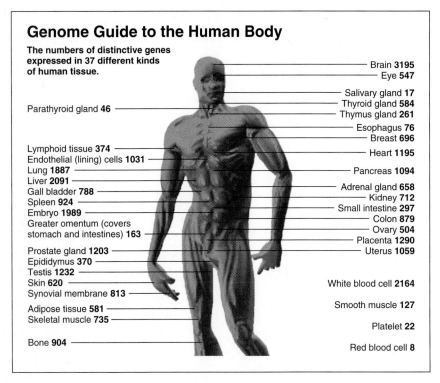

Genome Guide to the Human Body

The numbers of distinctive genes expressed in 37 different kinds of human tissue.

Brain **3195**
Eye **547**

Salivary gland **17**
Thyroid gland **584**
Thymus gland **261**

Parathyroid gland **46**

Esophagus **76**
Breast **696**

Lymphoid tissue **374**
Endothelial (lining) cells **1031**
Lung **1887**
Liver **2091**
Gall bladder **788**
Spleen **924**
Embryo **1989**
Greater omentum (covers stomach and intestines) **163**

Heart **1195**

Pancreas **1094**

Adrenal gland **658**
Kidney **712**
Small intestine **297**
Colon **879**
Ovary **504**
Placenta **1290**
Uterus **1059**

Prostate gland **1203**
Epididymus **370**
Testis **1232**
Skin **620**
Synovial membrane **813**

White blood cell **2164**

Smooth muscle **127**

Adipose tissue **581**
Skeletal muscle **735**

Platelet **22**

Bone **904**

Red blood cell **8**

Source: Copyright © 1996 by The New York Times Co. Reprinted by permission.

CONNECTIONS TO *GEOMETRY*

The Pythagorean Theorem

The ancient Egyptians knew that if the sides of a triangle are in the ratio of 3 to 4 to 5, then that triangle contains a right angle (of measure 90°). They also knew that the right angle is always opposite the longest side, the **hypotenuse.** They used this idea in a variety of ways in everyday life—for example, to ensure that vertical walls were actually vertical.

Later, in the sixth century B.C., the Greek mathematician Pythagoras showed that the 3:4:5 triangle contains a right angle because

$$3^2 + 4^2 = 5^2$$

More important, he proved that a relationship holds among the sides of every right triangle:

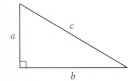

> **The Pythagorean Theorem**
>
> In a right triangle with legs of lengths a and b and hypotenuse of length c, $a^2 + b^2 = c^2$.
>
> The sum of the squares of the lengths of the two legs of a right triangle is equal to the square of the length of the hypotenuse.

We can find the third side of any right triangle if we know the other two sides by using this theorem or its variations, $c^2 - a^2 = b^2$ and $c^2 - b^2 = a^2$.

▪ ▪ ▪
EXAMPLE

Find the length of the third side of this triangle.

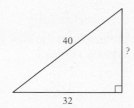

SOLUTION

$$c^2 - a^2 = b^2 \quad \text{Pythagorean theorem}$$
$$40^2 - 32^2 = b^2 \quad \text{Substituting known values}$$
$$576 = b^2 \quad \text{Simplifying}$$
$$\sqrt{576} = b$$
$$\sqrt{576} = 24 = b \quad \text{Finding the square root}$$

So the length of the third side is 24 units.

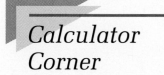
Calculator Corner

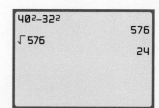

As shown at the right, you can use your calculator to do the computation for the previous example.

▪ ▪ ▪
EXAMPLE

A wooden brace 1.5 meters in length is used to secure the top of a vertical post that is 0.9 meter tall. How far from the base of the post is the base of the brace?

SOLUTION

We will first make a sketch of the situation and label the brace and the post.

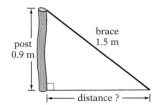

From the diagram, we can see that the brace, the post, and the ground form a right triangle. We are looking for the length of one leg. We substitute $a = 0.9$ and $c = 1.5$ into the formula and simplify.

$$a^2 + b^2 = c^2$$
$$(0.9)^2 + b^2 = 1.5^2$$
$$0.81 + b^2 = 2.25$$
$$b^2 = 1.44$$
$$b = 1.2$$

So the base of the brace is 1.2 meters from the base of the post. (The other theoretical answer, -1.2 meters, makes no sense.)

▪ ▪ ▪

EXAMPLE

Two children walk diagonally across a square lawn. If the perimeter of the lawn is 140 meters, about how far did they walk?

SOLUTION

First we draw and label a diagram.

Note that the diagonal forms two right triangles. We want to find the length of the hypotenuse of either of these triangles. Because the perimeter of the square is 140 meters, the length of each side is

$$\frac{140}{4} = 35 \text{ meters}$$

We substitute $a = b = 35$ in the Pythagorean theorem to find the length of the diagonal.

$$a^2 + b^2 = c^2$$
$$35^2 + 35^2 = c^2$$
$$1225 + 1225 = c^2$$
$$2450 = c^2$$
$$49.50 \approx c$$

So the children walked between 49 and 50 meters.

The Distance Formula

The distance between two points in the coordinate plane is the length of the line segment connecting them. Thus the distance between points A and B in the figure is the length of the line segment AB. We write $\overline{AB}$ to indicate this line segment.

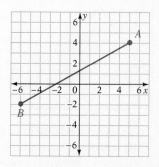

> **Distance Formula**
>
> The distance d between two points with coordinates (x_1, y_1) and (x_2, y_2) is
> $$d = \sqrt{(x_1 - x_2)^2 + (y_1 - y_2)^2}$$

It does not matter which of the coordinates we call (x_1, y_1) and which we call (x_2, y_2). We must, however, be careful about the signs of the coordinates.

▪▪▪
EXAMPLE

Find the distance between the points $(3, -2)$ and $(-1, 1)$.

SOLUTION

First we must identify the coordinates. It's a good idea to write down the identification:

$$\begin{array}{cc} (x_1, y_1) & (x_2, y_2) \\ \downarrow\ \ \downarrow & \downarrow\ \ \downarrow \\ (3, -2) & (-1, 1) \end{array}$$

Next we substitute these values in the distance formula.

$$d = \sqrt{(x_1 - x_2)^2 + (y_1 - y_2)^2}$$
$$= \sqrt{[3 - (-1)]^2 + (-2 - 1)^2}$$

Then we solve the equation by simplifying the radical.

$$d = \sqrt{(3 + 1)^2 + (-2 - 1)^2}$$
$$= \sqrt{(4)^2 + (-3)^2} = \sqrt{16 + 9} = \sqrt{25} = 5$$

So the distance from $(3, -2)$ to $(-1, 1)$ is 5 units.

◀

Calculator Corner

You can use your graphing calculator's Home Screen to evaluate the distance formula. In the previous example, use (a, b) and (c, d) on your Home Screen to stand for (x_1, y_1) and (x_2, y_2) as follows. Then rewrite the distance formula using these new variables. (Most graphing calculators will not let you use subscripts. Check the manual of your graphing calculator.)

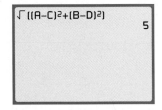

You can now **STO**re new values in *a*, *b*, *c*, and *d* and use the **RECALL** (2nd EN-TER) feature of your graphing calculator to recall the distance formula and find the distance between a new pair of points.

Try finding the distance between the points $(-4, 6)$ and $(8, -12)$.

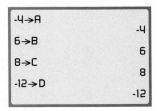 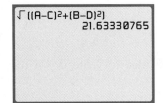

PRACTICE

Solve the following word problems. Sketch a diagram where none is given.

1. *Height of a Pole* A guy wire stretches from the top of a pole to the ground. The wire is 20 feet long and is attached to a spike 12 feet from the base of the pole. How tall is the pole?

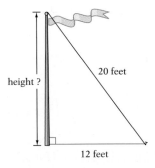

2. *Flight Distance* A bird's nest is 15 feet up in a tree, and the tree is 50 feet from a bird bath. If the mother bird flies directly from the nest to the base of the bird bath, about how far does she fly?

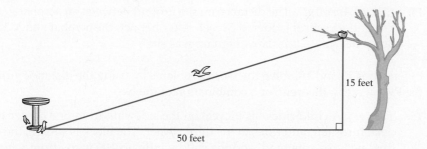

3. *Geometry*   Find the perimeter of the triangle with vertices $(0, 4)$, $(-3, 7)$, and $(-10, -1)$.

4. *Geometry*   Is the quadrilateral with vertices $(4, 6)$, $(4, 8)$, $(6, 6)$, and $(6, 8)$ a square? Justify your answer. (*Hint:* Show that the sides are equal. Because the Pythagorean theorem only works for right triangles, you can use it to prove that one of the angles is a right angle. You will have to find the length of the diagonal also.)

5. *Area of a Triangle*   What is the area of a silver pin that is shaped like a right triangle and has two legs equal to 2.1 inches and 2.8 inches, respectively?

6. *Soccer Field*   A standard soccer field measures 110 yards by 75 yards. Find the length of the diagonal across the field to the nearest hundredth of a yard.

7. *Length of a Diagonal*   A yard in the shape of a rectangle has a perimeter of 98 yards and a width of 21 yards. How much rope would be needed to reach from one corner diagonally across to the other?

8. *Area of a Rectangle*   Find the area of the rectangle with vertices at points $(1, 3)$, $(1, 8)$, $(5, 3)$, and $(5, 8)$.

9. *Difference in Area*   Television screens are measured diagonally. If the screens are square, what is the difference in area between a 21-inch screen and a 24-inch screen?

10. *Right Triangles*   Are $(-2, -2)$, $(2, 14)$, and $(8, 4)$ the vertices of a right triangle?

11. *Perimeter*   Find the perimeter of the four-sided figure that has vertices at $(-2, 4)$, $(0, 3)$, $(5, 5)$, and $(3, 6)$.

12. *Television Screen Size*   A rectangular television screen has a width of 24 inches and a diagonal of 37 inches. Find the length of the TV screen to the nearest inch, and then determine its perimeter.

13. *Rugby*   A standard rugby field is a rectangle 160 yards by 75 yards. Find the length of the diagonal of the field to the nearest yard.

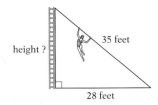

height ?
35 feet
28 feet

14. *Circus Acrobat*   The distance on the ground between an acrobat's safety line and a vertical ladder is 28 feet. After her act, the acrobat slides 35 feet down her safety line. How high up was she?

In Exercises 15 and 16, solve the word problems by using the distance formula, the Pythagorean theorem, or a combination of the two.

15. *Travel*   A child rides his bicycle on the sidewalk from one corner to the next, turns left, and rides to the third corner. The block is square, the child does not cross the road, and the child winds up 100 yards from where he started (diagonally). What are the dimensions of the block?

16. *Construction*   A board is placed on a stairway, just touching four steps, as shown in the figure. If the stairway has the measurements shown, what is the length of the board?

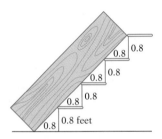

0.8   0.8
0.8   0.8
0.8   0.8
0.8   0.8
0.8   0.8 feet

CONNECTIONS TO *STATISTICS*

Finding the Variance and Standard Deviation

In an earlier statistics Connection, we discussed one measure of variation, or spread—the range. Recall that the range is the difference between the upper and lower limits of the data. While this is important, it does have one major disadvantage. It does not describe the variation among the variables. For instance, both of these sets of data have the same range, yet their values are definitely different.

$$90, 90, 90, 98, 90 \quad \text{Range} = 8$$
$$1, 6, 8, 1, 9, 5 \quad \text{Range} = 8$$

To better describe the variation, we will introduce two other measures of variation—*variance* and *standard deviation* (the variance is the square of the standard deviation). These measures tell us how much the actual values differ from the mean. The larger the standard deviation, the more spread out the values. The smaller the standard deviation, the less spread out the values. This measure is particularly helpful to teachers as they try to find whether their students' scores on a certain test are closely related to the class average.

> **To find the standard deviation of a set of values:**
>
> **a.** Find the mean of the data.
> **b.** Find the difference (deviation) between each of the scores and the mean.
> **c.** Square each deviation.
> **d.** Sum the squares.
> **e.** Dividing by one less than the number of values, find the "mean" of this sum (the **variance***).
> **f.** Find the square root of the variance (the **standard deviation**).
>
> *This formula is for the **sample variance**.

Follow along in the first example.

▪ ▪ ▪

EXAMPLE

Find the variance and standard deviation of the following scores on an exam:

$$92, 95, 85, 80, 75, 50$$

SOLUTION

First we find the mean of the data:

$$\text{Mean} = \frac{92 + 95 + 85 + 80 + 75 + 50}{6} = \frac{477}{6} = 79.5$$

Then we find the difference between each score and the mean (deviation).

Score	Score − Mean	Difference from mean
92	92 − 79.5	+12.5
95	95 − 79.5	+15.5
85	85 − 79.5	+5.5
80	80 − 79.5	+0.5
75	75 − 79.5	−4.5
50	50 − 79.5	−29.5

Next we square each of these differences and then sum them.

Difference	Difference squared
+12.5	156.25
+15.5	240.25
+5.5	30.25
+0.5	0.25
−4.5	20.25
−29.5	870.25
	1317.50 ⟵——— sum of the squares

The sum of the squares is 1317.50.

Next we find the "mean" of this sum (the variance).

$$\frac{1317.50}{5} = 263.5$$

Finally, we find the square root of this variance.

$$\sqrt{263.5} \approx 16.2$$

So, the standard deviation of the scores is 16.2; the variance is 263.5.

◢

▪▪▪

EXAMPLE

Find the standard deviation of the average temperatures recorded over a five-day period last winter:

$$18, 22, 19, 25, 12$$

SOLUTION

This time we will use a table for our calculations.

Temp.	Temp. − mean = deviation	Deviation squared
18	18 − 19.2 = −1.2	1.44
22	22 − 19.2 = 2.8	7.84
19	19 − 19.2 = −0.2	0.04
25 mean	25 − 19.2 = 5.8	33.64
12 ↓	12 − 19.2 = −7.2	51.84
96 ÷ 5 = 19.2		94.80 ◄——sum of squares

To find the variance, we divide by $5 - 1 = 4$.

$$\frac{94.8}{4} = 23.7$$

Finally we find the square root of this variance.

$$\sqrt{23.7} \approx 4.9$$

So the standard deviation for the temperatures recorded is 4.9; the variance is 23.7.

◢

Note that the values in the second example were much closer to the mean than those in the first example. This resulted in a smaller standard deviation.

We can write the formula for the standard deviation as

$$s = \sqrt{\frac{\Sigma(x_i - \bar{x})^2}{n - 1}}$$

where

      Σ means "the sum of."
      x_i represents each value x in the data.
      $\bar{x}$ is the mean of the x_i values.
      n is the total number of x_i values.

PRACTICE

1. The five states with the most covered bridges are

    Oregon: 106
    Vermont: 121
    Indiana: 152
    Ohio: 234
    Pennsylvania: 347

    Find the variance and standard deviation for the number of covered bridges in these states.

2. The five tallest skyscrapers in the United States are

    Sears Tower: 1454 feet
    World Trade Center: 1353 feet
    Empire State Building: 1250 feet
    Standard Oil Building: 1136 feet
    Hancock Insurance Building: 1127 feet

    Find the variance and standard deviation of the heights of these five sky-scrapers.

3. Four fast-food restaurants serve a fish sandwich. The caloric count for each is as follows: 486, 468, 402, 445. Find the variance and standard deviation of these caloric counts.

4. Scores on the most recent reading test are 7.7, 7.4, 7.3, and 7.9. Find the variance and standard deviation of these scores.

5. The highest temperatures recorded in eight specific states are 112, 100, 127, 120, 134, 118, 105, and 110. Find the variance and standard deviation of these scores.

6. TV viewing time during a certain week in millions was 95.2, 94.8, 91.7, 97.7, and 92.4. Find the variance and standard deviation of the numbers of viewers during this week.

7. Using data from Practice Exercises 1 through 6, ask and answer four more questions.

8. Which data from Practice Exercises 1 through 6 had the greatest variation?

SECTION GOALS

- *To use product and quotient properties to simplify square roots*
- *To multiply and divide radicals*
- *To add and subtract radical expressions*

9.2 Operations on Radicals

SECTION LEAD-IN*

Potters use radicals when they glaze their pottery in an oven, called a kiln (pronounced "kil"). The following formula is used to calculate the gas flow needed for heating a pottery kiln:

$$D = V \times A \times \overline{K} \times 1655\sqrt{\frac{H}{G}}$$

where

D = flow output in BTUs per hour
A = area of orifice opening in square inches
$\overline{K}$ = orifice coefficient
H = pressure in units WC (inches of water in a tube)
G = specific gravity
V = gas flow input

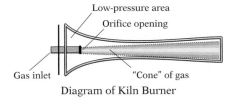

Diagram of Kiln Burner

- How does the value for D change as A or V or H get larger?

* From Mark Ward, "The Mysterious Hole," *Ceramics Monthly*, 43:5 (May 1995), pp. 50–53.

Radicals appear in equations that describe many situations, both scientific and otherwise. You must be able to compute with radicals to solve such equations.

Multiplying Radicals

We can write

$$\left(\sqrt{4}\right)\left(\sqrt{9}\right) = \left(\sqrt{2^2}\right)\left(\sqrt{3^2}\right) = (2)(3) = 6$$

and also

$$\sqrt{(4)(9)} = \sqrt{36} = \sqrt{6^2} = 6$$

Therefore, $\left(\sqrt{4}\right)\left(\sqrt{9}\right) = \sqrt{(4)(9)}$. This is true for all real numbers.

Product Property for Radicals

If a and b are positive real numbers, then
$$\left(\sqrt{a}\right)\left(\sqrt{b}\right) = \sqrt{ab}$$

The **product property** is very useful in simplifying radicals—that is, in removing all squares from under the square root sign. A radical that has no squares is in simplest form. For the problems in this chapter, we will assume that all variables are positive real numbers.

▪▪▪
EXAMPLE 1
Simplify: $\sqrt{125x^2y^5}$

SOLUTION

First we factor all the squares out of the radicand $125x^2y^5$, obtaining

$$(5^2)(5)(x^2)(y^2)(y^2)(y)$$

Applying the product rule, we get

$$\sqrt{(5^2)(5)(x^2)(y^2)(y^2)(y)}$$
$$= \left(\sqrt{5^2}\right)\left(\sqrt{5}\right)\left(\sqrt{x^2}\right)\left(\sqrt{y^2}\right)\left(\sqrt{y^2}\right)\left(\sqrt{y}\right)$$
$$= 5\sqrt{5} \cdot x \cdot y \cdot y\sqrt{y}$$
$$= 5xy^2\sqrt{5y}$$

▶ CHECK **Warm-Up 1**

▪▪▪
EXAMPLE 2
Multiply: $(2\sqrt{6xy})(\sqrt{30xy^2})$

SOLUTION

We use the product property to write the multiplication under one radical sign. This gives us

$$2\sqrt{(6xy)(30xy^2)}$$

Then we multiply and simplify the resulting radical.

$$2\sqrt{(6xy)(30xy^2)} = 2\sqrt{(6)(30)x^{1+1}y^{1+2}} \quad \text{Product property}$$
$$= 2\sqrt{(180x^2y^3)} \quad \text{Multiplying out}$$
$$= 2\sqrt{(36)(5)x^2y^2y} \quad \text{Factoring out squares}$$
$$= (2)(6)xy\sqrt{5y} \quad \text{Simplifying}$$

So $\left(2\sqrt{6xy}\right)\left(\sqrt{30xy^2}\right) = 12xy\sqrt{5y}$.

▶ CHECK **Warm-Up 2**

> Note that we always write the radicand last, as in
> $$3\sqrt{a}$$
> to make it perfectly clear what is under the radical sign.

! ! !
ERROR ALERT

Identify the error and give a correct answer.

Simplify: $3\sqrt{4a^4b^5}$

Incorrect Solution:
$3\sqrt{4a^4b^5} = 3\sqrt{2^2a^4b^4b^1}$
$= a^2b^2\sqrt{b}$

STUDY HINT

Perfect squares can be factored out of a radicand in different ways. They all lead to the same result once you apply the product property for radicals.

WRITER'S BLOCK

Describe the difference between the *square* of 9 and the *square root* of 9.

Dividing Radicals

The **quotient property** for radicals is used to divide radicals or to simplify ratios that involve radicals.

Quotient Property for Radicals

If a and b are real numbers, then

$$\frac{\sqrt{a}}{\sqrt{b}} = \sqrt{\frac{a}{b}}$$

▪▪▪

EXAMPLE 3

Simplify: $\dfrac{\sqrt{2000}}{\sqrt{50}}$

SOLUTION

Using the quotient property, we see that

$$\frac{\sqrt{2000}}{\sqrt{50}} = \sqrt{\frac{2000}{50}} \qquad \text{Quotient property}$$

$$= \sqrt{40} \qquad \text{Dividing out}$$

$$= \sqrt{(4)(10)} \qquad \text{Factoring out squares}$$

$$= \sqrt{2^2(10)} \qquad \text{Rewriting}$$

$$= 2\sqrt{10} \qquad \text{Simplifying}$$

So $\dfrac{\sqrt{2000}}{\sqrt{50}} = 2\sqrt{10}$.

▶ CHECK **Warm-Up 3**

Calculator Corner

Use your graphing calculator's Home Screen to evaluate square roots as for Example 3, illustrating the quotient property for radicals. You can verify that $\frac{\sqrt{a}}{\sqrt{b}} = \sqrt{\frac{a}{b}}$, as shown at the left in the following figure. It is also possible on many graphing calculators to "ask" the calculator if two numerical statements are correct, as shown at the right in the following figure. The calculator will return an answer of 1 if the statement is true and an answer of 0 if the statement is false. (Note: See the Calculator Corner on page 269 to review how to test statements.)

```
√2000/√50
           6.32455532
√(2000/50)
           6.32455532
```

```
√2000/√50=√(2000/
50)
                     1
```

▪▪▪

EXAMPLE 4

Simplify: $\dfrac{\sqrt{56xy^2}}{\sqrt{7x}}$

SOLUTION

Using the quotient property, we find that

$$\frac{\sqrt{56xy^2}}{\sqrt{7x}} = \sqrt{\frac{56xy^2}{7x}} \qquad \text{Quotient property}$$

$$= \sqrt{8y^2} \qquad \text{Dividing out}$$

$$= \sqrt{(4y^2)(2)} \qquad \text{Factoring out squares}$$

$$= \sqrt{2^2 y^2 (2)} \qquad \text{Rewriting}$$

$$= 2y\sqrt{2} \qquad \text{Simplifying}$$

So $\dfrac{\sqrt{56xy^2}}{\sqrt{7x}} = 2y\sqrt{2}$.

▶ CHECK **Warm-Up 4**

■■■
WRITER'S BLOCK
In your own words, explain why $\sqrt{4}$ is a *real number* but $\sqrt{-4}$ is not.

▪▪▪

EXAMPLE 5

Simplify: $\sqrt{\dfrac{36r^4 t^3 v^6}{25x^2}}$

SOLUTION

Because we cannot simplify the expression within the radical, we begin by applying the quotient property to write it as the quotient of two radicals. We will then simplify each one.

$$\sqrt{\frac{36r^4 t^3 v^6}{25x^2}} = \frac{\sqrt{36r^4 t^3 v^6}}{\sqrt{25x^2}} \qquad \text{Quotient property}$$

$$= \frac{\sqrt{6^2 r^4 t^2 v^6}}{\sqrt{5^2 x^2}} \qquad \text{Factoring out squares}$$

$$= \frac{\sqrt{6^2 (r^2)^2 t^2 \cdot t (v^3)^2}}{\sqrt{5^2 \cdot x^2}} \qquad \text{Rewriting}$$

This can be written as

$$\frac{6r^2 t v^3 \sqrt{t}}{5x}$$

▶ CHECK **Warm-Up 5**

Adding and Subtracting Radicals

Like radicals are terms that contain the same radical, such as $5\sqrt{3}$ and $2\sqrt{3}$. We combine like radicals by using the distributive property. For example,

$$5\sqrt{3} + 2\sqrt{3} = (5 + 2)\sqrt{3} = 7\sqrt{3}$$

We cannot combine unlike radicals. However, radicals that appear to be unlike can often be simplified and then combined.

▪▪▪
EXAMPLE 6

Combine the following radicals. Simplify your answers as much as possible.

a. $3\sqrt{5} + 2\sqrt{5} - \sqrt{5}$ **b.** $2\sqrt{5} + 5\sqrt{10}$

SOLUTION

a. These are like radicals. We combine the coefficients.

$$3\sqrt{5} + 2\sqrt{5} - \sqrt{5} = (3 + 2 - 1)\sqrt{5} = 4\sqrt{5}$$

b. This radical expression cannot be simplified further.

▶ *CHECK* **Warm-Up 6**

▪▪▪
EXAMPLE 7

Subtract $2\sqrt{2x} - \left(-2\sqrt{12x^2y}\right)$ from $\sqrt{18x} - 4\sqrt{3y}$. Assume all variables are positive real numbers.

SOLUTION

First we write the problem as a subtraction.

$$\sqrt{18x} - 4\sqrt{3y} - \left[2\sqrt{2x} - \left(-2\sqrt{12x^2y}\right)\right]$$

We remove the parentheses and then the brackets to get

$$= \sqrt{18x} - 4\sqrt{3y} - \left[2\sqrt{2x} + 2\sqrt{12x^2y}\right]$$
$$= \sqrt{18x} - 4\sqrt{3y} - 2\sqrt{2x} - 2\sqrt{12x^2y}$$

Next we simplify where possible, removing all squares from under radical signs.

$$\sqrt{18x} - 4\sqrt{3y} - 2\sqrt{2x} - 2\sqrt{12x^2y}$$
$$= \sqrt{2 \cdot 3^2 x} - 4\sqrt{3y} - 2\sqrt{2x} - 2\sqrt{2^2 x^2 \cdot 3y}$$
$$= 3\sqrt{2x} - 4\sqrt{3y} - 2\sqrt{2x} - 4x\sqrt{3y}$$

Finally, we combine like radicals.

$$= 3\sqrt{2x} - 2\sqrt{2x} - 4\sqrt{3y} - 4x\sqrt{3y} \quad \text{Regrouping}$$
$$= (3 - 2)\sqrt{2x} - (4 + 4x)\sqrt{3y} \quad \text{Distributive property}$$

We cannot combine 4 and $4x$, so we either leave the answer in this form or write it as $\sqrt{2x} - 4\sqrt{3y} - 4x\sqrt{3y}$. Also, because we are told to assume that all variables are positive real numbers, we do not need absolute value signs around x.

▶ CHECK **Warm-Up 7**

Practice what you learned.

SECTION FOLLOW-UP

The value for D gets larger as each of the variables A, V, or H gets larger. The flow of gas in BTUs must be carefully controlled, however, because the temperature of the kiln and how fast the kiln reaches the maximum desired temperature affects the color of the glazes.

9.2 WARM-UPS

Work these problems before you attempt the exercises.

1. Find the square root of $180t^5w^4$.

2. Multiply: $(\sqrt{27x^5y^3})(\sqrt{18xy})$

3. Simplify: $\dfrac{\sqrt{300}}{\sqrt{50}}$

4. Simplify: $\dfrac{\sqrt{30x^2y}}{\sqrt{5x^2y}}$

5. Simplify: $\sqrt{\dfrac{325x^5t^4}{25xy^4}}$

6. Simplify: $2\sqrt{27} - \sqrt{300}$

7. Subtract $-5\sqrt{44w^2x^3y} - 3\sqrt{99w^2xy}$ from $2\sqrt{11w^2x^3y}$.

9.2 EXERCISES

Note: Use your graphing calculator to check your results whenever possible.

In Exercises 1 through 36, simplify each radical or radical expression. Assume all variables represent positive real numbers.

1. $\sqrt{120x^9}$

2. $\sqrt{168y^3}$

3. $\sqrt{124x^2}$

4. $\sqrt{180y^5}$

5. $\sqrt{(2x)(6x^5)}$

6. $\sqrt{(8y^9)(2y)}$

7. $\sqrt{(5x)(15x^2)}$

8. $\sqrt{(32y^3)(2y)}$

9. $\dfrac{\sqrt{40}}{\sqrt{32}}$

10. $\dfrac{\sqrt{30}}{\sqrt{40}}$

11. $\dfrac{\sqrt{260}}{\sqrt{20}}$

12. $\dfrac{\sqrt{169}}{\sqrt{196}}$

13. $\sqrt{39y}(\sqrt{65y})$

14. $\sqrt{16y}(-\sqrt{72y})$

15. $(\sqrt{15x})(\sqrt{20x})$

16. $(\sqrt{35t})(\sqrt{15t})$

17. $\sqrt{(4x^2y)(8x^3)}$

18. $\sqrt{(18x^3y)(9x^3y)}$

19. $\sqrt{(-4x^2y)(-8x^3y^2)}$

20. $\sqrt{(-14x^3y^3)(-7x^5y)}$

21. $(\sqrt{4r^2t^3})(\sqrt{2r^4t^3})$

22. $(\sqrt{3r^2t^2})(\sqrt{42r^5t^6})$

23. $(\sqrt{20r^4})(\sqrt{24rt^3})$

24. $(\sqrt{8x^3y})(\sqrt{12x^4y^7})$

25. $\sqrt{24rt^8}(-\sqrt{8rt^3})$

26. $\sqrt{180tx^4}(\sqrt{12tx^2})$

27. $\sqrt{30rt^3}(\sqrt{60rt^2})$

28. $-\sqrt{9tx^3}(-\sqrt{135tx^4})$

29. $\dfrac{\sqrt{12r^2t^4}}{\sqrt{3r^8t^6}}$

30. $\dfrac{\sqrt{16rt}}{\sqrt{4rt}}$

31. $\dfrac{\sqrt{100rt^3}}{\sqrt{9rt}}$

32. $\dfrac{\sqrt{48x^5y}}{\sqrt{3xy}}$

33. $-\sqrt{60r}(-\sqrt{39r^2t})$

34. $-\sqrt{6t}(\sqrt{96t^2})$

35. $(\sqrt{42r^2})(9\sqrt{12r})(-\sqrt{2})$

36. $(\sqrt{4t^2})(-\sqrt{18t})(\sqrt{9})$

In Exercises 37 through 68, perform the indicated operations and simplify all radicals. Assume all variables represent positive real numbers.

37. $\sqrt{12} - \sqrt{75}$

38. $2\sqrt{28} - \sqrt{112}$

39. $\sqrt{45} + \sqrt{125}$

40. $\sqrt{40} + 3\sqrt{90}$

41. $\sqrt{52} - 2\sqrt{117}$

42. $\sqrt{16} + \sqrt{64} - \sqrt{100}$

43. $3\sqrt{27} + \sqrt{12} - 2\sqrt{18}$

44. $2\sqrt{24} - \sqrt{81} + 3\sqrt{24}$

45. $-\sqrt{8} - \sqrt{64} - \sqrt{8}$

46. $5\sqrt{48} + 2\sqrt{36} + \sqrt{48}$

47. $\sqrt{128} - 2\sqrt{250} - 5\sqrt{128}$

48. $\sqrt{27} + 2\sqrt{12} - \sqrt{75}$

49. $\sqrt{24} - \sqrt{81} + \sqrt{3}$

50. $-\dfrac{\sqrt{8}}{3} - \dfrac{2\sqrt{40}}{6} + \dfrac{\sqrt{8}}{12}$

51. $\dfrac{40\sqrt{8x^2}}{12} - \dfrac{10\sqrt{32x^2}}{15}$

52. $\dfrac{6\sqrt{8t^2}}{9} - \dfrac{8\sqrt{64t^2}}{12}$

53. $\dfrac{3t\sqrt{12t^2}}{8} - \dfrac{3t\sqrt{27t^2}}{12}$

54. $\dfrac{9x\sqrt{48x^2}}{15} - \dfrac{x\sqrt{75x^2}}{3}$

55. $\dfrac{t\sqrt{18t^2}}{5} - \dfrac{4t\sqrt{9t^2}}{20}$

56. $\dfrac{9x\sqrt{147x^2}}{6} - \dfrac{2x\sqrt{75x^2}}{18}$

57. $\sqrt{90} - 2\sqrt{40} + \sqrt{8}$

58. $4\sqrt{18x^2} - 10\sqrt{32x^2}$

59. $2\sqrt{24x^3y^4} + 3y\sqrt{24x^3y^2} - 4x\sqrt{81y^4}$

60. $3\sqrt{32t^2} - 2t\sqrt{2} + \sqrt{128t^2}$

61. $5\sqrt{3x^3y} + 2x\sqrt{27xy} - 4x\sqrt{27xy}$

62. $5v\sqrt{27v^3} - 3v^2\sqrt{3v}$

63. $\sqrt{4t} - \sqrt{9t} + \sqrt{16t}$

64. $\sqrt{20} - (\sqrt{10} + \sqrt{80} - \sqrt{100})$

65. $\sqrt{28} - (2\sqrt{63} + \sqrt{112} + \sqrt{56})$

66. $\sqrt{27} - (\sqrt{54} + 2\sqrt{48})$

67. Subtract $5\sqrt{18} - 2\sqrt{8}$ from $3\sqrt{12}$.

68. Subtract $6\sqrt{63} - \sqrt{28}$ from $\sqrt{56} - \sqrt{112}$.

MIXED PRACTICE

By doing these exercises, you will practice the topics up to this point in the chapter. Assume all variables represent positive real numbers.

69. Approximate the square root of 330 to two decimal places.

70. Find the length of a diagonal of a rectangle with sides of 15 and 20.

71. Simplify: $\dfrac{20\sqrt{40}}{5\sqrt{8}}$

72. Approximate the value of $-\sqrt{5}$.

73. A playing field has a diagonal measure of 98 yards. If it is 50 yards long, how wide is it to the nearest yard?

74. Multiply: $\sqrt{7x}(\sqrt{x})(\sqrt{4})$

75. Is the square of 8 real or not real and (if it is real) rational or irrational?

76. Is a triangle with sides of 5 inches, 8 inches, and 13 inches a right triangle?

77. Is $-\sqrt{38}$ real or not real? If real, is it rational or irrational?

78. The radical $2\sqrt{3}$ lies between what two multiples of 2?

79. Is the square of -7 real or not real, and (if real) is it rational or irrational?

80. Find all square roots of 64, both positive and negative.

81. Find the diagonal of a square that is 4 inches on one side.

82. Approximate the value of $\sqrt{93}$.

EXCURSIONS

Exploring Geometry

1. In these diagrams, the distance between any two dots horizontally or vertically is one unit. Use the distance formula and the Pythagorean theorem to label all the sides with their lengths.

a.    b.   c. 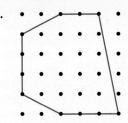   d.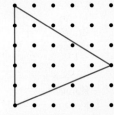

e. 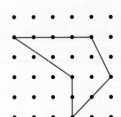   f. 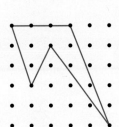   g. 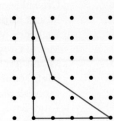   h.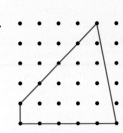

Exploring Numbers

2. a. Between any two fractions on a number line, there is another fraction that is halfway between them. We call this property **betweenness**. Use the idea of the mean to find the fraction that lies halfway between each of the following pairs of fractions.

   i. $\frac{1}{5}$ and $\frac{1}{125}$ ii. $\frac{1}{7}$ and $\frac{1}{8}$ iii. $\frac{1}{100}$ and $\frac{1}{10}$ iv. $\frac{11}{13}$ and $\frac{12}{13}$

b. This same property (betweenness) holds for decimal fractions. By finding the mean of the pair, find the decimal fraction that lies halfway between each of the following pairs of decimal fractions.

i. 0.21 and 0.22   **ii.** 0.004 and 0.005   **iii.** 0.231 and 0.232   **iv.** 0.10097 and 0.10098

Posing Problems

3. Ask and answer four questions using the following data.

Highest, Lowest, and Mean Elevations in the United States

State	Mean elevation (feet)	Highest point	Elevation (feet)	Lowest point	Elevation (feet)
Alabama	500	Cheaha Mountain	2,405	Gulf of Mexico	Sea level
Alaska	1,900	Mount McKinley	20,320	Pacific Ocean	Sea level
Arizona	4,100	Humphreys Peak	12,633	Colorado River	70
Arkansas	650	Magazine Mountain	2,753	Ouachita River	55
California	2,900	Mount Whitney	14,494	Death Valley	282
Colorado	6,800	Mount Elbert	14,433	Arkansas River	3,350
Connecticut	500	Mount Frissell, on south slope	2,380	Long Island Sound	Sea level
Delaware	60	On Ebright Road	448	Atlantic Ocean	Sea level
D.C.	150	Tenleytown, at Reno Reservoir	410	Potomac River	1
Florida	100	Sec. 30, T6N, R20W	345	Atlantic Ocean	Sea level
Georgia	600	Brasstown Bald	4,784	Atlantic Ocean	Sea level
Hawaii	3,030	Puu Wekiu, Mauna Kea	13,796	Pacific Ocean	Sea level
Idaho	5,000	Borah Peak	12,662	Snake River	710
Illinois	600	Charles Mound	1,235	Mississippi River	279
Indiana	700	Franklin Township, Wayne County	1,257	Ohio River	320
Iowa	1,100	Sec. 29, T100N, R41W	1,670	Mississippi River	480
Kansas	2,000	Mount Sunflower	4,039	Verdigris River	679
Kentucky	750	Black Mountain	4,139	Mississippi River	257
Louisiana	100	Driskill Mountain	535	New Orleans	8
Maine	600	Mount Katahdin	5,267	Atlantic Ocean	Sea level
Maryland	350	Backbone Mountain	3,360	Atlantic Ocean	Sea level
Massachusetts	500	Mount Greylock	3,487	Atlantic Ocean	Sea level
Michigan	900	Mount Arvon	1,979	Lake Erie	572
Minnesota	1,200	Eagle Mountain	2,301	Lake Superior	600
Mississippi	300	Woodall Mountain	806	Gulf of Mexico	Sea level
Missouri	800	Taum Sauk Mountain	1,772	St. Francis River	230
Montana	3,400	Granite Peak	12,799	Kootenai River	1,800
Nebraska	2,600	Johnson Township, Kimball County	5,424	Missouri River	840
Nevada	5,500	Boundary Peak	13,140	Colorado River	479
New Hampshire	1,000	Mount Washington	6,288	Atlantic Ocean	Sea level
New Jersey	250	High Point	1,803	Atlantic Ocean	Sea level
New Mexico	5,700	Wheeler Peak	13,161	Red Bluff Reservoir	2,842
New York	1,000	Mount Marcy	5,344	Atlantic Ocean	Sea level
North Carolina	700	Mount Mitchell	6,684	Atlantic Ocean	Sea level
North Dakota	1,900	White Butte	3,506	Red River	750

Ohio	850	Campbell Hill	1,549	Ohio River	455
Oklahoma	1,300	Black Mesa	4,973	Little River	289
Oregon	3,300	Mount Hood	11,239	Pacific Ocean	Sea level
Pennsylvania	1,100	Mount Davis	3,213	Delaware River	Sea level
Rhode Island	200	Jerimoth Hill	812	Atlantic Ocean	Sea level
South Carolina	350	Sassafras Mountain	3,560	Atlantic Ocean	Sea level
South Dakota	2,200	Harney Peak	7,242	Big Stone Lake	966
Tennessee	900	Clingmans Dome	6,643	Mississippi River	178
Texas	1,700	Guadalupe Peak	8,749	Gulf of Mexico	Sea level
Utah	6,100	Kings Peak	13,528	Beaverdam Wash	2,000
Vermont	1,000	Mount Mansfield	4,393	Lake Champlain	95
Virginia	950	Mount Rogers	5,729	Atlantic Ocean	Sea level
Washington	1,700	Mount Rainier	14,410	Pacific Ocean	Sea level
West Virginia	1,500	Spruce Knob	4,861	Potomac River	240
Wisconsin	1,050	Timms Hill	1,951	Lake Michigan	579
Wyoming	6,700	Gannett Peak	13,804	Belle Fourche River	3,099
United States	**2,500**	**Mount McKinley** (Alaska)	**20,320**	**Death Valley** (California)	**282**

4. Study the following table and then answer the questions.

NEW YORK GIANTS STARTING LINEUP (1996)

O F F E N S E	Position	Player	Age	Years in NFL	Position	Player	Age	Years in NFL	D E F E N S E
	WR	Thomas Lewis	24	2	LE	Jamal Duff	24	1	
	LT	Greg Bishop	25	3	LT	Keith Hamilton	25	4	
	LG	Ron Stone	25	3	RT	Robert Harris	27	4	
	C	Brian Williams	30	7	RE	Michael Strahan	24	3	
	RG	Rob Zatechka	24	1	LB	Jessie Armstead	25	3	
	RT	Scott Gragg	24	1	LB	Corey Widmer	27	4	
	TE	Howard Cross	28	7	LB	Corey Miller	27	5	
	WR	Chris Calloway	28	6	CB	Phillippi Sparks	27	4	
	QB	Dave Brown	26	4	CB	Thomas Randolph	25	2	
	RB	Rodney Hampton	27	6	SS	Jesse Campbell	27	5	
	FB	Charles Way	23	1	FS	Tito Wooten	24	2	

a. What is the mean age of the offense? the defense?

b. What was the mean "starting age" of the offense? the defense?

c. What is the range of the starting ages?

d. What is the range of experience?

e. Describe the "average" player. Justify your choice.

f. Ask and answer four other questions using this data.

Data Analysis

5. Use the graph on the following page to answer the questions that follow. Determine the validity of each statement. Justify your answers and show your calculations.

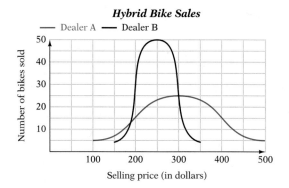

Hybrid Bike Sales

a. Dealer A had a greater variability in selling price than Dealer B.

b. Dealer B's median selling price was higher than Dealer A's.

c. The mode of the selling prices at Dealer A was equal to that at Dealer B.

d. Dealer A's mean selling price was lower than Dealer B's.

CONNECTIONS TO *GEOMETRY*

Calculating Volume

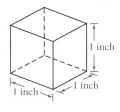

The volume of a three-dimensional object is equal to the amount of space it encloses. The units of volume are cubic units: cubic inches, cubic feet, and so on. A cubic inch is a cube with its length, width, and height all equal to 1 inch; a cubic foot is a cube with its length, width, and height all equal to 1 foot.

> **Cube**
>
> The geometric figure called a **cube** is a solid whose length, width, and height are equal in measure. Each side of a cube is called a face. Two faces meet in an **edge.**

To find the volume of any cube, we multiply its length, width, and height together. Because they all have the same measure:

> **To find the volume of a cube**
>
> If s is the length of an edge, then
>
> $$V = s^3$$

▪ ▪ ▪
EXAMPLE

Find the volume of a cube with an edge of $4\frac{1}{4}$ feet.

SOLUTION

We use the formula $V = s^3$.

$$V = s^3 \qquad \text{The formula}$$
$$= \left(4\tfrac{1}{4}\right)^3 \qquad \text{Substituting } s = 4\tfrac{1}{4}$$
$$= \tfrac{17}{4} \times \tfrac{17}{4} \times \tfrac{17}{4} \qquad \text{Multiplying out}$$
$$= \tfrac{4913}{64} \quad \text{or} \quad 76\tfrac{49}{64} \qquad \text{Simplifying}$$

So the volume is $76\tfrac{49}{64}$ cubic feet.

◢

A cube has six square surfaces that are called its **faces.** A solid whose faces are rectangles is called a **rectangular solid** (sometimes called a **rectangular prism**). A rectangular solid looks somewhat like a cube, but its dimensions—length, width, and height—are not necessarily all equal.

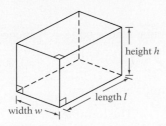

To find the volume of a rectangular solid, we multiply its length, width, and height. In symbols,

---

To find the volume of a rectangular solid

If ℓ is the length, w is the width, and h is the height, then
$$V = \ell \times w \times h$$

▪▪▪

EXAMPLE

Find the volume of a rectangular solid with length 14 inches, width 13 inches, and height 8 inches.

SOLUTION

We substitute the values for ℓ, w, and h in the formula and solve.

$$V = \ell \times w \times h \qquad \text{The formula}$$
$$= 14 \times 13 \times 8 \qquad \text{Substituting}$$
$$= 1456 \qquad \text{Multiplying}$$

So the volume is 1456 cubic inches.

◢

Calculating Surface Area

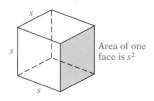

Area of one
face is s^2

Next we will find the surface area of a cube and a rectangular solid.

A cube has six faces, each in the shape of a square.

To find the surface area of a cube, we simply find the area of one of the square faces and multiply it by 6.

To find the surface area of a cube

If s is the length of an edge of a cube, then
$$SA = 6s^2$$

▪ ▪ ▪

EXAMPLE

Find the surface area of a cube whose edge is $3\frac{1}{2}$ inches long.

SOLUTION

We substitute into the formula and solve.

$$SA = 6s^2 \qquad \text{The formula}$$
$$= 6\left(3\tfrac{1}{2}\right)^2 \qquad \text{Substituting}$$
$$= 6\left(\tfrac{7}{2}\right)^2 \qquad \text{Changing to improper fraction}$$
$$= 6\left(\tfrac{49}{4}\right) \qquad \text{Removing the exponent}$$
$$= 73\tfrac{1}{2} \qquad \text{Multiplying and simplifying}$$

The surface area of the cube with an edge of $3\frac{1}{2}$ inches is $73\frac{1}{2}$ square inches.

◀

Notice that surface areas are given in square units.

To find the surface area of a rectangular solid, we find the areas of the faces and add these areas together. The solid shown in the following figure has two faces of area ℓh, two faces of area wh, and two faces of area $w\ell$.

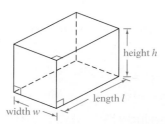

height h

length l

width w

> **To find the surface area of a rectangular solid**
>
> If ℓ is the length, w is the width, and h is the height, then
>
> $$SA = 2(\ell h + wh + w\ell)$$

▪ ▪ ▪

EXAMPLE

Find the surface area of a rectangular solid with a length of $2\frac{1}{2}$ inches, a width of 1 inch, and a height of $1\frac{1}{4}$ inches.

SOLUTION

We substitute in the formula and solve.

$$SA = 2(\ell h + wh + w\ell)$$ The formula

$$SA = 2\left[\left(2\tfrac{1}{2}\right)\left(1\tfrac{1}{4}\right) + (1)\left(1\tfrac{1}{4}\right) + (1)\left(2\tfrac{1}{2}\right)\right]$$ Substituting

$$= 2\left[\left(\tfrac{5}{2}\right)\left(\tfrac{5}{4}\right) + \tfrac{5}{4} + \tfrac{5}{2}\right]$$ Changing to improper fractions

$$= 2\left[\left(\tfrac{25}{8}\right) + \tfrac{5}{4} + \tfrac{5}{2}\right]$$ Multiplying

$$= 2\left[\tfrac{25}{8} + \tfrac{10}{8} + \tfrac{20}{8}\right]$$ Rewriting with LCD

$$= 2\left(\tfrac{55}{8}\right)$$ Adding

$$= \tfrac{110}{8} = 13\tfrac{3}{4}$$ Multiplying and simplifying

Thus the surface area of the rectangular solid is $13\frac{3}{4}$ square inches.

◢

PRACTICE

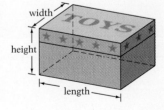

1. A toy box is advertised as holding 40 cubic feet of toys. If the length of the toy box is 5 feet and its height is 3 feet, how wide must the toy box be? Use the formula $w = V/(\ell h)$. If you paint the toy box on the outside, what surface area must the paint cover? Include the top.

2. Find the number of cubic feet in a freezer that has a height of $3\frac{1}{2}$ feet, a length of 8 feet, and a width of $2\frac{1}{4}$ feet.

3. To find the required capacity of an air conditioner (in BTUs), we multiply the volume of the room it is to cool by 3. How much capacity is needed for a room that measures $17\frac{1}{3}$ feet long, $11\frac{1}{2}$ feet wide, and 9 feet high?

4. A mini-storage facility advertises bins that are 15 feet wide, $10\frac{1}{2}$ feet long, and $8\frac{1}{4}$ feet wide. What is the volume of each bin?

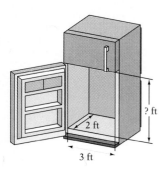

5. Find the volume of a cube that has an edge of $1\frac{1}{2}$ inches. What is its surface area?

6. A refrigerator is said to have a volume of 32 cubic feet. If the length of the interior of the refrigerator is 3 feet and the interior width is 2 feet, how high must the interior of the refrigerator be? Use the formula $h = V/w\ell$.

7. Find three things in your environment and calculate their surface area and volume.

8. ✐ Do surface area and volume have a relationship to each other that is constant? Justify your answer with data. You may use your results from Exercises 1 through 6 or make up other cubes and rectangular solids to test your ideas.

SECTION GOALS

- *To use the distributive property to multiply and simplify radical expressions*

- *To simplify an expression by rationalizing the denominator*

9.3 Simplifying Radical Expressions

SECTION LEAD-IN*

The size of the orifice (round opening) used for supplying gas to a kiln must be converted from a drill size or pipe size to an area given in square inches. The openings are small, usually less than $\frac{1}{2}$ inch. How do the areas vary? Write a formula to find the diameter of an orifice, given its area in square inches.

*From: Marc Ward, "The Mysterious Hole," *Ceramics Monthly,* 43:5 (May 1995), pp. 50–53.

The distributive property holds true for any real numbers.

Distributive Property

For real numbers a, b, and c,

$$a(b + c) = ab + ac$$

and

$$(b + c)a = ba + ca$$

We can use this to simplify radical expressions.

• • •

EXAMPLE 1

Simplify: $\sqrt{32}\left(\sqrt{24} - \sqrt{56}\right)$

SOLUTION

We begin by simplifying all radicals.

$$\sqrt{32}\left(\sqrt{24} - \sqrt{56}\right) = \sqrt{16 \cdot 2}\left(\sqrt{4 \cdot 6} - \sqrt{4 \cdot 2 \cdot 7}\right) \quad \text{Rewriting}$$
$$= \sqrt{4^2 \cdot 2}\left(\sqrt{2^2 \cdot 6} - \sqrt{2^2 \cdot 14}\right) \quad \text{Factoring}$$
$$= 4\sqrt{2}\left(2\sqrt{6} - 2\sqrt{14}\right) \quad \text{Simplifying}$$

Next we multiply, using the distributive property.

$$= \left(4\sqrt{2}\right)\left(2\sqrt{6}\right) - \left(4\sqrt{2}\right)\left(2\sqrt{14}\right)$$

We multiply out and simplify, combining the like radicals.

$$= 8\sqrt{12} - 8\sqrt{28}$$
$$= 8\sqrt{2^2 \cdot 3} - 8\sqrt{2^2 \cdot 7}$$
$$= 16\sqrt{3} - 16\sqrt{7} \quad \text{Multiplying}$$

So $\sqrt{32}\left(\sqrt{24} - \sqrt{56}\right) = 16\sqrt{3} - 16\sqrt{7}$.

▶ *CHECK* **Warm-Up 1**

■ ■ ■
EXAMPLE 2

Multiply and simplify: $\left(2\sqrt{3} - \sqrt{6}\right)\left(4\sqrt{2} - 1\right)$

SOLUTION

As we do when multiplying binomials, we use FOIL.

$$\left(2\sqrt{3} - \sqrt{6}\right)\left(4\sqrt{2} - 1\right)$$
$$\overset{F}{=} \left(2\sqrt{3}\right)\left(4\sqrt{2}\right) + \overset{O}{\left(2\sqrt{3}\right)\left(-1\right)} + \overset{I}{\left(-\sqrt{6}\right)\left(4\sqrt{2}\right)} + \overset{L}{\left(-\sqrt{6}\right)\left(-1\right)}$$
$$= 8\sqrt{6} - 2\sqrt{3} - 4\sqrt{12} + \sqrt{6}$$

Then we simplify all radicals and combine where possible.

$$= 8\sqrt{6} - 2\sqrt{3} - 4\sqrt{2^2 \cdot 3} + \sqrt{6}$$
$$= 8\sqrt{6} - 2\sqrt{3} - 8\sqrt{3} + \sqrt{6} \quad \text{Simplifying}$$
$$= 8\sqrt{6} + \sqrt{6} - 2\sqrt{3} - 8\sqrt{3} \quad \text{Rewriting}$$
$$= 9\sqrt{6} - 10\sqrt{3} \quad \text{Combining like terms}$$

So $\left(2\sqrt{3} - \sqrt{6}\right)\left(4\sqrt{2} - 1\right) = 9\sqrt{6} - 10\sqrt{3}$.

▶ *CHECK* **Warm-Up 2**

We can apply the product and quotient properties together when necessary.

■ ■ ■
EXAMPLE 3

Divide: $\dfrac{\sqrt{3x^3} \cdot \sqrt{6x}}{\sqrt{15x^2y^2}} \div \sqrt{\dfrac{3}{20w^2}}$

SOLUTION

First we use the quotient property to write the first fraction as one radical.

$$\frac{\sqrt{3x^3} \cdot \sqrt{6x}}{\sqrt{15x^2y^2}} \div \sqrt{\frac{3}{20w^2}} = \sqrt{\frac{3x^3 \cdot 6x}{15x^2y^2}} \div \sqrt{\frac{3}{20w^2}}$$

We then simplify each radical as much as possible.

$$= \sqrt{\frac{6x^2}{5y^2}} \div \sqrt{\frac{3}{20w^2}}$$

We rewrite the problem as a product. And we use the product property to combine the two radicals.

$$= \sqrt{\frac{6x^2}{5y^2}} \cdot \sqrt{\frac{20w^2}{3}}$$

$$= \sqrt{\frac{(6x^2)(20w^2)}{(5y^2)(3)}}$$

Finally, we multiply and simplify.

$$= \sqrt{\frac{120x^2w^2}{15y^2}} \qquad \text{Multiplying}$$

$$= \sqrt{\frac{8x^2w^2}{y^2}} \qquad \text{Dividing}$$

$$= \frac{2xw\sqrt{2}}{y} \qquad \text{Simplifying}$$

▶ *CHECK* **Warm-Up 3**

Rationalizing the Denominator

An expression is not considered simplified if it has a radical in the denominator. (Denominator radicals can lead to larger-than-usual rounding errors.) Removing a radical from the denominator of a fraction is **rationalizing the denominator.** We use a procedure that is similar to the way we write an equivalent fraction. That is, we multiply the numerator and denominator by the same number.

To rationalize a denominator

For a fraction of the form $\frac{a}{\sqrt{b}}$, multiply numerator and denominator by $\sqrt{b}$ to obtain

$$\frac{a}{\sqrt{b}} \cdot \frac{\sqrt{b}}{\sqrt{b}} = \frac{a\sqrt{b}}{b}$$

▪▪▪

EXAMPLE 4

Simplify: $\frac{\sqrt{12}}{\sqrt{45}}$

SOLUTION

First we simplify the radicals (if possible).

$$\frac{\sqrt{12}}{\sqrt{45}} = \frac{\sqrt{3 \cdot 4}}{\sqrt{9 \cdot 5}} = \frac{2\sqrt{3}}{3\sqrt{5}}$$

Then, to rationalize the denominator, we multiply numerator and denominator by $\sqrt{5}$. (We need not be concerned about the 3.) We get

$$\frac{2\sqrt{3}\sqrt{5}}{3\sqrt{5}\sqrt{5}} = \frac{2\sqrt{15}}{3\sqrt{5^2}} = \frac{2\sqrt{15}}{(3)(5)} = \frac{2\sqrt{15}}{15}$$

▶ *CHECK* **Warm-Up 4**

! ! !
ERROR ALERT

Identify the error and give a correct answer.

Rationalize the denominator:

$$\frac{3}{\sqrt{2}}$$

Incorrect Solution:

$$\frac{3}{\sqrt{2}} = \frac{3}{\sqrt{2}} \cdot \frac{1}{\sqrt{2}} = \frac{3}{2}$$

We would have eventually obtained the same result by multiplying the numerator and denominator by $3\sqrt{5}$, but there was no need to do that multiplication.

▪▪▪
EXAMPLE 5

Simplify this rational expression by rationalizing the denominator:

$$\frac{\sqrt{6} + 2\sqrt{15}}{\sqrt{3}}$$

SOLUTION

We multiply both numerator and denominator by $\sqrt{3}$ and then simplify.

$$\frac{(\sqrt{6} + 2\sqrt{15})\sqrt{3}}{\sqrt{3}\sqrt{3}} = \frac{\sqrt{6}\sqrt{3} + 2\sqrt{15}\sqrt{3}}{3}$$ Distributive property

$$= \frac{\sqrt{18} + 2\sqrt{45}}{3}$$

$$= \frac{\sqrt{9 \cdot 2} + 2\sqrt{9 \cdot 5}}{3}$$

$$= \frac{3\sqrt{2} + 6\sqrt{5}}{3} = \sqrt{2} + 2\sqrt{5}$$

▶ CHECK **Warm-Up 5**

> ▪▪▪
> **WRITER'S BLOCK**
> What do we do when we *rationalize* a fraction?

Two expressions $a + b$ and $a - b$ are said to be **conjugates** of each other, that is, the sum and difference of two numbers. When we multiply these conjugates, this gives us another way to rationalize some denominators.

> **To rationalize a denominator using a conjugate**
>
> For a fraction of the form $\frac{c}{a + b}$, where c can be any real number and either a or b or both are radicals, multiply numerator and denominator by the conjugate $a - b$.

▪▪▪
EXAMPLE 6

Simplify: $\frac{2 + \sqrt{15}}{\sqrt{5} + \sqrt{3}}$

SOLUTION

We multiply both numerator and denominator by the conjugate of $\sqrt{5} + \sqrt{3}$, which is $\sqrt{5} - \sqrt{3}$.

$$\frac{(2 + \sqrt{15})(\sqrt{5} - \sqrt{3})}{(\sqrt{5} + \sqrt{3})(\sqrt{5} - \sqrt{3})}$$ Multiplying numerator and denominator by $\sqrt{5} - \sqrt{3}$

$$= \frac{(2 + \sqrt{15})(\sqrt{5} - \sqrt{3})}{(\sqrt{5})^2 - (\sqrt{3})^2}$$ Using the special product in the denominator

$$= \frac{2\sqrt{5} - 2\sqrt{3} + \sqrt{75} - \sqrt{45}}{5 - 3}$$ Multiplying with FOIL; simplifying the denominator

$$= \frac{2\sqrt{5} - 2\sqrt{3} + 5\sqrt{3} - 3\sqrt{5}}{5 - 3}$$ Simplifying the radical terms

$$= \frac{3\sqrt{3} - \sqrt{5}}{2}$$ Combining like terms

> ❗❗❗
> *ERROR ALERT*
>
> Identify the error and give a correct answer.
>
> Rationalize the denominator:
>
> $$\frac{3}{2 + \sqrt{3}}$$
>
> *Incorrect Solution:*
>
> $$\frac{3}{2 + \sqrt{3}} \cdot \frac{2 + \sqrt{3}}{2 + \sqrt{3}}$$
>
> $$= \frac{6 + 3\sqrt{3}}{4 + 3}$$
>
> $$= \frac{6 + 3\sqrt{3}}{7}$$

Thus

$$\frac{2 + \sqrt{15}}{\sqrt{5} + \sqrt{3}} = \frac{3\sqrt{3} - \sqrt{5}}{2}$$

▶ CHECK **Warm-Up 6**

SECTION FOLLOW-UP

The area (in inches) of an orifice formed by a $\frac{1}{8}$-inch drill is

$$A = \pi\left(\frac{\frac{1}{8}}{2}\right)^2 = \pi\left(\frac{1}{16}\right)^2 = 0.0123$$

To find a formula for the area in terms of the diameter of an orifice, we first solve for r. Because $d = 2r$ implies $r = \frac{d}{2}$ (diameter $= d$) and $A = \pi r^2$ we have ·

$$A = \pi\left(\frac{d}{2}\right)^2 = \frac{\pi d^2}{4}$$

for the area written in terms of the diameter.

9.3 WARM-UPS

Work these problems before you attempt the exercises.

1. Simplify: $\sqrt{7}\left(\sqrt{28} - 2\sqrt{21}\right)$

2. Multiply and simplify: $\left(\sqrt{2} - 2\sqrt{3}\right)^2$

3. Multiply: $\left(\sqrt{\frac{125x^3y}{5xy}}\right)\left(\sqrt{\frac{3x^2y}{12y}}\right)$

4. Simplify: $\frac{\sqrt{120}}{\sqrt{50}}$

5. Simplify: $\frac{\sqrt{2} - 2\sqrt{3}}{\sqrt{6}}$

6. Simplify: $\frac{2 - 3\sqrt{5}}{\sqrt{3} + \sqrt{2}}$

9.3 EXERCISES

Note: Use your graphing calculator to check your results whenever possible.

In Exercises 1 through 46, simplify.

1. $\left(\sqrt{3v}\right)\left(\sqrt{3v} + \sqrt{27v}\right)$

2. $\left(\sqrt{12v}\right)\left(\sqrt{12v} - \sqrt{48v}\right)$

3. $\left(\sqrt{2w}\right)\left(\sqrt{2w} - 9\right)$

4. $\sqrt{2}\left(5\sqrt{8t^2} - 10\sqrt{32t^2}\right)$

5. $-\sqrt{3}\left(8t\sqrt{12t^2} - 3t\sqrt{27t^2}\right)$

6. $2\sqrt{3}\left(9x\sqrt{48x^2} - x\sqrt{75x^2}\right)$

7. $3\sqrt{2}\left(\sqrt{44} + 2\sqrt{8} + 6\sqrt{11}\right)$

8. $\left(\sqrt{3v}\right)\left(\sqrt{3v} + \sqrt{24v}\right)$

9. $\left(\sqrt{6v}\right)\left(\sqrt{12v} - \sqrt{54v}\right)$

10. $\left(3\sqrt{2} + 2\sqrt{10}\right)^2$

11. $\left(\sqrt{2} + 3\right)^2$

12. $\left(\sqrt{3} - 5\right)^2$

13. $\left(\sqrt{5} - 7\right)^2$

14. $\left(\sqrt{2} - 1\right)\left(\sqrt{2} + 1\right)$

15. $\left(\sqrt{3} - 2\right)\left(\sqrt{3} + 2\right)$

16. $\left(\sqrt{5} + 3\right)\left(\sqrt{5} - 3\right)$

17. $\left(\sqrt{10} + 5\right)\left(\sqrt{10} - 5\right)$

18. $\left(4\sqrt{2} - 3\right)^2$

19. $\left(4\sqrt{3} - 3\sqrt{6}\right)^2$

20. $\left(5\sqrt{2} + 1\right)^2$

21. $\left(\sqrt{2} + 2\right)^2$

22. $\left(\sqrt{1} - \sqrt{2}\right)\left(\sqrt{1} + \sqrt{2}\right)$

23. $\dfrac{3}{\sqrt{5}}$

24. $\dfrac{5}{\sqrt{2}}$

25. $\dfrac{10}{\sqrt{10}}$

26. $\dfrac{12\sqrt{10}}{10\sqrt{3}}$

27. $\dfrac{9\sqrt{15}}{6\sqrt{2}}$

28. $\dfrac{-2\sqrt{12}}{\sqrt{18}}$

29. $\dfrac{-7\sqrt{3}}{\sqrt{21}}$

30. $\dfrac{5}{-2\sqrt{75}}$

31. $\dfrac{12}{2\sqrt{72}}$

32. $\dfrac{7\sqrt{15}}{\sqrt{5}}$

33. $\dfrac{\sqrt{30}}{5\sqrt{3}}$

34. $\dfrac{9\sqrt{5}}{\sqrt{45}}$

35. $\dfrac{6\sqrt{3x}}{8\sqrt{24x}}$

36. $\dfrac{\sqrt{27y}}{2\sqrt{30y}}$

37. $\dfrac{\sqrt{18t}}{\sqrt{2t} - 1}$

38. $\dfrac{3 + \sqrt{2}}{\sqrt{2} + 1}$

39. $\dfrac{6 + \sqrt{5}}{\sqrt{5} + 2}$

40. $\dfrac{\sqrt{21}}{2\sqrt{3} - \sqrt{7}}$

41. $\dfrac{\sqrt{16} - \sqrt{25}}{\sqrt{12} - 2}$

42. $\dfrac{2\sqrt{5} - \sqrt{7}}{\sqrt{35} + 5}$

43. $\dfrac{\sqrt{120}}{\sqrt{3}(\sqrt{30} + \sqrt{24})}$

44. $\dfrac{2\sqrt{21}}{\sqrt{7}(\sqrt{63} - \sqrt{21})}$

45. $\dfrac{\sqrt{27}}{\sqrt{2}(\sqrt{54} - \sqrt{66})}$

46. $\dfrac{\sqrt{30}}{\sqrt{3}(\sqrt{15} + 2\sqrt{30})}$

In Exercises 47 through 72, multiply or divide as indicated. Write all answers without radicals in the denominator.

47. $\dfrac{\sqrt{8} + \sqrt{18}}{\sqrt{3}} \div \dfrac{\sqrt{3} + \sqrt{6}}{\sqrt{2}}$

48. $\dfrac{3\sqrt{5} + \sqrt{20}}{\sqrt{6}} \cdot \dfrac{5\sqrt{8} + \sqrt{3}}{\sqrt{5}}$

49. $\dfrac{2\sqrt{9} + \sqrt{18}}{\sqrt{3}} \cdot \dfrac{8\sqrt{3} + \sqrt{6}}{\sqrt{2}}$

50. $\dfrac{2\sqrt{15} - \sqrt{40}}{\sqrt{3}} \cdot \dfrac{2\sqrt{45} - \sqrt{20}}{\sqrt{10}}$

51. $\dfrac{x\sqrt{24x^3y}\sqrt{180xyx^3}}{x\sqrt{24x^2y}\sqrt{10xyx^2}\sqrt{18y}}$

52. $\left(\sqrt{\dfrac{28x^3y}{2x^2y}}\right)\left(\sqrt{\dfrac{180xyx^3}{48y}}\right)$

53. $\left(\sqrt{\dfrac{120yx^3}{10y}}\right)\left(\sqrt{\dfrac{15xy}{x}}\right)$

54. $\left(\sqrt{\dfrac{160tx^5}{8t}}\right)\left(\sqrt{\dfrac{16tx}{t}}\right)$

55. $\left(\sqrt{\dfrac{150wx^5}{10w}}\right) \div \left(\sqrt{\dfrac{15wy}{w}}\right)$

56. $\left(\sqrt{\dfrac{24x^3y}{24x^2y}}\right) \div \left(\sqrt{\dfrac{18xy}{180x^3y}}\right)$

57. $\left(\sqrt{\dfrac{50x^6}{5}}\right) \div \left(\sqrt{10x}\right)$

58. $\left(\sqrt{\dfrac{24x^2y}{4x^3y}}\right)\left(\sqrt{\dfrac{18xy}{3x^3y}}\right)$

59. $\dfrac{\sqrt{(5t)(10)(20y)}}{\sqrt{(7t)(49)(63y)}}$

60. $\dfrac{(\sqrt{300xy})}{(\sqrt{121xy})}$

61. $\dfrac{(\sqrt{256t})(\sqrt{220st})}{(\sqrt{81})(\sqrt{3s})(\sqrt{3})}$

62. $\dfrac{\sqrt{(18)(20t)}}{\sqrt{21t}} \cdot \sqrt{\dfrac{15t^2}{(14t)(25)}}$

63. $\dfrac{\sqrt{(28r)(11r)}}{\sqrt{(4r)(11rt)}} \cdot \sqrt{\dfrac{6rt^3}{(42rt)}}$

64. $\dfrac{-6\sqrt{(20x)(13)}}{2\sqrt{(13x)(7x)}} \cdot 2\sqrt{\dfrac{28x}{(5)}}$

65. $\dfrac{\sqrt{(10y)(8)(28y)}}{12\sqrt{(3y)(9)}} \cdot \sqrt{\dfrac{1}{(6y)}}$

66. $\dfrac{x\sqrt{48x^3y}\sqrt{180xy}}{x(\sqrt{x^2})(\sqrt{2y})(x^2\sqrt{8y})}$

67. $\sqrt{\dfrac{24x^3y}{24x^2y}} \cdot \sqrt{\dfrac{18xyx^3}{18y}}$

68. $\sqrt{\dfrac{120x^2}{4y}} \cdot \sqrt{\dfrac{15xy}{x}}$

69. $\sqrt{\dfrac{320tx^5}{8t}} \cdot \sqrt{\dfrac{16tx}{t}}$

70. $\sqrt{\dfrac{150wx^5}{6y}} \div \sqrt{\dfrac{1}{20wy}}$

71. $\sqrt{\dfrac{24}{7x^2y}} \div \sqrt{\dfrac{1}{140x^3y}}$

72. $\sqrt{\dfrac{50x^6}{5}} \div \dfrac{1}{\sqrt{10x}}$

MIXED PRACTICE

By doing these exercises, you will practice the topics up to this point in the chapter. Assume all variables represent positive real numbers.

73. Simplify: $\sqrt{90xyz^6}$

74. Simplify: $\sqrt{882}$

75. The radical $3\sqrt{23}$ lies between what two multiples of 3?

76. Find the diagonal of a square that has a side of 50 units. Express your result in simplest radical form.

77. Find, to the nearest hundredth, the length of the hypotenuse of a right triangle whose legs measure 75 and 110 units.

78. Multiply: $\left(\sqrt{15t} + 4\right)\left(\sqrt{7t} - 11\right)$

79. Between what two integers does the square root of 750 lie?

80. Simplify: $5\sqrt{18} + 2\sqrt{8} - \left(6\sqrt{32} - \sqrt{2}\right)$

81. Subtract $5\sqrt{10} - 2\sqrt{1000}$ from $\sqrt{500}$.

82. Subtract $\sqrt{108} - \sqrt{75} + 2\sqrt{27}$ from $3\sqrt{48}$.

EXCURSIONS

Class Act

1. Motorcycles often have a two-cylinder engine. To determine the engine size, multiply the displacement of the cylinder by the number of cylinders. The displacement (D) of one cylinder is found by using the formula

$$\text{Displacement} = \pi \times \left(\frac{\text{bore}}{2}\right)^2 \times \text{stroke}$$

where

$\pi \approx 3.14$
bore = diameter of the cylinder in centimeters
stroke = distance a piston moves in centimeters
displacement = volume of a cylinder in cubic centimeters

a. Solve this equation for the value of the bore. It may be easier if you rewrite the equation using variables instead of words.

b. *Research* What is a "bore" and what numbers are reasonable "bores"?

Exploring Geometry

The following problem is from an algebra text published in 1887.*

2. If a carriage wheel $14\frac{2}{3}$ feet in circumference takes one second more to revolve, the rate of the carriage per hour will be $2\frac{2}{3}$ miles less. How fast is the carriage traveling?

*H.S. Hall and S.R. Knight, *Elementary Algebra*, 4th Ed., London: Macmillan and Co., 1887.

3. The maximum distance you can see from a tall building is given by the equation

$$D \approx 69.4\sqrt{0.0003h}$$

where D is the distance in miles and h is the height of the building in feet.

a. Suppose the height of the building is given in meters. How would you have to change the formula so you can substitute and solve it?

b. **Research** The world's tallest building is *not* the Sears Tower in Chicago. What is? Where is it? How far can you see from the world's tallest building?

CONNECTIONS TO *GEOMETRY*

Cylinders

Volume of a Cylinder

A right circular cylinder is shown here. (We shall simply call it a cylinder.) Its bases are circles with the same radius, r.

Cylinders are solid figures. (Think of them as being closed on the ends.) The volume of a cylinder is equal to the area of one of its bases times its height. Because the area of a base is πr^2, we have the following formula:

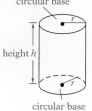

Right Circular Cylinder

circular base

height h

circular base

To find the volume of a cylinder

If r is the radius of the base and h is the height, then

$$V = \pi r^2 h$$

▪▪▪

EXAMPLE

Find the volume of a can that has a radius of 8 inches and a height of 12 inches.

SOLUTION

First we substitute the values into the formula for the volume of a cylinder.

$$V = \pi r^2 h = (3.14)(8^2)(12)$$

Then we simplify the resulting expression.

$$V = (3.14)(64)(12) = 2411.52$$

The volume of the cylinder is 2411.52 cubic inches.

Surface Area of a Cylinder

The following figure shows that we can think of the cylinder as being made up of three parts: two circular ends and a rectangle rolled up to form the side. The length of the rectangle is the circumference of a base, or $2\pi r$; its width is the height h. The surface area of the cylinder can be found by adding together the areas of the two end circles (πr^2 each) and the area of the rectangular part ($h \times 2\pi r$).

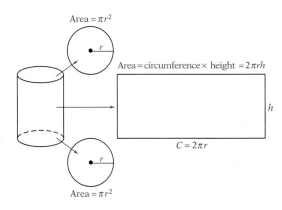

Area = πr^2

Area = circumference × height = $2\pi rh$

h

$C = 2\pi r$

Area = πr^2

To find the surface area of a cylinder

If r is the radius and h is the height, then

$$SA = \pi r^2 + \pi r^2 + 2\pi rh \quad \text{or} \quad 2\pi r^2 + 2\pi rh \quad \text{or} \quad 2\pi r(r + h)$$

▪▪▪

EXAMPLE

Find the surface area of a cylinder with a radius of 6 inches and a height of 12 inches.

SOLUTION

We substitute into the formula to find the surface area.

$$SA = 2\pi r(r + h)$$
$$= 2(3.14)(6)(6 + 12)$$
$$= 678.24$$

The surface area of the cylinder is 678.24 square inches.

◢

PRACTICE

Give your answers to the nearest tenth.

1. A can of soup is emptied into a pan. Both the can and the pan are shaped like cylinders. The soup can is 4.5 inches tall with a diameter of 2.5 inches.

The pan is 3 inches tall with a diameter of 5 inches. Is there enough space in the pan for a second can of soup?

2. Find the volume of a soda can that has a radius of $1\frac{1}{4}$ inches and a height of $4\frac{7}{8}$ inches. Use $\frac{22}{7}$ as an approximation for π.

3. You are cutting a label for a can. The can has a diameter of 2.5 inches and is 5.5 inches tall. What is the area of the label that fits exactly on this can?

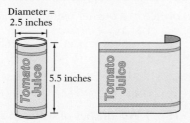

4. A container shaped like a cylinder holds glue for woodworking projects. If the container sits 9.5 inches high and has a diameter of 3 inches, how much glue can it hold?

5. What is the surface area of a cylinder with a radius of 3.5 inches and a height of 2.5 inches?

6. Modeling clay is shaped into a long cylinder with a diameter of 4 inches. A piece 6 inches long is cut off. What volume of clay is in this piece?

7. What is the surface area of a cylinder with a radius of 0.5 unit and a height of 2.5 units?

8. Find the volume of a pipe that is cylindrical in shape and has a length of 8 feet and a radius of $4\frac{1}{2}$ feet.

9. A can shaped like a cylinder is filled with water. If it has a radius of 3 inches and a height of 10 inches, how much water can it hold?

10. Water weighs 62.4 pounds per cubic foot. You have a bucket that is shaped like a cylinder with a diameter of 1 foot and a height of 1 foot, and it is filled with water. How much does the water weigh?

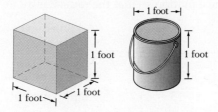

9.4 Solving Equations that Contain Radical Expressions

SECTION LEAD-IN

Use the formula described in the Section 9.2 Lead-In on page 496 to answer this question.

Find the number of BTUs that are produced per hour with a $\frac{1}{8}$-inch (diameter) orifice in a burner using natural gas:

$$D = V \times A \times \overline{K} \times 1655\sqrt{\frac{H}{G}}$$

where

$\overline{K} = 0.80$
$H = 7$ WC (inches of water in a tube)
$G = 0.65$
$V = 1000$ cubic feet per hour (for natural gas)

In previous chapters we have solved equations such as $7x = 49$ and $3x - 2 = 7$. But how would you solve an equation in the form of $\sqrt{x} = 5$?

To solve such equations, we must use the **equality property of squaring.**

The Equality Property of Squaring
If $\sqrt{x} = a$, then $x = a^2$.

In words, if two real numbers or expressions are equal, then their squares are also equal.

▪▪▪

EXAMPLE 1

Find the solution of the following equation.

$$\sqrt{x + 3} - 8 = 0$$

SOLUTION

First we rewrite the equation so that the radical is alone on one side.

$$\sqrt{x + 3} = 8 \qquad \text{Adding 8 to both sides}$$

Then we apply the squaring property.

$$\left(\sqrt{x + 3}\right)^2 = 8^2 \qquad \text{Squaring both sides}$$
$$x + 3 = 64 \qquad \text{Multiplying out}$$
$$x = 64 - 3 \quad \text{Adding } -3 \text{ to both sides}$$
$$x = 61 \qquad \text{Simplifying}$$

Check: Substituting in the original equation, we get

$$\sqrt{x + 3} - 8 = 0$$
$$\sqrt{61 + 3} - 8 = 0$$
$$\sqrt{64} - 8 = 0$$
$$8 - 8 = 0 \quad \text{True}$$

So when $\sqrt{x + 3} - 8 = 0$, $x = 61$.

▶*CHECK* **Warm-Up 1**

> **STUDY HINT**
>
> *Check every solution by substituting it into the original equation.*

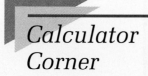

Calculator Corner

Let's examine and solve the equation $\sqrt{x + 3} - 8 = 0$ in three ways. This is often referred to as the **rule of three.**

a. symbolically   **b.** numerically   **c.** graphically

i. Example 1 solved the equation *symbolically.*

ii. You can use your graphing calculator to examine and solve the equation for *x numerically* by using the **TABLE** feature of the caculator. (Note: See the Calculator Corner on pages 189 and 191 to review how to set up tables.)

```
Y₁ ▋√(X+3)−8
Y₂
Y₃=
Y₄=
Y₅=
Y₆=
Y₇=
Y₈=
```

```
TABLE SETUP
 TblMin=58
 ΔTbl=1
Indpnt:  Auto   Ask
Depend:  Auto   Ask
```

```
  X      Y₁
 58    -.1898
 59    -.126
 60    -.0627
 61     0
 62     .06226
 63     .12404
 64     .18535
X=61
```

You can see from the **TABLE** that when $x = 61$, $y = 0$.

It is also possible to use the **TABLE** to examine another form of the equation, namely, $\sqrt{x + 3} = 8$. Now use the **TABLE** to find what *x*-value results in a *y*-value of 8.

```
Y₁ ▋√(X+3)
Y₂=
Y₃=
Y₄=
Y₅=
Y₆=
Y₇=
Y₈=
```

```
  X      Y₁
 58    7.8102
 59    7.874
 60    7.9373
 61    8
 62    8.0623
 63    8.124
 64    8.1854
X=61
```

This gives an *x*-value of 61 when $y = 8$.

iii. There are two methods for solving an equation *graphically.* Let's solve the equation in Example 1 by *both* methods and see how they compare. We will first graph the equation exactly as it is written: $\sqrt{x+3} - 8 = 0$. After obtaining the solution using the original equation, let's then <u>solve it</u> by rewriting the equation so that the radical is alone on one side: $\sqrt{x+3} = 8$.

Question: Will the graphs for the two methods *look the same*? If you work the problem correctly, how many solutions should you obtain?

Some people refer to this method as the **x-intercept method** because you are looking for the x-value when $y = 0$. In other words, where does the graph cross the x-axis? Use the **TRACE** feature of your graphing calculator to estimate where the graph crosses the x-axis.

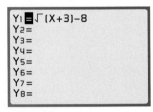
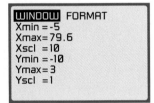
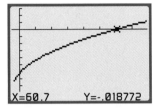

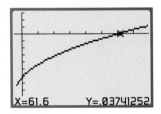

It appears that the x-value is somewhere between $x = 60.7$ and $x = 61.6$.

If you have a **CALCULATE** and **ROOT/ZERO** utility on your graphing calculator, use it to find the exact solution. (Note: See the Calculator Corner on page 365 to review how to find the roots of an equation.)

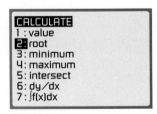
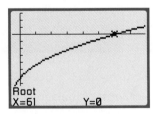

The solution is $x = 61$.

Now find the solution using the method sometimes referred to as the **multigraph method**. Where does $\sqrt{x+3} = 8$?

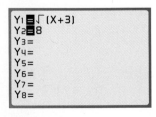

TRACE to estimate what the x-value at the point of intersection would be.

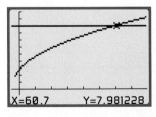

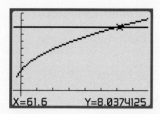

It appears that the correct answer for x is somewhere between 60.7 and 61.6. Now use the **CALCULATE** and **INTERSECT** feature of your graphing calculator to find the exact solution. (Note: See the Calculator Corner on page 125 to review how to find the points of intersection of two graphs.)

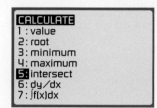 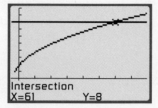

Once again you get the result $x = 61$.

Your graphing calculator can also be used to check your algebraic solution in several ways. One of these would be to use *function notation* (see pages 254–256).

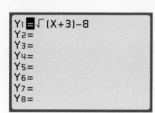 Go to your Home Screen and evaluate Y₁ at $x = 61$.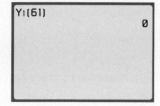

Now you have confirmed that when $x = 61$, Y₁ is indeed equal to 0.

▪▪▪
EXAMPLE 2

Solve the equation: $\sqrt{x + 5} - \sqrt{x} = 1$

SOLUTION

We want to rewrite the equation so that it has just one radical on a side:

$$\sqrt{x + 5} = \sqrt{x} + 1 \qquad \text{Adding } \sqrt{x} \text{ to both sides}$$

We apply the equality property of squaring, being careful to differentiate between the terms inside and outside the radical signs.

$$\left(\sqrt{x + 5}\right)^2 = \left(\sqrt{x} + 1\right)^2$$
$$x + 5 = x + 2\sqrt{x} + 1 \quad \text{Squaring}$$

We rewrite with the remaining radical alone on one side of the equation.

$$x - x + 5 - 1 = 2\sqrt{x}$$ Adding $-x$ and -1 to both sides

$$4 = 2\sqrt{x}$$ Combining like terms

$$2 = \sqrt{x}$$ Multiplying both sides by $\frac{1}{2}$

We again apply the squaring property.

$$(2)^2 = \left(\sqrt{x}\right)^2$$ Squaring

$$4 = x$$ Simplifying

Check: $\sqrt{x + 5} - \sqrt{x} = 1$ Original equation

$$\sqrt{4 + 5} - \sqrt{4} = 1$$ Substituting in the original equation

$$\sqrt{9} - \sqrt{4} = 1$$ Adding

$$3 - 2 = 1$$ Taking square roots

So when $\sqrt{x + 5} - \sqrt{x} = 1$, $x = 4$.

▶ *CHECK* **Warm-Up 2**

▪ ▪ ▪

EXAMPLE 3

The maximum distance you can see from a tall building is given by the following equation

$$D \approx 69.4\sqrt{0.0003h}$$

where D is the distance in miles and h is the height of the building in feet.

a. How far can you see from the top of the World Trade Center in New York City, which is 1353 feet tall?

b. How high must a building be for a person to see 100 miles from the top?

SOLUTION

a. To find the maximum seeing distance, we substitute $h = 1353$ into the equation and solve.

$$D \approx 69.4\sqrt{0.0003h}$$
$$\approx 69.4\sqrt{(0.0003)(1353)}$$
$$\approx 69.4\sqrt{0.4059}$$

We estimate the square root to two decimal places and multiply.

$$D \approx (69.4)(0.64) \approx 44.41$$

So from the top of the World Trade Center, one can see about 44 miles.

b. To answer the second question, we substitute $D = 100$ and use the equality property of squaring, being careful to differentiate between terms inside and outside the radical signs.

$$D \approx 69.4\sqrt{0.0003h}$$

$$100 \approx 69.4\sqrt{0.0003h} \qquad \text{Substituting}$$

$$(100)^2 \approx (69.4)^2\left(\sqrt{0.0003h}\right)^2 \qquad \text{Squaring}$$

$$10000 \approx 4816.36(0.0003h) \qquad \text{Simplifying}$$

$$10000 \approx 1.445h$$

$$6920 \approx h \qquad \text{Multiplying by } \tfrac{1}{1.445}$$

An observer would have to stand on top of a building 6920 feet tall to see 100 miles.

▶ *CHECK* **Warm-Up 3**

Practice what you learned.

SECTION FOLLOW-UP

When we complete our calculations, using 3.14 for π, we obtain 53,293 BTUs.

If we change to a high-pressure propane line at 3 psi (equivalent to 83.1 WC), we replace $V = 2500$ and $G = 1.52$. We substitute in the formula

$$D = V \times A \times \overline{K} \times 1655\sqrt{\frac{H}{G}}$$

$$D = 2500 \times 0.0122654 \times 0.80 \times 1655\sqrt{\frac{83.1}{1.52}}$$

$$D = 300{,}185 \text{ BTUs}$$

9.4 WARM-UPS

Work these problems before you attempt the exercises.

1. Solve for y: $\sqrt{y + 5} - 9 = 0$

2. Solve for x: $2\sqrt{x - 5} = 3\sqrt{x - 7}$

3. *Falling Objects*   Galileo discovered that the speed v of an object that has fallen d centimeters is $v = \sqrt{1960d}$, where v is in centimeters per second. A ball bearing is dropped from a height of 200 centimeters. How fast is it traveling when it reaches the ground?

9.4 EXERCISES

Note: Use your graphing calculator to check your results whenever possible.

In Exercises 1 through 24, solve the equation and check your answer. Assume all variables represent positive real numbers.

1. $\sqrt{p} = 6$

2. $\sqrt{m} = 9$

3. $\sqrt{y} - 5 = 0$

4. $\sqrt{x} - 11 = 0$

5. $\sqrt{x - 8} = 10$

6. $\sqrt{x + 6} = 19$

7. $\sqrt{t - 12} = 14$

8. $\sqrt{t - 17} = 4$

9. $\sqrt{5t + 6} = 3\sqrt{t}$

10. $\sqrt{3n + 4} = 2\sqrt{5}$

11. $\sqrt{y} + \sqrt{y + 5} = 5$

12. $\sqrt{t} - \sqrt{t + 9} = -1$

13. $2\sqrt{r} = \sqrt{8r - 16}$

14. $\sqrt{5y - 4} = \sqrt{4y + 1}$

15. $\sqrt{x + 2} = \sqrt{2x - 6}$

16. $\sqrt{3m + 8} = \sqrt{3m} - 1$

17. $\sqrt{m + 7} = 5$

18. $k = \sqrt{k^2 - 8k + 10}$

19. $p = \sqrt{p^2 - 3p - 16}$

20. $\sqrt{4x - 9} = \sqrt{x}$

21. $\sqrt{t} + 1 = \sqrt{t + 5}$

22. $3t = \sqrt{9t^2 - 6t + 32}$

23. $2r = \sqrt{4r^2 + 6r - 30}$

24. $\sqrt{t^2 - 5} = t - 1$

In Exercises 25 through 30, solve each word problem.

25. **Price of a Pizza**   At a certain pizza parlor, the price of a pizza is described by the equation

$$C = \frac{d^2 + 54}{42}$$

where C is the price in dollars and d is the diameter of the pizza.

   a.   What is the price of a 14-inch pizza?

   b.   A pizza is advertised at $10.00. According to the price equation, what diameter should this super-pizza have?

26. **Skidding Car**   The formula $s \approx \sqrt{30fd}$ is used by police to estimate the speed s in miles per hour that a car was initially traveling if it skidded d feet on a road. In this formula, $f = 0.4$ if the road is wet concrete and $f = 0.8$ if the road is dry concrete.

   a.   A car skids 40 feet on a dry concrete road. How fast was it initially traveling?

b. A car is moving at 60 miles per hour on a wet road when the driver slams on the brakes. How far can we expect this car to skid?

a.

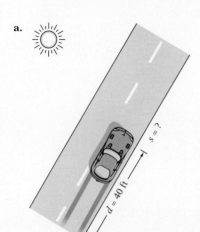

d = 40 ft
s = ?

b.

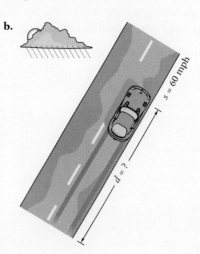

s = 60 mph
d = ?

27. *Filtered Light* The amount of light allowed through by n filters is $L \approx T^n$, where T is the amount of light allowed through by one filter and n is the number of filters.

   a. If a filter allows $\frac{2}{5}$ of the light through, what fraction of the light is allowed by 2 filters?

   b. One filter should block what percent of the light so that only 50% of the light gets through 2 filters?

28. *Population Growth* The population (P) after a given number of years (n) can be described by

$$(1 + r)^n \approx \sqrt{P/a}$$

   where a is the original population and r is the growth rate written as a decimal.

   a. What will the new population be after 2 years if the growth rate is 70% and the original population is 1000?

   b. What was the original population if the growth rate is 20% and, after 2 years, the population is 253?

29. *Luggage Size* The inside of a piece of luggage measures 36 inches by 24 inches by 8 inches. What is the maximum length of an umbrella that can be placed in this luggage? (*Hint:* Think what right triangles could be formed by the umbrella and the suitcase.)

30. *Growing Fruit Trees* A certain kind of fruit tree grows best when the trees are at least 6 feet apart. If the trees are grown in parallel rows, what is the smallest possible distance between two rows? (*Hint:* The answer will be a distance less than 6 feet.)

MIXED PRACTICE

By doing these exercises, you will practice the topics up to this point in the chapter. Assume all variables represent positive real numbers.

31. *Tile Size*　A rectangular tile has sides of 4 inches and 9 inches. What is the length of its diagonal to the nearest inch?

32. Solve for x: $\sqrt{4x - 13} = \sqrt{2x - 1}$

33. Simplify: $\sqrt{169x^2yz^2}$

34. Divide $(\sqrt{543})(\sqrt{18})$ by $2\sqrt{27}$.

35. Rationalize the denominator: $\dfrac{\sqrt{6} + 3\sqrt{3}}{\sqrt{3}}$

36. Multiply: $(\sqrt{6} + \sqrt{3} - \sqrt{12})\sqrt{8}$

37. Solve for x: $2\sqrt{x} - 1 = \sqrt{x}$

38. Multiply: $(\sqrt{12} - \sqrt{8})(-\sqrt{6} - 5)$

39. Solve for x: $\sqrt{3x - 10} = \sqrt{2x + 5}$

40. Multiply: $(\sqrt{35} + \sqrt{7} - \sqrt{21})\sqrt{7}$

41. Multiply: $\sqrt{5x} \cdot \sqrt{20x^2y}$

42. Multiply: $\sqrt{24}(\sqrt{32} + \sqrt{42} - \sqrt{3})$

43. Subtract $\sqrt{75} + 3\sqrt{27}$ from $\sqrt{108} - 2\sqrt{48}$.

44. Subtract $6\sqrt{1000} - 5\sqrt{10}$ from $3\sqrt{500}$.

45. Multiply: $(\sqrt{63} + \sqrt{54} - \sqrt{8})\sqrt{54}$

46. Divide: $\dfrac{\sqrt{(2x)(8)(28y)}}{\sqrt{(3x)(27y)(9)}}$

EXCURSIONS

Exploring Geometry

1.

> **To find the volume of a pyramid**
>
> If B is the area of the base and h is the height of the pyramid, its volume is
> $$V = \tfrac{1}{3}Bh$$

a. The Transamerica Tower in San Francisco is a pyramid with a square base. Each side of the base has a length of 52.13 meters. The height of the building is 259.8 meters. Find the volume of the building.

b. Another formula for the volume of a pyramid is

$$V = \tfrac{1}{3}a^2h$$

where a is the length of a side and h is the height.

Pyramid

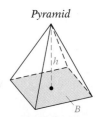

B is the area of the base.

The great pyramid of Egypt stands on a square base each side of which is 764 feet long; and its height is 480 feet. Find the number of cubic feet of stone used in its construction. Refer to part (a) and the first pyramid volume formula. What is the relationsip between a and B?

Exploring Numbers

2. 232, 233, and 234 are all hypotenuses of right triangles with integer sides. Find the sides of each of these triangles.

CHAPTER LOOK-BACK

Triangles are used as braces because, given three sides of a triangle, one and only one such figure exists. Its shape is, therefore, rigid. A rhombus can be formed whose sides are identical to a given square; a parallelogram can have sides identical in length with those of a rectangle.

Other shapes you will find in building structures are hexagons (honeycombs—a building of sorts), pentagons (geodesic domes), or trapezoids (found on certain roofs).

CHAPTER 9
REVIEW PROBLEMS

The following exercises will give you a good review of the material presented in this chapter. Assume all variables represent positive real numbers.

SECTION 9.1

1. Between what two integers does $\sqrt{150}$ lie?

2. Approximate the value of $3\sqrt{22}$.

3. Approximate the value of $6\sqrt{27}$.

4. Find the square root of 324.

5. Is $\sqrt{-26}$ real or not real and (if it is real) rational or irrational?

6. Simplify: $\sqrt{220x^2y^2}$

SECTION 9.2

Simplify each expression.

7. $\sqrt{(2x)(6x^5)}$

8. $\sqrt{(8x^3y)(4x^3y)}$

9. $\sqrt{(8y^9)(2y)}$

10. $\sqrt{(-4x^2y)(-8x^3y^2)}$

11. $-\sqrt{80} - 20\sqrt{40} + \sqrt{800}$

12. $2\sqrt{12} - \sqrt{75}$

SECTION 9.3

Rationalize each denominator and leave your answer in simplest radical form.

13. $\dfrac{3}{2\sqrt{5}}$

14. $\dfrac{3 + 4\sqrt{20}}{\sqrt{12} + 3\sqrt{5}}$

15. $\dfrac{6\sqrt{15}}{2 - \sqrt{5}}$

16. $\dfrac{6}{2\sqrt{15}}$

17. $\sqrt{\dfrac{160tx^5}{8t}} \cdot \sqrt{\dfrac{16tx}{t}}$

18. $\left[\dfrac{\sqrt{96}\,(\sqrt{84})}{(\sqrt{24})}\right]\left[\dfrac{(\sqrt{18})(\sqrt{36})}{(\sqrt{4})}\right]$

19. $\dfrac{[(\sqrt{12})(2)]}{(\sqrt{3})} - \dfrac{(7\sqrt{2})}{\sqrt{8}}$

20. $\left(3\sqrt{2} + 2\sqrt{10}\right)\left(\sqrt{5} - \sqrt{20}\right)$

21. $\left(3\sqrt{2} + 2\sqrt{10}\right)^2$

22. $\left(4\sqrt{3} - 3\sqrt{6}\right)^2$

23. $\dfrac{\sqrt{8}}{2\sqrt{12} - 1}$

24. $\dfrac{\sqrt{12} - 5\sqrt{2}}{2\sqrt{2} + \sqrt{12}}$

SECTION 9.4

Solve each equation.

25. $3m = \sqrt{9m^2 - 6m + 32}$

26. $\sqrt{m - 12} = 14$

27. $\sqrt{t + 2} = \sqrt{2t - 6}$

28. $2r = \sqrt{4r^2 + 6r - 30}$

29. $\sqrt{3m + 8} = \sqrt{5m - 1}$

30. $\sqrt{m - 17} = -4$

MIXED REVIEW

31. Between what two multiples of 6 does the number $-6\sqrt{37}$ lie?

32. Between what two integers does $\sqrt{173}$ lie?

33. Subtract $5\sqrt{18} - 2\sqrt{8}$ from $3\sqrt{12}$.

34. Rationalize the denominator: $\dfrac{12\sqrt{10}}{12\sqrt{3} - \sqrt{10}}$

35. Solve for t: $2\sqrt{t - 7} = \sqrt{t + 7}$

36. What is the area of a rectangle that has a diagonal of 52 and one side that measures 20?

CHAPTER 9 TEST

This exam tests your knowledge of the material in Chapter 9. Assume all variables represent positive real numbers.

1. **a.** Approximate the value of $\sqrt{2550}$.

  b. Between what two multiples of 7 does the number $7\sqrt{63}$ lie?

  c. Simplify: $\sqrt{5^2 r^{33} t^{52}}$

2. **a.** Multiply and simplify: $\left(\sqrt{18xy^2 t}\right)\left(\sqrt{180x^2 y^2 t^2}\right)$

  b. Simplify: $\dfrac{2\sqrt{3} + 5\sqrt{12}}{\sqrt{3}}$

  c. Multiply and simplify: $\left(\sqrt{\dfrac{120yx^3}{10y}}\right)\left(\sqrt{\dfrac{15xy}{x}}\right)$

3. Simplify:

  a. $2\sqrt{12} - \left(\sqrt{24} - 2\sqrt{54}\right)$

  b. $3\sqrt{2}\left(6\sqrt{18x^2} - 10\sqrt{32x^2}\right)$

  c. $\sqrt{72p} + 20\sqrt{8p} - \sqrt{16p}$

4. Rationalize the denominator:

   a. $\dfrac{2}{\sqrt{12}}$

   b. $\dfrac{15\sqrt{3} + 3\sqrt{5}}{\sqrt{5}}$

   c. $\dfrac{\sqrt{18}}{\sqrt{3} + \sqrt{2}}$

5. Solve these equations for x.

   a. $\sqrt{x + 5} = \sqrt{4x - 2}$

   b. $\sqrt{x - 6} + 9 = 12$

   c. $\sqrt{x + 10} = 20$

CUMULATIVE REVIEW

CHAPTERS 1–9

The following exercises will help you maintain the skills you have learned in this and previous chapters.

1. Factor: $5x^2 - 2x - 24$

2. Factor: $9x^2 - 121$

3. Multiply: $(2x - 7)(3x + 11)$

4. Add: $\frac{1}{3}x^2 + \frac{2}{4}x + \frac{3}{5}(x^2 - x)$

5. Solve: $\frac{2}{3} + \frac{x}{4} = 90$

6. Simplify: $\dfrac{\frac{3x}{x - 1}}{\frac{1}{x^2 + 3x - 4}}$

7. Solve by substitution:
   $7y - 3x = 22$
   $2x + 4 = 2y$

8. Divide $2x^3 + 4x^2 - 12x + 8$ by $x - 2$.

9. Divide: $\dfrac{-24y^2 - 36x^2}{-3xy}$

10. Solve by addition: $4x + 6y = -8$
    $-2x + 4y = 18$

11. Find the slope of the line whose equation is $-4x + 8y = 9$.

12. Tickets to a show are $5 for children and $9 for adults. How many tickets of each kind were sold if the total revenue for the show was $3500 and the same number of adults' and children's tickets were sold?

13. Write the equation of the line that passes through the points $(2, 3)$ and $\left(\frac{1}{2}, 6\right)$.

14. Write the equation of the line that passes through the origin and has slope -4.

15. Evaluate $2xy + 3x^2y^4 - 10x + 4y$ when $x = -1$ and $y = 2$.

16. List all the whole-number factors of 364.

17. Factor 4000 into prime factors.

18. Write as a whole number: $(3^2 2^3 5^2 7^4)^0$

19. Find all the real numbers that solve the inequality $-3x + 46 > 4$.

20. If $2n + 1$ is a positive odd integer,

    a. what is three times this integer?

    b. what is the next odd integer?

    c. what is the integer that appears immediately to the left of this integer on the number line?

QUADRATIC EQUATIONS

Quadratic equations are extremely useful in a variety of fields, but especially when working with moving objects. For example, the equation $d = v_0t + \frac{1}{2}at^2$ can be used to determine the distance traveled by a falling object; $v_0 =$ the object's initial speed, $t =$ the time during which the object traveled, and $a =$ the acceleration due to gravity, which is 32 ft/s^2. For a competition diver, such information is useful in determining the height required for certain dives, or perhaps in determining the amount of time available before hitting the water.

■ *What moving objects have you used or encountered whose motion could be described using quadratic equations?*

SKILLS CHECK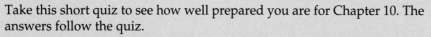

Take this short quiz to see how well prepared you are for Chapter 10. The answers follow the quiz.

1. Factor: $x^2 + 4x - 5$

2. Factor: $(2ab^2)^2 - 9x^2y^2$

3. Factor: $x^3 - 3x^2 - 4x$

4. Find the product of $(3x + 8)(x - 6)$.

5. Factor $10a^2 + 16ab - 5ab - 8b^2$ by grouping.

6. A formula for finding the area of a triangle is

$$A = \sqrt{s(s - a)(s - b)(s - c)}$$

where a, b, and c are sides of a triangle and s is half its perimeter. Find A when $a = 3$, $b = 4$, and $c = 5$.

7. What is the square of 3.2?

8. Find $\sqrt{441}$.

9. Multiply: $(3 + \sqrt{2})(3 - \sqrt{2})$

10. Multiply: $\left(\dfrac{2 - 3\sqrt{5}}{2}\right)^2$

ANSWERS: **1.** $(x + 5)(x - 1)$ [Section 6.2]   **2.** $(2ab^2 - 3xy)(2ab^2 + 3xy)$ [Section 6.3]
3. $x(x - 4)(x + 1)$ [Section 6.2]   **4.** $3x^2 - 10x - 48$ [Section 5.4]   **5.** $(5a + 8b)(2a - b)$ [Section 6.1]
6. $A = 6$ square units [Section 2.2]   **7.** 10.24 [Section 1.4]   **8.** 21 [Section 1.4]   **9.** 7 [Section 9.2]
10. $\dfrac{49 - 12\sqrt{5}}{4}$ [Section 9.2]

CHAPTER LEAD-IN

The Summer Olympic Games typically include events such as the

1. hammer throw

2. high jump

3. javelin throw

Sketch the "shape" of each of these events. What do the shapes have in common?

10.1 Using Factoring and Square Roots to Solve Quadratic Equations

SECTION LEAD-IN

A diver dives off a platform and reaches the water in 2.0 seconds. Describe the shape of the path of the dive.

SECTION GOALS

- *To solve quadratic equations by factoring*
- *To solve equations using the principle of zero products*
- *To find an equation with given roots*
- *To solve equations using the equality property of square roots*
- *To work word problems that involve quadratic relationships*

Second-degree equations in one variable are known as **quadratic equations.** The **degree of an equation in one variable** is the highest power of its variable. So the highest power in a quadratic equation is 2.

> The **standard form of a quadratic equation in one variable** is
> $$ax^2 + bx + c = 0$$
> where a, b, and c are real numbers and $a \neq 0$.

An equation is a quadratic equation if it can be put in this standard form. The variable need not be x.

Some examples of quadratic equations are

$$2x^2 - 4x = 4 \qquad 4y^2 - 36 = 0 \qquad 10z^2 - 5z = 0$$

When a quadratic equation is in standard form, the polynomial on the left is called a **quadratic polynomial.** We can solve the equation if we can factor the quadratic polynomial. We do so by applying the **principle of zero products.**

Principle of Zero Products

> **Principle of Zero Products**
>
> If a and b represent real numbers, and
> $$ab = 0$$
> then $a = 0$ or $b = 0$ or both.

To use the principle, we set each factor of the quadratic polynomial equal to zero and solve. We obtain one solution (to the quadratic equation) for each factor. A quadratic equation has two solutions, or *roots*. The two roots may not necessarily be real or distinct.

■ ■ ■

EXAMPLE 1

Solve: $6x^2 + x - 15 = 0$

SOLUTION

This is a quadratic equation in standard form. The terms of the quadratic polynomial do not have a common factor. It turns out that we can factor the given quadratic polynomial and write it as

$$(2x - 3)(3x + 5) = 0$$

We now apply the principle of zero products by setting each factor equal to zero and solving the resulting equations.

$$
\begin{array}{c|c}
2x - 3 = 0 & 3x + 5 = 0 \\
2x = 3 & 3x = -5 \\
x = \dfrac{3}{2} & x = \dfrac{-5}{3}
\end{array}
$$

We check by substituting both solutions into the original quadratic equation. For $x = \frac{3}{2}$, we get

$$6x^2 + x - 15 = 0$$
$$6\left(\frac{3}{2}\right)^2 + \left(\frac{3}{2}\right) - 15 = 0$$
$$6\left(\frac{9}{4}\right) + \frac{3}{2} - 15 = 0$$
$$\frac{54}{4} + \frac{6}{4} - \frac{60}{4} = 0$$
$$0 = 0$$

So $x = \frac{3}{2}$ is a solution. Checking that $x = \frac{-5}{3}$ is a solution is left for you to do.
So $x = \frac{3}{2}$ and $x = \frac{-5}{3}$ are solutions for $6x^2 + x - 15 = 0$.

▶ CHECK **Warm-Up 1**

Calculator Corner

Use your graphing calculator to graph the equation $Y_1 = 6x^2 + x - 15$ on a "friendly" grid or screen. Check your graphing calculator manual to see how to do this on your particular model. The screens that follow are from the *TI-82/83.*

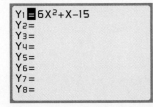 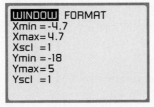 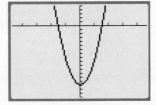

This graph is called a **parabola.**

Now **TRACE** on the graph to find the two points where the parabola crosses the *x*-axis.

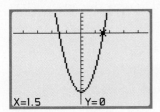 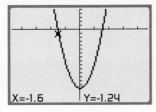 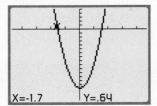

The first screen above shows that the parabola crosses the *x*-axis at the point $(1.5, 0)$. The second and third screens above show that it also crosses the *x*-axis somewhere between the point $(-1.7, 0.64)$ and the point $(-1.6, -1.24)$. When you found the points where the parabola crossed the *x*-axis by the paper-and-pencil method, you got the point $\left(-\frac{5}{3}, 0\right)$, which can also be written as $(-1.667, 0)$.

Some graphing calculators have a **CALC**ulation utility that can calculate the *x*-intercept directly on the graph screen. The following screens are for the *TI-82/83* graphing calculators. (Note: See the Calculator Corner on pages 365 and 366 to review how to find the roots of an equation.)

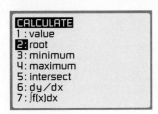 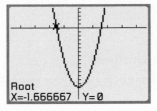

Because $x = -\frac{5}{3}$ equals -1.666667 when it is rounded off, you have obtained the same point as you did when you used the paper-and-pencil method.

The solutions of an equation are also called its **roots.** All quadratic equations have two roots. When the roots are identical, we say that the root is a **double root.** When we know the roots of a quadratic equation, we can use the principle of zero products to write a quadratic equation with those roots.

EXAMPLE 2

Find a quadratic equation that has the roots -3 and 4.

SOLUTION

STUDY HINT

The x^2 term in a quadratic equation tells you that it has exactly two roots.

Because the roots of the equation are -3 and 4, we know that

$$x = -3 \quad \text{and} \quad x = 4$$

Therefore,

$$x + 3 = 0 \quad \text{and} \quad x - 4 = 0 \quad \text{Adding to both sides}$$

Thus the factors of the equation we seek are $x + 3$ and $x - 4$. We multiply these two factors together and set their product equal to zero. (Because each is equal to zero, their product is equal to zero.) We get

$$(x + 3)(x - 4) = x^2 - 4x + 3x - 12 = 0$$

So an equation with the roots -3 and 4 is $x^2 - x - 12 = 0$.

▶ *CHECK* **Warm-Up 2**

The solution for Example 2 is not unique. In fact, any multiple of the equation we found will also give the roots -3 and 4. That is, equations such as

$$2x^2 - 2x - 24 = 0 \quad \text{and} \quad 100x^2 - 100x - 1200 = 0$$

all have the same solutions as the equation in Example 2.

Solving Quadratic Equations: Equality Property of Square Roots

Here is a useful property that can be used to solve equations in the form

$$x^2 = \text{a number} \quad \text{or} \quad (x + b)^2 = \text{a number}$$

> **Equality Property of Square Roots**
>
> For real numbers a and c, where c is non-negative, if $a^2 = c$, then $a = \sqrt{c}$ or $a = -\sqrt{c}$.

We often write "$a = \sqrt{c}$ or $a = -\sqrt{c}$" as

$$a = \pm\sqrt{c}$$

We read this as "*a* equals plus or minus the square root of *c*." In an equation, *a* can be a variable or any expression that represents a real number.

▪ ▪ ▪

EXAMPLE 3

Solve: $y^2 - 36 = 0$

SOLUTION

We could solve this equation by factoring it into the difference of two squares. However, we will use this new method instead.

The given equation can be written in the form $y^2 = 36$. Because 36 is a non-negative real number, we can apply the equality property of square roots.

$y^2 = 36$		Adding 36 to both sides
$y = \sqrt{36}$ or	$y = -\sqrt{36}$	Equality property of square roots
$y = 6$	$y = -6$	Simplifying

We check in the original equation.

$$
\begin{array}{c|c}
y^2 - 36 = 0 & y^2 - 36 = 0 \\
(6)^2 - 36 = 0 & (-6)^2 - 36 = 0 \\
36 - 36 = 0 & 36 - 36 = 0 \\
0 = 0 & 0 = 0
\end{array}
$$

So 6 and -6 are both solutions of the equation $y^2 - 36 = 0$.

▶ CHECK **Warm-Up 3**

> **WRITER'S BLOCK**
>
> Contrast the *equality property of squaring* and the *equality property of square roots*. When is each used?

In the next example, we solve an equation that involves the square of an expression.

▪ ▪ ▪

EXAMPLE 4

Solve: $(2x + 3)^2 = 16$

SOLUTION

We have "perfect" squares on both sides of this equation, so we can apply the equality property immediately.

$(2x + 3)^2 = 16$		
$2x + 3 = \sqrt{16}$ or	$2x + 3 = -\sqrt{16}$	By the equality property
$2x + 3 = 4$	$2x + 3 = -4$	Taking the square root
$2x = 1$	$2x = -7$	Adding -3
$x = \frac{1}{2}$	$x = \frac{-7}{2}$	Multiplying by $\frac{1}{2}$

Both of these solutions check in the original equation (try them), so $x = \frac{1}{2}$ and $x = \frac{-7}{2}$ are both solutions for the equation $(2x + 3)^2 = 16$.

> **❗❗❗**
>
> *ERROR ALERT*
>
> Identify the error and give a correct answer.
>
> Solve: $(x + 3)^2 = 16$
>
> *Incomplete Solution:*
>
> $(x + 3)^2 = 16$
>
> $x + 3 = 4$
>
> $x = 1$

 CHECK **Warm-Up 4**

Calculator Corner

It is possible to solve $(2x + 3)^2 = 16$ by graphing each side of the equation. This method is called the *multi-graph method* because you are graphing more than one equation in order to find the point(s) where the two functions are *equal (intersect).*

Use the following steps to find the point(s) of intersection on your graphing calculator.

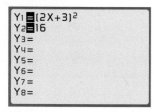 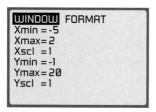 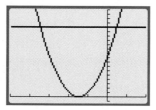

Now you can use **TRACE** to estimate the two points of intersection, *or,* if your graphing calculator has the capability, you could **CALC**ulate the two points of intersection. (Note: See the Calculator Corner on page 125 to review how to find the points of intersection of two graphs.)

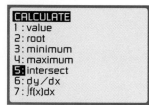 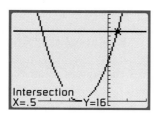 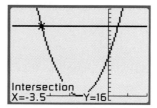

STUDY HINT

Before you work a problem, take a few seconds to write the problem in your notebook and record its page number. If you do this, you can always find the section in the text to go back to for review.

...

EXAMPLE 5

Solve: $9(2x - 5)^2 = 16$

SOLUTION

We begin by rewriting the equation in the form $a^2 = c$.

$$(2x - 5)^2 = \frac{16}{9} \quad \text{Multiplying by } \frac{1}{9}$$

Then we continue.

$$2x - 5 = \sqrt{\frac{16}{9}} \quad \text{or} \quad 2x - 5 = -\sqrt{\frac{16}{9}} \qquad \text{Equality property of square roots}$$

$$\begin{array}{c|c}
2x - 5 = \frac{4}{3} & 2x - 5 = \frac{-4}{3} \\
2x = \frac{4}{3} + 5 & 2x = \frac{-4}{3} + 5 \\
2x = \frac{19}{3} & 2x = \frac{11}{3} \\
x = \frac{19}{6} & x = \frac{11}{6}
\end{array}$$

$2x - 5 = \frac{4}{3}$	Taking the square root
$2x = \frac{4}{3} + 5$	Adding $+5$
$2x = \frac{19}{3}$	Adding
$x = \frac{19}{6}$	Multiplying by $\frac{1}{2}$

Both solutions check.

▶ *CHECK* **Warm-Up 5**

Applying Quadratic Equations

Quadratic equations in two variables provide very powerful tools for solving practical (word) problems. Many actual occurrences are described by quadratic equations.

• • •

EXAMPLE 6

The path of a toy rocket is described by the equation

$$h = v_0 t - 16t^2 + h_0$$

Its initial (launch) velocity (v_0) is 64 feet per second, and its initial height (h_0) is 4 feet.

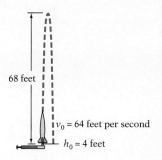

68 feet

v_0 = 64 feet per second

h_0 = 4 feet

a. At what time does it reach its maximum height of 68 feet?

b. When does it return to its launch height of $h_0 = 4$ feet?

SOLUTION

a. We first substitute $h = 68$, $v_0 = 64$, and $h_0 = 4$ into the equation:

$$h = v_0 t - 16t^2 + h_0 \qquad \text{The given equation}$$
$$68 = 64t - 16t^2 + 4$$

We then rewrite the equation in standard quadratic form.

$$0 = 64t - 16t^2 + 4 - 68 \qquad \text{Adding } -68 \text{ to both sides}$$
$$0 = -16t^2 + 64t - 64 \qquad \text{The equation in standard form}$$

The right side of this equation can be factored.

$$0 = -16(t - 2)(t - 2)$$

The factors are identical; setting $t - 2$ equal to zero then gives

$$t - 2 = 0$$
$$t = 2$$

The projectile reaches its maximum height at $t = 2$ seconds after launch.

b. We substitute $h = 4$, $v_0 = 64$, and $h_0 = 4$ into the original equation, and we obtain

$$4 = 64t - 16t^2 + 4$$

Adding -4 to both sides yields

$$0 = 64t - 16t^2$$
$$0 = 4t - t^2 \qquad \text{Multiplying by } \tfrac{1}{16}$$
$$0 = t(4 - t) \qquad \text{Factoring}$$

We set both factors equal to zero and find

$$t = 0 \qquad \Big| \qquad 4 - t = 0$$
$$4 = t$$

The solution $t = 0$ gives us the launch time. The solution $t = 4$ seconds is the one that gives us the time the projectile takes to return to its launch height of 4 feet.

▶ *CHECK* **Warm-Up 6**

Practice what you learned.

SECTION FOLLOW-UP

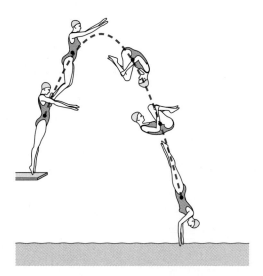

The path of her dive is in the shape of a **parabola.** The equation of this path is, roughly, quadratic.

10.1 WARM-UPS

Work these problems before you attempt the exercises.

1. Factor and solve using the principle of zero products:

$$4r^2 + 12r = 72$$

2. Find an equation with the roots $-\frac{2}{5}$ and 10.

3. Solve: $x^2 - 144 = 0$

4. Solve:
$$(x + 100)^2 - 36 = 0$$

5. Solve: $25(x - 10)^2 - 49 = 0$

6. *Launching a Toy Rocket* Suppose the rocket in Example 6 is launched from ground level ($h_0 = 0$). When does it return to the ground?

10.1 EXERCISES

Note: Use your graphing calculator to check your results whenever possible.

In Exercises 1 through 24, factor and solve using the principle of zero products.

1. $6n^2 + 30n - 36 = 0$

2. $x^2 + 2x + 1 = 0$

3. $r^2 + 20r + 100 = 0$

4. $y^2 + 11y + 18 = 0$

5. $x^2 + 13x + 40 = 0$

6. $20x^2 + 27x + 9 = 0$

7. $16r^2 - 24r + 8 = 0$

8. $4r^2 + 16r - 180 = 0$

9. $5t^2 - 90t + 385 = 0$

10. $t^2 - 3t + 2 = 0$

11. $r^2 + 8r + 16 = 0$

12. $t^2 - 20t + 100 = 0$

13. $64x^2 - 36 = 0$

14. $y^2 - 4 = 0$

15. $3t^2 - 39t + 120 = 0$

16. $2y^2 + 46y + 264 = 0$

17. $t^2 + 3t - 10 = 0$

18. $x^2 - 10x + 21 = 0$

19. $y^2 - 24y + 144 = 0$

20. $5n^2 - 2000 = 0$

21. $5x^2 = 125$

22. $2r^2 - 6r - 8 = 0$

23. $4 + 11t^2 + 24t = 0$

24. $81t^2 + 144t + 64 = 0$

In Exercises 25 through 48, the roots of a quadratic equation are given. Find an equation that has these roots.

25. -4 and 5

26. 10 and 21

27. 2 and -7

28. -2 and -1

29. 10 and $\frac{2}{5}$

30. 10 and 100

31. -3 and -7

32. -2 and 24

33. a double root of -5　　**34.** -10 and 10　　**35.** -14 and -1　　**36.** -2 and -18

37. $\frac{1}{5}$ and $-\frac{2}{3}$　　**38.** $-\frac{1}{3}$ and $\frac{4}{7}$　　**39.** $-\frac{2}{3}$ and $\frac{1}{4}$　　**40.** $\frac{2}{3}$ and -6

41. -12 and 12　　**42.** -27 and -1　　**43.** $\sqrt{2}$ and $-\sqrt{2}$　　**44.** $\sqrt{3}$ and $-\sqrt{3}$

45. $-2\sqrt{3}$ and $2\sqrt{3}$　　**46.** $-\sqrt{5}$ and $\sqrt{5}$　　**47.** $-2\sqrt{6}$ and $2\sqrt{6}$　　**48.** $4\sqrt{5}$ and $-4\sqrt{5}$

In Exercises 49 through 74, solve by using the equality property of square roots.

49. $x^2 - 4 = 0$　　　　　　**50.** $r^2 - 81 = 0$　　　　　　**51.** $t^2 - 9 = 0$

52. $y^2 - 100 = 0$　　　　**53.** $r^2 - 169 = 0$　　　　**54.** $t^2 - 25 = 0$

55. $y^2 - 256 = 0$　　　　**56.** $y^2 - 225 = 0$　　　　**57.** $n^2 - 49 = 0$

58. $49t^2 - 9 = 0$　　　　**59.** $64x^2 - 36 = 0$　　　　**60.** $4y^2 - 25 = 0$

61. $25x^2 - 100 = 0$　　　**62.** $100x^2 - 81 = 0$　　　**63.** $81t^2 - 49 = 0$

64. $(5 + 16n)^2 - 125 = 0$　**65.** $(5 - n)^2 - 200 = 0$　**66.** $(22 - y)^2 = 3600$

67. $(4x - 7)^2 = 196$　　　**68.** $(4x - 12)^2 - 484 = 0$　**69.** $(2 - 3y)^2 - 289 = 0$

70. $(x + 1)^2 = 16$　　　　**71.** $(x - 100)^2 = 144$　　　**72.** $49(6t - 2)^2 = 144$

73. $4(2x - 8)^2 - 121 = 0$　**74.** $36(5 - 2y)^2 - 25 = 0$

In Exercises 75 through 80, solve the word problems.

75. *Football*　Big Dave kicked a football at an angle. To find how long it took to hit the ground, solve

$$12t - \frac{1}{2}(9.80)t^2 = 0$$

The answer will be in seconds.

76. *Football*　To find how far the football (in Exercise 75) traveled, use the formula

$$d = 65t - 4.9t^2$$

Replace t with your solution from Exercise 75 and solve for d, the distance traveled. Your answer is in feet.

77. Suppose a rock is dropped off a cliff into a stream 80 meters below. How far will it have fallen after 1, 2, 3, and 4 seconds? Use

$$y = \frac{1}{2}(9.8)s^2$$

where y represents the distance the rock falls in s seconds. Let $s = 1.00$, $s = 2.00$, and so on. Your answer for y will be in meters.

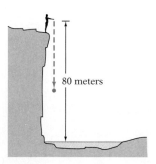

80 meters

78. Use the information in Exercise 77 and find out how long it will take for the rock to hit the water.

79. ✏ "I threw a ball into the air; it fell to Earth I know not where. . . ."

But, I can tell how long it is in the air by using

$$0 = 15.0t + \frac{1}{2}(-9.80)t^2$$

The answer t will be in seconds. This problem has only one "useful" answer. Explain why.

80. ✏ We can tell when the ball in Exercise 79 passed a height of 8 meters by solving

$$8.0 = 15.0t + \frac{1}{2}(-9.8)t^2$$

This problem has two answers. Both are correct. Explain why.

EXCURSIONS

Exploring Problem Solving

1. A football is kicked and leaves the punter's foot at a height of 1 meter above the ground. Let

y_0 be the original position
y be the final position
t be the time (in seconds) in the air
x be the distance (in meters) traveled

and use these equations:

$$y = y_0 + 12.0t - 4.90t^2$$
$$x = 16.0t$$

a. How long is the football in the air?

b. How far does it travel before hitting the ground?

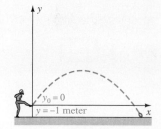

$y_0 = 0$
$y = -1$ meter

Class Act

2. A square sheet of aluminum with sides of 6.4 feet is being altered to form a sand box. A square is cut out of each corner of the sheet. The remaining aluminum is folded to form an open box. The new box is 0.7 foot tall and can hold 17.5 cubic feet of sand.

a. ✏ Explain how the equation

$$(6.4 - 2x)^2 = \frac{17.5}{0.7}$$

can be used to solve for the amount cut from the length and from the width.

b. ✏ Explain how $4x^2$ gives the total area "cut away."

Exploring Geometry

3. The following problem is from an algebra text published in 1887.*

   The perimeter of one square exceeds that of another by 100 feet; and the area of the larger square exceeds three times the area of the smaller by 325 square feet. Find the length of each square's sides.

   Source: H. S. Hall and S. R. Knight, *Elementary Algebra,* 4th Ed, London: Macmillan and Co., 1887.

Exploring Numbers

4. Find the replacements for a, b, c, d, and e that make these two statements true at the same time. Use only the digits 1, 2, 3, 4, and 6. Each letter stands for a unique digit.

$$a^b + b^b + c^b = dee$$
$$d^b + e^b + e^b = abc$$

SECTION GOAL

▪ *To solve quadratic equations by completing the square*

10.2 Solving Quadratic Equations by Completing the Square

SECTION LEAD-IN

A volleyball player serves the ball. Describe the path of the volleyball.

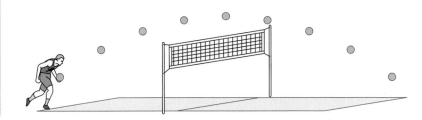

Most quadratic equations are not easy to factor. If we could transform such equations so that one side was the square of a binomial, as in the equation $(2y + 3)^2 = 16$, then we could solve it by taking square roots. Actually, we can do exactly that, and the method we use is called **completing the square.**

To complete the square for a quadratic equation

1. Write the equation in the form $x^2 + bx = c$, where b and c are real numbers.
2. Find $\frac{b}{2}$, square it, and add the result to both sides of the equation.
3. Write the left side of the equation as $\left(x + \frac{b}{2}\right)^2$, and simplify the right side of the equation.
4. Take the square root of each side of the equation to complete the solution.

We shall apply this procedure to several different equations. It is an important skill.

▪ ▪ ▪
EXAMPLE 1

Solve by completing the square: $x^2 - x - 6 = 0$

SOLUTION

Step 1:   This equation is in standard form. But we want the terms that contain the variable to be on the left and the constant on the right. So we add 6 to both sides, obtaining

$$x^2 - x = 6$$

The equation is now in the proper form for completing the square.

Step 2:   Because b (the coefficient of x) is -1,

$$\tfrac{b}{2} \text{ is } -\tfrac{1}{2} \quad \text{and} \quad \left(\tfrac{b}{2}\right)^2 \text{ is } \left(-\tfrac{1}{2}\right)^2$$

We add this value, $\left(-\tfrac{1}{2}\right)^2$, to both sides of the equation.

$$x^2 - x + \left(-\tfrac{1}{2}\right)^2 = 6 + \left(-\tfrac{1}{2}\right)^2$$

Step 3:   We have transformed the left side of the equation into the square of the binomial $\left[x + \left(-\tfrac{1}{2}\right)\right]^2$. To see that this is so, multiply it out. You get

$$\left[x + \left(-\tfrac{1}{2}\right)\right]^2 = x^2 + 2(x)\left(-\tfrac{1}{2}\right) + \left(-\tfrac{1}{2}\right)^2$$
$$= x^2 + \left(-\tfrac{2x}{2}\right) + \left(-\tfrac{1}{2}\right)^2$$
$$= x^2 - x + \left(-\tfrac{1}{2}\right)^2$$

We have "completed the square" of the left side.

We write the left side as $\left[x + \left(-\tfrac{1}{2}\right)\right]^2$ and simplify the right side, obtaining

$$\left[x + \left(-\tfrac{1}{2}\right)\right]^2 = 6 + \left(-\tfrac{1}{2}\right)^2$$
$$= 6 + \tfrac{1}{4} \qquad \text{Squaring}$$
$$= \tfrac{25}{4} \qquad \text{Combining and writing as an improper fraction}$$

> ▪ ▪ ▪
> **WRITER'S BLOCK**
>
> Why do we want to "complete the square" to solve a quadratic equation?

Step 4:   To solve

$$\left[x + \left(-\tfrac{1}{2}\right)\right]^2 = \tfrac{25}{4}$$

we take the square root of both sides of the equation.

$$x + \left(-\tfrac{1}{2}\right) = \tfrac{5}{2} \quad \text{or} \quad x + \left(-\tfrac{1}{2}\right) = -\tfrac{5}{2}$$

Thus x has two values, which we find as follows:

$$x - \frac{1}{2} = \frac{5}{2} \qquad \qquad x - \frac{1}{2} = -\frac{5}{2}$$

$$= \frac{5}{2} + \frac{1}{2} \qquad \qquad = -\frac{5}{2} + \frac{1}{2} \quad \text{Adding } \tfrac{1}{2} \text{ to both sides}$$

$$= \frac{6}{2} \qquad \qquad = -\frac{4}{2} \quad \text{Combining like terms}$$

$$= 3 \qquad \qquad = -2 \quad \text{Simplifying fractions}$$

So $x = 3$ and $x = -2$ are solutions to the equation $x^2 - x - 6 = 0$. You should check both solutions to make sure they satisfy the original equation.

▶*CHECK* **Warm-Up 1**

Calculator Corner

In step 4 of Example 1, we found the equation of the parabola in the form $\left[x + \left(-\frac{1}{2}\right)\right]^2 = \frac{25}{4}$. Set this equation equal to zero in order to obtain:

$$\left[x + \left(-\frac{1}{2}\right)\right]^2 - \frac{25}{4} = 0$$

Now graph this parabola on your graphing calculator using a "friendly" window. **TRACE** to estimate the *vertex* of the parabola.

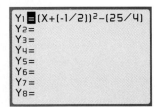 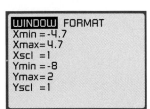 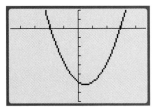

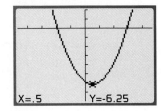

Question: Is $\left(\frac{1}{2}, -\frac{25}{4}\right)$ the same as $(0.5, -6.25)$?

Now graph the following parabolas on your graphing calculator using "friendly" windows.

a. $(x + 4)^2 - 7 = 0$ **b.** $(x + 3)^2 - (7/2) = 0$ **c.** $[x + (-3/2)]^2 + (5/2) = 0$

Can you state a conjecture about this form of the equation of a parabola and the vertex of the parabola?

■ ■ ■

EXAMPLE 2

Solve: $n^2 - 6n - 10 = 0$

SOLUTION

We must first rewrite the equation in such a way that the variable terms appear on one side of the equation and the constant term on the other side. To do so, we add 10 to both sides.

$$n^2 - 6n = 10$$

To complete the square, we need to add $\left(\frac{b}{2}\right)^2$ to both sides. Because b in this equation is -6, $\frac{b}{2}$ is $\frac{-6}{2} = -3$. Then we add $(-3)^2$ to both sides.

$$n^2 - 6n + (-3)^2 = 10 + (-3)^2$$

We have completed the square, so the left side of this equation is equivalent to $[n + (-3)]^2$. We must also simplify the right side.

$$[n + (-3)]^2 = 10 + (-3)^2$$
$$= 10 + 9$$
$$= 19$$

Taking the square roots of both sides of $[n + (-3)]^2 = 19$, we obtain

$$n - 3 = \sqrt{19} \quad \text{or} \quad n - 3 = -\sqrt{19}$$

Adding 3 to both sides of each equation gives

$$n = 3 + \sqrt{19} \quad \text{or} \quad n = 3 - \sqrt{19}$$

We will check $n = 3 - \sqrt{19}$ in the original equation; you should check $n = 3 + \sqrt{19}$ yourself.

$n^2 - 6n - 10 = 0$	Original equation
$\left(3 - \sqrt{19}\right)^2 - 6\left(3 - \sqrt{19}\right) - 10 = 0$	Substituting $n = 3 - \sqrt{19}$
$9 - (2)\left(3\sqrt{19}\right) + 19 - 18 + 6\sqrt{19} - 10 = 0$	Multiplying out
$9 - 6\sqrt{19} + 19 - 18 + 6\sqrt{19} - 10 = 0$	Simplifying
$28 - 28 - 6\sqrt{19} + 6\sqrt{19} = 0$	Combining like terms
$0 = 0$	True

▶ CHECK **Warm-Up 2**

> *Note:* The squared term *must* have a coefficient of 1 before you can complete the square. If it does not, divide *each* term by the coefficient of the squared term before you complete the square.

The technique of completing the square will be very useful to you if you study conic sections in a later course.

Practice what you learned.

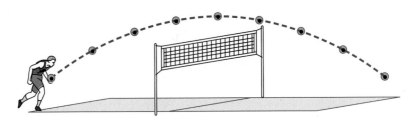

SECTION FOLLOW-UP

The track of the volleyball is a parabola. The equation of this path is quadratic.

10.2 WARM-UPS

Work these problems before you attempt the exercises.

1. Solve by completing the square: $x^2 - 3x = 0$

2. Solve by completing the square: $x^2 = 20x + 21$

10.2 EXERCISES

Note: Use your graphing calculator to check your results whenever possible.

For Exercises 1 through 12, write each equation in the form $x^2 + bx = c$. Then find $\frac{b}{2}$ and $\left(\frac{b}{2}\right)^2$.

1. $x^2 - 24x + 40 = 0$

2. $y^2 + 12y - 72 = 0$

3. $y^2 + 30y + 27 = 0$

4. $y^2 + 64y + 125 = 0$

5. $x^2 - 30x - 200 = 0$

6. $3x^2 - 45x + 162 = 0$

7. $10y^2 - 110y + 300 = 0$

8. $-3y^2 + 3y = 168$

9. $7x^2 - 35x = 98$

10. $2y^2 - 2y - 122 = 0$

11. $5y^2 - 50y - 140 = 0$

12. $4x^2 + 24x = 364$

In Exercises 13 through 30, solve each equation by completing the square.

13. $x^2 - 2x - 15 = 0$

14. $y^2 - 3y - 10 = 0$

15. $y^2 - 26y + 25 = 0$

16. $x^2 + 6x - 7 = 0$

17. $4x^2 + 36x - 88 = 0$

18. $9x^2 - 18x + 8 = 0$

19. $y^2 + 8y - 20 = 0$

20. $x^2 - 10x - 24 = 0$

21. $x^2 - 18x - 19 = 0$

22. $y^2 + 10y - 11 = 0$

23. $y^2 - 8y + 12 = 0$

24. $x^2 - 22x + 21 = 0$

25. $x^2 + 12x - 13 = 0$

26. $y^2 - 6y - 40 = 0$

27. $y^2 + 22y + 120 = 0$

28. $x^2 + 14x + 48 = 0$

29. $x^2 - 4x - 5 = 0$

30. $y^2 - 24y + 140 = 0$

In Exercises 31 through 48, solve the word problems. Be sure to read each problem carefully before starting.

31. *Electricity*   The current I (in amperes) in a certain circuit at any time t (in seconds) is given by the equation

$$I = t^2 - 7t + 12$$

a.   What is the current after 10 seconds?

b.   When is the current equal to 6 amperes?

32. *Hang Gliding*   The area of a hang-gliding sail can be computed by the formula

$$A = \frac{k^2}{\sqrt{2}}$$

where k is the length of the keel. $\left(\text{Use } \sqrt{2} \approx 1.4.\right)$

a.   Find A when k is $4\frac{1}{2}$ feet.

b.   Find k when A is 18 square feet.

33. *Geometry*   Triangles, squares, pentagons, and hexagons are all examples of polygons. A straight line that joins any two non-adjacent vertices is called a diagonal. The number of diagonals D of any polygon with n sides is given by

$$D = \tfrac{1}{2}n^2 - \tfrac{3}{2}n$$

Complete the square to find the number of sides a figure with 35 diagonals has.

34. *Roller Coaster*   At Knotts Berry Farm in California, a roller coaster has a 70-foot vertical drop. The height y of its last car t seconds after it starts its descent can be approximated by

$$y_{height} = -0.8t^2 - 10t + 70$$

Complete the square to find when the car is 17 feet from the ground. Round your answer to the nearest whole second.

35. *Number Problem*   Find two positive numbers whose difference is 6 and whose product is 40.

36. *Garden Fencing*   Lydia, an avid gardener, wants to use 131 feet of fencing to make two rectangular gardens as shown in the figure. The area enclosed is 713 square feet. What are the dimensions of the outside perimeter if the dimensions are whole numbers?

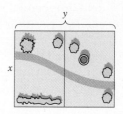

37. *Dropping a Ball*   The Empire State Building is 1472 feet tall.

    a. If a ball is dropped from the top, how fast will it be traveling the instant before it hits the ground? Use the formula $v = 5.45\sqrt{d}$, where v is in miles per hour and d is in feet.

    b. From how high would you have to drop a ball for it to be traveling 545 miles per hour the instant before it hits the ground?

38. *Geometry*   A 5-inch square is cut from each corner of a square piece of cardboard. If the sides are then turned up and taped to form an open-topped box, whose volume is 720 cubic inches, find the length of the sides of the original piece of cardboard.

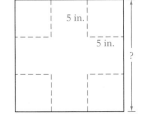

39. *Number Problem*   The product of two consecutive positive odd integers is 899. What are the integers?

40. *Geometry*   Two cubes have the same surface area. One is closed and one has an open top. Each edge of the open cube is $\frac{2}{3}$ inch longer than each edge of the closed cube. Find the dimensions of each.

41. *Shock Waves*   The formula $d = t^2 + 2t$ describes the distance d in miles that shock waves from an explosion travel in t seconds. How long will it take for shock waves to travel 15 miles?

42. *Number Problem*   Find two numbers whose sum is 16 and whose product is 15.

43. *Viral Infection*   The formula $N = 40x - x^2$ describes the number of cases N of a virus reported x days after exposure. How long does it take for 400 people to get the virus?

44. *Geometry*   The area of a trapezoid is 49.5 square inches. The shorter base is 3 inches longer than the height. The longer base is 3.5 inches shorter than 4 times the height. Find the dimensions.

45. *Office Renovation*   In the planning stages of the renovation of an 1800-square-foot space, the amount of office space allocated per person is increased by 25 square feet when 6 people are removed to an adjoining building. How many people share this renovated space after the move? How much space does each person have?

46. *Lab Fee*   A mathematics class must split a $100 computer fee. During add/drop, 15 more students signed up for the class, and the share per student dropped $1.50. How many students were in the class before and after add/drop?

47. *Nuclear Waste*   The amount of nuclear waste, in gallons, flowing into a stream t days after a leak in the cooling system is $w = 8t^2 + 4t$. After how many days is the waste level 840 gallons?

48. *Bacterial Growth* The formula $C = 20t^2 - 200t + 600$ describes the concentration C of bacteria per cubic centimeter in a body of water t days after treatment to slow bacterial growth. After how many days will the concentration be 100 bacteria per cubic centimeter?

MIXED PRACTICE

By doing these exercises, you will practice the topics up to this point in the chapter.

49. Solve by completing the square:
$36x^2 - 48x + 7 = 0$

50. Solve by using the equality property of square roots: $(7 - 2x)^2 = 64$

51. Solve by factoring: $49x^2 - 9 = 0$

52. Solve by using the equality property of square roots: $(3x - 17)^2 - 81 = 0$

53. Solve by factoring: $4x^2 - 484 = 0$

54. Solve by using the equality property of square roots: $(13 - x)^2 - 9 = 0$

55. Solve: $9(16x - 18)^2 = 1$

56. Solve: $x^2 - 13x - 14 = 0$

57. Solve: $x^2 = 5$

58. Solve: $24r^2 = 36$

59. Solve: $9y^2 - 18y + 9 = 0$

60. Solve: $y^2 + 14y = 120$

61. Solve: $9t^2 = 16$

62. Solve: $9y^2 + 18y - 16 = 0$

EXCURSIONS

Posing Problems

1. Ask and answer four questions using the following data.

World Population, Land Areas, and Elevations

Area	Estimated population, mid-1955	Approximate land area (sq. mi.)	Percent of total land area	Population density per sq. mi.	Elevation (feet)	
					Highest	Lowest
WORLD	5,734,106,000	58,433,000	100.0	98.1[1]	Mt. Everest, Asia, 29,028	Dead Sea, Asia, 1290 below sea level
ASIA, incl. Philippines, Indonesia, and European and Asiatic Turkey; excl. Asiatic former U.S.S.R.	3,403,451,000	10,644,000	18.2	319.7	Mt. Everest, Tibet-Nepal, 29,028	Dead Sea, Israel-Jordan, 1290 below sea level

AFRICA	721,472,000	11,707,000	20.0	61.2	Mt. Kilimanjaro, Tanzania, 19,340	Lake Assal, Djibouti, 571 below sea level
NORTH AMERICA, including Hawaii, Central America, and Caribbean region	454,187,000	9,360,000	16.0	48.5	Mt. McKinley, Alaska, 20,320	Death Valley, Calif., 282 below sea level
SOUTH AMERICA	319,553,000	6,883,000	11.8	46.2	Mt. Aconcagua, Arg.-Chile, 23,034	Valdes Peninsula, 131 below sea level
ANTARCTICA	—	6,000,000	10.3	—	Vinson Massif, Sentinel Range, 16,863	Sea level
EUROPE, incl. Iceland; excl. European former U.S.S.R. and European Turkey	509,254,000	1,905,000	3.3	267.3	Mont Blanc, France, 15,781	Sea level
OCEANIA, incl. Australia, New Zealand, Melanesia, Micronesia, and Polynesia[2]	28,680,000	3,284,000	5.6	8.7	Wilhelm, Papua, New Guinea, 14,793	Lake Eyre, Australia, 38 below sea level
Former U.S.S.R., both European and Asiatic	297,508,000	8,647,000	14.8	34.4	Communism Peak, Pamir, 24,590	Caspian Sea, 96 below sea level

1. In computing density per square mile, the area of Antarctica is omitted. 2. Although Hawaii is geographically part of Oceania, its population is included in the population figure for North America. *Note:* The land area of Asia including the Asiatic portion of the former U.S.S.R. is 17,240,000 sq. miles. *Source:* U.S. Bureau of the Census, International Data Base, as cited in the *1996 Information Please® Almanac* (©1995 Houghton Mifflin Co.), p. 481. All rights reserved. Used with permission by Information Please LLC.

2. Ask and answer four questions using the following data.

Gestation, Incubation, and Longevity of Certain Animals

Animal	Gestation or incubation, in days (& average)	Longevity, in years (& record exceptions)	Animal	Gestation or incubation, in days (& average)	Longevity, in years (& record exceptions)
Ass	365	18–20 (63)	Horse	329–345 (336)	20–25 (50+)
Bear	180–240[1]	15–30 (47)	Kangaroo	32–39[1]	4–6 (23)
Cat	52–69 (63)	10–12 (26+)	Lion	105–113 (108)	10 (29)
Chicken	22	7–8 (14)	Man	253–303	([2])
Cow	c. 280	9–12 (39)	Monkey	139–270[1]	12–15[1](29)
Deer	197–300[1]	10–15 (26)	Mouse	19–31[1]	1–3 (4)
Dog	53–71 (63)	10–12 (24)	Parakeet (Budgerigar)	17–20 (18)	8 (12+)

Duck	21–35[1](28)	10 (15)	Pig	101–130 (115)	10 (22)
Elephant	510–730 (624)[1]	30–40 (71)	Pigeon	11–19	10–12 (39)
Fox	51–63[1]	8–10 (14)	Rabbit	30–35 (31)	6–8 (15)
Goat	136–160 (151)	12 (17)	Rat	21	3 (5)
Groundhog	21–32	4–9	Sheep	144–152 (151)[1]	12 (16)
Guinea pig	58–75 (68)	3 (6)	Squirrel	44	8–9 (15)
Hamster, golden	15–17	2 (8)	Whale	365–547[1]	—
Hippopotamus	220–255 (240)	30 (49+)	Wolf	60–63	10–12 (16)

1. Depending on kind. 2. Life expectancy varies. *Source:* James G. Doherty, General Curator, The Wildlife Conservation Society, as cited in the *1996 Information Please® Almanac* (©1995 Houghton Mifflin Co.), p. 573. All rights reserved. Used with permission by Information Please LLC.

Data Analysis

3. ✏ What can you learn from the following data?

Getting on the 1996 Presidential Ballot

Each state had a different requirement for getting on the 1996 Presidential ballot with a party affiliation. Here are the deadlines and signatures that were required, using the easiest method in each state.

State	Deadline	Signatures required	% of registered voters
California	Oct. 24, 1995	89,006	0.60
Ohio	Nov. 20	33,463	0.54
South Dakota	April 2, 1996	7,792	0.72
North Dakota	April 12	7,000	1.51
Tennessee	May 1	37,179	1.39
Wyoming	May 1	8,000	3.36
South Carolina	May 5	10,000	0.67
Texas	May 13	43,963	0.51
Arizona	May 21	15,062	0.73
North Carolina	June 1	51,904	1.43
Kansas	June 1	16,418	1.25
Maine	June 4	4,000	0.44
Washington	July 6	200	0.01
Georgia	July 9	30,036	1.00
New Mexico	July 9	2,339	0.33
Nevada	July 11	3,761	0.60
Florida	July 15	65,596	1.00
Oklahoma	July 15	41,711	2.04
Indiana	July 15	29,822	0.97
Colorado	July 16	0[1]	0.00
Michigan	July 18	30,891	0.50
New Jersey	July 29	800	0.02
Massachusetts	July 30	10,000	0.32
Montana	July 31	10,471	2.04
Pennsylvania	Aug. 1	30,000	0.51
West Virginia	Aug. 1	6,837	0.77

State	Deadline	Signatures required	% of registered voters
Nebraska	Aug. 1	5,741	0.62
Illinois	Aug. 5	25,000	0.41
Maryland	Aug. 5	10,000	0.42
Missouri	Aug. 5	10,000	0.34
Connecticut	Aug. 7	7,500	0.40
New Hampshire	Aug. 7	3,000	0.45
Alaska	Aug. 7	2,586	0.77
Iowa	Aug. 16	1,500	0.09
Delaware	Aug. 17	180	0.06
New York	Aug. 20	15,000	0.17
District of Columbia	Aug. 20	3,500	0.97
Virginia	Aug. 23	16,000	0.53
Oregon	Aug. 27	18,316	1.02
Kentucky	Aug. 29	5,000	0.23
Louisiana	Aug. 29	0[1]	0.00
Idaho	Aug. 31	9,644	1.54
Utah	Sept. 1	300	0.03
Arkansas	Sept. 1	0[2]	0.00
Alabama	Sept. 2	35,973	1.62
Wisconsin	Sept. 3	2,000	0.06
Hawaii	Sept. 6	3,829	0.78
Rhode Island	Sept. 6	1,000	0.18
Mississippi	Sept. 6	0[3]	0.00
Minnesota	Sept. 10	2,000	0.07
Vermont	Sept. 19	1,000	0.27

1 No signatures required. Must pay $500 to get on ballot.
2 Just hold a meeting.
3 Just be organized.

Class Act

4. A storage bin that is 5 feet tall has a length and width of equal size. Its volume is 320 cubic feet. The length and width are increased by the same amount. The new volume is 551.25 cubic feet.

   a. What was the original length and width?

   b. By how much were the length and width increased?

   c. Explain how the equation $y^2 = \frac{320}{5}$ represents the length and width of the original bin.

   d. Explain how the equation $(x + 8)^2 = \frac{551.25}{5}$ can be used to find the size of the increase.

10.3 Solving Quadratic Equations by Using the Quadratic Formula

SECTION GOALS

▪ *To solve quadratic equations using the quadratic formula*

▪ *To use the discriminant to describe the roots of a quadratic equation*

SECTION LEAD-IN

The shot put is an Olympic event. A men's shot is a 16-pound ball (women's = 7.6 lbs) made of iron or brass. It is put, not thrown, from shoulder level and can travel almost 80 feet. Its path of flight is a parabola. The equation of that flight is quadratic. Find another "throwing" sport with a quadratic relationship.

We can solve any quadratic equation by completing the square. However, the method is complicated and often involves tricky fractions and radicals. In this section, we discuss another method that works for all quadratic equations but is almost as simple as substituting numbers into a formula. The particular formula we use is called the **quadratic formula.** You should memorize this formula.

Quadratic Formula

The quadratic equation

$$ax^2 + bx + c = 0$$

where a, b, and c are real numbers and $a \neq 0$, has the solutions

$$x = \frac{-b \pm \sqrt{b^2 - 4ac}}{2a}$$

To use the formula, you must be able to identify the constants a, b, and c in the equation.

■ ■ ■

EXAMPLE 1

Write each quadratic equation in standard form, and identify the constants a, b, and c for use in the quadratic formula.

a. $4x^2 - 6x = 9$ **b.** $2x = 3x^2 - 2$ **c.** $x^2 - 25 = 0$

SOLUTION

a. We add -9 to both sides of the equation, obtaining the standard form

$$4x^2 - 6x - 9 = 0$$

Then

$$a = 4 \qquad b = -6 \qquad c = -9$$

We *do not* need to divide by the coefficient of x^2 to use the quadratic formula.

b. If we move the non-zero terms to the left side, we get the standard form

$$-3x^2 + 2x + 2 = 0$$

Then

$$a = -3 \qquad b = 2 \qquad c = 2$$

But if we move $2x$ to the right side, we get

$$0 = 3x^2 - 2x - 2$$

so

$$a = 3 \qquad b = -2 \qquad c = -2$$

Both sets of constants give *the same solutions* in the quadratic formula.

c. The equation $x^2 - 25 = 0$ is in standard form, but the x-term has zero as its coefficient. Thus

$$a = 1 \qquad b = 0 \qquad c = -25$$

▶ *CHECK* **Warm-Up 1**

> **■■■**
>
> **WRITER'S BLOCK**
>
> **What is meant by the** *standard form* **of an equation? What is useful about the standard form?**

> **STUDY HINT**
>
> *Write all quadratic equations in standard form before you try to use the quadratic formula.*

Using the Formula

Now let's apply the quadratic formula.

▪▪▪

EXAMPLE 2

Solve: $4x^2 + 5x - 6 = 0$

SOLUTION

This equation is in standard form, and

$$a = 4 \qquad b = 5 \qquad c = -6$$

> **STUDY HINT**
>
> *Make sure you always use parentheses when you substitute values for variables.*

We substitute these values into the quadratic formula and simplify, getting

$$x = \frac{-b \pm \sqrt{b^2 - 4ac}}{2a} = \frac{-(5) + \sqrt{(5)^2 - 4(4)(-6)}}{2(4)}$$

$$= \frac{-(5) \pm \sqrt{25 - (16)(-6)}}{8} = \frac{-(5) \pm \sqrt{25 + 96}}{8}$$

$$= \frac{-5 \pm \sqrt{121}}{8} = \frac{-5 \pm 11}{8}$$

We have simplified the expression on the right as much as possible. It is time now to separate the two solutions.

$$x = \frac{-5 + 11}{8} \qquad \text{or} \qquad x = \frac{-5 - 11}{8}$$

$$= \frac{6}{8} = \frac{3}{4} \qquad \Bigg| \qquad = \frac{-16}{8} = -2$$

So $x = \frac{3}{4}$ and $x = -2$ when $4x^2 + 5x - 6 = 0$. Always check your results.

▶ *CHECK* **Warm-Up 2**

Calculator Corner

You can use your graphing calculator's Home Screen to verify the numerical results you obtained using the quadratic formula. Verify the results for Example 2 with your graphing calculator. First store the values of a, b, and c. (Note: See the Calculator Corner on page 85 to review how to store values for variables.) Then type in the quadratic formula on the Home Screen (see the first screen at the right).

Use the calculator's **2nd ENTRY** feature to recall the previous line and change the $+$ sign after the $-$B to a $-$ sign (see the second screen at the right).

The results $x = 0.75$ and $x = -2$ verify the results obtained using the paper-and-pencil method of calculation.

```
4→A
                    4
5→B
                    5
-6→C
                   -6
(-B+√(B²-4AC))/2A
                  .75
```

```
(-B+√(B²-4AC))/2A
                  .75
(-B-√(B²-4AC))/2A
                   -2
```

Remember that you can now **STO**re new values for a, b, and c and then use **2nd ENTRY** to recall the two forms of the quadratic formula in order to work more problems using the quadratic formula.

▪▪▪

EXAMPLE 3

Solve: $3n^2 - 6n - 10 = 0$

SOLUTION

We identify

$$a = 3 \qquad b = -6 \qquad c = -10$$

Here we are solving for n, not x. In the formula, however, they are interchangeable. Using the quadratic formula, we substitute as follows:

$$n = \frac{-b \pm \sqrt{b^2 - 4ac}}{2a} = \frac{-(-6) \pm \sqrt{(-6)^2 - 4(3)(-10)}}{2(3)} = \frac{6 \pm \sqrt{36 + 120}}{6}$$

$$= \frac{6 \pm \sqrt{156}}{6} = \frac{6 \pm \sqrt{2^2 \cdot 39}}{6} = \frac{6 \pm 2\sqrt{39}}{6}$$

So

$$n = \frac{6 + 2\sqrt{39}}{6} \qquad \text{or} \qquad n = \frac{6 - 2\sqrt{39}}{6}$$

$$= \frac{3 + \sqrt{39}}{3} \qquad \qquad \qquad = \frac{3 - \sqrt{39}}{3}$$

Because 39 is positive, these are real-number solutions. We always check the results. We will check $n = \frac{3 - \sqrt{39}}{3}$ and leave $\frac{3 + \sqrt{39}}{3}$ for you to do.

$$3n^2 - 6n - 10 = 0$$

$$3\left(\frac{3 - \sqrt{39}}{3}\right)^2 - 6\left(\frac{3 - \sqrt{39}}{3}\right) - 10 = 0$$

$$3\left(\frac{9 - 6\sqrt{39} + 39}{9}\right) - 2\left(3 - \sqrt{39}\right) - 10 = 0$$

$$3\left(\frac{48 - 6\sqrt{39}}{9}\right) - 6 + 2\sqrt{39} - 10 = 0$$

$$16 - 2\sqrt{39} - 6 + 2\sqrt{39} - 10 = 0$$

$$16 - 6 - 10 - 2\sqrt{39} + 2\sqrt{39} = 0$$

$$0 = 0$$

The solution checks.

In the next example, we again must substitute a value and solve the resulting quadratic equation to answer a question.

STUDY HINT

To check a solution that includes a square root of a number that is not a perfect square, leave the answer in simplest radical form. Do not estimate the value of the radical.

10.3 EXERCISES

Note: Use your graphing calculator to check your results whenever possible.

In Exercises 1 through 12, write each quadratic equation in standard form, and identify a, b, and c.

1. $3x^2 - 45x = 152$

2. $6x^2 - 36x + 30 = 0$

3. $-0.4x^2 + 1.6x = 1.8$

4. $-0.5x^2 = 9.0x + 3.85$

5. $2x^2 + 22x = 84$

6. $10x^2 = 110x + 300$

7. $-\frac{3}{5}x^2 + \frac{3}{8}x = 168$

8. $-\frac{5}{7}x^2 = 85x + 260$

9. $7x^2 - 35x = 98$

10. $2x^2 - 2x - 122 = 0$

11. $4.5x^2 - 5.08x = 1.40$

12. $4x^2 + 2.4x = 36.9$

In Exercises 13 through 54, solve each quadratic equation using the quadratic formula. Identify those that have no real-number solution.

13. $3x^2 + 30x + 27 = 0$

14. $5x^2 - 30x - 80 = 0$

15. $x^2 - 16x + 64 = 0$

16. $x^2 - 13x - 169 = 0$

17. $x^2 + 12x + 35 = 0$

18. $x^2 + 22x + 72 = 0$

19. $x^2 - 169 = 0$

20. $x^2 - 25 = 0$

21. $4x^2 + 20x + 24 = 0$

22. $9x^2 - 15x - 6 = 0$

23. $2x^2 + 22x - 84 = 0$

24. $10x^2 - 110x = -300$

25. $x^2 - x = 30$

26. $x^2 - 5x - 6 = 0$

27. $x^2 - 18x + 81 = 0$

28. $x^2 + 10x + 25 = 0$

29. $25x^2 = 100$

30. $9x^2 - 16 = 0$

31. $5x^2 - 22x = 15$

32. $16x^2 - 16x + 4 = 0$

33. $2x^2 - 8x + 6 = 0$

34. $6x^2 + 78x + 180 = 0$

35. $x^2 + 16x + 63 = 0$

36. $x^2 - 9x + 18 = 0$

37. $x^2 + 16x + 64 = 0$

38. $x^2 - 2x + 1 = 0$

39. $2x^2 + 14x + 88 = 0$

40. $15x^2 + 48x + 33 = 0$

41. $x^2 - 361 = 0$

42. $3x^2 - 108 = 0$

43. $3x^2 - 6x + 9 = 0$

44. $49x + 17x^2 + 90 = 0$

45. $4x^2 + 8x + 4 = 0$

46. $16x^2 + 40x + 25 = 0$

47. $x^2 - 18x + 77 = 0$

48. $x^2 + 21x + 104 = 0$

49. $0.5x^2 + 2.5x + 2 = 0$

50. $x^2 + x + 0.25 = 0$

51. $\frac{1}{3}x^2 - \frac{8}{3}x + 5\frac{1}{3} = 0$

52. $x^2 - 2\frac{1}{3}x - 2 = 0$

53. $x(x + 5) + 4 = 0$

54. $x(x + 11) = 0$

In Exercises 55 through 66, use the discriminant $b^2 - 4ac$ to determine which equations have real-number solutions.

55. $x^2 = 1 - 5x$

56. $-4x^2 = 27x + 3$

57. $x^2 + 64 = 17x$

58. $x^2 + 125x = 14$

59. $6x^2 + 24x + 24 = 0$

60. $9x^2 - 54x + 81 = 0$

61. $25x^2 + 1 = 10x$

62. $x^2 + 625 = 50x$

63. $x^2 + 8 = -18x$

64. $-3x^2 - 729x = 2$

65. $x^2 + 512 = 13x$

66. $-2x^2 - 216 = 16x$

Solve the following application problems.

67. *Manufacturing*   A manufacturing company produces n items per week at a cost C of

$$C = (n - 100)^2 + 400$$

It also sells every item it produces at $27.16 each. From these sales, the company makes an amount P according to the equation

$$P = 27.16n$$

Find, to the nearest whole number, the values of n that cause P and C to be equal.

68. *Pollution*   The number of tons of pollutants P dumped into a flowing stream by a factory in its tth year of operation is represented by the equation $(P - 12)^3 = 8^3 t$. What year is the first in which 36 tons are dumped?

69. *Family Income*   The equation

$$I = \frac{3t^2 - 13t + 505}{100}$$

describes the median income for married couples in the United States t years after 1955, where I is in thousands of dollars. Find the year in which the median income was $10,000.

70. *Temperature*   The temperature F in degrees Fahrenheit t hours after 8 A.M. in a mountain resort is given by

$$\frac{5t^2 - 48t + 288}{36} = F$$

At what time was the temperature 5 degrees?

71. *Bacterial Growth*   The formula

$$N = \frac{-14400 + 120t + 100t^2}{144 + t^2}$$

gives the number of bacteria present in a culture t hours after 100 bacteria are treated with an antibacterial agent. How many hours will it take before the population is 0?

72. ***Roller Coaster***   A roller coaster at an amusement park in California has an initial 10-foot vertical drop. The speed v of the last car t seconds after it starts this descent can be approximated as $v = 7.2t + 4.6$, where v is in feet per second. The height of the last car from the bottom of this drop, in feet, can be estimated by $h = -0.8t^2 - 10t + 70$. At what time does the last car reach the bottom of the vertical drop? What is the speed of the last car at this time?

73. ***Hang Gliding***   The proper length keel k of a hang glider for light wind conditions can be computed by the formula

$$k \approx \sqrt{1.19(35 + w)}$$

where w is the rider's weight in pounds.

 a.   Find the keel length for a 180-pound rider.

 b.   A 14-foot keel will support a rider of what weight?

74. ***Boiling Temperature***   The temperature T at which water will boil decreases as the elevation E above sea level increases. The formula is

$$E \approx 1000(100 - T) + 580(100 - T)^2$$

where E is measured in meters, and T is measured in degrees Celsius.

 a.   What is the altitude if water can boil at 99.2°C? (100°C is the boiling point at sea level.)

 b.   At what temperature would water boil at 6125 meters above sea level?

75. ***Number Problem***   The two roots

$$\frac{-b + \sqrt{b^2 - 4ac}}{2a} \quad \text{and} \quad \frac{-b - \sqrt{b^2 - 4ac}}{2a}$$

have a sum of $-\frac{b}{a}$ and a product of $\frac{c}{a}$. Verify that fact and use it to answer the following questions.

 a.   One solution of a quadratic equation $6x^2 - 22x + n = 0$ is $\frac{17}{3}$. What is the other solution, and what is n in the quadratic equation?

 b.   A quadratic equation $4x^2 - bx + 25 = 0$ has been solved, and one solution is 25. What is the other solution, and what is b in the quadratic equation?

76. ***Stopping Distance***   The formula $10D = 7x^2 + 7.5x$ describes the distance D in feet that it takes to stop a vehicle traveling x miles per hour on dry pavement.

 a.   How fast can you drive if you wish to be able to stop your car within 65 feet?

 b.   On black ice, a truck's stopping distance is 3 times its stopping distance on dry pavement. A truck traveling 5 miles per hour applies the brakes, on black ice, at a distance of 65 feet in front of a rubber traffic cone. Will the truck hit the cone?

77. **Velocity** The World Trade Center Buildings in New York City are 1353 feet tall. A ball is thrown upward from the side of the rooftop observation deck, at an initial velocity of 128 feet per second, at the same time that a ball is dropped from the same spot. Which ball will hit the ground first, and how many seconds later will the second ball hit the ground? The equation giving the height h of the ball that was tossed upward is

$$h = -16t^2 + 128t + 1353$$

The equation for the falling ball is

$$h = 1353 - 16t^2$$

In both equations, t is the time in seconds. Use $h = 0$ for the ground.

78. **Well Depth** Ahmed has a well on his land. He drops a stone over the opening and 2.86 seconds later hears the stone hit the water. He wants to find the depth of the well. An engineer gives him the formula

$$T = \frac{\sqrt{d}}{4} + \frac{d}{1100}$$

where d is the depth of the well and T is the time between dropping a stone and hearing it hit the water. How deep is Ahmed's well?

79. **Moon Rocks** A rock is thrown straight upward, on Earth and on the moon, with an initial velocity of 50 feet per second from an initial height of 5.5 feet. The equations describing the resulting motion are

On Earth: $h_E = -16t^2 + v_0t + h_0$

On the moon: $h_m = -2.7t^2 + v_0t + h_0$

a. By how much do the heights differ after 3 seconds?

b. How long does each rock take to return to its initial height?

80. **Stopping Distance** A quadratic equation for estimating highway stopping distance is

$$y = x + \frac{x^2}{20}$$

where y is the number of feet it takes to stop if a car is traveling x miles per hour. The formula, based on an old Highway Code in Great Britain, takes into account the driver's "thinking" distance and the actual braking distance.

a. What is the difference in stopping distance between a car traveling at 55 miles per hour and a car traveling at 60 miles per hour?

b. How fast could a car be traveling and still stop within 120 feet?

MIXED PRACTICE

By doing these exercises, you will practice the topics up to this point in the chapter.

81. Solve by completing the square:
$4x^2 + 36x + 81 = 0$

82. Solve: $6(2 - x)^2 = 54$

83. Solve by factoring: $2x^2 - 19x - 21 = 0$

84. Solve by using the quadratic formula:
$x^2 - 13x + 40 = 0$

85. Solve by completing the square:
$-56 + x^2 + x = 0$

86. Solve: $(2x - 17)^2 = 121$

87. Solve by factoring: $x^2 + 22x + 121 = 0$

88. Solve by using the quadratic formula:
$-24 + x^2 = 2x$

89. Solve by using the quadratic formula:
$7x^2 - 5x = 2$

90. Solve: $(3 - x)^2 - 9 = 0$

EXCURSIONS

Class Act

1. a. A certain mathematics text uses

$$x = -\frac{b}{2a} \pm \sqrt{\frac{b^2}{4a^2} - \frac{c}{a}}$$

as its quadratic formula. Is this formula equivalent to

$$x = \frac{-b \pm \sqrt{b^2 - 4ac}}{2a}$$

Justify your answer.

b. ✏ Which formula is easiest to use and why?

2. Graphs take many forms. Ask four questions that can be answered by the graph at the top of the following page. Answer your own questions or swap and answer someone else's questions.

3. A company produces n items per week that cost c dollars to produce. The relationship between the cost and the number produced is given by the equation

$$\frac{1}{n^2} = \frac{1 - \frac{200}{n}}{c - 10,400}$$

This equation, rational in form, is actually a quadratic equation. Verify this.

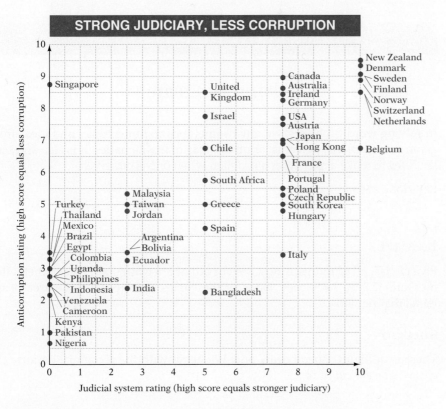

STRONG JUDICIARY, LESS CORRUPTION

Source: Reprinted by permission of *FORBES Magazine* © Forbes Inc., 1996.

Posing Problems

4. Ask and answer four questions about the following data.

Per Capita Consumption of Principal Foods[1]

Food	1993	1992	1991
Red meat	111.9	114.1	111.9
Poultry[3]	61.1	60.0	58.0
Fish and shellfish[2]	14.9	14.7	14.8
Eggs	30.1	30.2	30.0
Fluid milk and cream[3]	226.7	230.8	233.1
Ice cream	16.1	16.3	16.3
Cheese (excluding cottage)	26.2	26.0	25.0
Butter (actual weight)	4.5	4.2	4.2
Margarine (actual weight)	10.8	11.0	10.6
Total fats and oils[4]	65.0	65.6	63.8
Fruits (farm weight)[5]	308.7	291.7	289.5
Peanuts (shelled)	6.0	6.6	6.5
Vegetables (farm weight)	396.6	391.7	388.5
Sugar (refined)	64.2	64.5	63.7
Corn sweeteners (dry weight)	81.5	77.9	75.4
Flour and cereal products	189.2	187.0	185.4
Soft drinks (gal)	46.6	45.4	44.9
Coffee bean equivalent	10.0	10.3	10.4
Cocoa (chocolate liquor equivalent)	4.6	4.6	4.6

1. As of August 1994. Except where noted, consumption is from commercial sources and is in pounds retail weight. 2. Boneless, trimmed equivalent. 3. Includes milk and cream produced and consumed on farms. 4. Fat-content basis. 5. Excludes wine grapes. Note: Data are latest available. *Source: 1996 Information Please® Almanac* (©1995 Houghton Mifflin Co.), p. 71. All rights reserved. Used with permission by Information Please LLC.

CONNECTIONS TO *GEOMETRY*

Using Diagrams

In solving word problems, we must sometimes combine topics and ideas from geometry, fractions, percents, or other areas of mathematics. It is also an excellent idea to make a sketch for a problem when one is not given.

Let's look at some examples.

▪ ▪ ▪

EXAMPLE

Five circles, all of radius 1 inch, are cut out of a rectangular piece of metal that measures 2.3 by 9.8 inches. None of the circles overlap. Find the area of the metal that remains.

SOLUTION

There is no figure, so we draw one. We begin with the rectangle, because circles will be *cut out* of it.

Next we draw in the five circles. We know that they are all circles and that they *don't overlap.* But we can place them wherever we like. Finally, we shade in the area we must find, as a reminder.

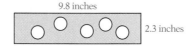

We can find the area of the metal that remains by subtracting the area of the five circles from the area of the rectangle.

$$\text{Area(shaded)} = \text{area(rectangle)} - \text{area(five circles)}$$

First we find the area of the rectangle.

$$A = \ell \times w = 2.3 \times 9.8 = 22.54 \text{ square inches}$$

Then we find the area of one circle.

$$A = \pi r^2$$

Because the radius of the circle is 1 inch, the area is

$$A = 3.14 \times 1^2 = 3.14 \text{ square inches}$$

And the area of five circles is 5 times 3.14.

$$5 \times 3.14 = 15.70 \text{ square inches}$$

So the area of the remaining metal is $22.54 - 15.70 = 6.84$ square inches.

In the next example, we are given information about a geometric figure along with a percent, and we are asked to find another piece of information.

• • •

EXAMPLE

A rectangular area rug is 4 feet wide by 6 feet long. A rectangular piece equivalent to 40% of the area of the original rug is cut off its length. If the width remains 4 feet, what is the length of the cut piece? What are the dimensions of the remaining portion of the original rug?

SOLUTION

This problem will definitely be easier to solve if we first make a sketch. We draw a rectangle, label one side 4 feet and one 6 feet, and draw a dashed line to show where the cut is made. The sketch also shows what we know about the percents. See the accompanying figure.

Let's first find the area of the cut-off piece. To do that, we must find the area of the original rug.

$$\text{Area} = \ell w = 6 \times 4 = 24 \text{ square feet}$$

We know that 40% of the rug has been cut off, so we ask (and then must answer) the question "40% of 24 square feet is what?"

The percent is 40, the base is 24, and the amount is the unknown value.

$$\text{Percent (as a decimal)} \times \text{base} = \text{amount}$$
$$0.40 \times 24 = n$$
$$9.60 = n \qquad \text{Multiplying out}$$

So 9.60 square feet of the rug has been cut off. We can use this area to find the length that was cut off, knowing its width is 4 feet. We have

$$A = \ell \times w$$
$$9.6 = \ell \times 4 \qquad \text{Substituting for } A \text{ and } w$$
$$\frac{9.6}{4} = \frac{\ell \times 4}{4} \qquad \text{Dividing both sides by 4}$$
$$2.4 = \ell \qquad \text{Dividing out}$$

So the length of the cut-off piece is 2.4 feet.

To find the length of the remaining rug, we subtract.

$$6 - 2.4 = 3.6 \text{ feet}$$

Its "width" is still 4 feet, so the dimensions of the remainder of the original rug are 3.6 feet by 4 feet.

PRACTICE

1. *Quilt Panels*   A quilt has panels shaped as shown in the figure at the left. The inner panel is a square $9\frac{1}{2}$ square inches in area. The perimeter of the large square is 21 inches. What is the total area of the trapezoids?

2. *Chicken Wire*   Chicken wire is to be placed along a straight fence. There are 21 fence poles that are 27 inches apart. The poles are 3 inches in diameter. How much chicken wire is needed to reach exactly from the first pole to the last?

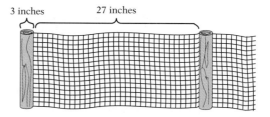

3 inches   27 inches

3. *Geometry*   A cube with an edge of 3 inches has a spherical ball placed inside of it. The ball has a diameter of 2 inches. The cube is then filled with sand right up to and level with the top, covering the ball. What volume of sand is in the cube? Use 3.14 for π and round to the nearest hundredth.

4. *Geometry*   A square card table that is $3\frac{1}{4}$ feet on a side is serving as the workspace for a round jigsaw puzzle with a radius of $18\frac{1}{4}$ inches. What percent of the card table will be exposed when the puzzle is complete? Round to the nearest whole percent.

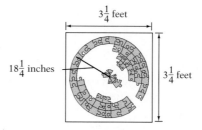

$3\frac{1}{4}$ feet

$18\frac{1}{4}$ inches

$3\frac{1}{4}$ feet

5. *Area of a Table*   The diameter of a round tablecloth is 20.6 centimeters. If $15\frac{1}{2}$% of the tablecloth is hanging off the table, what is the actual area of the top of the table in square centimeters? Round to the nearest tenth.

6. *Stone Border*   Stones are to be placed around a rectangular swimming pool. There will be 19 stones on each side along the longer sides of the pool, and 14 stones on each side along the shorter ends. How many stones will there be in all?

7. *Liquid in a Container*   A cylindrical container is 12 inches tall, and its round base has a diameter of 7.35 inches. If it is filled to 80% of its capacity, how much liquid is it holding? Round to the nearest inch.

8. *Number of Skeins*   A knitted afghan has 24 panels, each measuring $8\frac{1}{2}$ inches by $4\frac{1}{4}$ inches. If each square foot of knitted panel takes 1 skein of yarn, how many skeins will you need for this project?

9. *Geometry*   Find the area of the shaded region of the figure at the right. The radius of the large circle is 3. The radius of the small circle is $1\frac{1}{2}$. Use 3.14 for π and round to the nearest tenth of a unit.

10. *Patio Space*   A circular pool with a diameter of 21.08 feet is built inside a square patio with a side of 25.6 feet. How much patio space is there around the pool?

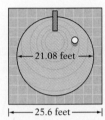

10.4 Graphing Quadratic Equations in Two Variables

SECTION LEAD-IN

A long jumper runs between 40 and 45 meters and then jumps. The world record distance is about 30 feet. The path of the jump is a parabola. The equation of that path is quadratic. Find another "jumping" sport with a quadratic relationship.

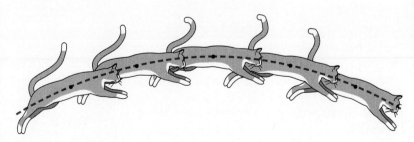

Note that this leaping cat's center of mass moves in a parabolic arc.

SECTION GOALS

- To graph quadratic equations in two variables
- To find the vertex of a parabola
- To find the x- and y-intercepts
- To graph a parabola using the vertex and intercepts.

The standard form of a quadratic equation in one variable is

$$ax^2 + bx + c = 0$$

It consists of a quadratic polynomial that is set equal to zero.

If we set the polynomial equal to a second variable instead, as in

$$ax^2 + bx + c = y$$

we have a **quadratic equation in two variables.** Such an equation can be plotted on the coordinate plane. In this section, we consider the graphs of quadratic equations in which a is 1 or -1, so that

$$y = x^2 + bx + c \quad \text{or} \quad y = -x^2 + bx + c$$

The graphs of two equations of this type are shown in the accompanying figure. These graphs are called **parabolas.** Note that they open upward when the coefficient of x^2 is positive and that they open downward when the coefficient of x^2 is negative.

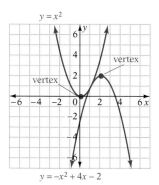

$y = x^2$

$y = -x^2 + 4x - 2$

Calculator Corner

GRAPH each of the following equations on your graphing calculator. What do these four parabolas have in common?

GRAPH each of the following equations on your graphing calculator. What do these four parabolas have in common?

GRAPH each of the following equations on your graphing calculator. What do these four parabolas have in common?

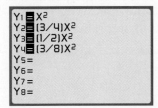

GRAPH each of the following equations on your graphing calculator. What do these four parabolas have in common?

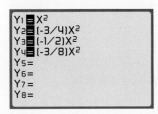

State a conjecture about how the value of a affects the graph of a parabola.

▪▪▪

EXAMPLE 1

Graph $y = x^2 + 2$ and $y = -x^2$ on the same graph.

SOLUTION

We proceed just as though we were graphing a linear equation. We set up a table of values, choose several values for x, and compute the corresponding values of y. We obtain

$$y = x^2 + 2 \qquad\qquad y = -x^2$$

x	y
0	2
1	3
−1	3
2	6
−2	6

x	y
0	0
1	−1
−1	−1
2	−4
−2	−4

Now we plot pairs of these values on the coordinate plane. Then we connect the points with a smooth curve—all parabolas have shapes similar to those in the figure at the beginning of this section. The result is the two curves at the right.

Check by using your graphing calculator.

▶ *CHECK* **Warm-Up 1**

> **WRITER'S BLOCK**
>
> What is the relationship between graphing a parabola and graphing a quadratic equation?

The Vertex of a Parabola

The lowest point on a parabola that opens upward, or the highest point on a parabola that opens downward, is called its **vertex.** The vertices of the parabolas in the previous figure are at the points (0, 2) and (0, 0), respectively.

Vertex of a Parabola

The parabola with equation

$$y = ax^2 + bx + c$$

has its vertex at the point whose x-coordinate is $\dfrac{-b}{2a}$.

•••

EXAMPLE 2

Find the coordinates of the vertex of the parabola whose equation is

$$y = -x^2 + 4x - 3$$

SOLUTION

We compute $\dfrac{-b}{2a}$. In this equation, $b = 4$ and $a = -1$. Then

$$\frac{-b}{2a} = \frac{-(4)}{[2(-1)]} = \frac{-4}{-2} = 2$$

Thus the x-coordinate of the vertex is 2. To find the y-coordinate, we substitute 2 into the original equation.

$$y = -x^2 + 4x - 3$$
$$= -(2)^2 + 4(2) - 3 = -4 + 8 - 3 = 1$$

So the coordinates of the vertex of the graph of $y = -x^2 + 4x - 3$ are (2, 1).

▶ CHECK **Warm-Up 2**

The Intercepts of a Parabola

Recall that the y-intercept of a line is the point at which it intersects the y-axis. A parabola may or may not have a y-intercept. If it does have one, we find it just as we do for a line.

---

y-Intercept of a Parabola

Every parabola with the equation

$$y = ax^2 + bx + c$$

has a y-intercept; it is at the point on the parabola whose x-coordinate is zero.

An x-intercept of a graph is a point at which the graph intersects the x-axis. A parabola like those in the figure in Example 1 can have two x-intercepts, or none, or only one.

> **x-Intercepts of a Parabola**
>
> If a parabola with the equation
>
> $$y = ax^2 + bx + c$$
>
> has x-intercepts, they are at the points on the parabola whose y-coordinates are zero.

You can use the following procedure for locating the x- and y-intercepts.

> **To find the x- and y-intercepts of a parabola**
>
> To find the x-intercept(s), if they exist,
>
> $$\text{set } y = 0 \text{ and solve for } x$$
>
> To find the y-intercept,
>
> $$\text{set } x = 0 \text{ and solve for } y$$

▪ ▪ ▪

EXAMPLE 3

Find the x- and y-intercepts of the parabola that has the equation

$$y = -x^2 + 4x - 3$$

SOLUTION

First we set $y = 0$ and solve for x. We use the quadratic formula to obtain

$$0 = -x^2 + 4x - 3 \qquad \text{Setting } y = 0$$

$$a = -1 \qquad b = 4 \qquad c = -3 \quad \text{The values of } a, b, c$$

$$x = \frac{-b \pm \sqrt{b^2 - 4ac}}{2a} \qquad \text{The quadratic formula}$$

$$= \frac{-(4) \pm \sqrt{(4)^2 - 4(-1)(-3)}}{2(-1)} \qquad \text{Substituting for } a, b, c$$

$$= \frac{-4 \pm \sqrt{16 - 12}}{-2}$$

$$= \frac{-(4) \pm \sqrt{4}}{-2} = \frac{-4 \pm 2}{-2}$$

So

$$x = \frac{-4 + 2}{-2} \qquad \text{or} \qquad x = \frac{-4 - 2}{-2}$$

$$x = \frac{-2}{-2} = 1 \qquad \Big| \qquad x = \frac{-6}{-2} = 3$$

The x-intercepts are (1, 0) and (3, 0). (*Note:* We could have solved this equation by factoring.)

To find the y-intercept, we set $x = 0$ and solve for y.

$$y = -x^2 + 4x - 3$$

$$= -(0)^2 + 4(0) - 3 = -3$$

The y-intercept is (0, −3).

> ▪▪▪
>
> **WRITER'S BLOCK**
>
> Define *intercept* in your own words. Can a graph of a parabola have three x-intercepts? Justify your answer.

▶ CHECK **Warm-Up 3**

Graphing a Parabola

Once we know the locations of the vertex and intercepts of a parabola, we can graph it fairly easily. The next example shows how.

▪ ▪ ▪

EXAMPLE 4

Determine the locations of the vertex and intercepts of the parabola that has the equation $y = x^2 - 2x - 8$. Then graph the parabola.

SOLUTION

We find the vertex of the parabola by identifying

$$\frac{-b}{2a} = \frac{-(-2)}{2(1)} = \frac{2}{2} = 1$$

We substitute $x = 1$ into the original equation.

$$y = x^2 - 2x - 8$$
$$= (1)^2 - 2(1) - 8 = 1 - 2 - 8 = -9$$

The vertex, then, is at $(1, -9)$.

To find the x-intercepts of this graph, we let $y = 0$ and solve for x.

$$0 = x^2 - 2x - 8$$

We solve by factoring.

$$0 = (x - 4)(x + 2) \qquad \text{Factoring}$$
$$x - 4 = 0 \quad \text{or} \quad x + 2 = 0$$
$$x = 4 \qquad\qquad x = -2$$

The x-intercepts are $(4, 0)$ and $(-2, 0)$.

We set $x = 0$ to find the y-intercept.

$$y = (0)^2 - 2(0) - 8 = -8$$

So the y-intercept is $(0, -8)$.

Recall that when the coefficient of the x^2 term is positive, the graph opens upward; when that coefficient is negative, the graph opens downward. The coefficient here is 1, so the graph opens upward.

Before we graph, we will find two more points by choosing two values for x and solving for y. We will choose $x = 3$ and $x = -3$.

When $x = 3$,

$$y = x^2 - 2x - 8$$
$$= (3)^2 - 2(3) - 8 = 9 - 6 - 8 = -5$$

When $x = -3$,

$$y = x^2 - 2x - 8$$
$$= (-3)^2 - 2(-3) - 8 = 9 + 6 - 8 = 7$$

So our two additional points are

$$(3, -5) \quad \text{and} \quad (-3, 7)$$

We graph the vertex, intercepts, and these two points, as shown at the left below. Then we connect the points with a smooth curve to graph the parabola, as shown at the right below.

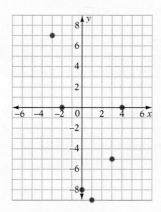

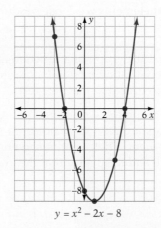

$$y = x^2 - 2x - 8$$

▶ CHECK Warm-Up 4

Note: If the equation you are graphing has no real-number solutions when $y = 0$, then the graph has no x-intercepts. You can still graph it, but you will have to find several more points that are on the graph.

Optional Topic: Quadratic Functions and Their Graphs

A function f given by

$$f(x) = ax^2 + bx + c$$

where a, b, and c are real numbers and a is not equal to zero, is called a **quadratic function.** (It is also a polynomial function of degree 2.)

To determine ordered pairs $(x, f(x))$ of a quadratic function, we substitute values for x and find corresponding values of $f(x)$.

▪ ▪ ▪

EXAMPLE 5

Find six ordered pairs $(x, f(x))$ of the function f, given $f(x) = x^2 - 2x - 1$. Sketch the graph.

SOLUTION

Let us find the ordered pairs in which x is $0, 1, -1, 2, -2$, and 3. We first substitute 0 for x in the function equation and find

$$f(0) = 0^2 - 2(0) - 1 = -1$$

Thus $(0, -1)$ is one ordered pair of the function. You should verify that the other first elements we chose produce the ordered pairs

$$(1, -2), (-1, 2), (2, -1), (-2, 7), (3, 2)$$

Recall that the graph of a function is the graph of all its ordered pairs. We can graph a quadratic function by graphing several ordered pairs and then sketching in the curve that they suggest.

We first graph the six ordered pairs we computed, as shown at the left below. The points are obviously not on a straight line, so we draw a smooth curve through them. The result is shown at the right below. Note the arrowheads, which we add to suggest that the graph goes on indefinitely.

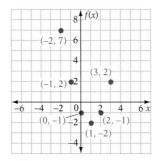 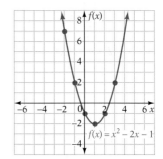

The curve sketched in the figure at the right above is a vertical parabola, and every quadratic function has such a parabola as its graph.

▶ CHECK **Warm-Up 5**

Practice what you learned.

SECTION FOLLOW-UP

Research Which of these sports have quadratic relationships?

    a. pole vault

    b. hurdles

    c. ski jump

10.4 WARM-UPS

Work these problems before you attempt the exercises.

1. Graph: $y = -3x^2 - 1$

2. Find the vertex of the parabola $y = -x^2 - 5$.

3. Find the x- and y-intercepts of the graph of
$$9x^2 - 36x + 20 = y$$
Does the parabola open upward or downward?

4. Find the x- and y-intercepts, the vertex, and two additional points on the parabola
$$x^2 - 4x - 1 = y$$
Then graph it.

5. Find six ordered pairs $(x, f(x))$ of the function f, given $f(x) = 2x^2 - 2x + 1$.

10.4 EXERCISES

Note: Use your graphing calculator to check your results whenever possible.

In Exercises 1 through 18, find the vertex of the graph of each parabola, and determine whether the graph opens upward or downward. Do not graph.

1. $13 - y = x^2 + 14x$

2. $y - 3 = 6x - x^2$

3. $-y = 2x + x^2 - 1.7$

4. $-4x + 7 + x^2 + 3 = y$

5. $105 - 20x + x^2 = y$

6. $-y = x^2 - 16x - 67$

7. $y - 8 - (x + 5)^2 = 0$

8. $(22 - x)^2 + 6 = y$

9. $-y = (x + 3)^2 - 3$

10. $-y = x^2 - 3$

11. $y - 10 = x^2$

12. $-y = x^2 + 12$

13. $\frac{1}{7} + y = x^2$

14. $x^2 - y = 7.8$

15. $y - 12 = x^2$

16. $x^2 - y = -18$

17. $y = -\left(x + \frac{1}{2}\right)^2 - 3$

18. $\frac{1}{6}x - y = x^2$

In Exercises 19 through 30, find the x- and y-intercepts if they exist.

19. $4x^2 + 16x - 180 = y$

20. $5x^2 - 90x + 385 = y$

21. $x^2 + 10x + 25 = y$

22. $x^2 + 8x + 16 = y$

23. $x^2 - 20x + 100 = y$

24. $x^2 - 18x + 81 = y$

25. $(4x - 7)^2 = y + 118$

26. $81x^2 + 144x - 49 = y$

27. $9x^2 - 18x + 8 = y$

28. $(22 - x)^2 = y + 360$

29. $(4x - 12)^2 - 484 = y$

30. $(2 - 3x)^2 - 324 = y$

In Exercises 31 through 42, find the vertex, the x- and y-intercepts (if they exist), and two additional points. Do not graph.

31. $2x^2 + 7x + 3 = y$

32. $x^2 - 16 = y$

33. $y = 3x^2 - 5x - 2$

34. $2x^2 - 3x + 1 = y$

35. $x^2 - 3x - 10 = y$

36. $x^2 - 3x + 2 = y$

37. $x^2 - 11x - 12 = y$

38. $30x^2 + 9x - 3 = y$

39. $x^2 - 16x - 36 = y$

40. $x^2 - x = y + 30$

41. $x^2 - 2x + 6 = y$

42. $x^2 + x + 18 = y$

In Exercises 43 through 48, graph each parabola, finding the vertex, the x- and y-intercepts (if they exist), and two additional points.

43. $x^2 - 4 = y$

44. $x^2 + 4 = y$

45. $-x^2 + 4 = y$

46. $-x^2 - 4 = y$

47. $x^2 - 9 = y$

48. $-x^2 + 9 = y$

OPTIONAL EXERCISES

In Exercises 49 through 54, find six ordered pairs $(x, f(x))$ of the given function.

49. $f(x) = 2x^2 - 3x + 1$

50. $f(x) = 3x^2 + x - 2$

51. $f(x) = -x^2 + 3x + 1$

52. $f(x) = -x^2 - 2x + 3$

53. $f(x) = 8x^2 - 5x - 2$

54. $f(x) = 2x^2 + 8x - 5$

In Exercises 55 through 66, graph each function by locating and graphing six points $(x, f(x))$ of the function.

55. $y = -x^2$

56. $y = x^2$

57. $f(x) = x^2 - 2$

58. $f(x) = -x^2 - 2$

59. $y = -x^2 + 3$

60. $y = x^2 - 3$

61. $f(x) = x^2 - x + 1$

62. $f(x) = 4x^2 - x + 1$

63. $f(x) = x^2 - 3x + 5$

64. $f(x) = x^2 + 3x + 5$

65. $f(x) = -x^2 + 3x + 2$

66. $f(x) = x^2 + 3x + 2$

For Exercises 67 through 78, graph each function and calculate the roots of $f(x)$.

67. $f(x) = x^2 - 3$

68. $f(x) = x^2 - 5$

69. $y = x^2 - 2x - 2$

70. $y = x^2 - 3x + 3$

71. $f(x) = x^2 + 4x - 1$

72. $f(x) = x^2 - 5x - 1$

73. $f(x) = 2x^2 + 2x + 2$

74. $f(x) = -x^2 + 2x - 2$

75. $f(x) = 2x^2 - x - 1$

76. $f(x) = 3x^2 + 3x - 2$

77. $f(x) = x^2 - 6x + 5$ **78.** $f(x) = x^2 + 5x - 5$

MIXED PRACTICE

By doing these exercises, you will practice the topics up to this point in the chapter.

79. Solve by using the quadratic formula: $26x^2 + 13 = -2x$

80. Find the vertex, x- and y-intercepts, and two additional points on the graph of the equation $y = -x^2 + 1$. Then graph it.

81. Solve: $(5x + 7)^2 = 64$

82. Solve by completing the square: $x^2 - 3x - 1\frac{3}{4} = 0$

83. Solve by using the quadratic formula: $2x^2 - 8 = 0$

84. Solve by factoring: $x^2 + 14x - 32 = 0$

85. Solve: $(x - 12)^2 = 121$

86. Solve: $2x^2 + 8x - 1 = 0$

87. Solve by using the quadratic formula: $7x^2 + 6x + 8 = 0$

88. Find the vertex, x- and y-intercepts, and two additional points on the graph of the equation $2x^2 + 8x + 5 = y$. Then graph it.

89. Find the vertex, x-, and y-intercepts, and two additional points on the graph of the equation $y = -x^2 + 2x$.

90. Give the vertex and the direction of the graph of $18 + y = x^2$.

91. Give the vertex and the direction of the graph of $-y = (x - 5)^2 + 3$.

92. Describe the roots of $-7x^2 + 13x - 5 = 0$.

93. The output of a machine is given as $x_0 = t^2 + 3t - 21$, where x_0 is the output in gallons and t is the amount of time in seconds that the machine has been running. How long has the machine been running when the output is 90 gallons?

94. Solve by completing the square: $x^2 + 4x = 0$

EXCURSIONS

Data Analysis

1. a. Graph each of the following pairs of equations.

i. $y = 2x^2 + 3x + 1$
   $y = 3x + 1$

ii. $y = x^2 - 2x + 2$
   $y = -2x + 2$

iii. $-2x^2 + 8x - 5 = y$
   $8x - 5 = y$

iv. $2x^2 - 3x - 2 = y$
   $-3x - 2 = y$

b. Describe how the equations in each pair are related to each other.

c. Make a rule that you believe will always result in the same mathematical relationship.

d. *Research*   What is the second equation in each pair as described mathematically?

Class Act

2. 👉   A student claims that if a quadratic equation has two intercepts, the distance between those two intercepts is the same as the distance from the vertex to the x-axis.

Is the student correct? Justify your answer. Show any work you did to prove or disprove this idea.

3. In Experiments 1–3, we will graph data from the following table and predict whether the result is linear or quadratic. The data is from a football kicking experiment.

F O O T B A L L   K I C K I N G   D A T A

VELOCITY (feet per second)	HANG TIME (seconds)	DISTANCE (feet)	HEIGHT (feet)
65	2.9	132	33.0
66	2.9	136	34.0
67	2.95	140	35.1
68	3.0	145	36.1
69	3.05	149	37.2
70	3.1	153	38.2
71	3.15	158	39.4
72	3.2	162	40.0
73	3.2	167	41.6
74	3.3	171	42.8
75	3.3	176	43.9
76	3.4	181	45.1
77	3.4	185	46.3
78	3.45	190	47.5
79	3.5	195	48.7

In all the equations that follow, use

$$g = 32.15 \text{ feet per second}$$

and

$$V_{x_0} = V_{y_0} = (\text{velocity})\left(\frac{\sqrt{2}}{2}\right)$$

a. *Experiment 1*   Verify the relationship

$$\text{Maximum height} = V_{y_0}t - \tfrac{1}{2}gt^2 \quad (\text{for } t = \tfrac{V_{y_0}}{g})$$

by following these steps.

i. Choose a velocity in the table, then calculate

$$t = \frac{V_{y_0}}{g}$$

ii. Substitute t in the following equation and solve for y (the height)

$$y = V_{y_0}t - \tfrac{1}{2}gt^2$$

iii. Check to see if y is equal to the height in the table that corresponds to your chosen velocity.

iv. Is the relation between maximum height and velocity linear or quadratic? Why?

b. *Experiment 2*   Verify the relationship

$$0 = V_{y_0}(\text{hang time}) - \tfrac{1}{2}g\,(\text{hang time})^2$$

Use t to stand for hang time.

i. Choose a velocity from the table and solve the equation for t.

$$0 = V_{y_0}t - \tfrac{1}{2}gt^2$$

ii. Compare the calculated hang time with that in the table.

iii. Is the relationship linear or quadratic? Justify your answer.

c. *Experiment 3*   Verify the relationship

$$\text{Distance} = (V_{x_0})\,(\text{hang time})$$

i. Choose a velocity from the table with its hang time (t), then calculate

$$d = V_{x_0}t$$

ii. Compare the calculated d with the distance in the table.

iii. Is this relationship linear or quadratic?

d. *Discussion*

i. Did your graphs give you enough information to decide what kind of relationship there was between these variables? Why or why not?

ii. Just because your data fits certain kinds of equations, can you conclude that the equation explains the relationship? Why or why not?

◤ **CHAPTER LOOK-BACK**

The path followed by any projectile (something thrown, shot, or propelled in some way) is a parabola. The equation that represents the path followed by a

1. thrown hammer

2. high jumper

3. thrown javelin

is a quadratic equation.

CHAPTER 10
REVIEW PROBLEMS

SECTION 10.1

Use the principle of zero products to solve the following equations.

1. $81t^2 - 49 = 0$

2. $8x^2 + 2x - 45 = 0$

Use the equality property of square roots to solve the following equations.

3. $49x^2 - 9 = 0$

4. $4x^2 - 25 = 0$

5. $(5 - x)^2 - 400 = 0$

6. $(5 + 16x)^2 = 125$

SECTION 10.2

Solve the following equations by completing the square.

7. $-40 + n^2 + 3n = 0$

8. $t^2 - 4t + \frac{32}{9} = 0$

9. $n^2 - 3n - 108 = 0$

10. $x^2 - 2x + 1 = 0$

11. $x^2 - 20x - 100 = 0$

12. $x^2 - 16x + 64 = 0$

Endangered Animals Many endangered animal species are the subjects of protection programs. The equation

$$C = 1010 + 475t - 10t^2$$

describes the crocodile population C after t years of such a program, where t is positive and less than 26 years.

13. How many crocodiles existed when the program began, at time $t = 0$?

14. How many years will it take for C to reach 3500?

SECTION 10.3

Solve the following equations by using the quadratic formula.

15. $49x^2 - 14x + 1 = 0$

16. $16y^2 - 20y - 6 = 0$

17. $4x^2 - 16x - 9 = 0$

18. $-30r + 189 + r^2 = 0$

19. $25 + t^2 + 26t = 0$

20. $36n^2 - 48n + 16 = 0$

21. *Thrown Projectile* The equation $h = vt - 16t^2 + h_0$ represents the height h of an object t seconds after it is given an upward velocity of v feet per second from an initial altitude of h_0. An object is thrown upward at a velocity of 64 feet per second, from an initial height of 10 feet. How long will it take to hit the ground?

22. *Profit from Stocks* The profit earned by a particular fund in dollars is given by $P = 4x^2 - 12x$, where x is the number of shares of stock sold. To the nearest half share, how many shares must be sold to obtain a profit of $10,000? (Use the quadratic formula to solve this problem. Estimate the value of your radical to the nearest hundred.)

SECTION 10.4

In Exercises 23 and 24, find the vertex, the x- and y-intercepts, and two additional points on the graph of the given equation. Do not graph.

23. $7x^2 + 14x + 21 = y$

24. $6x^2 + 2x = y$

In Exercises 25 through 28, find the vertex, the x- and y-intercepts, and two additional points on the graph of the given equation. Then graph it.

25. $-7 - x^2 + 2x = y$

26. $y = x^2 - 8$

27. $-x^2 + y = 2x$

28. $-y - 6 - (x - 1)^2 = 0$

MIXED REVIEW

29. Solve: $x^2 - 26x + 169 = 0$

30. Solve: $x^2 + 225 = 0$

31. Solve: $56 + n^2 - 15n = 0$

32. Solve: $9x^2 + 18x - 27 = 0$

33. Solve: $4t^2 - 4t - 80 = 0$

34. Solve by factoring: $x^2 - 36x + 324 = 0$

35. Solve by using the equality property of square roots: $(9x + 8)^2 = 121$

36. *Geometry* The area of a certain circle and that of a certain square are the same. If the circle has a radius of 8 inches, what are the dimensions of the square? (Use 3.14 for π.)

37. Solve by using the quadratic formula: $3x^2 - 14x = 0$

38. **Profit** The profit obtained from doing a job is found by using the formula $2.6t = p^2$, where t is the amount of time in seconds spent on the job, and p is the number of dollars of profit. Give the amount of profit to the nearest dollar when $t = 3\frac{1}{2}$ hours.

39. Solve by using the quadratic formula: $x^2 - 8x + 15 = 0$

40. Find the vertex, x- and y-intercepts, and two additional points on the graph of the equation $9 - x^2 = y$. Then graph it.

CHAPTER 10 TEST

This exam tests your knowledge of the material in Chapter 10.

1. Factor and solve.

   a. $4x^2 + 6x + 2 = 0$ **b.** $18r^2 - 6r - 4 = 0$ **c.** $8x^2 + 64x + 128 = 0$

2. Solve using the equality property of square roots.

   a. $x^2 = 900$ **b.** $(x + 1)^2 = 64$ **c.** $(x - 5)^2 = 81$

3. Solve each of the following by completing the square.

   a. $x^2 + 3x - 40 = 0$ **b.** $8x^2 - 2x - 1 = 0$ **c.** $x^2 + 12x + 11 = 0$

4. Solve by using the quadratic formula.

   a. $x^2 - 20x + 100 = 0$ **b.** $x^2 + 5x - 14 = 0$ **c.** $35x^2 - 21x + 56 = 0$

5. Find the vertex, the x- and y-intercepts, and two additional points on the graph of each of the following equations. Then graph it.

   a. $y = -3x^2 + 1$

   b. $y = 4x^2 - 4x + 4$

   c. $y = (1 - x)^2$

6. Solve each of the following problems.

   a. A rectangular lot that measures 20 yards by 30 yards is to be reduced by the same amount all the way around to allow for a sidewalk. If the area of the new lot is to be 504 square yards, how wide will the sidewalk be? (See the accompanying figure.)

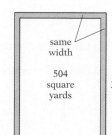

same width

504 square yards

30 yards

20 yards

b. A storage cabinet is to be built into a workshop. The length of the cabinet is 5 feet less than three times its width. The height is 4 feet and the volume is 200 cubic feet. Find the length and the width.

c. The height of a triangular sail exceeds the base by 4 feet. The area is 48 square feet, find the dimensions.

CUMULATIVE REVIEW

CHAPTERS 1–10

The following exercises will help you maintain the skills you have learned in this and previous chapters.

1. Write $\dfrac{(15 - y)}{21}$ in words.

2. Evaluate: $-3.4(-6.8) + (-2.4 + 1.2)^3 - 5^2$

3. Evaluate: $\dfrac{t}{r} + rt^2 - 4r^2t$ when $r = 0.2$ and $t = 1.4$.

4. Solve: $9x + 5 = 12$

5. Simplify: $-2y(x + y) + y^2$

6. Which is larger, $-|2.5|$ or -2.6?

7. Solve for x: $-3(x + 4) = -9 + 5(x - 5)$

8. Simplify: $[-(-2^3)]^2$

9. Find the solution set: $0.8x - 2.7 \geq 1.6$

10. Write in symbols and simplify: the difference between a number and -212

11. Multiply: $4(3y - 8)(y - 2)$

12. Find the slope of the line whose equation is $-4x + 2y = 9$.

13. Solve: $3y^2 = 147$

14. The sum of two numbers is 107 and their difference is 2. Find the numbers.

15. Solve: $\dfrac{9}{x - 7} - \dfrac{3}{x + 6} = \dfrac{12}{x^2 - x - 42}$

16. Solve using addition: $\dfrac{1}{3}x - \dfrac{2}{3}y = 9$
$3x - 4y = 9$

17. Multiply: $\left(\sqrt{12y} - \sqrt{2}\right)\left(-\sqrt{6y} - \sqrt{8}\right)$

18. Multiply: $(-0.04x^2y^3)^2(-0.3x^3y^{-2})^3$

19. Len is 5 years older than Connie.

    a. If Connie is n years old, how old is Len?

    b. If Len is $n + 8$ years old, how old is Connie?

20. Solve: $x^2 + 4x = 60$

ANSWERS to Warm-Ups

SECTION 1.1 *(page 11)*

1. all are rational **2.** $5; -42$ **3.** **4.** $10 > 4 > -3 > -34$ **5.** $-97; -7$

SECTION 1.2 *(page 23)*

1. -1409 **2.** $-1\frac{5}{42}$ **3.** 1.68 **4.** $3, -2\frac{1}{3}, 3.6$ **5.** 80 **6.** $-62\frac{5}{6}$ **7.** -2

SECTION 1.3 *(page 35)*

1. -7395 **2.** $33\frac{3}{4}$ **3.** 2.61 **4.** $-\frac{1}{4}, \frac{3}{7}, -\frac{10}{32}$ **5.** -2 **6.** $-\frac{39}{76}$ **7.** 0.06

SECTION 1.4 *(page 49)*

1. $2; 1.9; 1.935$ **2.** \$4.6 thousand, \$6.5 thousand, \$8.2 thousand, \$8.6 thousand **3.** $\frac{27}{343}$ **4.** $256; 9$ **5.** $24;$
31 **6.** $2{,}000{,}000$ **7.** $6{,}010{,}000{,}300$ **8.** -5.31 **9.** -42.6

SECTION 2.1 *(page 70)*

1. $\{-1, 0, 1, 2, 3, 4\}$; $\{x \mid x \text{ is an integer greater than } -2 \text{ and less than } 5\}$ **2.** $A \cup B = \{1, 2, 3, 4, 5, 7\}$; $(A \cup B) \cup C =$
$\{1, 2, 3, 4, 5, 7\}$ **3.** $A' = \{6, 7, 8, 9, 10\}$; $B' = \{2, 4, 6, 7, 8, 9, 10\}$; $(A \cap B)' = B'$ **4.** 16 times **5.** about 3 minutes
6. 30% **7.** 117 people

SECTION 2.2 *(page 87)*

1. $-4y^3, -2y^2, 5$ **2.** coefficients: $3, 4, -3$; variables: x, x, xy **3.** 49.5 **4.** -70.25

SECTION 2.3 *(page 97)*

1. commutative property of addition; associative property of addition **2.** -4 **3.** $2r^2t$ **4.** $a - 20$

SECTION 2.4 *(page 111)*

1. $n + 5; n - 12$ **2.** $\frac{3}{4}(n \div 5)$ **3.** $4(5.50 + m) + 5m = 22 + 9m$ **4.** $7y$ **5.** The difference of the
product of 14 and b, and 8; or 8 less than the product of fourteen and b.

SECTION 3.1 *(page 135)*

1. $n = -13$ **2.** $m = 43$ **3.** $n = -1.2$ **4.** $x = 12$ **5.** $t = -605$ **6.** $y = 1$ **7.** $n = -8$
8. $x = 12$ **9.** $y = \frac{8}{13}$ **10.** $t = -1\frac{1}{4}$ **11.** $r = -7$ **12.** $n = -3$

SECTION 3.2 *(page 146)*

1. $t = j - yx$ **2.** $t = \frac{3r - 15}{2}$ **3.** $D = O - S$ **4.** $D = 2262$ miles **5.** $F = 172.4°$

SECTION 3.3 *(page 161)*

1. $n = 41$ **2.** Nancy is 64 years old. Susan is 16 years old. **3.** $3, 5, 7$ **4.** 3 nickels, 6 dimes, 12 quarters

SECTION 3.4 *(page 179)*

1. $x > \frac{14}{9}$

2. $x < 14.1$

3. $x < 2$

4. $x \le \frac{1}{4}$

5. $x > -13$

6. $x < 3$

7. $x < -\frac{13}{7}$

8. $x \le -\frac{1}{14}$

SECTION 3.5 *(page 190)*

1. a. yes **b.** yes **c.** no **d.** yes **2. a.** no **b.** no **c.** no **d.** yes **3.** $x = -4.5$

4.

x	y
-2	-1
7	0
16	1

5. $(-11, 3), (-8, 0), (-4, -4)$

SECTION 4.1 *(page 211)*

1.

2. $A: (-4, -3)$
$B: (-2, 1)$
$C: (3, 2)$
$D: (2, -2)$

3.

4. x-intercept: $(-2, 0)$;
y-intercept: $(0, 6)$

4. x-intercept: $(3, 0)$;
y-intercept: $\left(0, \frac{9}{4}\right)$

5.

SECTION 4.2 *(page 222)*

1. $m = -1$

2. $m = \frac{7}{11}$ **3.** $m = 0$ **4.** undefined **5.** They do *not* lie on the same line. **6.** a triangle

SECTION 4.3 *(page 238)*

1. $m = -2$; y-intercept: $(0, 3)$ **2.** **3.** slope: $-\frac{1}{2}$; y-intercept: $\left(0, \frac{3}{2}\right)$

4. 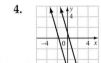 **5.** **6.** parallel **7.** $y = 4x + 5$ **8.** $y = -3x - 9$ **9.** $y = x - 2$

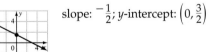

SECTION 4.4 *(page 247)*

1.   **2.**

SECTION 4.5 *(pages 256–257)*

1. $\left\{\left(\frac{1}{2}, 1\right), (1, 2), \left(1\frac{1}{2}, 3\right)\right\}$; Range: $\{1, 2, 3\}$ **2. a.** a function   **b.** not a function   **c.** a function   **d.** not a function
3. $-2; 10; 3a^2 - 2$ **4. a.** function   **b.** function

SECTION 5.1 *(pages 276–277)*

1. 25 **2.** $-625; 25$ **3.** 121 **4.** $\frac{1}{8}; \frac{-1}{1296}$ **5.** $64z^3$ **6.** $\frac{1}{x'}; -\frac{1}{x^5}$ **7.** $-\frac{216}{125}$ **8.** $-10\frac{1}{8}$ **9.** r^6s^3
10. $-81x^{10}y^{16}$ **11.** $\frac{25r^6s^4}{81t^{10}}$ **12.** $\frac{1}{625r^8t^{16}}$ **13.** $\frac{x^2}{9r^2t^6}$

SECTION 5.2 *(page 287)*

1. $8x^5y$ **2.** $\frac{3x^8y^4}{4}$ **3.** $-132x^9y^6$ **4.** $-\frac{37.4}{xy^2}$ **5.** $\frac{1}{2r}$ **6.** $\frac{3}{2x^7}$ **7.** $\frac{-26m^7}{5n^2t^3}$

SECTION 5.3 *(page 298)*

1. binomial; trinomial   **2.** $15x^2 + 2x + 9$ **3.** $11x^2 + 15x + 2$ **4.** -7 **5.** $8n^2 - 9n + 8$

SECTION 5.4 *(page 313)*

1. $18x^2 - 30x$ **2.** $-2x^4 + 2x^3 - 6x^2$ **3.** $56x^3 - 11x^2 - 75x + 36$ **4.** $3r^2 - 22r - 16$ **5.** $r^2 - 16$
6. $z^2 + 6z + 9$ **7.** $35x^2 - 2x + 3$ **8.** $-6x^2 - 7x - 13$

SECTION 5.5 *(page 323)*

1. $-\frac{x^3}{2} - \frac{3x}{2} + \frac{1}{x}$ **2.** $\frac{2}{3xy} - \frac{3}{x} + \frac{1}{2x^2y}$ **3.** $x + 4$ **4.** $3x^2 + x + 4 + \frac{12}{x - 2}$
5. $4x^2 - 8x + 20 + \frac{-35}{x + 2}$

SECTION 6.1 *(page 341)*

1. $2^2 \cdot 3 \cdot 5$ **2.** $3r \cdot rt$ and $3t \cdot r^2$; Answers may vary.   **3.** $7rt^2$ **4.** $2x^2y(11 - 5xy^4)$
5. $x(6y^6 - 20y^4 + x^5)$ **6.** $4x(3x^3y^2 - 5y^3 + 4x)$ **7.** $(x + 5)(x + 2y)$ **8.** $(r - 3)(r + 4)$
9. $(2)(y - 7)(y + 3)$

SECTION 6.2 *(page 350)*

1. $(n + 5)(n + 1)$ **2.** $(n - 10)(n - 2)$ **3.** $4(x - 6)(x + 3)$ **4.** $(x - 3y)(x + 5y)$
5. $(2x + 1)(3x + 20)$

SECTION 6.3 *(page 361)*

1. $(x + 7)(x - 7)$ **2.** $2(8x + 3)(8x - 3)$ **3.** $(x + 4)^2$ **4.** $4y(m - 8)^2$

SECTION 6.4 *(page 370)*

1. $x = 0$ or $x = 3$ **2.** $r = -6$ or $r = 3$ **3.** $n = 0, n = 3$, or $n = \frac{-7}{2}$ **4.** 10 sides
5. at 2 seconds and at 4 seconds

SECTION 7.1 *(page 384)*

1. all real numbers except $x = \frac{2}{3}$ **2.** $5\frac{1}{3}$ **3.** x **4.** $-3n - 5$

SECTION 7.2 *(page 394)*

1. $\frac{t(t - x)}{t + x}$ **2.** $r(r + 3)$ **3.** 9 **4.** $\frac{y}{3x}$ **5.** $x = 8$ **6.** $2x(x + 2)^2(x - 2)$

SECTION 7.3 *(page 403)*

1. $\frac{14}{18(x - 3y)}$ and $\frac{3}{18(x - 3y)}$ **2.** $\frac{2y^2 - 11y + 24}{y(y - 3)}$ **3.** $\frac{x^2 + 3x - 6}{(x - 2)(x + 1)}$ **4.** $\frac{-(y - 6)(y + 1)}{2(y - 2)(y - 3)}$ **5.** $t = -\frac{249}{99}$

SECTION 7.4 *(pages 414–415)*

1. 180 mph **2.** $1\frac{7}{8}$ hours **3.** 14 drawings

SECTION 8.1 *(page 437)*

1. yes **2.**  $(0, -4)$ **3.** **4.** **5.**

SECTION 8.2 *(page 448)*

1. $\left(1, \frac{2}{3}\right)$ **2.** $(1, -7)$ **3.** $(-4, 10)$ **4.** These represent the same line. All points on the line are solutions.
5. no solution **6.** $(1, 2)$ **7.** $(2, 4)$

SECTION 8.3 *(page 463)*

1. $45.35 and $78.65 **2.** car, 50.5 mph; train, 74.5 mph **3.** 15 mg of 4% mixture; 5 mg of 8% mixture

SECTION 9.1 *(page 483)*

1. 25 and -25 **2.** $13|y|$; 13 **3.** real and rational **4.** 31 and 32

SECTION 9.2 *(page 501)*

1. $6t^2w^2\sqrt{5t}$ **2.** $9x^3y^2\sqrt{6}$ **3.** $\sqrt{6}$ **4.** $\sqrt{6}$ **5.** $\frac{x^2t^2\sqrt{13}}{y^2}$ **6.** $-4\sqrt{3}$ **7.** $12wx\sqrt{11xy} + 9w\sqrt{11xy}$

SECTION 9.3 *(page 514)*

1. $14 - 14\sqrt{3}$ **2.** $14 - 4\sqrt{6}$ **3.** $\frac{5x^2}{2}$ **4.** $\frac{2\sqrt{15}}{5}$ **5.** $\frac{\sqrt{3} - 3\sqrt{2}}{3}$
6. $2\sqrt{3} - 2\sqrt{2} - 3\sqrt{15} + 3\sqrt{10}$

SECTION 9.4 *(page 525)*

1. $y = 76$ **2.** $x = \frac{43}{5}$ **3.** $280\sqrt{5}$ centimeters per second

SECTION 10.1 *(page 543)*

1. $r = -6$ or $r = 3$ **2.** $5x^2 - 48x - 20 = 0$ **3.** $x = 12$ or $x = -12$ **4.** $x = -94$ or $x = -106$
5. $x = 11\frac{2}{5}$ or $x = 8\frac{3}{5}$ **6.** $t = 4$ seconds after launch

SECTION 10.2 *(page 550)*

1. $x = 3$ or $x = 0$ **2.** $x = 21$ or $x = -1$

SECTION 10.3 *(page 561)*

1. $3t^2 + (-6)t + (-9) = 0; a = 3, b = -6, c = -9$ **2.** $x = -1$ or $x = -2$ **3.** $v = -3 \pm \sqrt{2}$ **4.** probably; the results show that this answer may be correct within rounding error   **5.** discriminant is -4; no real-number solution

SECTION 10.4 *(page 579)*

1. 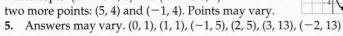   **2.** $(0, -5)$ **3.** x-intercepts: $\left(\frac{10}{3}, 0\right)$ and $\left(\frac{2}{3}, 0\right)$; y-intercept: $(0, 20)$; opens upward

4. Vertex: $(2, -5)$
y-intercept: $(0, -1)$
x-intercepts: $\left(2 + \sqrt{5}, 0\right)$ and $\left(2 - \sqrt{5}, 0\right)$
two more points: $(5, 4)$ and $(-1, 4)$. Points may vary.

5. Answers may vary. $(0, 1), (1, 1), (-1, 5), (2, 5), (3, 13), (-2, 13)$

ANSWERS to Chapter 1 Odd-Numbered Exercises

SECTION 1.1 *(pages 12–15)*

1. R, I **3.** R, I **5.** R **7.** neither **9.** -6 **11.** -1933 **13.** -5.86 **15.** $0.9\overline{99}$ **17.** 6
19. -18 **21.** 6 **23.** $26\frac{7}{8}$ **25.** 2 **27.** -142 **29.** 47 **31.** 8 **33.** 122 **35.** -355

37. -18 **39.** $-6 \quad 0 \ |-2| \ 3$ **41.** $-(-1)$; $-1 \ 0 \quad 3$ **43.** $-(-90)$; $-1 \ 0 \ 1 \ -(-3)$

45. $-(-|54|)$; $0 \quad 48 \ |-49| \ 50$ **47.** $0 \quad 53 \quad 55$ **49.** $0 \quad 89 \quad 91$ **51.** $-400\frac{2}{5} \quad -3\frac{2}{3}$; $-401 \quad -67.98 \quad -3 \ 0$

53. $4 > -2 > -5$ **55.** $|43| > -25 > -|-43|$ **57.** $|-35| > |25| > -|65|$
59. $|-104| > -|102| > -103$ **61.** $<$ **63.** $<$ **65.** $>$ **67.** $-|-11| < 15 < 26$
69. $-45 < -|-13| < 27$ **71.** $-23\frac{1}{3} < 19\frac{2}{5} < |41|$ **73.** $-52.7 < -\left|38\frac{2}{5}\right| < 29\frac{3}{8}$
75. $|-|-123|| > |-9| > -|-1465|$ **77.** $-|-564| > -|766| > -2637$
79. $\left|-13\frac{2}{5}\right| > -3\frac{1}{2} > -\left|-30\frac{4}{7}\right|$ **81. a.** roaster **b.** fryer **83.** $200,000 < 292,000,000,000$
85. $298,624 > 27,472$ **87.** $1985, 1987, 1983, 1986$ **89.** D, A, B, C **91. a.** violet **b.** yellow **c.** red **d.** violet
e. blue **f.** orange **g.** yellow **h.** red

SECTION 1.2 *(pages 23–25)*

1. -90 **3.** 221 **5.** 156 **7.** -93 **9.** $-10\frac{31}{77}$ **11.** $-24\frac{5}{24}$ **13.** $\frac{-47}{77}$ **15.** $-3\frac{1}{8}$
17. $-14\frac{1}{2}$ **19.** $-44\frac{20}{77}$ **21.** 1.14 **23.** -5.8 **25.** -9.76 **27.** -16.26 **29.** 21 **31.** -5.68
33. -81 **35.** -23 **37.** -190 **39.** 75 **41.** -103 **43.** -125 **45.** $-11\frac{1}{3}$ **47.** 1.2
49. -145.6 **51.** -8.4 **53.** -7.3 **55.** -9.46 **57.** $-5\frac{1}{4}$ **59.** 3 **61.** 7.9 **63.** $-3 + 2$
65. $-4 + -|-9|$ **67.** 15 **69.** -63 **71.** $7\frac{5}{6}$ **73.** $17\frac{5}{7}$ **75.** $0; 1$ **77.** $2; 7$ **79.** $1; 6; 7$
81. $3; 6; 5$ **83.** $29\frac{1}{2}$ **85.** $-76\frac{5}{6}$ **87.** $63\frac{3}{4}$ **89.** -107.45 **91.** $-94\frac{41}{42}$

SECTION 1.3 *(pages 36–37)*

1. 1×-7 or -1×7 **3.** -1 **5.** -6 **7.** 0 **9.** 6.7 **11.** 0 **13.** -35 **15.** -54
17. -784 **19.** -370 **21.** $5\frac{1}{3}$ **23.** $-\frac{2}{3}$ **25.** $-20\frac{1}{2}$ **27.** $30\frac{1}{3}$ **29.** -5.92 **31.** 0.0084
33. 300 **35.** 352 **37.** 1080 **39.** $-30,132$ **41.** $-\frac{4}{15}$ **43.** $\frac{1}{9}$ **45.** -0.06 **47.** -2.8
49. -0.008 **51.** $\frac{7}{2}$ **53.** $-\frac{5}{19}$ **55.** $-\frac{1}{6}$ **57.** $\frac{3}{10}$ **59.** undefined **61.** 0 **63.** -1
65. -7.6 **67.** $-\frac{6}{73}$ **69.** $12\frac{1}{2}$ **71.** -23 **73.** $4\frac{34}{65}$ **75.** $-\frac{6}{13}$ **77.** $-1\frac{1}{7}$ **79.** $-0.0\overline{3}$
81. -16 **83.** $4\frac{5}{13}$ **85.** $-\frac{11}{16}$ **87.** $-\frac{10}{43}$ **89.** $-\frac{3}{8}$ **91.** -2050 **93.** $\frac{1}{7}$
95. $-\frac{5}{6} \quad \frac{4}{9} -\left(-\frac{7}{8}\right)$; $-1 \quad 0 \quad 1$ **97.** $-55\frac{1}{4}$ **99.** $-(-2.64) > -2\frac{6}{7} > -12\frac{5}{8}$

SECTION 1.4 *(pages 50–55)*

1. a. 260; 262.0; 262.002 **b.** 80; 77.0; 76.963 **c.** 980; 984.3; 984.339 **d.** 20; 24.6; 24.591 **3. a.** $16.0 thousand
b. $14,500 **c.** $13.8 thousand **d.** $15.9 thousand **e.** $20,000 **5.** 1870 **7.** 1880 **9.** 5,698,000,000
11. 1,400,000 **13.** 62,100,000,000 **15.** 815,000,000,000 **17.** 34,900,000 **19.** 2,286,000 **21.** 13,000

23. $1\frac{11}{25}$ **25.** -2.197 **27.** -0.0625 **29.** 3.6963 **31.** 0.064 **33.** 3.61 **35.** 0 **37.** $12\frac{19}{27}$

39. a. $6.85, 6.94999, 6.9049$ **b.** $7.23, 7.2349875$ **c.** $1.99, 1.957867$ **41. a.** $20,736$ **b.** 6 **c.** 2 **43. a.** 1000

b. 256 **c.** 512 **45. a.** 66 **b.** 63 **c.** 71 **d.** 64 **47. a.** 84 **b.** 87 **c.** 89 **d.** 81 **49. a.** no **b.** no

51. a. no **b.** no **53.** $10,000,000$ vehicles **55.** $20,000$ graduates **57.** $1,800,000$ people **59.** $\frac{5204}{25,797}$;

the fraction of women finishers was larger in 1991, so it increased. **61.** tallest: $14,999$ feet; shortest: $5,000$ feet

63. 20 times **65.** 4000 times **67.** $1,598,762 \div 18,212$ **69.** $883,260 \div 32,597$ **71.** -28 **73.** -5

75. -9.644 **77.** -1 **79.** 2.85 **81.** -2.91 **83.** 6.02 **85.** $8\frac{1}{9}$ **87.** -1 **89.** -1 **91.** $8\frac{1}{2}$

93. $16.8; 16.78; 16.779$ **95.** 77 **97.** $-(-3) > -2\frac{1}{2} > -\left|3\frac{1}{2}\right|$ **99.** $2\frac{17}{30}$

Chapter 1 Review Problems *(pages 58–59)*

1. 34 **2.** 25 **3.** 24 **4.** [number line: $-8\ -7$ ... $0\ \ 2$] **5.** $-\frac{23}{4} < \frac{13}{5} < \left|-\frac{19}{6}\right|$ **6.** $5.128; 5.12; -5.1$

7. [number line: $\left|-1\frac{1}{3}\right|$ 0 $\left|-\frac{9}{5}\right|$ 2.4] **8.** between -10 and -9 **9.** -6.7 **10.** 1.404 **11.** -6.3 **12.** $-25\frac{21}{40}$

13. -5 **14.** $-7\frac{19}{20}$ **15.** 54 **16.** $11\frac{16}{35}$ **17.** -79.6 **18.** $-4\frac{5}{28}$ **19.** $-31\frac{1}{3}$ **20.** $\frac{4}{5}$

21. -53.75 **22.** -307.5 **23.** -8892 **24.** $-18\frac{3}{8}$ **25.** $9\frac{3}{11}$ **26.** 193.45 **27.** -4540

28. $-1\frac{19}{85}$ **29.** $\frac{5}{41}$ **30.** -100 **31.** 204.6 **32.** $-4\frac{8}{23}$ **33.** 0.000000125 **34.** -216 **35.** $1\frac{1}{8}$

36. -49 **37.** 37 **38.** 81 **39.** 7000 **40.** 0.003 **41.** $\left|-\left|\frac{8}{3}\right|\right| > -|2.4| > -|-9.5|$

42. -1 **43.** $-4\frac{31}{60}$ **44.** $-3\frac{1}{4}$ **45.** -0.012 **46.** $-\frac{115}{147}$ **47.** 1 **48.** $23\frac{2}{3}$ **49.** $3\frac{1}{14}$

50. $-33\frac{1}{4}$ **51.** 22.398 **52.** between -14 and -13 **53.** 2865 **54.** -3504

Answers to Chapter 2 Odd-Numbered Exercises

Section 2.1 *(pages 70-78)*

1. Not empty. 0 is an element. **3.** Empty. No number is both even and odd. **5.** $\{0, 2, 4, 6, \ldots\}$

7. $\{2, 3, 5, 7, 11, 13, 17, 19, 23, 29\}$ **9.** $\{1, 2, 3, 4, 6, 12\}$ **11.** $\{-4, -3, -2, -1, 0, 1, 2, 3, 4, 5, 6, 7, 8, 9\}$ **13.** true

15. true **17.** $\{0, 2, 3, 4, 5, 9\}$ **19.** $\{0\}$ **21.** $A \cup B = \{\text{integers}\}$ or $A \cup B = B$; $A \cap B = \{\text{whole numbers}\}$

or $A \cap B = A$ **23.** $A \cup B = \{n \mid n \text{ is a whole number}\}$; $A \cap B = \varnothing$ **25.** $2\frac{3}{4}$ inches **27.** $181\frac{2}{5}$ miles

29. 25 steps **31.** $666,000$ bills **33.** 37 touchdowns **35.** 13 feet $2\frac{1}{16}$ inches **37.** 36 times longer

39. 12.35% **41.** 100 bpm **43.** $125°F$ **45.** $9\frac{5}{8}$ **47.** 24 less than at the start **49.** 5.5 meters

51. $9,269,000$ members **53.** $\$6.87$ **55.** 24% **57.** 22.8 ounces **59.** $75\%; 25\%$

61. $19,775\frac{59}{144}$ ft 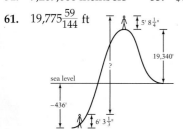 **63. a.** $\$66.08$ **b.** Answers may vary. **65. a.** $\frac{1}{5}; \frac{1}{7}; \frac{2}{7}; \frac{1}{7}, \frac{1}{5}, \frac{2}{7}$

b. Answers may vary. **67.** $\$62.10$ **69.** 1337 cubic centimeters

71. Answers may vary. **73.** Answers may vary. **75.** 15 semesters

77. 6 trips **79.** $\$533$ **81.** 2290 feet higher **83.** 550

SECTION 2.2   *(pages 87-89)*

1. terms: $3x$, $-2x$, $3y$, $8x$; coefficients: 3, -2, 3, 8; variable parts: x, x, y, x **3.** terms: $-9y^3$, $7y^2$, $7y$, 4; coefficients: $-9, 7, 7, 4$; variable parts: y^3, y^2, y, none   **5.** terms: $-7ef$, $-8ef$, $7ef$, $7ef^2$; coefficients: $-7, -8, 7, 7$; variable parts: ef, ef, ef, ef^2 **7.** terms: $7fg$, $-8fg$, $6fg$, $-6f$, $7f^2g^2$; coefficients: $7, -8, 6, -6, 7$; variable parts: fg, fg, fg, f, f^2g^2
9. 0   **11.** 1   **13.** 0   **15.** -12 **17.** -2 **19.** -5 **21.** -4 **23.** -16 **25.** -4 **27.** -3
29. -3 **31.** 36   **33.** 0   **35.** 3   **37.** 9   **39.** 101   **41.** -210 **43.** 256   **45.** 281.62
47. -1579.75 **49.** 38.83   **51.** 512   **53.** 10,816   **55.** -428.76 **57.** $\frac{5}{48}$ **59.** -24 **61.** 9

SECTION 2.3   *(pages 97-100)*

1. true; associative property of addition   **3.** true; commutative property of addition   **5.** false   **7.** false
9. associative property of addition   **11.** commutative property of multiplication
13. commutative property of addition   **15.** $\left(\frac{86}{5} + 38\right) + 5$ **17.** $97 + [3 + 4(876) + 4(723)]$
19. $45 - [8(24)](23)$ **21.** $27 + [(-18) + 3(6)(-2)]$ **23.** -126 **25.** -75 **27.** -90 **29.** 10,005
31. false   **33.** false   **35.** true   **37.** true   **39.** false   **41.** true   **43.** false   **45.** true
47. $14x^2y$ and $-11x^2y$, $8xy^2$ and $-2xy^2$ **49.** $3rt^2$ and $4rt^2$ **51.** $10u + 6v$ **53.** $-y$ **55.** $2n^3 + 5n^2 - n$
57. $13x^2 - 10x$ **59.** $2s^2 + 7s$ **61.** $8x^2 - 8x$ **63.** $-6t + 45$ **65.** $16a - 20$ **67.** $3x + 20$
69. $3.2x - 51$ **71.** $-1\frac{13}{20}x^2 + 2\frac{1}{20}y$ **73.** $5nt^2 - 2.9n^2t + 9$ **75.** $-252n^2y - 15ny^2 - 9$
77. $2xy - 7y$ **79.** commutative property of addition
81. $-3 - 6 \div 5 = -3 - \frac{6}{5} = (-1)(3) + (-1)\left(\frac{6}{5}\right)$ property of multiplication by -1

$$= (-1)\left(3 + \frac{6}{5}\right) \qquad \text{distributive property}$$
$$= -\left(3 + \frac{6}{5}\right) \qquad \text{property of multiplication by } -1$$
$$= -\left(\frac{6}{5} + 3\right) \qquad \text{commutative property of addition}$$

83. distributive property   **85.** $74t - 36c - 41$ **87.** $-11a - 20$ **89.** Expressions (a) and (b) are equivalent.
91. $5 + 6 \times 7 = 5 + 7 \times 6$ commutative property of multiplication
   $= 7 \times 6 + 5$ commutative property of addition
93. No. Justifications may vary.

SECTION 2.4   *(pages 111-115)*

1. $13 - f$ **3.** $n - 4$ **5.** t^2 **7.** $\$0.10d$ **9.** $15n$ cents   **11.** $2(8 - x)$ **13.** $n + y$ **15.** $n + 8$
17. $\frac{n}{12}$ **19.** $1.5n$ **21.** $\frac{1}{2}y$ **23.** $n + y$ **25.** $x + y$ **27.** $11x + 4$ **29.** $n + (n - 2)$
31. $n - \$1.94$ **33.** $n + 126$ **35.** $n - 285$ **37.** $\frac{1}{3}(n + 4)n$ **39.** $n \div (n - 4)$ **41.** $\frac{1}{4}n - 19$
43. $2(n + 0.02) + 3n$ **45.** $\frac{n}{2} + n$ **47.** $n - 10$ **49.** $n \div 54$ **51.** $n - \$.94$ **53.** $2(n + 3)$
55. $n + \left(n - \frac{1}{5}n\right)$ **57.** $n^2 + n + 2$ **59.** $n^2 \div 23$ **61.** $1\frac{1}{4}n$ **63.** $n - (n + 6)$ **65.** $(n - 8) + n$
67. $n - (n + 30)$ **69.** the sum of n and 6   **71.** the quotient of n and 7   **73.** the product of 6 and n
75. the difference of n and 10   **77.** $155r$ dollars   **79.** $d - 4.15$ dollars   **81.** $50n + 6p$ cents   **83.** nh
85. c stamps   **87.** np people   **89.** $6n + 80d$ dollars   **91.** $[n + 2] + [(n - d) + 2]$ years   **93.** $36 - A$
95. 24 more than the product of 16 and a **97.** 76   **99.** $(6)(n)(56) \div 12$

CHAPTER 2 REVIEW PROBLEMS   *(pages 117-119)*

1. false   **2.** true   **3.** true   **4.** $\{5, 6\}$ **5.** $\{5, 6\}$ **6.** $\{0, 5, 6, 13\}$ **7.** variable parts: v; t; rv; tr
coefficients: 5; -1; -1; -1
8. -2201 **9.** 6   **10.** $-16,811$ **11.** 33   **12.** -51 **13.** $2(-1) = 6 + (-8)$
$-2 = -2$
14. associative property of multiplication   **15.** yes   **16.** -128 **17.** $-36y$ **18.** $9x - 2$
19. $6t^2 - 5t + 4tx$ **20.** $4a - 72$ **21.** $-6x + 15$ **22.** $-14x + 22y$ **23.** the difference between 4
times a number and 7 times a number   **24.** $n - 12n$ **25.** $q + (q \div 10)$ **26.** the opposite of 9 times the
sum of 18 and a **27.** $x + (x - 14)$ or $y + (y + 14)$ **28.** $x - 1$; $x - 2$ **29.** 10,985   **30.** no

31. $n + 10$ **32.** -7 **33.** $-3a^2 + 11a + 36$ **34.** $a - 4$ **35.** $n + 1; n + 2$ **36.** 334   **37.** $67n$
38. $0.4x + 34.32y$ **39.** 7 less than x **40.** $3p + 2(p + 0.50)$ **41.** $-11x - 18$ **42.** $4 + 3y$
43. $4 \div n$ **44.** $(n + 10) \div 25$ **45.** the product 3 and x **46.** $1\frac{1}{2}n$

CUMULATIVE REVIEW   *(page 120)*

1. -0.1944 **2.** $-5\frac{10}{19}$ **3.** $4\frac{38}{75}$ **4.** $25mt + 21$ **5.** $25 \div t$ **6.** 57.3   **7.** -5.5 **8.** 1020
9. $-35\frac{7}{30}$ **10.** $-63\frac{2}{3}$ **11.** $-x + 22$ **12.** $-y + 64$ **13.** $-|23.9| < -(-25.7) < |-26.3|$
14. $754\frac{1}{6}$ **15.** $m - 77$ **16.** 13   **17.** $-305\frac{1}{4}$ **18.** 288   **19.** -81 **20.** $(n + 5) \div 11$

ANSWERS to Chapter 3 Odd-Numbered Exercises

SECTION 3.1   *(pages 136–137)*

1. $x = 6$ **3.** $y = 99$ **5.** $t = 11$ **7.** $n = 110$ **9.** $x = -480$ **11.** $x = -6.2$ **13.** $n = \frac{-1}{2}$
15. $t = -10.7$ **17.** $t = 3$ **19.** $t = 1\frac{1}{3}$ **21.** $n = 40$ **23.** $t = 32$ **25.** $t = 70$ **27.** $t = -726$
29. $x = 1584$ **31.** $r = 16{,}200$ **33.** $x = 1$ **35.** $y = 3$ **37.** $x = -0.06$ **39.** $m = 1$ **41.** $x = 9$
43. $x = -16$ **45.** $x = -27$ **47.** $x = -0.475$ **49.** $m = -20$ **51.** $x = -280$ **53.** $r = 1.35$
55. $y = 0.4$ **57.** $y = 102$ **59.** $x = 1$ **61.** $x = 3$ **63.** $x = 8.4$ **65.** $x = -30$ **67.** $r = -3\frac{3}{5}$
69. $x = \frac{-4}{5}$ **71.** $x = 28$ **73.** $x = \frac{1}{3}$ **75.** $x = \frac{-1}{3}$ **77.** $v = -4.25$ **79.** $x = \frac{-59}{43}$ **81.** $x = \frac{43}{39}$
83. $x = \frac{-4}{17}$

SECTION 3.2   *(pages 147–151)*

1. $c = e - h$ **3.** $c = e + h$ **5.** $c = eh$ **7.** $c = \frac{h}{e}$ **9.** $c = \frac{e + h}{2}$ **11.** $c = \frac{h + 6}{2}$ **13.** $x = 3e + 24$
15. $c = t + \frac{x}{h}$ **17.** $c = hk - fk$ **19.** $c = \frac{hr - hj}{2}$ **21.** $c = \frac{e}{a + h}$ **23.** $c = \frac{h}{j - r}$ **25.** $C = 122.4$ calories
per serving   **27.** $C = \$54.13$ final cost   **29.** 12.6 amperes   **31.** $S = 2926$ **33.** $d = 4$ grams per cubic
centimeter   **35.** $S = \$2370$ **37.** $P = 1200$ **39.** $F = 31{,}640$ grams per unit   **41.** $F = 41$ gallons
43. $d = 765$ miles   **45.** $S = 86$ **47.** $e = \frac{c - f}{r}$ **49.** $x = -2\frac{3}{8}$ **51.** $y = 1\frac{19}{63}$ **53.** 900   **55.** $x = 9$
57. $t = -16$ **59.** $x = 2\frac{1}{3}$ **61.** $m = \frac{12}{37}$

SECTION 3.3   *(pages 161–165)*

1. $n = -7$ **3.** $n = 5$ **5.** $\$0.89$ **7.** $x = 318$ **9.** $n = 1026$ **11.** $x = -422$ **13.** 17, 19, 21
15. $n = 193$ **17.** $n = 25$ **19.** 0.8 kilogram   **21.** 30 inches   **23.** 75 miles per hour   **25.** 75 passengers
27. 1200 millimeters   **29.** $-2, -1, 0, 1$ **31.** 3.5 hours   **33.** 27 inches   **35.** $-321, -319, -317$
37. Sophia works $14\frac{1}{3}$ hours. Sherly works $25\frac{2}{3}$ hours.   **39.** First sister is 14 years, 8 months old. Second sister is
9 years, 8 months old. Third sister is 38 years, 8 months old.   **41.** 4 weeks   **43.** 840 calories   **45.** Answers
may vary.   **47.** Answers may vary.   **49.** $x = \frac{-9}{14}$ **51.** $t = \frac{52}{37}$ **53.** $x = -83$ **55.** $t = 333$
57. $34.50 + (4.50)(n - 3)$

SECTION 3.4   *(pages 180–181)*

1. no   **3.** yes   **5.** no   **7.** no   **9.** $x > 33$ **11.** $y \geq -8$ **13.** $r \geq -7$ **15.** $t < 20$ **17.** $m > -3$
19. $x \geq 17$ **21.** $x \geq -7\frac{3}{4}$ **23.** $y < -11\frac{1}{2}$ **25.** $r > 2\frac{6}{17}$ **27.** $t > 45$ **29.** $y \leq \frac{-5}{12}$ **31.** $x \leq 5.7$
33. $y > -1.625$ **35.** $t \geq 4\frac{58}{125}$ **37.** $t > \frac{3}{8}$ **39.** $x \leq -1\frac{3}{7}$ **41.** $y > 14.4$ **43.** $r < 2\frac{1}{3}$ **45.** $y \geq \frac{6}{7}$
47. $r \geq 6$ **49.** $r < 11$ **51.** $r \geq -1$ **53.** $x < -1\frac{5}{9}$ **55.** $r > 9$ **57.** $n \geq -45$ **59.** $t < \frac{4}{5}$
61. $r > \frac{-21}{2}$ **63.** $r < \frac{2}{3}$ **65.** $y \leq \frac{-3}{10}$

67. $y \le \frac{25}{7}$ **69.** $t < \frac{-15}{7}$ **71.** $x > 4$

73. $y > \frac{-15}{13}$ **75.** $t \ge \frac{6}{5}$ **77.** $y > -5$ **79.** $t \ge \frac{-79}{30}$ **81.** $x \le \frac{-26}{11}$ **83.** $x = \frac{v-y}{8}$ **85.** $v = \frac{x}{7}$

87. $y > \frac{113}{10}$ **89.** $y = 1$ **91.** $t = -42$ **93.** $w = 31$ **95.** $c = \frac{f+x}{e}$ **97.** $m = -2184$

99. The longer piece is 15.6 feet.

SECTION 3.5   *(pages 190–193)*

1. yes **3.** yes **5.** no **7.** no **9.** yes; yes **11.** no; no **13.** yes; no **15.** no; yes

17. yes; no **19.** yes; yes **21.** (5, 17) **23.** $\left(2, 3\frac{1}{2}\right)$ **25.** $\left(1\frac{1}{3}, -2\right)$ **27.** $(5, -5)$ **29.** not possible

31. $A = 3$ **33.** $A = -0.4$ **35.** $A = -\frac{1}{2}$

37.

x	y
−4	11
2	2
$\frac{8}{3}$	1

39.

x	y
0	−2
−5	0
−10	2

41.

x	y
2	5
2	3
2	$-\frac{15}{16}$

43.

x	y
0	$\frac{4}{3}$
10	$\frac{5}{3}$
−30	$\frac{1}{3}$

45.

x	y
1	−3
0	−4
−1	−5

47.

x	y
1	14
0	10
−1	6

49.

x	y
1	−2
0	3
−1	8

51.

x	y
1	2
0	0
−1	−2

53.

x	y
1	4
0	2
−1	0

55.

x	y
1	4
0	1
−1	−2

57.

x	y
1	$-\frac{2}{5}$
0	$\frac{3}{5}$
−1	$1\frac{3}{5}$

59.

x	y
1	1
0	$\frac{1}{2}$
−1	0

61. When Victor was 2, Marcy was 6. When Victor was 15, Marcy was 19. **63.** 65 miles; 110 miles; 50 miles

65. $y = \frac{11}{3}$ **67.** $t = 1$ **69.** $w = 1.6$ feet, $l = 0.8$ feet **71.** $-41, -39, -37$ **73.** $A = 3x^2$

CHAPTER 3 REVIEW PROBLEMS   *(pages 197-198)*

1. $x = -42$ **2.** $x = -12.5$ **3.** $x = \frac{7}{3}$ **4.** $x - \frac{-153}{7}$ **5.** $x = \frac{1}{8}$ **6.** $x \le \frac{-1}{8}$ **7.** $x = \frac{-1}{2}$

8. $x = \frac{16}{5}$ **9.** $r = v(m - x)$; or $vm - vx$ **10.** $p = \frac{t(s-r)}{m}$ **11.** $x = \frac{e}{f-r}$ **12.** $t = \frac{11r}{s-4}$ **13.** 8 amperes

14. $S = 351$ **15.** -44 and -42 **16.** 43 dimes, 27 nickels **17.** Bimla is 32. Her sister is 22. **18.** 59

19. -36 **20.** $-13; -12; -11$ **21.** $x < \frac{-8}{3}$ **22.** $x < -6$ **23.** $x > \frac{-1}{12}$

24. $x < \frac{-1}{2}$ **25.** $x < 0.5$ **26.** $x \ge 9.2$ **27.** $x \le 14.1$ **28.** $x < -3.1$

29.

x	y
−1	−5
−2	−7
3	3

30. $A = 4$ **31.** $\left(-\frac{2}{5}, -\frac{6}{5}\right)$ **32.** (10, 4) **33.** $x < 11$ **34.** $x < \frac{31}{6}$ **35.** $r = \frac{wx - v}{t}$

36. $n = 102$ **37.** $c = t + w$ **38.** $x \le 0.2$ **39.** 15

40. One piece will be 12 inches long; the other will be 24 inches long.

CUMULATIVE REVIEW   *(pages 199–200)*

1. $x = \frac{7}{2}$ **2.** $t \le -1$ **3.** 304 **4.** $x = 18$

5. Yes. Explanations may vary. **6.** -36 **7.** $n \ge \frac{6}{5}$ **8.** -52.44 **9.** 0 **10.** 23,010

$$-(5 - y) = -1(5 - y)$$
$$= (-1)(5) + (-1)(-y)$$
$$= -5 + y$$
$$= y - 5$$

11. $6.7r^2t + 4.9rt - 5.8r$ **12.** $-30\frac{37}{60}$ **13.** $-\frac{3}{7}$ **14.** $-|18.6|$

15. -125 **16.** 1 **17.** -6 **18.** -6 **19.** $-1\frac{23}{40}$

20. $-85n < 42.5$

Answers to Chapter 4 Odd-Numbered Exercises

SECTION 4.1 *(pages 212–215)*

1. quadrant III **3.** quadrant II **5.** quadrant III **7.** quadrant IV **9.** $(4, 4)$ **11.** $(-6, 6)$
13. $(-5, 0)$ **15.** $(-6, -1)$ **17.** Answers may vary. One answer is $(-3, 3)$. **19.** $(-2, -3)$ **21.** $(1, 0)$
23. Answers may vary. One answer is $(3, -1)$.

25.

x	y
0	-4
1	1
2	6

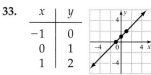

27.

x	y
-2	-4
-1	0
0	4

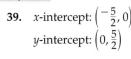

29.

x	y
-1	$\frac{8}{5}$
0	$\frac{3}{5}$
1	$-\frac{2}{5}$

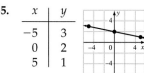

31.

x	y
-1	-3
0	0
1	3

33.

x	y
-1	0
0	1
1	2

35.

x	y
-5	3
0	2
5	1

37. x-intercept: $(4, 0)$
y-intercept: $\left(0, -\frac{4}{5}\right)$

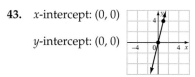

39. x-intercept: $\left(-\frac{5}{2}, 0\right)$
y-intercept: $\left(0, \frac{5}{2}\right)$

41. x-intercept: $\left(-\frac{3}{5}, 0\right)$
y-intercept: $\left(0, \frac{3}{5}\right)$

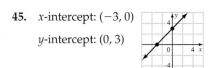

43. x-intercept: $(0, 0)$
y-intercept: $(0, 0)$

45. x-intercept: $(-3, 0)$
y-intercept: $(0, 3)$

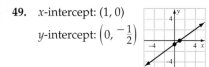

47. x-intercept: $(3, 0)$
no y-intercept

49. x-intercept: $(1, 0)$
y-intercept: $\left(0, -\frac{1}{2}\right)$

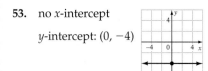

51. x-intercept: $(-1, 0)$
no y-intercept

53. no x-intercept
y-intercept: $(0, -4)$

55. x-intercept: $\left(\frac{7}{2}, 0\right)$
y-intercept: $\left(0, \frac{7}{3}\right)$

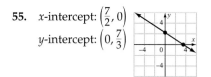

57. a.

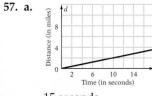

15 seconds

b. 6 miles

SECTION 4.2 *(pages 222–224)*

1. -1 **3.** 1 **5.** $\frac{3}{5}$ **7.** 0 **9.** 1 **11.** 5 **13.** undefined **15.** $-\frac{2}{3}$

Answers may vary for Exercises 17 through 23.

17. $(4, 3)$ **19.** $(6, -5)$ **21.** $(1, -3)$ **23.** $(1, 4)$ **25.** $m = 0$ **27.** $m = -\frac{3}{2}$ **29.** collinear
31. not collinear **33.** no **35.** There are no right angles. This is not a rectangle. **37. a.** negative slope
b. positive slope **c.** neither, slope $= 0$ **d.** positive slope **39.** Answers may vary; may include $(0, 2)$, $(1, 5)$, $(2, 8)$,
$(-1, -1)$, and $(-2, -4)$ **41.** $x = -\frac{2}{3}$, so the point is $\left(-\frac{2}{3}, -3\right)$. **43.** The y-intercept is $\left(0, -\frac{11}{4}\right)$. The
x-intercept is $\left(\frac{11}{6}, 0\right)$. **45.** yes

47.
slope $= \frac{3}{2}$

49.

x	y
-4	$-3\frac{1}{2}$
2	1
-6	-5

51. no **53.** $x = 10\frac{1}{3}$

SECTION 4.3 *(pages 239–241)*

1. $y = \frac{2}{3}x + \frac{1}{3}$ **3.** $y = \frac{7}{3}x - \frac{5}{3}$ **5.** $y = -\frac{3}{11}x$ **7.** $y = 4x - 5$ **9.**
  slope $= \frac{2}{3}$ slope $= \frac{7}{3}$ slope $= -\frac{3}{11}$ slope $= 4$
  y-intercept: $\left(0, \frac{1}{3}\right)$ y-intercept: $\left(0, -\frac{5}{3}\right)$ y-intercept: $(0, 0)$ y-intercept: $(0, -5)$

11.   **13.**   **15.**   **17.**   **19.** neither

21. perpendicular **23.** perpendicular **25.** perpendicular **27.** parallel **29.** parallel

31. a. neither **b.** parallel **c.** perpendicular **33. a.** neither **b.** neither **c.** parallel **35.** $y = -3x + \frac{1}{2}$
37. $y = \frac{4}{5}x + \frac{2}{5}$ **39.** $y = \frac{3}{4}x + 8$ **41.** $y = 9x - 5$ **43.** $y = 4x + 6$ **45.** $y = 2x - 1$
47. $y = -5x - 12$ **49.** $y = -7x + 4$ **51.** $y = -2$ **53.** $y = -4x - 21$ **55.** $y = \frac{3}{4}x$
57. $y = \frac{1}{5}x - 5$ **59.** $y = x + 3$ **61.** $y = \frac{10}{7}x + \frac{18}{7}$ **63.** $y = x + 2$ **65.** $y = 2x - 4$
67. $y = 2x + 5$ **69.** $y = -4x + 1$ **71.** $y = \frac{1}{5}x + \frac{9}{5}$ **73.** $y = \frac{3}{2}x - \frac{5}{2}$ **75.** $y = -3x + 1$
77. $y = \frac{1}{2}x - \frac{17}{2}$ **79.** $y = x + 5$ **81.** $y = x - 6$ **83.** $y = x$ **85.** $y = -\frac{1}{2}x - \frac{13}{2}$ **87.** $y = -\frac{1}{3}x - 1$
89. $y = \frac{1}{2}x + 2$ **91.**   **93.** yes **95.**   **97.** $2x - 2y = 4$

x	y
0	-2
2	0
4	2

99. x-intercept: $\left(-\frac{12}{5}, 0\right)$

Tables may vary.

SECTION 4.4 *(pages 248–250)*

1.   **3.**   **5.**   **7.**   **9.**   **11.**

13. **15.** **17.** **19.** **21.** **23.**

In Exercises 25 through 42, the three representative points may vary.

25. **27.** **29.** **31.** **33.** **35.**

37. **39.** **41.**

43. $y = \dfrac{7}{10}x - \dfrac{11}{5}$ **45.** $y = -4x + 12$ **47.** -1 **49.** x-intercept: $(9, 0)$; y-intercept: $\left(0, \dfrac{-9}{8}\right)$

51. Answers may vary; they could include $(0, -9)$, $(1, -6)$, or $(2, -3)$ **53.**

SECTION 4.5 *(pages 257–258)*

1. $f(x) = 3x$; possible ordered pairs are $\{(0, 0), (1, 3), (2, 6), (3, 9)\}$ **3.** $f(x) = x - 3$; possible ordered pairs are $\{(0, -3), (1, -2), (2, -1), (3, 0)\}$ **5.** $f(x) = 2x - 4$; possible ordered pairs are $\{(0, -4), (1, -2), (2, 0), (3, 2)\}$
7. $f(x) = x \cdot (x - 1)$; possible ordered pairs are $\{(0, 0), (1, 0), (2, 2), (3, 6)\}$ **9.** Domain: $\{0\}$; Range: $\{3, 5, 7\}$; no
11. Domain: $\{3, 6\}$; Range: $\{2, 7\}$; yes **13.** Domain: $\{0, 5, 6\}$; Range: $\{3, 0\}$; yes **15.** Domain: $\{5, 6, 8\}$;
Range: $\{2, 7\}$; yes **17.** $7; 11; 5$ **19.** $-2; 14; 7$ **21.** $1; -2a + 1; -2x - 3$ **23.** $3; 3 - 2a; 5 - 2x$
25. function **27.** not a function **29.** function **31.** not a function

CHAPTER 4 REVIEW PROBLEMS *(pages 261–263)*

1. x-intercept: $(1, 0)$; y-intercept: $(0, -1)$ **2.**   **3.** no x-intercept; y-intercept: $(0, 4)$

4. x-intercept: $(0, 0)$; y-intercept: $(0, 0)$ **5.** $y = 12x - 6$ **6.** $m = \dfrac{9}{2}$ **7.** $(3, -5)$

8. $m = \dfrac{2}{3}$ **9.** $(-1, -3)$ and $(5, -7)$ **10.** $(5, 4)$ and $(2, 3)$ **11.** $m = 0$ **12.** $m = \dfrac{1}{0} =$ undefined

13. collinear **14.** $m = \dfrac{1}{2}$ **15.** slope $= \dfrac{3}{2}$, y-intercept: $\left(0, -\dfrac{11}{2}\right)$ **16.** neither **17.**

18. $y = -x + 5$ **19.** perpendicular **20.** $y = -5x - 2$ **21.** $y = -\dfrac{5}{4}x$ **22.** $y = x - 1$

23. $y = -x - 2$ **24.** $y = \dfrac{8}{3}x$ **25.** Answers may vary; they may include $(0, 0)$ and $(4, 2)$.

26. **27.** **28.** **29.** x-intercept: $(-2, 0)$ **30.** y-intercept: $(0, 7)$

31. parallel **32.** $y = -6x - 4$ **33.** $y = \dfrac{1}{2}x - \dfrac{1}{2}$ **34.** $y = 5x - 12$

CUMULATIVE REVIEW *(page 264)*

1. $x = 9\dfrac{2}{3}$ **2.** $x \geq -\dfrac{1}{60}$ **3.** $3 \div n$ **4.** $-\left|-\dfrac{7}{8}\right| < -|-0.5| < |-0.56|$ **5.** x-intercept: $(-5, 0)$; y-intercept: $(0, -5)$ **6.** 0 **7.** $x \leq 17$ **8.** 32 **9.** $t = -125.5$ **10.** -2.11

11.

x	y
1	-3
0	-5
$\dfrac{5}{2}$	0

12. **13.** 16 **14.** 325 **15.** $12xv^2 - 5x^2v + 11v^3 + 12x^2$ **16.** $m = \dfrac{1}{3}$ **17.** $r = \dfrac{t}{x - 4}$

18. -2 **19.** **20.** neither

Answers to Chapter 5 Odd-Numbered Exercises

SECTION 5.1 *(pages 277–278)*

1. 1 **3.** 3 **5.** x **7.** 1 **9.** -49 **11.** -243 **13.** $\dfrac{64}{729}$ **15.** -0.64 **17.** 49 **19.** 729
21. 10.48576 **23.** 103.823 **25.** $-\dfrac{1}{2}$ **27.** -3 **29.** $\dfrac{1}{m^4}$ **31.** $\dfrac{1}{p^2}$ **33.** $\dfrac{-1}{t}$ **35.** $\dfrac{1}{36y^2}$
37. $-\dfrac{3}{x^3}$ **39.** $-x^3$ **41.** $\dfrac{25}{9}$ **43.** $-\dfrac{6561}{256}$ **45.** $\dfrac{2048}{2401}$ **47.** 64 **49.** $x = \dfrac{at^2}{2}$ **51.** $\dfrac{I}{pr} = t$
53. $16x^2$ **55.** $125n^3$ **57.** $64x^2$ **59.** $81n^2$ **61.** $\dfrac{1}{w^{12}x^3y^3}$ **63.** $n^{32}y^4z^4$ **65.** $625x^{12}y^8$ **67.** $\dfrac{1}{64n^6y^9}$
69. $4x^{10}y^6$ **71.** $16n^{12}y^{20}$ **73.** $\dfrac{x^{10}}{32y^{15}}$ **75.** $\dfrac{27n^3}{y^6}$ **77.** $\dfrac{-27y^6}{x^9}$ **79.** $36n^2y^4$ **81.** $121x^4y^{10}$ **83.** 1
85. $\dfrac{9}{x^2}$ **87.** $\dfrac{n^4}{16}$ **89.** $\dfrac{32w^{15}x^{20}}{243y^{20}}$ **91.** $\dfrac{9n^4z^6}{25y^8}$ **93.** $\dfrac{64x^6}{81y^8}$ **95.** $\dfrac{49y^{14}z^{12}}{25n^8}$ **97.** $\dfrac{n^2p^2s^4}{m^6r^6t^2}$
99. $\dfrac{2187n^{14}p^{35}x^{14}y^7z^7}{m^{21}}$

SECTION 5.2 *(pages 287–289)*

1. $-m^{10}$ **3.** $12x^3$ **5.** $\dfrac{-10}{x^8}$ **7.** $\dfrac{46}{t^2}$ **9.** $15r^5t^2$ **11.** $-12xy^3$ **13.** $\dfrac{224t^4}{r^2}$ **15.** $\dfrac{6ty^2}{x^8}$ **17.** $96r^2u$

19. $280r^6u^3$ **21.** $\dfrac{18y}{r^2x^3}$ **23.** $\dfrac{1008m^2}{t^4}$ **25.** $\dfrac{3}{2}x^7y^4$ **27.** $\dfrac{-5x^5y^7}{2}$ **29.** $\dfrac{-8v^5}{t}$ **31.** $\dfrac{6}{5}x^6y^3$ **33.** $\dfrac{1}{5}x^9y^2$

35. $-9.03x^5y^7$ **37.** $0.3x^4y$ **39.** $\dfrac{-48m^2t^3}{n}$ **41.** w^3 **43.** n^8 **45.** $\dfrac{-1}{n^7}$ **47.** $\dfrac{-1}{w^7}$ **49.** $27w^{13}$

51. $48n^5$ **53.** $\dfrac{1}{w^3x^2}$ **55.** $\dfrac{3w^{11}}{8}$ **57.** $\dfrac{2}{7n^4}$ **59.** $\dfrac{-2}{n^8}$ **61.** $\dfrac{x}{w^6y}$ **63.** $\dfrac{27}{w^6x}$ **65.** $\dfrac{2}{3nty}$ **67.** $\dfrac{9}{wx^5y^4}$

69. $\dfrac{-9}{n^3t^3y}$ **71.** $9wx^6$ **73.** $\dfrac{-3n^2}{t^9y^4}$ **75.** $\dfrac{55w^2x^7}{y^3}$ **77.** $\dfrac{y^3z^5}{2x^7}$ **79.** $\dfrac{7n^{13}}{10m^{10}t^5}$ **81.** $a = 76{,}032$ miles/hour^2

83. 64 **85.** $72u^2$ **87.** $4x^2y^6$ **89.** $\dfrac{1}{r^{10}t^{20}v^{25}}$ **91.** $\dfrac{-48y^5}{x}$ **93.** -32 **95.** $\dfrac{36}{x^7y^5}$ **97.** $\dfrac{m^2}{4r^8}$

99. $-\dfrac{1}{9t^3}$

SECTION 5.3 *(pages 298–301)*

1. degree 2 **3.** degree 4 **5.** monomial **7.** polynomial; trinomial **9.** polynomial; binomial

11. polynomial; trinomial **13.** $5x + 3$ **15.** $9x - 6$ **17.** $\left(\dfrac{1}{3}\right)x^2 - \left(\dfrac{2}{3}\right)x - 2$ **19.** $\dfrac{7}{6}x - \dfrac{7}{6}$

21. $3.7x^2 - 2.1x$ **23.** $-1.1x^2 + 2.6x$ **25.** $6r^2 - 3$ **27.** $-2t^2 + 4t + 1$

29. $1.2n^3 - 1.8n^2 - 0.6n + 0.2$ **31.** $14n^2 - 23n + 8$ **33.** $6x^2 - 4x - 6$ **35.** $x^3 + 2x^2 + 3x$

37. $-x^2 - x + 11$ **39.** $9x^3 - 2x^2 + 9x - 17$ **41.** $x + 13$ **43.** $7x + 7$ **45.** $\dfrac{-1}{5}x^2 + x + 1$

47. $-2\dfrac{1}{2}x^2 + \dfrac{2}{3}x$ **49.** $3.5x^2 - 3.3x + 4$ **51.** $-1.1x^3 + 3.1x^2 + x + 3.6$ **53.** $9x^3 + 11x^2 - 10x + 12$

55. $-z^4 - 18z^2 + 11z + 8$ **57.** $-27x^2 + x + 4$ **59.** $-25x^2 - 33x + 35$ **61.** $-10t^2 + 40t + 6$

63. $7x^2 + 10x - 16$ **65.** $4x^2 - 8$ **67.** $-16x^3 + 33x^2 - 4x + 12$ **69.** $16x^2 - 24xy + 29y^2$

71. $9r^2 + 28r - 16$ **73.** $-13m^3 - 7m^2 - 11m - 35$ **75.** $-x^3 + x^2 - 8x + 12$

77. $-6x^2 - 3x - 3xy - 8$ **79.** $\dfrac{5}{12}x^2 - \dfrac{7}{15}x + \dfrac{11}{72}$ **81.** Answers may vary. **83.** $N_A - N_L = 10x - 0.1x^2$

85. $\dfrac{3}{8}rt^2y^2$ **87.** $10t^2 + 7t - 3$ **89.** $\dfrac{-5x}{9y}$ **91.** $\dfrac{216y^{21}}{x^3}$ **93.** t^7 **95.** $\dfrac{3v^6}{t^5}$ **97.** $\dfrac{1}{81}$

99. $7.31x^2 - 1.62x - 11.13$ **101.** $-48s^5t^{20}$

SECTION 5.4 *(pages 313–316)*

1. $3n^2 + 15n$ **3.** $84y^3 + 108y^2$ **5.** $21n^3 + 35n^2 - 56n$ **7.** $2x^4 - 10x^3 + 5x^2 - 28x + 15$

9. $80n^4 - 56n^3 - 55n^2 + 28n - 35$ **11.** $-132n^5 + 96n^4 + 92n^3 - 164n^2 + 80n + 48$ **13.** $21x^2 + 20x + 4$

15. $35t^2 + 23t - 4$ **17.** $4y^2 - 32y + 55$ **19.** $12r^2 - 11rt + 2t^2$ **21.** $21s^2 - 23ms + 6m^2$

23. $-6x^4 - 8x^2 - 15x^3 - 20x$ **25.** $n^2 + 18n + 81$ **27.** $9x^2 - 16$ **29.** $x^2 - 4$

31. $36x^2 - 24xy + 4y^2$ **33.** $16x^2 - 26x - 10$ **35.** $-2x^4 + 5x^3 + 2x^2 + 9x$ **37.** $3x^3 - 7x^2 + 25x$

39. $4x^3 - 7x^2 + 3x - 10$ **41.** $24x^2 + 55x - 92$ **43.** $10x^3 - 2x^2 - 32x - 32$ **45.** $3x^3 - 5x^2 + 9x$

47. $2x^3 - 3x^2y + 2xy^2$ **49.** $-3x^3 - 4x^2 - 13x + 16$ **51.** $-7x^2 - 3xy - 9x - 7$ **53.** $-10x^2 + 25x - 12$

55. $-82x^2 + 19x + 37$ **57.** $n^2 + n - 12$ **59.** $n^2 + 24n + 144$ **61.** $V = 2n^3 - 7n^2 + 3n$ **63.** $\dfrac{9v^5}{10rt^2}$

65. $4x^2 - 2x - 12$ **67.** $\dfrac{256}{81}x^8y^8$ **69.** $-x^4 - 2x^3 + 10x^2 - 25x$ **71.** $560r^6y^{12}$ **73.** $\dfrac{14r}{15s^2}$

SECTION 5.5 *(pages 323–325)*

1. $2x + 4$ **3.** $3x^2 - \dfrac{3}{2}$ **5.** $-2 - x$ **7.** $-3 + r^2$ **9.** $\dfrac{4}{x^7} + 1$ **11.** $\dfrac{5}{r^2} - 1$ **13.** $5x + \dfrac{3x^2}{y}$

15. $7r^2 - \dfrac{3r}{u}$ **17.** $\dfrac{1}{y} + 4x^2$ **19.** $\dfrac{1}{u} - 3r^2$ **21.** $\dfrac{13y}{x^2} + \dfrac{11z}{x^2y}$ **23.** $\dfrac{-8u^5}{r^6} + \dfrac{6}{r^6}$ **25.** $\dfrac{-9x^2}{y} - 8xy - 7$

27. $\dfrac{9r}{u} - 6u - \dfrac{4}{r}$ **29.** $-y^2 + 3y^4 - y^3$ **31.** $\dfrac{11}{u} - 8 - 6u$ **33.** $-1 + \dfrac{1}{3r^4}$ **35.** $\dfrac{-2}{x} + \dfrac{2}{7y} - 3$

37. $\dfrac{-3}{r^2u} + \dfrac{1}{2u^4} - \dfrac{4r}{u}$ **39.** $x - 13$ **41.** $4x - 21 + \dfrac{54}{x + 3}$ **43.** $x + 2$ **45.** $x + 1$ **47.** $y + 2$

49. $m - 12 + \dfrac{-26}{m - 1}$ **51.** $3x + 13 + \dfrac{139}{x - 13}$ **53.** $t^2 + t + 2$ **55.** $t^2 + 2t + 4$ **57.** $t^2 - 4t + 16$

59. $x - 8$ **61.** $x^3 + x^2 + x + 1$ **63.** $8x^3 + 12x^2 + 18x + 27$ **65.** $A = 220 - \dfrac{10}{7}H$; numerical answers

may vary. **67.** $\dfrac{-1}{8^4}$ **69.** $-55r^3 - 75r^2v + 25rv$ **71.** $x^2 + 4$ **73.** $117r^4t^2y^2$

75. $-55x^3 + 54x^2 - 63x + 10$ **77.** $21x^2 - 43x - 14$ **79.** $5n^2 + n + 15$

CHAPTER 5 REVIEW PROBLEMS   *(pages 328–329)*

1. 36   **2.** -125 **3.** $\frac{1}{1024}$ **4.** $-(2^4)$ **5.** They are equal.   **6.** $\frac{16y^6}{9x^4}$ **7.** $4x^4y^6$ **8.** $\frac{1}{9r^6t^8y^2}$

9. $\frac{s^{12}}{512r^{18}t^3}$ **10.** $\frac{625m^{12}}{n^8}$ **11.** $6.08x^4y^4$ **12.** $\frac{14}{5}x^5y^4$ **13.** $-210m^7n$ **14.** $62.4m^2n$ **15.** $\frac{4r^3}{t}$

16. $\frac{-7n^3}{m^2}$ **17.** $4xy^2$ **18.** $\frac{3}{2txy}$ **19.** $\frac{7}{m^2n^{10}r^5}$ **20.** $-30r^6t^4$ **21.** $8x-1$ **22.** $-3x^3-x^2-3x$

23. $-12y^3+y^2-4y+5$ **24.** $12t^3-10t^2+2t-1$ **25.** $10x^2-26x+12$

26. $7x^2+59xy+\frac{6x}{y^2}-36y^2+\frac{54}{y}$ **27.** $-12y^2-36y-15$ **28.** $3r^2t-2$ **29.** $\frac{-7}{2t^3}+\frac{4r^2}{t}-\frac{9}{2}$

30. x^2+5x-3 **31.** x^2+4x **32.** $3x^2-3x$ **33.** $-\frac{mn^7}{9}$ **34.** $x^2+6x+23+\frac{84}{x-3}$ **35.** $\frac{-x^6y^9z^{15}}{729}$

36. $\frac{2r^7}{3t^6}$ **37.** $21t^4-12t^3+39t^2-36t-72$

CUMULATIVE REVIEW   *(page 330)*

1. -1.64 **2.** $-13txy^2-6tx^2y+8ty^3+12x^2$ **3.** $x=-10\frac{1}{2}$ **4.** 6.0×10^{-14} **5.** $m=\frac{7}{12}$

6. 23.86881 **7.** $x\le-0.9$ **8.** $t=-10$ **9.** 17 **10.** $-4r^2+2r^3t$

11. $-16\frac{1}{6}$ **12.** $-\frac{1}{25}$ **13.** 3.3009 **14.** $216x^4y^5t^2$ **15.** $6x^2+4x-10$

16. $-216t^{12}y^6$ **17.** -56 **18.** $18x^2y^3+18rx^2y^3$ **19.** Tables may vary.   **20.**

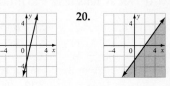

ANSWERS to Chapter 6 Odd-Numbered Exercises

SECTION 6.1   *(pages 341–342)*

1. 1, ②, ③, 4, ⑤, 6, 10, 12, 15, 20, 30, 60   **3.** 1, ⑤, ⑦, 35   **5.** $2^5\cdot7$ **7.** $2\cdot3^2\cdot7$ **9.** $6m$ **11.** $5mt$

13. $3y$ **15.** $3xy^3$ **17.** $4mn$ **19.** $4xy$ **21.** true   **23.** true   **25.** $6xy(3y+4z)$

27. $3rs^2(-rs-2)$ **29.** $4t^2y(tx+2)$ **31.** $3rt(5+3t)$ **33.** $2xy^2(-9+14x)$ **35.** $mn(2m^3n-m-t)$

37. $(5x+4)(2-x)$ **39.** $(n-1)(16n-1)$ **41.** $(3x-5)(18x+7)$ **43.** $(x+4)(x+2)$

45. $(r+3)(r+2)$ **47.** $2(3t-5)(t-2)$ **49.** $2(2+x)(5-3x)$ **51.** $(5x+6)(2x-3)$

53. $(4-5x)(6x-1)$ **55.** $2(v+2)(3v+5)$ **57.** $(y-2)(y-3)$ **59.** $7(y-1)(y-5)$

SECTION 6.2   *(pages 350–352)*

1. $(y+2)(y+9)$ **3.** $(x+4)(x-3)$ **5.** $(r-5)(r-8)$ **7.** $(r-2)(r+1)$ **9.** $(m+8)(m-5)$

11. $(n-2)(n-1)$ **13.** $3(r-4)(r-13)$ **15.** $2(x-10)(x-2)$ **17.** $3(t+9)(t+1)$

19. $4(x-3)(x+2)$ **21.** $5(x-10)(x+4)$ **23.** $10u(t-6)(t-5)$ **25.** $3(t+8)(t-7)$

27. $(y-11)(y+3)$ **29.** $(t+7)(t-6)$ **31.** $(y+8)(y-7)$ **33.** $5(x^2-10x-28)$

35. $3(x+3y)(x-y)$ **37.** $2(9r-1)(r-1)$ **39.** $5(3y-1)(y+1)$ **41.** $8(2n-1)(n-1)$

43. $4(4v-5)(v+1)$ **45.** $t(3r+2)(r+4)$ **47.** $(7y+2)(y-1)$ **49.** $5n(x-6)(x+1)$

51. $3(x-3)(x-1)$ **53.** $(7v-10)(8v+3)$ **55.** $3t(4x+1)(x-4)$ **57.** $16(4t+1)(t-1)$

59. $3(5w+11)(w+1)$ **61.** $6(x+3)(x-1)$ **63.** $9n(x-2)(x-3)$ **65.** $2(2m-3)(m+5)$

67. $6(n-1)(n+6)$ **69.** $2m(v-3)(2v+9)$ **71.** $2(y-8)(y-4)$ **73.** $7(5n-8)(n+1)$

75. $28(x+1)(x+1)$ **77.** $3t(x+5)(x-3)$ **79.** $6(x+10)(x-5)$ **81.** $-8(3r+5)(r-1)$

83. $x(t-11)(t-6)$ **85.** $(7x-4)(x+2)$ **87.** $(y+2t)(3y-7)$ **89.** $9x^2y(1-3y^2)$ **91.** $5\cdot53$

SECTION 6.3   *(pages 361–362)*

1. $(t+3)(t-3)$ **3.** $(2y+5)(2y-5)$ **5.** $(t^2+4)(t+2)(t-2)$ **7.** $5(x+5)(x-5)$

9. $(9t^8+7t^4)(9t^8-7t^4)$ **11.** $(x-3+y)(x-3-y)$ **13.** $(t-8)^2$ **15.** $(y+2)^2$ **17.** $(t-11)^2$

19. $(y+5)^2$ **21.** $(2n-11)^2$ **23.** $(5y-12)^2$ **25.** $x(r+30)(r-30)$ **27.** $3z(y+6)(y-6)$

29. $(2x+3y)^2$ **31.** $9(y-1)^2$ **33.** $2(5n+3)^2$ **35.** $4(3n-2)^2$ **37.** $x^2-⑧x+16=(x-4)^2$

39. $\boxed{169} - 26x + x^2 = (x - 13)^2$ **41.** $25x^2 - \boxed{30x} + 9 = (5x - 3)^2$ **43.** $49x^2 - 42x + \boxed{9} = (7x - 3)^2$
45. $4 - 72x + \boxed{324}x^2 = (18x - 2)^2$ **47.** $\boxed{16}x^2 + 48x + 36 = (4x + 6)^2$ **49.** $(t - 8)^2$
51. $3(4 + t)(4 - t)$ **53.** $9x^2y^2(3 - x^3y - 9y)$ **55.** $3(t - 7)(t - 6)$

SECTION 6.4 *(pages 370–373)*

1. $x = 0$ or $x = -1$ **3.** $n = -\frac{1}{2}$ or $n = 65$ **5.** $x = -\frac{3}{2}$ or $x = 3$ **7.** $r = 4$ or $r = 5$ **9.** $r = 3$ or
$r = -28$ **11.** $y = \frac{1}{2}$ or $y = -\frac{3}{4}$ **13.** $t = 0$ or $t = 3$ **15.** $x = 0$ or $x = -13$ **17.** $t = 0$ or $t = 4$
19. $n = 0$ or $n = 5$ **21.** $t = -10$ or $t = 10$ **23.** $t = -\frac{8}{9}$ **25.** $y = -12$ **27.** $t = -\frac{10}{3}$ or $t = \frac{10}{3}$
29. $r = -4$ **31.** $y = 1$ **33.** $r = -9$ or $r = 5$ **35.** $t = 50$ or $t = 2$ **37.** $t = -\frac{4}{9}$ or $t = \frac{4}{7}$
39. $n = \frac{3}{2}$ or $n = -\frac{5}{3}$ **41.** $x = \frac{1}{2}$ **43.** 15 and 17 **45.** 11; -10 **47.** 13 **49.** 2 inches
51. 100 square inches and 225 square inches **53.** 22 and 23 **55.** 52 square inches **57.** 15; -10
59. $(2y + 1)(4a - 3)$ **61.** $6(x^2 + 11x - 5)$; cannot be factored further **63.** $y = -20$ or $y = 6$ **65.** -11
and -10 or 10 and 11 **67.** $r = \frac{1}{2}$ and $r = 4$ **69.** $(y + 22)(y - 22)$

CHAPTER 6 REVIEW PROBLEMS *(pages 374–375)*

1. $5rt(4rt - 2r + t)$ **2.** $3xy(6y - 2 + 3xy)$ **3.** $9xy(4xy + 2x - y)$ **4.** $4(t + 3m)(x + 2)$
5. $(6a + 7)(x + 5y)$ **6.** true **7.** $(x + 5)(x - 1)$ **8.** $(y + 4)(y - 3)$ **9.** $3(t - 3)(t - 4)$
10. $4(r + 6)(r - 4)$ **11.** $x(r - 8)(r + 5)$ **12.** $(5t - 3)(t - 1)$ **13.** cannot be factored
14. $4(y - 7)(2y + 3)$ **15.** $(3x + y)(9x - 5y)$ **16.** $2(4x - 7)(3x + 2)$ **17.** $-25(1 + 5x)(1 - 5x)$
18. $3(4x + 1)(4x - 1)$ **19.** $6(y + 2)^2$ **20.** $-(3x - 4)^2$ **21.** $(2x + 1)^2$ **22.** $3t(r - 8)^2$ **23.** $r = 7$ or
$r = 3$ **24.** $n = 20$ or $n = -20$ **25.** $x = 7$ or $x = -5$ **26.** $x = 8$ or $x = 3$ **27.** $r = -\frac{6}{5}$ or $r = 2$
28. $m = -\frac{7}{5}$ or $m = -2$ **29.** -22 and -20 or 20 and 22 **30.** -11 **31.** 6 inches by 18 inches
32. rectangle 1: $\ell = 10$ inches, $w = 13$ inches; rectangle 2: $\ell = 10$ inches, $w = 7$ inches **33.** $(4x - 5)(2x + 3)$
34. $(2r - 9)(2r + 41)$ **35.** $(8x - 3)(7x + 1)$ **36.** $-2(4w - 3)(2w - 1)$ **37.** 5 and -7 or -5 and 7
38. $(8x - 3)(3x - 4)$ **39.** $(6x - 1)(8x + 3)$ **40.** $xy(3x + 9y - xy)$ **41.** $(x + 5)(x - 3)$ **42.** $r = -9$
or $r = -13$ **43.** $y = 6$ or $y = -3$ **44.** $x = \frac{4}{3}$ or $x = -\frac{4}{3}$ **45.** $y = 12$ or $y = -8$ **46.** $(y + 2)(4y - x)$

CUMULATIVE REVIEW *(page 376)*

1. $-3(x + 1)(x - 1)$ **2.** $75r^3t^5v^4$ **3.** $2(x - 7)(x + 1)$ **4.** $y = \frac{1}{3}x + 6$ **5.** $\frac{1}{10}$ **6.** $81x^8y^4$ **7.** 90
8. $y = \frac{4}{3}x - \frac{17}{3}$ **9.** 270 **10.** $t \geq -\frac{1}{15}$ **11.** $x = -2$ or $x = 2$ **12.** $r = -0.2$
13. $(x + 12)(x - 12)$ **14.** $6x^2 - 15x + 6$ **15.** $8y - 12x$ **16.** $b = c(d - mx)$ or $b = cd - cmx$
17. $x = -10$ **18.** $x = 3$ or $x = 2$ **19.** Tables may vary.   **20.**

ANSWERS to Chapter 7 Odd-Numbered Exercises

SECTION 7.1 *(pages 384–387)*

1. all real numbers except $y = 0$ **3.** all real numbers except $x = -4$ **5.** all real numbers except $y = 3$ and
$y = -3$ **7.** all real numbers except $x = 2$ and $x = 1$ **9.** $\frac{10}{13}$ **11.** $\frac{25}{31}$ **13.** undefined
15. undefined **17.** $\frac{1}{y}$ **19.** $\frac{y}{r}$ **21.** $\frac{y}{x}$ **23.** $\frac{y^2}{t^2}$ **25.** $\frac{x}{x^2 + 1}$ **27.** $\frac{r^5}{r^2 - 3}$ **29.** $\frac{1}{5}$ **31.** 2
33. $\frac{1}{x - 1}$ **35.** $\frac{1}{x - 5}$ **37.** $\frac{x + 3}{x + 5}$ **39.** $\frac{x + 3}{x + 2}$ **41.** $y + 2$ **43.** $\frac{y + 2}{y - 2}$ **45.** 3 **47.** 3

49. $\dfrac{2x(7x-1)}{5}$ **51.** $\dfrac{2x+y}{3}$ **53.** $\dfrac{8+20+168}{40+100+840}=0.2$ **55.** $E=0.315$ **57.** $\dfrac{P}{v^3}=K;\ K\approx 0.0148$

$\dfrac{196}{980}=0.2$

59. 1 **61.** $\dfrac{2}{3}$ **63.** 4 **65.** $\dfrac{1}{4}$ **67.** -7 **69.** 0

71. $-\dfrac{7}{2}$ **73.** $-\dfrac{15}{13}$ **75.** $\dfrac{3}{2}$ **77.** $-\dfrac{1}{5}$ **79.** 12

SECTION 7.2 *(pages 394–396)*

1. $\dfrac{1}{x^3 n^6}$ **3.** $\dfrac{98r^{14}}{x^9}$ **5.** $\dfrac{55(x-3)}{2}$ **7.** 7 **9.** $\dfrac{2(y-1)}{y+9}$ **11.** $\dfrac{1}{4}$ **13.** 3 **15.** $\dfrac{y+9}{y-1}$ **17.** $\dfrac{r-2}{7r-1}$

19. $(y-9)^2$ **21.** $\dfrac{t+2}{t+9}$ **23.** 1 **25.** $\dfrac{y^2}{x^2}$ **27.** $\dfrac{2}{5}$ **29.** $\dfrac{m}{6(m-n)}$ **31.** 1 **33.** $\dfrac{7r}{(r+t)(7r+12t)}$

35. $x=2$ or $x=1$ **37.** $x=-\dfrac{2}{5}$ **39.** $\dfrac{25-n}{40-n}$ **41.** $456x^2 y$ **43.** $100x^2 y$ **45.** $42x^2 y$ **47.** $168x^2 y$

49. $160xy^2$ **51.** $52xy$ **53.** $82xy$ **55.** $116w^2 x^2 y^2$ **57.** $172x^2 y^5$ **59.** $\dfrac{x+7}{x-7}$ **61.** $\dfrac{1}{x-4}$

63. $\dfrac{x+4}{x}$ **65.** $\dfrac{(r+10)(y-10)}{(r+8)(y+8)}$

SECTION 7.3 *(pages 403–405)*

1. $\dfrac{5y}{5xy^2}$ and $\dfrac{3x}{5xy^2}$ **3.** $\dfrac{1}{x+1}$ and $\dfrac{x-1}{x+1}$ **5.** $\dfrac{x^3 y}{8x^2 y^2}$ and $\dfrac{6y}{8x^2 y^2}$ **7.** $\dfrac{16wy}{300wxy},\dfrac{210wx}{300wxy}$ and $\dfrac{255xy}{300wxy}$ **9.** $\dfrac{8y}{45}$

11. $\dfrac{14y}{45}$ **13.** $\dfrac{10+3y}{6y}$ **15.** $\dfrac{x-3}{3x}$ **17.** $\dfrac{6y-35}{18y}$ **19.** $\dfrac{43-20x}{15x}$ **21.** $\dfrac{5x-6}{6(4x+3)}$ **23.** $\dfrac{x-1}{x-2}$

25. $\dfrac{6}{x+2}$ **27.** $\dfrac{-2}{x-5}$ **29.** $\dfrac{3(y+18)}{2(y-36)}$ **31.** $\dfrac{7(r+27)}{3(r+56)}$ **33.** $\dfrac{80}{81}$ **35.** $\dfrac{8x}{3(3x-10)}$ **37.** $\dfrac{(12-x)(x-8)}{8x^2}$

39. $\dfrac{30}{47}$ **41.** $\dfrac{1}{2x}$ **43.** $r=-36$ **45.** $x=\dfrac{-18}{13}$ **47.** $y=6$ or $y=-2$ **49.** $y=2$ **51.** $x=-5$

53. $t=4$ **55.** $n=3$ **57.** $x=\dfrac{-23}{3}$ or $x=-4$ **59.** $r=-3$ **61.** $x=9$ **63.** $x=\dfrac{-5}{3}$

65. $\dfrac{16x^2+10x-9}{5(x-2)(x+2)}$ **67.** $\dfrac{2(x-1)}{(x-2)(x+1)}$ **69.** $\dfrac{t}{xy}$ **71.** has meaning except when $x=0$ and when $x=4$

73. $\dfrac{r}{2(r-2)}$ **75.** $\dfrac{7x-11}{10x}$ **77.** $(r+4)(r-2)$ **79.** $\dfrac{1}{x+4}$

SECTION 7.4 *(pages 415–418)*

1. 18 minutes **3.** 12 seconds **5.** $9\dfrac{1}{11}$ minutes **7.** hose A, 32 minutes; hose B, 96 minutes **9.** woman is 21; sister is 28 **11.** The pine board weighs more. **13.** 12 hours **15.** length, $1\dfrac{13}{20}$ feet; width, $\dfrac{11}{20}$ feet

17. 30 **19.** 120 flavors **21.** 38 minutes; 5053 meters per hour **23.** 28 miles per hour **25.** $7\dfrac{1}{2}$ hours

27. post, $4\dfrac{1}{2}$ feet; man, 6 feet **29.** $\dfrac{1}{6}$ cup **31.** $t=3$ **33.** $n=2$ **35.** $\dfrac{39y-74}{30(3y-2)}$ **37.** $\dfrac{20}{15}$

39. $\dfrac{2x+1}{x}$

CHAPTER 7 REVIEW PROBLEMS *(pages 420–422)*

1. $\dfrac{3}{2x^2}$ **2.** when $x=-\dfrac{3}{2}$ **3.** x **4.** 100 **5.** $\dfrac{y-4}{2}$ **6.** when $x=0$ **7.** $\dfrac{2}{(x-3)(2x+3)}$

8. $\dfrac{y(y-3)}{8}$ **9.** $\dfrac{1}{(y-5)^2}$ **10.** $\dfrac{2}{x+1}$ **11.** $\dfrac{(x+3)^2}{3}$ **12.** $\dfrac{2(x-3)}{x-5}$ **13.** $x=\dfrac{52}{7}$ **14.** $x=\dfrac{2}{21}$

15. $t=-8$ **16.** $x=5$ or $x=-2$ **17.** $\dfrac{220x^2 y^3}{300x^2 y^3},\dfrac{200y^2}{300x^2 y^3}$ and $\dfrac{27x}{300x^2 y^3}$ **18.** $(x+5)(x-5)$

19. $\dfrac{4(x-3)}{(x+1)(x-4)(x-3)}$ and $\dfrac{16(x+1)}{(x+1)(x-4)(x-3)}$ **20.** $48wx^3 y$ **21.** $400w^2 x^2 y^3$ **22.** $180w^2 x^2 y^2$

23. $\dfrac{4x^2-15x+25}{(x+5)(x-5)}$ **24.** $\dfrac{3(3x-1)}{10x^2}$ **25.** $\dfrac{1}{y}$ **26.** $\dfrac{6(8x-11)}{(3x-4)(2x-3)}$ **27.** $\dfrac{6x-13}{(x+3)(x-3)}$

28. $\dfrac{-9x+14}{(2x-7)(3x-4)}$ **29.** $\dfrac{3x^2+2}{4x+1}$ **30.** $\dfrac{y^2+5}{y+2}$ **31.** $\dfrac{y(2y+1)}{3x}$ **32.** $\dfrac{-(x-3)^2}{1+y}$ **33.** $x=-\dfrac{2}{3}$

34. $x = 2$ **35.** $x = \dfrac{-9}{2}$ **36.** $x = 8$ **37.** 24 feet smaller   **38.** $13\frac{1}{3}$ hours   **39.** Ms. Appel, 25 mph;

Ms. Allen, 10 mph   **40.** $\dfrac{30}{8}$ **41.** $y = \dfrac{28}{5}$ **42.** $x = 6$ or $x = 2$ **43.** $\dfrac{12.4}{9.4}$ **44.** $\dfrac{1}{25}$ **45.** $\dfrac{x}{x+1}$

46. -2 **47.** $\dfrac{48}{r^5}$ **48.** 2

CUMULATIVE REVIEW   *(pages 423–424)*

1. $x = 21$ **2.** slope: $\frac{1}{3}$; y-intercept: $(0, -2)$ **3.** $-6\frac{41}{63}$ **4.** $\dfrac{2}{x+5}$ **5.** -459 **6.** $y = \frac{1}{2}x + 4$

7. $y^2 - 11y + 28$ **8.** $2(x - 3)(x + 1)$ **9.** $y = \dfrac{-37}{10}$ **10.** $x = \dfrac{-100}{3}$ **11.** $\dfrac{rt(3 - rt)}{4}$

12. $x < -3\frac{1}{6}$ **13.** $\dfrac{8x^2y^3 + 10xy^3}{(x + 2)(x - 2)}$ **14.** $x = 9$ **15.** $\dfrac{2(2y + 35)}{(y - 7)(y + 7)}$ **16.** $x^2 + 4x + 2$ **17.** $\frac{1}{2}$

18. $4x^2 + 20x + 25$ **19.**   **20.**

ANSWERS to Chapter 8 Odd-Numbered Exercises

SECTION 8.1   *(pages 437–439)*

1. yes   **3.** yes   **5.** no   **7.** no   **9.** $(6, -3)$ **11.**   no solution

13.   all points on the line   **15.** $(1, 3)$ **17.** $(-2, -4)$

19. $(2, 3)$ **21.**   **23.**   **25.**   **27.**

29.   **31.**   **33.**   **35.**   **37.** B   **39.** D   **41.** G

43. H   **45.** L   **47.** K

SECTION 8.2   *(pages 449–451)*

1. $(3, -1)$ **3.** $(1, 2)$ **5.** $(5, 4)$ **7.** $\left(3, \frac{4}{9}\right)$ **9.** $(1, 4)$ **11.** $\left(1, \dfrac{-2}{9}\right)$ **13.** $\left(-2, \dfrac{3}{2}\right)$ **15.** $(2, 11)$

17. Same line. All points on this line are solutions.   **19.** These lines are parallel. There is no solution.

21. $(-6, 5)$ **23.** $(3, 5)$ **25.** $(1, 2)$ **27.** $(-2, -8)$ **29.** $(4, 4)$ **31.** $(5, 1)$ **33.** $(3, 4)$ **35.** $(1, 5)$

37. $(11, 7)$ **39.** $(2, 1)$ **41.** $(-4, -3)$ **43.** $(0, 1)$ **45.** $(1, -1)$ **47.** $(-3, -2)$ **49.** $(9, -10)$

51. $(-6, -5)$ **53.** $(1, 2)$ **55.** $(-3, 18)$ **57.** $(1, 2)$ **59.** No solution; the lines are parallel.

61. $(-48, -124)$ **63.** $\left(\frac{1}{2}, \frac{7}{2}\right)$ **65.** $(8, -9)$ **67.** $(-3, -1)$ **69.** 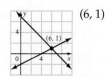 $(6, 1)$

71. $(2, 2)$ **73.** $(1, -2)$ **75.** No solution; the lines are parallel.

SECTION 8.3   *(pages 463–467)*

1. 59 nickels; 299 pennies   **3.** 30 ounces of 60% alcohol; 10 ounces of 40% alcohol   **5.** $700 was invested at 5%; $800 was invested at 8%   **7.** 6 hours; 30 miles from the lake   **9.** when one is mailing more than 4 items   **11.** 22 20¢-coupons; 8 25¢-coupons   **13.** 40° and 140°   **15.** 15 and 75   **17.** 15 liters of 25% solution; 35 liters of 55% solution   **19.** 18 mph   **21.** 9408 square inches   **23.** 72 million; 132 pounds   **25.** 21 and 26   **27.** 6 hours; 360 miles   **29.** 10 nickels; 9 quarters   **31.** $x = 430$ adults' tickets, $y = 195$ children's tickets   **33.** 17 and 21   **35.** $\left(-\frac{6}{7}, -\frac{36}{7}\right)$ **37.** yes   **39.** Same line. All points on this line are solutions.   **41.** $(2, 0)$ **43.** $(0, 5)$

CHAPTER 8 REVIEW PROBLEMS   *(pages 473–474)*

1. no   **2.** infinite number of solutions   **3.** $(3, 0)$ **4.** No solution; the lines are parallel.

5. $(0, -4)$ **6.** $(6, -1)$ **7.**   **8.**

9.   **10.**   **11.** no   **12.** Yes, $(-1, 3)$ is in the solution set.

13. $\left(-\frac{12}{5}, \frac{33}{5}\right)$ **14.** $(-2, 1)$ **15.** $\left(-\frac{9}{7}, -\frac{1}{2}\right)$ **16.** $\left(\frac{5}{7}, \frac{10}{7}\right)$ **17.** $\left(\frac{5}{8}, -\frac{11}{8}\right)$ **18.** $\left(\frac{10}{3}, -\frac{2}{3}\right)$

19. The lines are parallel. There is no solution.   **20.** $(-5, -1)$ **21.** 9.8 and 9.6   **22.** $\frac{33}{4}$ and $\frac{13}{4}$

23. 6 pounds   **24.** 6 hours   **25.** $\left(-\frac{2}{3}, \frac{7}{3}\right)$ **26.** $\left(-\frac{1}{6}, 2\right)$ **27.** $(-1, -1)$ **28.** $(50, 14)$

29. $\left(-\frac{17}{4}, \frac{3}{4}\right)$ **30.**

CUMULATIVE REVIEW   *(pages 475–476)*

1. $8(x - 2)(x - 1)$ **2.** $\dfrac{2(5x - 1)}{(x + 4)(3x - 5)}$ **3.** $3(x + 9)(x - 9)$ **4.** $\dfrac{-9}{x^3 y}$ **5.**   **6.** $t = \dfrac{5r}{3r - 4}$

7. $(3, 4)$ **8.** $-3x^2 - 10x + 6$ **9.** $48x^8y^4$ **10.** $y = -5x + 19$ **11.** $(37, 42)$ **12.** $\frac{1}{2}$ **13.** $\frac{3}{9 - x^2}$

14. $\frac{66.6t^5}{r}$ **15.** $29x^2 - 12x + 7$ **16.** $x = -7$ **17.** $r \le -0.0446$ **18.** $t = 1$ **19.** $5x^2 + 24x + 29$

20.

ANSWERS to Chapter 9 Odd-Numbered Exercises

SECTION 9.1   *(page 484)*

1. 18   **3.** 29   **5.** 63   **7.** 105   **9.** 12 and 13   **11.** 17 and 18   **13.** 33 and 34   **15.** 43 and 44
17. 37 and 38   **19.** 100 and 101   **21.** between 5 and 10   **23.** between -12 and -18 **25.** between
-10 and -12 **27.** between -69 and -72 **29.** not real   **31.** real and irrational   **33.** real and rational
35. real and rational   **37.** $2m^4n^4$ **39.** $2xy$ **41.** $30xy$ **43.** $30mn^4$ **45.** $0.45x^{22}y^{28}$ **47.** $1.03s^{17}t^{26}$

SECTION 9.2   *(pages 501–503)*

1. $2x^4\sqrt{30x}$ **3.** $2x\sqrt{31}$ **5.** $2x^3\sqrt{3}$ **7.** $5x\sqrt{3x}$ **9.** $\frac{\sqrt{5}}{2}$ **11.** $\sqrt{13}$ **13.** $13y\sqrt{15}$
15. $10x\sqrt{3}$ **17.** $4x^2\sqrt{2xy}$ **19.** $4x^2y\sqrt{2xy}$ **21.** $2r^3t^3\sqrt{2}$ **23.** $4r^2t\sqrt{30rt}$ **25.** $-8rt^5\sqrt{3t}$
27. $30rt^2\sqrt{2t}$ **29.** $\frac{2}{r^3t}$ **31.** $\frac{10t}{3}$ **33.** $6r\sqrt{65rt}$ **35.** $-108r\sqrt{7r}$ **37.** $-3\sqrt{3}$ **39.** $8\sqrt{5}$
41. $-4\sqrt{13}$ **43.** $11\sqrt{3} - 6\sqrt{2}$ **45.** $-4\sqrt{2} - 8$ **47.** $-32\sqrt{2} - 10\sqrt{10}$ **49.** $2\sqrt{6} - 9 + \sqrt{3}$
51. $4x\sqrt{2}$ **53.** 0   **55.** $\frac{3t^2\sqrt{2} - 3t^2}{5}$ **57.** $-\sqrt{10} + 2\sqrt{2}$ **59.** $10xy^2\sqrt{6x} - 36xy^2$ **61.** $-x\sqrt{3xy}$
63. $3\sqrt{t}$ **65.** $-8\sqrt{7} - 2\sqrt{14}$ **67.** $6\sqrt{3} - 11\sqrt{2}$ **69.** 18.17   **71.** $4\sqrt{5}$ **73.** 84 yards
75. real and rational   **77.** real and irrational   **79.** real and rational   **81.** $4\sqrt{2}$

SECTION 9.3   *(pages 514–516)*

1. $12v$ **3.** $2w - 9\sqrt{2w}$ **5.** $-21t^2$ **7.** $24\sqrt{22} + 24$ **9.** $6v\sqrt{2} - 18v$ **11.** $11 + 6\sqrt{2}$
13. $54 - 14\sqrt{5}$ **15.** -1 **17.** -15 **19.** $102 - 72\sqrt{2}$ **21.** $6 + 4\sqrt{2}$ **23.** $\frac{3\sqrt{5}}{5}$ **25.** $\sqrt{10}$
27. $\frac{3\sqrt{30}}{4}$ **29.** $-\sqrt{7}$ **31.** $\frac{\sqrt{2}}{2}$ **33.** $\frac{\sqrt{10}}{5}$ **35.** $\frac{3\sqrt{2}}{16}$ **37.** $\frac{6t + 3\sqrt{2t}}{2t - 1}$ **39.** $4\sqrt{5} - 7$
41. $\frac{-\sqrt{3} - 1}{4}$ **43.** $\frac{10\sqrt{3} - 4\sqrt{15}}{3}$ **45.** $\frac{-9 - 3\sqrt{11}}{4}$ **47.** $\frac{10\sqrt{2} - 10}{3}$ **49.** $27\sqrt{2} + 30$ **51.** $\frac{x\sqrt{y}}{y}$
53. $6x\sqrt{5xy}$ **55.** $\frac{x^2\sqrt{xy}}{y}$ **57.** $x^2\sqrt{x}$ **59.** $\frac{10\sqrt{10}}{147}$ **61.** $\frac{32t\sqrt{55}}{27}$ **63.** $\sqrt{t}$ **65.** $\frac{\sqrt{70}}{27}$
67. $x^2\sqrt{x}$ **69.** $8x^3\sqrt{10}$ **71.** $4\sqrt{30x}$ **73.** $3z^3\sqrt{10xy}$ **75.** 12 and 15   **77.** 133.14 units
79. 27 and 28   **81.** $10\sqrt{5} + 15\sqrt{10}$

SECTION 9.4   *(pages 526–528)*

1. $p = 36$ **3.** $y = 25$ **5.** $x = 108$ **7.** $t = 208$ **9.** $t = \frac{3}{2}$ **11.** $y = 4$ **13.** $r = 4$
15. $x = 8$ **17.** no real-number solution   **19.** no real-number solution   **21.** $t = 4$ **23.** $r = 5$
25. a. \$5.95   **b.** about 19 inches   **27. a.** $\frac{4}{25}$ **b.** 29%   **29.** 44 inches   **31.** 10 inches   **33.** $13xz\sqrt{y}$
35. $\sqrt{2} + 3$ **37.** $x = 1$ **39.** $x = 15$ **41.** $10x\sqrt{xy}$ **43.** $-16\sqrt{3}$ **45.** $9\sqrt{42} + 54 - 12\sqrt{3}$

CHAPTER 9 REVIEW PROBLEMS   *(pages 529–530)*

1. 12 and 13   **2.** between 12 and 15   **3.** between 30 and 36   **4.** 18   **5.** not real   **6.** $2xy\sqrt{55}$
7. $2x^3\sqrt{3}$ **8.** $4x^3y\sqrt{2}$ **9.** $4y^5$ **10.** $4x^2y\sqrt{2xy}$ **11.** $-4\sqrt{5} - 40\sqrt{10} + 20\sqrt{2}$ **12.** $-\sqrt{3}$
13. $\frac{3\sqrt{5}}{10}$ **14.** $\frac{6\sqrt{3} - 9\sqrt{5} + 16\sqrt{15} - 120}{-33}$ **15.** $-12\sqrt{15} - 30\sqrt{3}$ **16.** $\frac{\sqrt{15}}{5}$ **17.** $8x^3\sqrt{5}$
18. $36\sqrt{42}$ **19.** $\frac{1}{2}$ **20.** $-3\sqrt{10} - 10\sqrt{2}$ **21.** $58 + 24\sqrt{5}$ **22.** $102 - 72\sqrt{2}$ **23.** $\frac{8\sqrt{6} + 2\sqrt{2}}{47}$
24. $\frac{16 - 7\sqrt{6}}{2}$ **25.** $m = 5\frac{1}{3}$ **26.** $m = 208$ **27.** $t = 8$ **28.** $r = 5$ **29.** $m = \frac{9}{2}$ **30.** no

real-number solution   **31.** -36 and -42 **32.** 13 and 14   **33.** $6\sqrt{3} - 11\sqrt{2}$ **34.** $\frac{72\sqrt{30} + 60}{211}$

35. $t = \frac{35}{3}$ **36.** 960 square units

CUMULATIVE REVIEW   *(pages 531–532)*

1. $(5x - 12)(x + 2)$ **2.** $(3x - 11)(3x + 11)$ **3.** $6x^2 + x - 77$ **4.** $\frac{14}{15}x^2 - \frac{1}{10}x$ **5.** $x = 357\frac{1}{3}$
6. $3x^2 + 12x$ **7.** $x = 2; y = 4$ **8.** $2x^2 + 8x + 4 + \frac{16}{x - 2}$ **9.** $\frac{8y^2 + 12x^2}{xy}$ **10.** $x = -5; y = 2$
11. $\frac{1}{2}$ **12.** 250 children's tickets; 250 adults' tickets   **13.** $y = -2x + 7$ **14.** $y = -4x$ **15.** 62
16. 1, 2, 4, 7, 13, 14, 26, 28, 52, 91, 182, and 364   **17.** $2^5 5^3$ **18.** 1   **19.** $x < 14$ **20. a.** $6n + 3$ **b.** $2n + 3$
c. $2n$

ANSWERS to Chapter 10 Odd-Numbered Exercises

SECTION 10.1   *(pages 543–545)*

1. $n = -6$ or $n = 1$ **3.** $r = -10$ (double root)   **5.** $x = -5$ or $x = -8$ **7.** $r = \frac{1}{2}$ or $r = 1$
9. $t = 7$ or $t = 11$ **11.** $r = -4$ (double root)   **13.** $x = -\frac{3}{4}$ or $x = \frac{3}{4}$ **15.** $t = 5$ or $t = 8$
17. $t = -5$ or $t = 2$ **19.** $y = 12$ (double root)   **21.** $x = -5$ or $x = 5$ **23.** $t = -\frac{2}{11}$ or $t = -2$
25. $x^2 - x - 20 = 0$ **27.** $x^2 + 5x - 14 = 0$ **29.** $5x^2 - 52x + 20 = 0$ **31.** $x^2 + 10x + 21 = 0$
33. $x^2 + 10x + 25 = 0$ **35.** $x^2 + 15x + 14 = 0$ **37.** $15x^2 + 7x - 2 = 0$ **39.** $12x^2 + 5x - 2 = 0$
41. $x^2 - 144 = 0$ **43.** $x^2 - 2 = 0$ **45.** $x^2 - 12 = 0$ **47.** $x^2 - 24 = 0$ **49.** $x = 2$ or $x = -2$
51. $t = 3$ or $t = -3$ **53.** $r = 13$ or $r = -13$ **55.** $y = 16$ or $y = -16$ **57.** $n = 7$ or $n = -7$
59. $x = \frac{3}{4}$ or $x = -\frac{3}{4}$ **61.** $x = 2$ or $x = -2$ **63.** $t = \frac{7}{9}$ or $t = -\frac{7}{9}$ **65.** $n = 5 \pm 10\sqrt{2}$
67. $x = 5\frac{1}{4}$ or $x = -1\frac{3}{4}$ **69.** $y = -5$ or $y = 6\frac{1}{3}$ **71.** $x = 112$ or $x = 88$ **73.** $x = 6\frac{3}{4}$ or $x = 1\frac{1}{4}$
75. $t \approx 2.4$ seconds
77. $s = 1, y = 4.9$ meters
$s = 2, y = 9.8$ meters
$s = 3, y = 14.7$ meters
$s = 4, y = 19.6$ meters
79. $t \approx 3$ seconds; At $t = 0$, the ball is not yet in the air.

SECTION 10.2   *(pages 550–553)*

1. equation: $x^2 - 24x = -40$ **3.** equation: $y^2 + 30y = -27$ **5.** equation: $x^2 - 30x = 200$
$\frac{b}{2}$: -12 $\frac{b}{2}$: 15 $\frac{b}{2}$: -15
$\left(\frac{b}{2}\right)^2$: 144 $\left(\frac{b}{2}\right)^2$: 225 $\left(\frac{b}{2}\right)^2$: 225

7. equation: $y^2 - 11y = -30$ **9.** equation: $x^2 - 5x = 14$ **11.** equation: $y^2 - 10y = 28$

$\frac{b}{2}: -5\frac{1}{2}$ $\qquad$ $\frac{b}{2}: -\frac{5}{2}$ $\qquad$ $\frac{b}{2}: -5$

$\left(\frac{b}{2}\right)^2: \frac{121}{4}$ $\qquad$ $\left(\frac{b}{2}\right)^2: \frac{25}{4}$ $\qquad$ $\left(\frac{b}{2}\right)^2: 25$

13. $x = 5$ or $x = -3$ **15.** $y = 25$ or $y = 1$ **17.** $x = 2$ or $x = -11$ **19.** $y = 2$ or $y = -10$
21. $x = 19$ or $x = -1$ **23.** $y = 6$ or $y = 2$ **25.** $x = 1$ or $x = -13$ **27.** $y = -10$ or $y = -12$
29. $x = 5$ or $x = -1$ **31. a.** 42 amperes **b.** at 1 second and at 6 seconds **33.** $n = 10$ sides **35.** 4 and 10
37. a. 209 miles per hour **b.** 10,000 feet **39.** 29 and 31 **41.** $t = 3$ seconds **43.** $x = 20$ days
45. 18 people; 100 sq. ft/person **47.** 10 days **49.** $x = \frac{7}{6}$ and $x = \frac{1}{6}$ **51.** $x = \pm\frac{3}{7}$ **53.** $x = \pm 11$
55. $x = 1\frac{7}{48}$ or $x = 1\frac{5}{48}$ **57.** $x = \pm\sqrt{5}$ **59.** $y = 1$ (double root) **61.** $t = \pm\frac{4}{3}$

SECTION 10.3 *(pages 562–566)*

1. $3x^2 - 45x - 152 = 0$ **3.** $0.4x^2 - 1.6x + 1.8 = 0$ **5.** $2x^2 + 22x - 84 = 0$
$a = 3, b = -45, c = -152$ $\quad$ $a = 0.4, b = -1.6, c = 1.8$ $\quad$ $a = 2, b = 22, c = -84$

7. $\frac{3}{5}x^2 - \frac{3}{8}x + 168 = 0$ **9.** $7x^2 - 35x - 98 = 0$ **11.** $4.5x^2 - 5.08x - 1.40 = 0$
$a = \frac{3}{5}, b = -\frac{3}{8}, c = 168$ $\quad$ $a = 7, b = -35, c = -98$ $\quad$ $a = 4.5, b = -5.08, c = -1.40$

13. $x = -1$ or $x = -9$ **15.** $x = 8$ (double root) **17.** $x = -5$ or $x = -7$ **19.** $x = \pm 13$
21. $x = -2$ or $x = -3$ **23.** $x = -14$ or $x = 3$ **25.** $x = 6$ or $x = -5$ **27.** $x = 9$ (double root)
29. $x = \pm 2$ **31.** $x = 5$ or $x = \frac{-3}{5}$ **33.** $x = 3$ or $x = 1$ **35.** $x = -7$ or $x = -9$
37. $x = -8$ (double root) **39.** no real-number solutions **41.** $x = \pm 19$ **43.** no real-number solutions
45. $x = -1$ (double root) **47.** $x = 11$ or $x = 7$ **49.** $x = -1$ or $x = -4$ **51.** $x = 4$ (double root)
53. $x = -1$ or $x = -4$
55. discriminant: 29 **57.** discriminant: 33 **59.** discriminant: 0
two real-number solutions $\quad$ two real-number solutions $\quad$ one real-number solution
61. discriminant: 0 **63.** discriminant: 292 **65.** discriminant: -1879
one real-number solution $\quad$ two real-number solutions $\quad$ no real-number solutions
67. $n = 64$ and $n = 164$ **69.** 1970 **71.** approximately 11 hours **73. a.** 16 feet **b.** 130 pounds
75. a. The second root is -2. $n = -68$ **b.** The second root is $\frac{1}{4}$. $b = 101$ **77.** falling ball; 5 seconds later
79. a. 119.7 feet **b.** On Earth: 3.125 seconds; on the moon: 18.52 seconds **81.** $x = -4\frac{1}{2}$ (double root)
83. $x = 10\frac{1}{2}$ or $x = -1$ **85.** $x = -8$ or $x = 7$ **87.** $x = -11$ (double root) **89.** $x = -\frac{2}{7}$ or $x = 1$

SECTION 10.4 *(pages 579–582)*

1. vertex: $(-7, 62)$; opens downward **3.** vertex: $(-1, 2.7)$; opens downward **5.** vertex: $(10, 5)$; opens upward
7. vertex: $(-5, 8)$; opens upward **9.** vertex: $(-3, 3)$; opens downward **11.** vertex: $(0, 10)$; opens upward
13. vertex: $\left(0, -\frac{1}{7}\right)$; opens upward **15.** vertex: $(0, 12)$; opens upward **17.** vertex: $\left(-\frac{1}{2}, -3\right)$; opens
downward **19.** y-intercept: $(0, -180)$; x-intercepts: $(5, 0)$ and $(-9, 0)$ **21.** y-intercept: $(0, 25)$;
x-intercept: $(-5, 0)$ **23.** y-intercept: $(0, 100)$; x-intercept: $(10, 0)$ **25.** y-intercept: $(0, -69)$;
x-intercepts: $\left(\frac{7 \pm \sqrt{118}}{4}, 0\right)$ **27.** y-intercept: $(0, 8)$; x-intercepts: $\left(\frac{2}{3}, 0\right)$ and $\left(\frac{4}{3}, 0\right)$ **29.** y-intercept: $(0, -340)$;
x-intercepts: $\left(\frac{17}{2}, 0\right)$ and $\left(-\frac{5}{2}, 0\right)$ **31.** vertex: $\left(-\frac{7}{4}, -\frac{25}{8}\right)$; x-intercepts: $\left(-\frac{1}{2}, 0\right)$ and $(-3, 0)$; y-intercept: $(0, 3)$
33. vertex: $\left(\frac{5}{6}, -4\frac{1}{12}\right)$; x-intercepts: $\left(-\frac{1}{3}, 0\right)$ and $(2, 0)$; y-intercept: $(0, -2)$ **35.** vertex: $\left(1\frac{1}{2}, -12\frac{1}{4}\right)$;
x-intercepts: $(-2, 0)$ and $(5, 0)$; y-intercept: $(0, -10)$ **37.** vertex: $\left(\frac{11}{2}, -\frac{169}{4}\right)$; x-intercepts: $(12, 0)$ and $(-1, 0)$;
y-intercept: $(0, -12)$ **39.** vertex: $(8, -100)$; x-intercepts: $(18, 0)$ and $(-2, 0)$; y-intercept: $(0, -36)$
41. vertex: $(1, 5)$; x-intercepts: none; y-intercept: $(0, 6)$

43. x-intercepts: $(2, 0)$ and $(-2, 0)$
y-intercept: $(0, -4)$
vertex: $(0, -4)$
Additional points may vary.

45. x-intercepts: $(2, 0)$ and $(-2, 0)$
y-intercept: $(0, 4)$
vertex: $(0, 4)$
Additional points may vary.

47. x-intercepts: $(3, 0)$ and $(-3, 0)$
y-intercept: $(0, -9)$
vertex: $(0, -9)$
Additional points may vary.

49. $(0, 1), (-1, 6), (-2, 15), (1, 0), (2, 3), (3, 10)$ **51.** $(0, 1), (-1, -3), (-2, -9), (1, 3), (2, 3), (3, 1)$
53. $(0, -2), (-1, 11), (-2, 40), (1, 1), (2, 20), (3, 55)$ **55.** **57.** **59.**

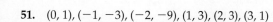

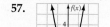

61. **63.** **65.** **67.** $\{(\pm\sqrt{3}, 0)\}$

69. $\{(1 \pm \sqrt{3}, 0)\}$ **71.** $\{(-2 \pm \sqrt{5}, 0)\}$ **73.** no real solutions

75. $\left\{(1, 0), \left(-\frac{1}{2}, 0\right)\right\}$ **77.** $\{(5, 0), (1, 0)\}$ **79.** There are no real-number solutions.

81. $x = \frac{1}{5}$ or $x = -3$ **83.** $x = 2$ or $x = -2$ **85.** $x = 23$ or $x = 1$ **87.** There are no real-number solutions.

89. x-intercepts: $(2, 0)$ and $(0, 0)$
y-intercept: $(0, 0)$
vertex: $(1, 1)$
Additional points may vary.

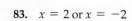

91. vertex: $(5, -3)$; opens downward **93.** 9.14 seconds

CHAPTER 10 REVIEW PROBLEMS *(pages 585–587)*

1. $t = \pm\frac{7}{9}$ **2.** $x = -2\frac{1}{2}$ or $x = 2\frac{1}{4}$ **3.** $x = \pm\frac{3}{7}$ **4.** $x = \pm 2\frac{1}{2}$ **5.** $x = -15$ or $x = 25$

6. $x = \frac{-5 \pm 5\sqrt{5}}{16}$ **7.** $n = 5$ or $n = -8$ **8.** $t = 2\frac{2}{3}$ or $t = 1\frac{1}{3}$ **9.** $n = 12$ or $n = -9$

10. $x = 1$ (double root) **11.** $x = 10 \pm 10\sqrt{2}$ **12.** $x = 8$ (double root) **13.** 1010 crocodiles

14. 6 years **15.** $x = \frac{1}{7}$ (double root) **16.** $y = 1\frac{1}{2}$ or $y = \frac{-1}{4}$ **17.** $x = 4\frac{1}{2}$ or $x = \frac{-1}{2}$

18. $r = 21$ or $r = 9$ **19.** $t = -1$ or $t = -25$ **20.** $n = \frac{2}{3}$ (double root) **21.** approximately 4 seconds

22. $51\frac{1}{2}$ shares

23. vertex: $(-1, 14)$
  y-intercept: $(0, 21)$
  x-intercept: none
  Additional points may vary.

24. vertex: $\left(-\frac{1}{6}, -\frac{1}{6}\right)$
  y-intercept: $(0, 0)$
  x-intercepts: $(0, 0)$ and $\left(-\frac{1}{3}, 0\right)$
  Additional points may vary.

25. vertex: $(1, -6)$
  y-intercept: $(0, -7)$
  x-intercepts: none
  Additional points may vary.

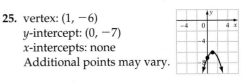

26. vertex: $(0, -8)$
  y-intercept: $(0, -8)$
  x-intercepts: $(\pm 2.8, 0)$
  Additional points may vary.

27. vertex: $(-1, -1)$
  y-intercept: $(0, 0)$
  x-intercepts: $(0, 0)$ and $(-2, 0)$
  Additional points may vary.

28. vertex: $(1, -6)$
  y-intercept: $(0, -7)$
  x-intercepts: none
  Additional points may vary.

29. $x = 13$ (double root)   **30.** no real-number solutions   **31.** $n = 7$ or $n = 8$ **32.** $x = 1$ or $x = -3$

33. $t = 5$ or $t = -4$ **34.** $x = 18$ (double root)   **35.** $x = \frac{1}{3}$ or $x = -\frac{19}{9}$ **36.** 14 inches by 14 inches

37. $x = \frac{14}{3}$ or $x = 0$ **38.** \$181   **39.** $x = 3$ or $x = 5$

40. vertex: $(0, 9)$
  y-intercept: $(0, 9)$
  x-intercepts: $(3, 0)$ and $(-3, 0)$
  Additional points may vary.

CUMULATIVE REVIEW   *(page 588)*

1. Answers may vary. One is "the difference of 15 and a number, divided by 21."   **2.** -3.608 **3.** 7.168

4. $x = \frac{7}{9}$ **5.** $-2xy - y^2$ **6.** $-|2.5|$ **7.** $x = 2\frac{3}{4}$ **8.** 64 **9.** $x \geq 5.375$ **10.** $n + 212$

11. $12y^2 - 56y + 64$ **12.** 2 **13.** $y = 7$ or $y = -7$ **14.** 54.5 and 52.5 **15.** $x = -10\frac{1}{2}$

16. $x = -45, y = -36$ **17.** $-6y\sqrt{2} - 4\sqrt{6y} + 2\sqrt{3y} + 4$ **18.** $0.0000432x^{13}$ **19. a.** $n + 5$ years
b. $n + 3$ years   **20.** $x = 6$ or $x = -10$

INDEX